Springer Tracts in Advanced Robotics

Volume 141

The Springer Tracts in Advanced Robotics (STAR) publish new developments and advances in the fields of robotics research, rapidly and informally but with a high quality. The intent is to cover all the technical contents, applications, and multidisciplinary aspects of robotics, embedded in the fields of Mechanical Engineering, Computer Science, Electrical Engineering, Mechatronics, Control, and Life Sciences, as well as the methodologies behind them. Within the scope of the series are monographs, lecture notes, selected contributions from specialized conferences and workshops, as well as selected PhD theses.

Special offer: For all clients with a print standing order we offer free access to the electronic volumes of the Series published in the current year.

Indexed by SCOPUS, DBLP, EI Compendex, zbMATH, SCImago.
All books published in the series are submitted for consideration in Web of Science.

More information about this series at http://www.springer.com/series/5208

Peter Corke

Robotics and Control

Fundamental Algorithms in MATLAB®

 Springer

Peter Corke
School of Electrical Engineering
and Robotics
Queensland University of Technology
Brisbane, QLD, Australia

ISSN 1610-7438   ISSN 1610-742X (electronic)
Springer Tracts in Advanced Robotics
ISBN 978-3-030-79178-0   ISBN 978-3-030-79179-7   (eBook)
https://doi.org/10.1007/978-3-030-79179-7

This Springer imprint is published by the registered company Springer Nature Switzerland AG
The registered company address is: Gewerbestrasse 11, 6330 Cham, Switzerland

To my family Phillipa, Lucy and Madeline for their indulgence and support;
my parents Margaret and David for kindling my curiosity;
and to Lou Paul who planted the seed that became this book.

Foreword

At the dawn of the century's third decade, robotics is reaching an elevated level of maturity and continues to benefit from the advances and innovations in its enabling technologies. These all are contributing to an unprecedented effort to bringing robots to human environment in hospitals and homes, factories and schools; in the field for robots fighting fires, making goods and products, picking fruits and watering the farmland, saving time and lives. Robots today hold the promise for making a considerable impact in a wide range of real-world applications from industrial manufacturing to healthcare, transportation, and exploration of the deep space and sea. Tomorrow, robots will become pervasive and touch upon many aspects of modern life.

The *Springer Tracts in Advanced Robotics (STAR)* is devoted to bringing to the research community the latest advances in the robotics field on the basis of their significance and quality. Through a wide and timely dissemination of critical research developments in robotics, our objective with this series is to promote more exchanges and collaborations among the researchers in the community and contribute to further advancements in this rapidly growing field.

This is a refined remake of the volume of the second edition of *Robotics, Vision and Control – Fundamental Algorithms in MATLAB®* by Peter Corke in 2017. The work now comes in two split volumes: one devoted to *Robotics and Control*, and the other to *Robotic Vision*. The first volume contains material from the first nine chapters of the previous single volume, covering: foundations on pose, time, and motion; mobile robots with navigation and localization; kinematics, dynamics, and control of robot manipulators. On the other hand, the second volume contains material from the first two chapters and the tenth to fourteenth chapters of the previous single volume, covering: foundations on pose, computer vision, image processing and feature extraction; image formation and multiple images for the geometry of vision.

The outcome is a two-volume handy set which is confirmed to be shining in our STAR series!

Naples, Italy and Stanford, USA
November 2020

Bruno Siciliano and *Oussama Khatib*
STAR Editors

Foreword
to the Second Edition

Once upon a time, a very thick document of a dissertation from a faraway land came to me for evaluation. *Visual robot control* was the thesis theme and *Peter Corke* was its author. Here, I am reminded of an excerpt of my comments, which reads, *this is a masterful document, a quality of thesis one would like all of one's students to strive for, knowing very few could attain – very well considered and executed.*

The connection between robotics and vision has been, for over two decades, the central thread of Peter Corke's productive investigations and successful developments and implementations. This rare experience is bearing fruit in this second edition of his book on *Robotics, Vision, and Control*. In its melding of theory and application, this second edition has considerably benefited from the author's unique mix of academic and real-world application influences through his many years of work in robotic mining, flying, underwater, and field robotics.

There have been numerous textbooks in robotics and vision, but few have reached the level of integration, analysis, dissection, and practical illustrations evidenced in this book. The discussion is thorough, the narrative is remarkably informative and accessible, and the overall impression is of a significant contribution for researchers and future investigators in our field. Most every element that could be considered as relevant to the task seems to have been analyzed and incorporated, and the effective use of Toolbox software echoes this thoroughness.

The reader is taken on a realistic walkthrough the fundamentals of mobile robots, navigation, localization, manipulator-arm kinematics, dynamics, and joint-level control, as well as camera modeling, image processing, feature extraction, and multi-view geometry. These areas are finally brought together through extensive discussion of visual servo system. In the process, the author provides insights into how complex problems can be decomposed and solved using powerful numerical tools and effective software.

The *Springer Tracts in Advanced Robotics (STAR)* is devoted to bringing to the research community the latest advances in the robotics field on the basis of their significance and quality. Through a wide and timely dissemination of critical research developments in robotics, our objective with this series is to promote more exchanges and collaborations among the researchers in the community and contribute to further advancements in this rapidly growing field.

Peter Corke brings a great addition to our STAR series with an authoritative book, reaching across fields, thoughtfully conceived and brilliantly accomplished.

Oussama Khatib
Stanford, California
October 2016

Preface

Tell me and I will forget.
Show me and I will remember.
Involve me and I will understand.
Chinese proverb

Simple things should be simple,
complex things should be possible.
Alan Kay

These are exciting times for robotics and we have seen much recent progress: the rise of the self-driving car, the Mars science laboratory rover making profound discoveries on Mars, the Philae comet landing attempt, and the DARPA Robotics Challenge. We have witnessed the drone revolution – flying machines that were once the domain of the aerospace giants can now be bought for just tens of dollars. All this has been powered by the continuous and relentless improvement in computer power and tremendous advances in low-cost inertial sensors – driven largely by consumer demand for better mobile phones and gaming experiences. It's getting easier for individuals to create robots – 3D printing is now very affordable, the Robot Operating System (ROS) is both capable and widely used, and powerful hobby technologies such as the Arduino, Raspberry Pi, Dynamixel servo motors and Lego's EV3 brick are available at low cost. This in turn has contributed to the rapid growth of the global maker community – ordinary people creating at home what would once have been done by a major corporation. We have also witnessed an explosion of commercial interest in robotics – many startups and a lot of acquisitions by big players in the field. Robotics even featured on the front cover of the Economist magazine in 2014!

So how does a robot work? Robots are data-driven machines. They acquire data, process it and take action based on it. The data comes from sensors measuring the velocity of a wheel, the angle of a robot arm's joint or the intensities of millions of pixels that comprise an image of the world that the robot is observing. For many robotic applications the amount of data that needs to be processed, in real-time, is massive. For a vision sensor it can be of the order of tens to hundreds of megabytes per second.

"Computers in the future may weigh no more than 1.5 tons." Popular Mechanics, forecasting the relentless march of science, 1949

Progress in robots has been, and continues to be, driven by more effective ways to process data. This is achieved through new and more efficient algorithms, and the dramatic increase in computational power that follows Moore's law.◀ When I started in robotics and vision in the mid 1980s, see Fig. 0.1, the IBM PC had been recently released – it had a 4.77 MHz 16-bit microprocessor and 16 kbytes (expandable to 256 k) of memory. Over the intervening 30+ years computing power has perhaps doubled 20 times which is an increase by a factor of one million.

Over the fairly recent history of robotics a very large body of algorithms has been developed to efficiently solve large-scale problems in perception, planning, control and localization – a significant, tangible, and collective achievement of the research community. However its sheer size and complexity presents a very real barrier to somebody new entering the field. Given so many algorithms from which to choose, a real and important question is:

What is the right algorithm for this particular problem?

One strategy would be to try a few different algorithms and see which works best for the problem at hand, but this is not trivial and leads to the next question:

How can I evaluate algorithm X on my own data without spending days coding and debugging it from the original research papers?

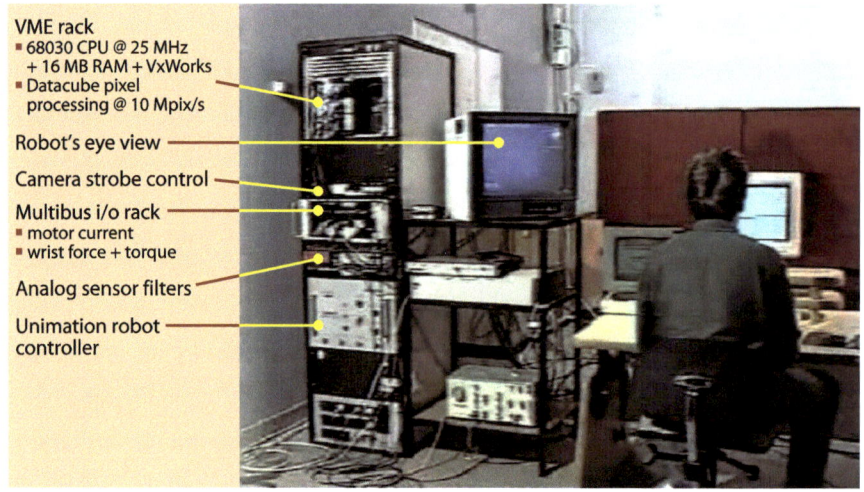

VME rack
- 68030 CPU @ 25 MHz
  + 16 MB RAM + VxWorks
- Datacube pixel
  processing @ 10 Mpix/s

Robot's eye view

Camera strobe control

Multibus i/o rack
- motor current
- wrist force + torque

Analog sensor filters

Unimation robot
controller

Fig. 0.1.
Once upon a time a lot of equipment was needed to do vision-based robot control. The author with a large rack full of real-time image processing and robot control equipment (1992)

Two developments come to our aid. The first is the availability of general purpose mathematical software which makes it easy to prototype algorithms. There are commercial packages such as MATLAB®, Mathematica®, Maple® and MathCad®▸, as well as open source projects include SciLab, Octave, and Matplotlib. All these tools deal naturally and effortlessly with vectors and matrices, can create complex and beautiful graphics, and can be used interactively or as a programming environment. The second is the open-source movement. Many algorithms developed by researchers are available in open-source form. They might be coded in one of the general purpose mathematical languages just mentioned, or written in a mainstream language like C, C++, Java or Python.

Respectively the trademarks of The Math-Works Inc., Wolfram Research, MapleSoft and PTC.

For more than twenty years I have been part of the open-source community and maintained two open-source MATLAB Toolboxes that date back to my own Ph.D. work and have evolved since then, growing features and tracking changes to the MATLAB language. One of these, the Robotics Toolbox has been translated into a number of different languages such as Python, SciLab and LabView and more recently some of its functionality is finding its way into the MATLAB Robotics System Toolbox™ published by The MathWorks. It forms the basis of this book.

These Toolboxes have some important virtues. Firstly, they have been around for a long time and used by many people for many different problems so the code can be accorded some level of trust. New algorithms, or even the same algorithms coded in new languages or executing in new environments, can be compared against implementations in the Toolbox.

» allow the user to work with real problems, not just trivial examples

Secondly, they allow the user to work with real problems, not just trivial examples. For real robots, those with more than two links, the computation required is beyond unaided human ability. Thirdly, they allow us to gain insight which can otherwise get lost in the complexity. We can rapidly and easily experiment, play *what if* games, and depict the results graphically using the powerful 2D and 3D graphical display tools of MATLAB. Fourthly, the Toolbox code makes many common algorithms tangible and accessible. You can read the code, you can apply it to your own problems, and you can extend it or rewrite it. It gives you a "leg up" as you begin your journey into robotics.

» a narrative that covers robotics and computer vision – both separately and together

This book takes a conversational approach, weaving text, mathematics and code examples into a narrative. I want to show how complex problems can be decomposed and solved using just a few simple lines of code. More formally this is an inductive learning approach, going from specific and concrete examples to the more general.

» consider it a grand tasting menu

The topics covered in this book are based on my own interests but also guided by real problems that I observed over many years as a practitioner of both robotics and computer vision. I want to give the reader a flavor of what robotics is about and what it can do – consider it a grand tasting menu. I hope that by the end of this book you will share my enthusiasm for these topics.

» software is a first-class citizen in this book

This book is unlike other text books, and deliberately so. Firstly, software is a first-class citizen in this book. Software is a tangible instantiation of the algorithms described – it can be read and it can be pulled apart, modified and put back together again. There are a number of classic books that use software in an illustrative fashion and which have influenced my approach, for example *LaTeX: A document preparation system* (Lamport 1994), *Numerical Recipes in C* (Press et al. 2007), *The Little Lisper* (Friedman et al. 1987) and *Structure and Interpretation of Classical Mechanics* (Sussman et al. 2001). Over 600 examples in this book illustrate how the Toolbox software can be used and generally provide *instant gratification* in just a couple of lines of MATLAB code.

» instant gratification in just a couple of lines of MATLAB code

Secondly, building the book around MATLAB and the Toolboxes means that we are able to tackle more realistic and more complex problems than other books.

» this book provides a complementary approach

The emphasis on software and examples does not mean that rigor and theory are unimportant – they are very important, but this book provides a complementary approach. It is best read in conjunction with standard texts which do offer rigor and theoretical nourishment. The end of each chapter has a section on further reading and provides pointers to relevant textbooks and key papers. I try hard to use the least amount of mathematical notation required, if you seek deep mathematical rigor this may not be the book for you.

The Toolboxes also include some great open-source software and I am grateful to the following for code that has been incorporated into the Robotics Toolbox: mobile robot localization and mapping by Paul Newman; a quadrotor simulator by Pauline Pounds; a Symbolic Manipulator Toolbox by Jörn Malzahn; pose-graph SLAM code by Giorgio Grisetti and 3D robot models from the ARTE Robotics Toolbox by Arturo Gil.

As I wrote I became fascinated by the mathematicians, scientists and engineers whose work, hundreds of years ago, underpins the science of robotics today. Some of their names have become adjectives like Coriolis, Gaussian, or Cartesian; nouns like Jacobian, or units like Newton and Coulomb. They are interesting characters from a distant era when science was a hobby and their day jobs were as doctors, alchemists, gamblers, astrologers, philosophers or mercenaries. In order to know whose shoulders we are standing on, I have included small vignettes about the lives of some of these people – a smattering of history as a backstory.

In my own career I have had the good fortune to work with many wonderful people who have inspired and guided me. Long ago at the University of Melbourne John Anderson fired my interest in control and Graham Holmes tried with mixed success to have me "think before I code". Early on, I spent a life-direction-changing ten months working with Richard (Lou) Paul in the GRASP laboratory at the University of Pennsylvania in the period 1988–1989. The genesis of the Toolboxes was my Ph.D. research (1991–1994) and my advisors Malcolm Good (University of Melbourne) and Paul Dunn (CSIRO) asked me good questions and guided my research. Laszlo Nemes (CSIRO) provided great wisdom about life and the ways of organizations, and encouraged me to publish and to open-source my software. Much of my career was spent at CSIRO where I had the privilege and opportunity to work on a diverse range of real robotics projects and to work with a truly talented set of colleagues and friends. Part way through writing the first edition I joined the Queensland University of Technology which made time available to complete that work, and in 2015 sabbatical leave to complete the second.

Many people have helped me in my endeavor and I thank them. I was generously hosted for periods of productive writing at Oxford (both editions) by Paul Newman, and at MIT (first edition) by Daniela Rus. Daniela, Paul and Cédric Pradalier made constructive suggestions and comments on early drafts of that edition. For the second edition I was helped by comments on draft chapters by: Tim Barfoot, Dmitry Bratanov, Duncan Campbell, Donald Dansereau, Tom Drummond, Malcolm Good, Peter Kujala, Obadiah Lam, Jörn Malzahn, Felipe Nascimento Martins, Ajay Pandey, Cédric Pradalier, Dan Richards, Daniela Rus, Sareh Shirazi, Surya Singh, Ryan Smith, Ben Talbot, Dorian Tsai and Ben Upcroft; and assisted with wisdom and content by: François Chaumette, Donald Dansereau, Kevin Lynch, Robert Mahony and Frank Park.

I have tried my hardest to eliminate errors but inevitably some will remain. Please email bug reports to me at rvc@petercorke.com as well as suggestions for improvements and extensions.

Writing the second edition was financially supported by EPSRC Platform Grant EP/M019918/1, QUT Science & Engineering Faculty sabbatical grant, QUT Vice Chancellor's Excellence Award 2015, QUT Robotics and Autonomous Systems discipline and the ARC Centre of Excellence for Robotic Vision (grant CE140100016).

Over both editions I have enjoyed wonderful support from MathWorks, through their author program, and from Springer. My editor Thomas Ditzinger has been a great supporter of this project and Armin Stasch, with enormous patience and dedication in layout and typesetting, has transformed my untidy ideas into a thing of beauty.

Finally, my deepest thanks are to Phillipa who has supported me and "the book" with grace and patience for a very long time and in many different places – without her this book could never have been written.

Peter Corke
Brisbane,
Queensland
March 2019

Note on the Second Edition

The revision principle was to keep the good (narrative style, code as a first-class citizen, soft plastic cover) and eliminate the bad (errors and missing topics). There were more errors than I would have liked and I thank everybody who submitted errata and suggested improvements.

New content includes matrix exponential notation; the basics of screw theory and Lie algebra; inertial navigation; differential steer and omnidirectional mobile robots; a deeper treatment of SLAM systems including scan matching and pose graphs; greater use of MATLAB computer algebra; operational space control; deeper treatment of manipulator dynamics and control; visual SLAM and visual odometry; structured light; bundle adjustment; and light-field cameras.

In the first edition I shied away from Lie algebra, matrix exponentials and twists but I think it's important to cover them. The topic is deeply mathematical and I've tried to steer a middle ground between hardcore algebraic topology and the homogenous transformation only approach of most other texts, while also staying true to the overall approach of this book.

All MATLAB generated figures have been regenerated to reflect recent improvements to MATLAB graphics and all code examples have been updated as required and tested, and are available as MATLAB Live Scripts.

The second edition of the book is matched by new major releases of my Toolboxes: Robotics Toolbox (release 10) and the Machine Vision Toolbox (release 4). These newer versions of the toolboxes have some minor incompatibilities with previous releases of the toolboxes, and therefore also with the code examples in the first edition of the book.

Note on this Edition

This book is essentially the first nine chapters of the second edition of Robotics, Vision & Control with all known errata incorporated. It omits all content related to computer vision and vision-based control.

Contents

1   **Introduction** 1
1.1   Robots, Jobs and Ethics 7
1.2   About the Book 8
     1.2.1   MATLAB Software and the Toolboxes 8
     1.2.2   Notation, Conventions and Organization 9
     1.2.3   Audience and Prerequisites 10
     1.2.4   Learning with the Book 11
     1.2.5   Teaching with the Book 11
     1.2.6   Outline 11
     Further Reading 12

     Part I   Foundations 13
2   **Representing Position and Orientation** 15
2.1   Working in Two Dimensions (2D) 20
     2.1.1   Orientation in 2-Dimensions 21
     2.1.2   Pose in 2-Dimensions 24
2.2   Working in Three Dimensions (3D) 29
     2.2.1   Orientation in 3-Dimensions 30
     2.2.2   Pose in 3-Dimensions 44
2.3   Advanced Topics 47
     2.3.1   Normalization 47
     2.3.2   Understanding the Exponential Mapping 48
     2.3.3   More About Twists 50
     2.3.4   Dual Quaternions 53
     2.3.5   Configuration Space 53
2.4   Using the Toolbox 54
2.5   Wrapping Up 56
     Further Reading 58
     Exercises 59

3   **Time and Motion** 61
3.1   Time-Varying Pose 61
     3.1.1   Derivative of Pose 61
     3.1.2   Transforming Spatial Velocities 62
     3.1.3   Incremental Rotation 64
     3.1.4   Incremental Rigid-Body Motion 65
3.2   Accelerating Bodies and Reference Frames 66
     3.2.1   Dynamics of Moving Bodies 66
     3.2.2   Transforming Forces and Torques 67
     3.2.3   Inertial Reference Frame 67
3.3   Creating Time-Varying Pose 68
     3.3.1   Smooth One-Dimensional Trajectories 68

3.3.2   Multi-Dimensional Trajectories 71
3.3.3   Multi-Segment Trajectories 72
3.3.4   Interpolation of Orientation in 3D 73
3.3.5   Cartesian Motion in 3D 75
3.4   Application: Inertial Navigation 77
3.4.1   Gyroscopes 77
3.4.2   Accelerometers 79
3.4.3   Magnetometers 83
3.4.4   Sensor Fusion 85
3.5   Wrapping Up 88
Further Reading 88
Exercises 89

Part II   Mobile Robots 91
4   Mobile Robot Vehicles 97
4.1   Wheeled Mobile Robots 97
4.1.1   Car-Like Mobile Robots 97
4.1.2   Differentially-Steered Vehicle 107
4.1.3   Omnidirectional Vehicle 110
4.2   Flying Robots 112
4.3   Advanced Topics 117
4.3.1   Nonholonomic
and Under-Actuated Systems 117
4.4   Wrapping Up 119
Further Reading 120
Toolbox and MATLAB Notes 121
Exercises 121

5   Navigation 123
5.1   Reactive Navigation 124
5.1.1   Braitenberg Vehicles 124
5.1.2   Simple Automata 126
5.2   Map-Based Planning 128
5.2.1   Distance Transform 128
5.2.2   D* 132
5.2.3   Introduction to Roadmap Methods 134
5.2.4   Probabilistic Roadmap Method (PRM) 135
5.2.5   Lattice Planner 138
5.2.6   Rapidly-Exploring Random Tree (RRT) 142
5.3   Wrapping Up 144
Further Reading 145
Resources 146
MATLAB Notes 146
Exercises 146

6   Localization 149
6.1   Dead Reckoning 153
6.1.1   Modeling the Vehicle 153
6.1.2   Estimating Pose 155
6.2   Localizing with a Map 158
6.3   Creating a Map 163
6.4   Localization and Mapping 165
6.5   Rao-Blackwellized SLAM 167
6.6   Pose Graph SLAM 168

6.7   Sequential Monte-Carlo Localization 173
6.8   Application: Scanning Laser Rangefinder 176
         Laser Odometry 177
         Laser-Based Map Building 179
         Laser-Based Localization 180
6.9   Wrapping Up 180
         Further Reading 181
         Toolbox and MATLAB Notes 183
         Exercises 183

Part III   Arm-Type Robots 187
7   Robot Arm Kinematics 191
7.1   Forward Kinematics 191
         7.1.1   2-Dimensional (Planar) Robotic Arms 192
         7.1.2   3-Dimensional Robotic Arms 194
7.2   Inverse Kinematics 203
         7.2.1   2-Dimensional (Planar) Robotic Arms 203
         7.2.2   3-Dimensional Robotic Arms 205
7.3   Trajectories 209
         7.3.1   Joint-Space Motion 209
         7.3.2   Cartesian Motion 212
         7.3.3   Kinematics in Simulink 212
         7.3.4   Motion through a Singularity 213
         7.3.5   Configuration Change 214
7.4   Advanced Topics 215
         7.4.1   Joint Angle Offsets 215
         7.4.2   Determining Denavit-Hartenberg Parameters 215
         7.4.3   Modified Denavit-Hartenberg Parameters 216
7.5   Applications 218
         7.5.1   Writing on a Surface 218
         7.5.2   A Simple Walking Robot 219
7.6   Wrapping Up 223
         Further Reading 224
         MATLAB and Toolbox Notes 225
         Exercises 225

8   Manipulator Velocity 227
8.1   Manipulator Jacobian 227
         8.1.1   Jacobian in the World Coordinate Frame 227
         8.1.2   Jacobian in the End-Effector Coordinate Frame 230
         8.1.3   Analytical Jacobian 230
8.2   Jacobian Condition and Manipulability 232
         8.2.1   Jacobian Singularities 232
         8.2.2   Manipulability 233
8.3   Resolved-Rate Motion Control 235
         8.3.1   Jacobian Singularity 238
8.4   Under- and Over-Actuated Manipulators 238
         8.4.1   Jacobian for Under-Actuated Robot 239
         8.4.2   Jacobian for Over-Actuated Robot 240
8.5   Force Relationships 242
         8.5.1   Transforming Wrenches to Joint Space 242
         8.5.2   Force Ellipsoids 242
8.6   Inverse Kinematics: a General Numerical Approach 243
         8.6.1   Numerical Inverse Kinematics 243

8.7   Advanced Topics 245
       8.7.1   Computing the Manipulator Jacobian Using Twists 245
8.8   Wrapping Up 245
       Further Reading 246
       MATLAB and Toolbox Notes 246
       Exercises 246

9   Dynamics and Control 249
9.1   Independent Joint Control 249
       9.1.1   Actuators 249
       9.1.2   Friction 250
       9.1.3   Effect of the Link Mass 251
       9.1.4   Gearbox 252
       9.1.5   Modeling the Robot Joint 253
       9.1.6   Velocity Control Loop 255
       9.1.7   Position Control Loop 259
       9.1.8   Independent Joint Control Summary 260
9.2   Rigid-Body Equations of Motion 261
       9.2.1   Gravity Term 262
       9.2.2   Inertia Matrix 264
       9.2.3   Coriolis Matrix 265
       9.2.4   Friction 266
       9.2.5   Effect of Payload 266
       9.2.6   Base Force 267
       9.2.7   Dynamic Manipulability 267
9.3   Forward Dynamics 269
9.4   Rigid-Body Dynamics Compensation 270
       9.4.1   Feedforward Control 271
       9.4.2   Computed Torque Control 272
       9.4.3   Operational Space Control 273
9.5   Applications 274
       9.5.1   Series-Elastic Actuator (SEA) 274
9.6   Wrapping Up 276
       Further Reading 276
       Exercises 278

Appendices 281
A   **Installing the Toolboxes** 283
B   **Linear Algebra Refresher** 287
C   **Geometry** 295
D   **Lie Groups and Algebras** 307
E   **Linearization, Jacobians and Hessians** 313
F   **Solving Systems of Equations** 317
G   **Gaussian Random Variables** 327
H   **Kalman Filter** 331
I   **Graphs** 337

     Bibliography 341

     Index 347
     Index of People 347
     Index of Functions, Classes and Methods 348
     General Index 351

Nomenclature

The notation used in robotics varies considerably across books and research papers. The symbols used in this book, and their units where appropriate, are listed below. Some symbols have multiple meanings and their context must be used to disambiguate them.

Notation	Description
x^*	desired value of x
x^+	predicted value of x
$x^{\#}$	measured, or observed, value of x
$\hat{x}$	estimated value of x
$\bar{x}$	mean of x or relative value
$x\langle k\rangle$	k^{th} element of a time series
v	a vector
$\hat{v}$	a unit-vector parallel to v
$\tilde{v}$	homogeneous representation of vector v
$v[i]$	i^{th} element of vector v
v_x	a component of a vector
A	a matrix
$A[i,j]$	the element (i,j) of A
A_{ij}	the element (i,j) of A
$f(x)$	a function of x
$F_x(x)$	the derivative $\partial f/\partial x$
$F_{xy}(x,y)$	the derivative $\partial^2 f/\partial x\partial y$
$\mathring{q}$	unit quaternion, $\mathring{q} \in \mathbb{S}^3$
$0_{m\times n}$	an $m \times n$ matrix of zeros
$1_{m\times n}$	an $m \times n$ matrix of ones

Symbol	Description	Unit
B	viscous friction coefficient	N m s rad^{-1}
B	magnetic field intensity (or magnetic flux density)	T
$C(q, \dot{q})$	manipulator centripetal and Coriolis term	$\text{kg m}^2\,\text{s}^{-1}$
$\mathcal{C}$	configuration space of a robot with N joints: $\mathcal{C} \subset \mathbb{R}^N$	
f	force	N
$F(\dot{q})$	friction torque	Nm
$G(q)$	manipulator gravity loading term	Nm
$\mathbb{H}$	the set of all quaternions (H for Hamilton)	
$I_{n \times n}$	$n \times n$ identity matrix	
J	inertia	kg m^2
J	inertia tensor, $J \in \mathbb{R}^{3 \times 3}$	kg m^2
J	Jacobian matrix	
$^A J_B$	Jacobian transforming velocities in frame B to frame A	
k, K	constant	
K_i	amplifier gain (transconductance)	A V^{-1}
K_{m}	motor torque constant	Nm A^{-1}
m	mass	kg
$\mathbf{M}(q)$	manipulator inertia matrix	kg m^2
$N(\mu, \sigma^2)$	a normal (Gaussian) distribution with mean μ and standard deviation σ	
q	generalized coordinates, configuration $q \in \mathcal{C}$	m, rad
Q	generalized force $Q \in \mathbb{R}^N$	N, Nm
R	an orthonormal rotation matrix, $R \in \mathbf{SO}(2)$ or $\mathbf{SO}(3)$	
$\mathbb{R}$	set of real numbers	
$\mathbb{R}^2$	set of all 2-D points	
$\mathbb{R}^3$	set of all 3-D points	
s	Laplace transform operator	
$\mathbb{S}^1$	unit circle, set of angles $[0, 2\pi)$	
$\mathbb{S}^n$	unit sphere embedded in $\mathbb{R}^{n+1}$	
$\mathrm{se}(n)$	Lie algebra for $\mathbf{SE}(n)$, an $\mathbb{R}^{(n+1) \times (n+1)}$ augmented skew-symmetric matrix	
$\mathrm{so}(n)$	Lie algebra for $\mathbf{SO}(n)$, an $\mathbb{R}^{n \times n}$ skew-symmetric matrix	
$\mathbf{SE}(n)$	special Euclidean group, the set of all poses in n dimensions, represented by an $\mathbb{R}^{(n+1) \times (n+1)}$ homogeneous transformation matrix	
$\mathbf{SO}(n)$	special orthogonal group, the set of all orientations in n dimensions, represented by an $\mathbb{R}^{n \times n}$ orthogonal matrix	
S	twist in 3 dimensions, $S \in \mathbb{R}^6$	
t	time	s
$\mathcal{T}$	task space of robot: $\mathcal{T} \subset \mathbf{SE}(3)$	K
T	sample interval	s
T	homogeneous transformation, $T \in \mathbf{SE}(2)$ or $\mathbf{SE}(3)$	
$^A T_B$	homogeneous transform representing frame $\{B\}$ with respect to frame $\{A\}$. If A is not given then assumed relative to world coordinate frame 0. Note that $^A T_B = (^B T_A)^{-1}$	
v	velocity	m s^{-1}
v	velocity vector	m s^{-1}
W	wrench, a vector of forces and moments $(f_x, f_y, f_z, m_x, m_y, m_z)$	N, Nm
X, Y, Z	Cartesian coordinates	
$\mathbb{Z}$	set of all integers	
$\mathbb{Z}^+$	the set of all integers greater than zero	

Symbol	Description	Unit
γ	robot steering angle	rad
$\boldsymbol{\Gamma}$	3-angle representation of rotation, $\boldsymbol{\Gamma} \in \mathbb{R}^3$	rad
$\boldsymbol{\Gamma}$	body torque, $\boldsymbol{\Gamma} \in \mathbb{R}^3$	Nm
θ	angle	rad
$\theta_r, \theta_p, \theta_y$	roll pitch yaw angles	rad
λ	an eigenvalue	
ν	innovation	
ν	spatial velocity, $\nu = (v_x, v_y, v_z, \omega_x, \omega_y, \omega_z) \in \mathbb{R}^6$	m s^{-1}, rad s^{-1}
ξ	abstract representation of Cartesian pose (pronounced ksi)	
$^A\xi_B$	abstract representation of relative pose, frame $\{B\}$ with respect to frame $\{A\}$ or rigid-body motion from frame $\{A\}$ to $\{B\}$	
π	mathematic constant	
$\boldsymbol{\pi}$	a plane	
σ	standard deviation	
σ	robot joint type, $\sigma = $ R for revolute and $\sigma = $ P for prismatic	
$\boldsymbol{\Sigma}$	Lie algebra $\boldsymbol{\Sigma} = [\cdot] \in \mathbf{se}(3)$	
τ	torque	N m
τ_C	Coulomb friction torque	N m
ω	rotational rate	rad s^{-1}
$\boldsymbol{\omega}$	angular velocity vector	rad s^{-1}
ϖ	rotational speed of a motor or propellor	rad s^{-1}
$\boldsymbol{\Omega}$	Lie algebra $\boldsymbol{\Omega} = [\cdot]_\times \in \mathbf{so}(3)$	

Operator	Description	MATLAB
$\|\cdot\|$	norm, or length, of vector: $\mathbb{R}^n \mapsto \mathbb{R}$	`norm, .norm`
$v_1 \cdot v_2$	dot, or inner, product, also $v_1^T v_2$: $\mathbb{R}^n \times \mathbb{R}^n \mapsto \mathbb{R}$	`dot`
$v_1 \times v_2$	cross, or vector, product: $\mathbb{R}^n \times \mathbb{R}^n \mapsto \mathbb{R}^n$	`cross`
A^{-1}	inverse of A: $\mathbb{R}^{n\times n} \mapsto \mathbb{R}^{n\times n}$	`inv`
A^+	pseudo-inverse of A: $\mathbb{R}^{n\times m} \mapsto \mathbb{R}^{m\times n}$	`pinv`
A^*	adjugate of $A \mapsto \det(A)A^{-1}$, $\mathbb{R}^{n\times n} \mapsto \mathbb{R}^{n\times n}$	
A^T	transpose of A: $\mathbb{R}^{n\times m} \mapsto \mathbb{R}^{m\times n}$	`'`
A^{-T}	transpose of inverse $A \mapsto (A^T)^{-1} = (A^{-1})^T$, $\mathbb{R}^{n\times n} \mapsto \mathbb{R}^{n\times n}$	
$\bullet$	transform a point (coordinate vector) by a relative pose: $SE(n) \times \mathbb{R}^n \mapsto \mathbb{R}^n$	`*`
$\oplus$	composition: $S_E^O(n) \times S_E^O(n) \mapsto S_E^O(n)$	`*`
$\ominus$	composition with inverse: $S_E^O(n) \times S_E^O(n) \mapsto S_E^O(n)$	`/`
$\ominus$	unary inverse: $S_E^O(n) \mapsto S_E^O(n)$	`.inv`
$\Delta(\cdot)$	maps incremental pose change to differential motion: $SE(3) \mapsto \mathbb{R}^6$	`tr2delta`
$\Delta^{-1}(\cdot)$	maps differential motion to incremental pose change: $\mathbb{R}^6 \mapsto SE(3)$	`delta2tr`
$\mathscr{R}_i(\theta)$	pure rotation about axis i: $\mathbb{R} \mapsto SE(3)$	`SE3.rotx|y|z`
$\mathscr{R}(\omega)$	pure rotation by $\|\omega\|$ about ω: $\mathbb{R}^3 \mapsto SE(3)$	`SE3.angvec`
$\mathscr{T}_i(d)$	pure translation along axis i: $\mathbb{R} \mapsto SE(2), SE(3)$	`SE2, SE3`
$\mathscr{T}(t)$	pure translation by vector: $\mathbb{R}^n \mapsto SE(n)$	`SE2, SE3`
$[\cdot]_t$	translational component of pose: $SE(n) \mapsto \mathbb{R}^n$	`.t`
$[\cdot]_R$	rotational component of pose: $SE(n) \mapsto \mathbb{R}^{n\times n}$	`.R`
$[\cdot]_\times$	skew-symmetric matrix: $\mathbb{R} \mapsto so(2), \mathbb{R}^3 \mapsto so(3)$	`skew`
$\vee_\times(\cdot)$	*unpack* skew-symmetric matrix: $so(2) \mapsto \mathbb{R}, so(3) \mapsto \mathbb{R}^3$	`vex`
$[\cdot]$	augmented skew-symmetric matrix: $\mathbb{R}^3 \mapsto se(2), \mathbb{R}^6 \mapsto se(3)$	`skewa`
$\vee(\cdot)$	*unpack* augmented skew-symmetric matrix: $se(2) \mapsto \mathbb{R}^3, so(3) \mapsto \mathbb{R}^6$	`vexa`
$\mathrm{Ad}(\cdot)$	adjoint representation: $SE(3) \mapsto \mathbb{R}^{6\times 6}$	`.Ad`
$\mathrm{ad}(\cdot)$	logarithm of adjoint representation: $SE(3) \mapsto \mathbb{R}^{6\times 6}$	`.ad`
$\circ$	quaternion (Hamiltonian) multiplication: $\mathbb{H} \times \mathbb{H} \mapsto \mathbb{H}$	`*`
$\mathring{v}$	pure quaternion: $\mathbb{R}^3 \mapsto \mathbb{H}$	`Quaternion.pure`
$\sim$	equivalence of representations	
$\simeq$	homogeneous coordinate equivalence	
$\ominus$	smallest angular difference between two angles on a circle: $\mathbb{S}^1 \times \mathbb{S}^1 \mapsto \mathbb{R}$	`angdiff`
$\mathcal{K}(\cdot)$	forward kinematics: $\mathcal{C} \mapsto \mathcal{T}$	`fkine`
$\mathcal{K}^{-1}(\cdot)$	inverse kinematics: $\mathcal{T} \mapsto \mathcal{C}$	`ikine`
$\mathcal{D}^{-1}(\cdot)$	manipulator inverse dynamics function: $\mathcal{C}, \mathbb{R}^N, \mathbb{R}^N \mapsto \mathbb{R}^N$	`rne`
$\{F\}$	coordinate frame F	
$[a, b]$	interval a to b inclusive	
(a, b)	interval a to b exclusive, not including a or b	
$[a, b)$	interval a to b, not including b	
$(a, b]$	interval a to b, not including a	

MATLAB® Toolbox Conventions

- A Cartesian coordinate, a point, is expressed as a column vector.
- A set of points is expressed as a matrix with columns representing the coordinates of individual points.
- A rectangular region is defined by two opposite corners $[x_{\min}\ x_{\max}; y_{\min}\ y_{\max}]$.
- A robot configuration, a set of joint angles, is expressed as a row vector.
- Time series data is expressed as a matrix with rows representing time steps.
- A MATLAB matrix has subscripts (i, j) which represent row and column respectively.

Common Abbreviations

2D	2-dimensional
3D	3-dimensional
DOF	Degrees of freedom
n-tuple	A group of n numbers, it can represent a point of a vector

Chapter

1 Introduction

The term robot means different things to different people. Science fiction books and movies have strongly influenced what many people expect a robot to be or what it can do. Sadly the practice of robotics is far behind this popular conception. One thing is certain though – robotics will be an important technology in this century. Products such as vacuum cleaning robots have already been with us for over a decade and self-driving cars are coming. These are the vanguard of a wave of smart machines that will appear in our homes and workplaces in the near to medium future.

In the eighteenth century the people of Europe were fascinated by automata such as Vaucanson's duck shown in Fig. 1.1a. These machines, complex by the standards of the day, demonstrated what then seemed *life-like* behavior. The duck used a cam mechanism to sequence its movements and Vaucanson went on to explore mechanization of silk weaving. Jacquard extended these ideas and developed a loom, shown in Fig. 1.1b, that was essentially a programmable weaving machine. The pattern to be woven was encoded as a series of holes on punched cards. This machine has many hallmarks of a modern robot: it performed a physical task and was reprogrammable.

The term robot first appeared in a 1920 Czech science fiction play "Rossum's Universal Robots" by Karel Čapek (pronounced Chapek). The term was coined by his brother Josef, and in the Czech language means serf labor but colloquially means hardwork or drudgery. The robots in the play were artificial people or androids and as in so many robot stories that follow this one, the robots rebel and it ends badly for humanity. Isaac Asimov's robot series, comprising many books and short stories written between 1950 and 1985, explored issues of human and robot interaction and morality. The robots in these stories are equipped with "positronic brains" in which the "Three laws of robotics" are encoded. These stories have influenced subsequent books and movies which in turn have shaped the public perception of what robots are. The mid twentieth century also saw the advent of the field of *cybernetics* – an uncommon term today but then an exciting science at the frontiers of understanding life and creating intelligent machines.

The first patent for what we would now consider a robot was filed in 1954 by George C. Devol and issued in 1961. The device comprised a mechanical arm with

Fig. 1.1.
Early programmable machines. **a** Vaucanson's duck (1739) was an automaton that could flap its wings, eat grain and defecate. It was driven by a clockwork mechanism and executed a single program; **b** The Jacquard loom (1801) was a reprogrammable machine and the program was held on punched cards (photograph by George P. Landow from www.victorianweb.org)

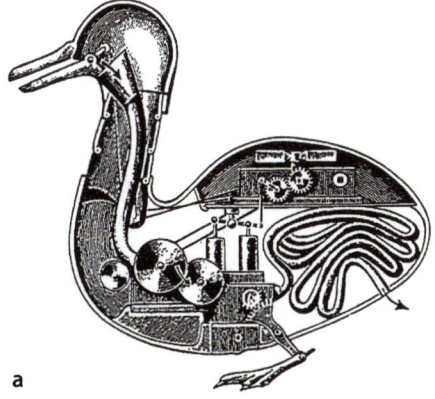

a

b

P. Corke, *Robotics and Control*, Springer Tracts in Advanced Robotics 141, https://doi.org/10.1007/978-3-030-79179-7_1

a gripper that was mounted on a track and the sequence of motions was encoded as magnetic patterns stored on a rotating drum. The first robotics company, Unimation, was founded by Devol and Joseph Engelberger in 1956 and their first industrial robot shown in Fig. 1.2 was installed in 1961. The original vision of Devol and Engelberger for robotic automation has become a reality and many millions of arm-type robots such as shown in Fig. 1.3 have been built and put to work at tasks such as welding, painting, machine loading and unloading, electronic assembly, packaging and palletizing. The use of robots has led to increased productivity and improved product quality. Today many products we buy have been assembled or handled by a robot.

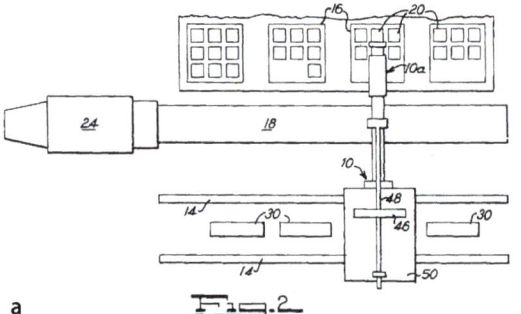

a

b

Fig. 1.2.
Universal automation. **a** A plan view of the machine from Devol's patent; **b** the first Unimation robot working at a General Motors factory (photo courtesy of George C. Devol)

Unimation Inc. (1956–1982). Devol sought financing to develop his unimation technology and at a cocktail party in 1954 he met Joseph Engelberger who was then an engineer with Manning, Maxwell and Moore. In 1956 they jointly established Unimation, the first robotics company, in Danbury Connecticut. The company was acquired by Consolidated Diesel Corp. (Condec) and became Unimate Inc. a division of Condec. Their first robot went to work in 1961 at a General Motors die-casting plant in New Jersey. In 1968 they licensed technology to Kawasaki Heavy Industries which produced the first Japanese industrial robot. Engelberger served as chief executive until it was acquired by Westinghouse in 1982. People and technologies from this company have gone on to be very influential on the whole field of robotics.

George C. Devol, Jr. (1912–2011) was a prolific American inventor. He was born in Louisville, Kentucky, and in 1932 founded United Cinephone Corp. which manufactured phonograph arms and amplifiers, registration controls for printing presses and packaging machines. In 1954, he applied for US patent 2,988,237 for Programmed Article Transfer which introduced the concept of Universal Automation or "Unimation". Specifically it described a track-mounted polar-coordinate arm mechanism with a gripper and a programmable controller – the precursor of all modern robots.

In 2011 he was inducted into the National Inventors Hall of Fame. (Photo on the right: courtesy of George C. Devol)

Joseph F. Engelberger (1925–2015) was an American engineer and entrepreneur who is often referred to as the "Father of Robotics". He received his B.S. and M.S. degrees in physics from Columbia University, in 1946 and 1949, respectively. Engelberger has been a tireless promoter of robotics. In 1966, he appeared on *The Tonight Show Starring Johnny Carson* with a Unimate robot which poured a beer, putted a golf ball, and directed the band. He promoted robotics heavily in Japan, which led to strong investment and development of robotic technology in that country.

Engelberger served as chief executive of Unimation until 1982, and in 1984 founded Transitions Research Corporation which became HelpMate Robotics Inc., an early entrant in the hospital service robot sector. He was elected to the National Academy of Engineering, received the Beckman Award and the Japan Prize, and has written two books: Robotics in Practice (1980) and Robotics in Service (1989). Each year the Robotics Industries Association presents an award in his honor to "persons who have contributed outstandingly to the furtherance of the science and practice of robotics."

These first-generation robots are fixed in place and cannot move about the factory – they are not mobile. By contrast mobile robots as shown in Figs. 1.4 and 1.5 can move through the world using various forms of mobility. They can locomote over the ground using wheels or legs, fly through the air using fixed wings or multiple rotors, move through the water or sail over it. An alternative taxonomy is based on the function that the robot performs. *Manufacturing* robots operate in factories and are the technological descendents of the first-generation robots. *Service robots* supply services to people such as cleaning, personal care, medical rehabilitation or fetching and carrying as shown in Fig. 1.5b. *Field robots*, such as those shown in Fig. 1.4, work outdoors on tasks such as environmental monitoring, agriculture, mining, construction and forestry. *Humanoid robots* such as shown in Fig. 1.6 have the physical form of a human being – they are both mobile robots and service robots. ◀

In practice the categorization of robots is not very consistently applied.

Fig. 1.3.
Manufacturing robots, technological descendants of the Unimate shown in Fig. 1.2. **a** A modern six-axis robot designed for high accuracy and throughput (image courtesy ABB robotics); **b** Baxter two-armed robot with built in vision capability and programmable by demonstration, designed for moderate throughput piece work (image courtesy Rethink Robotics)

Rossum's Universal Robots (RUR). In the introductory scene Helena Glory is visiting Harry Domin the director general of Rossum's Universal Robots and his robotic secretary Sulla.

Domin	Sulla, let Miss Glory have a look at you.
Helena	(stands and offers her hand) Pleased to meet you. It must be very hard for you out here, cut off from the rest of the world [the factory is on an island]
Sulla	I do not know the rest of the world Miss Glory. Please sit down.
Helena	(sits) Where are you from?
Sulla	From here, the factory
Helena	Oh, you were born here.
Sulla	Yes I was made here.
Helena	(startled) What?
Domin	(laughing) Sulla isn't a person, Miss Glory, she's a robot.
Helena	Oh, please forgive me …

The full play can be found at http://ebooks.adelaide.edu.au/c/capek/karel/rur. (Image on the left: Library of Congress item 96524672)

A manufacturing robot is typically an arm-type manipulator on a fixed base such as Fig. 1.3a that performs repetitive tasks within a local work cell. Parts are presented to the robot in an orderly fashion which maximizes the advantage of the robot's high speed and precision. High-speed robots are hazardous and safety is achieved by excluding people from robotic work places, typically placing the robot inside a cage. In contrast the Baxter robot shown in Fig. 1.3b is human safe, it operates at low speed and stops moving if it encounters an obstruction.

Field and service robots face specific and significant challenges. The first challenge is that the robot must operate and move in a complex, cluttered and changing environment. A delivery robot in a hospital must operate despite crowds of people and a time-varying configuration of parked carts and trolleys. A Mars rover as shown in Fig. 1.5a must navigate rocks and small craters despite not having an accurate local map in advance of its travel. Robotic, or self-driving cars, such as shown in Fig. 1.5c, must follow roads, avoid obstacles and obey traffic signals and the rules of the road. The second challenge for these types of robots is that they must operate safely in the presence of people. The hospital delivery robot operates among people, the robotic car contains people and a robotic surgical device operates *inside* people.

Fig. 1.4. Non-land-based mobile robots. **a** Small autonomous underwater vehicle (Todd Walsh © 2013 MBARI); **b** Global Hawk unmanned aerial vehicle (UAV) (photo courtesy of NASA)

Cybernetics, artificial intelligence and robotics. Cybernetics flourished as a research field from the 1930s until the 1960s and was fueled by a heady mix of new ideas and results from neurology, control theory and information theory. Research in neurology had shown that the brain was an electrical network of neurons. Harold Black, Henrik Bode and Harry Nyquist at Bell Labs were researching negative feedback and the stability of electrical networks, Claude Shannon's information theory described digital signals, and Alan Turing was exploring the fundamentals of computation. Walter Pitts and Warren McCulloch proposed an artificial neuron in 1943 and showed how it might perform simple logical functions. In 1951 Marvin Minsky built SNARC (from a B24 autopilot and comprising 3 000 vacuum tubes) which was perhaps the first neural-network-based learning machine as his graduate project. William Grey Walter's robotic tortoises showed life-like behavior. Maybe an electronic brain could be built!

An important early book was Norbert Wiener's *Cybernetics or Control and Communication in the Animal and the Machine* (Wiener 1965). A characteristic of a cybernetic system is the use of feedback which is common in engineering and biological systems. The ideas were later applied to evolutionary biology, psychology and economics.

In 1956 a watershed conference was hosted by John McCarthy at Dartmouth College and attended by Minsky, Shannon, Herbert Simon, Allen Newell and others. This meeting defined the term artificial intelligence (AI) as we know it today with an emphasis on digital computers and symbolic manipulation and led to new research in robotics, vision, natural language, semantics and reasoning. McCarthy and Minsky formed the AI group at MIT, and McCarthy left in 1962 to form the Stanford AI Laboratory. Minsky focused on artificially simple "blocks world". Simon, and his student Newell, were influential in AI research at Carnegie-Mellon University from which the Robotics Institute was spawned in 1979. These AI groups were to be very influential in the development of robotics and computer vision in the USA. Societies and publications focusing on cybernetics are still active today.

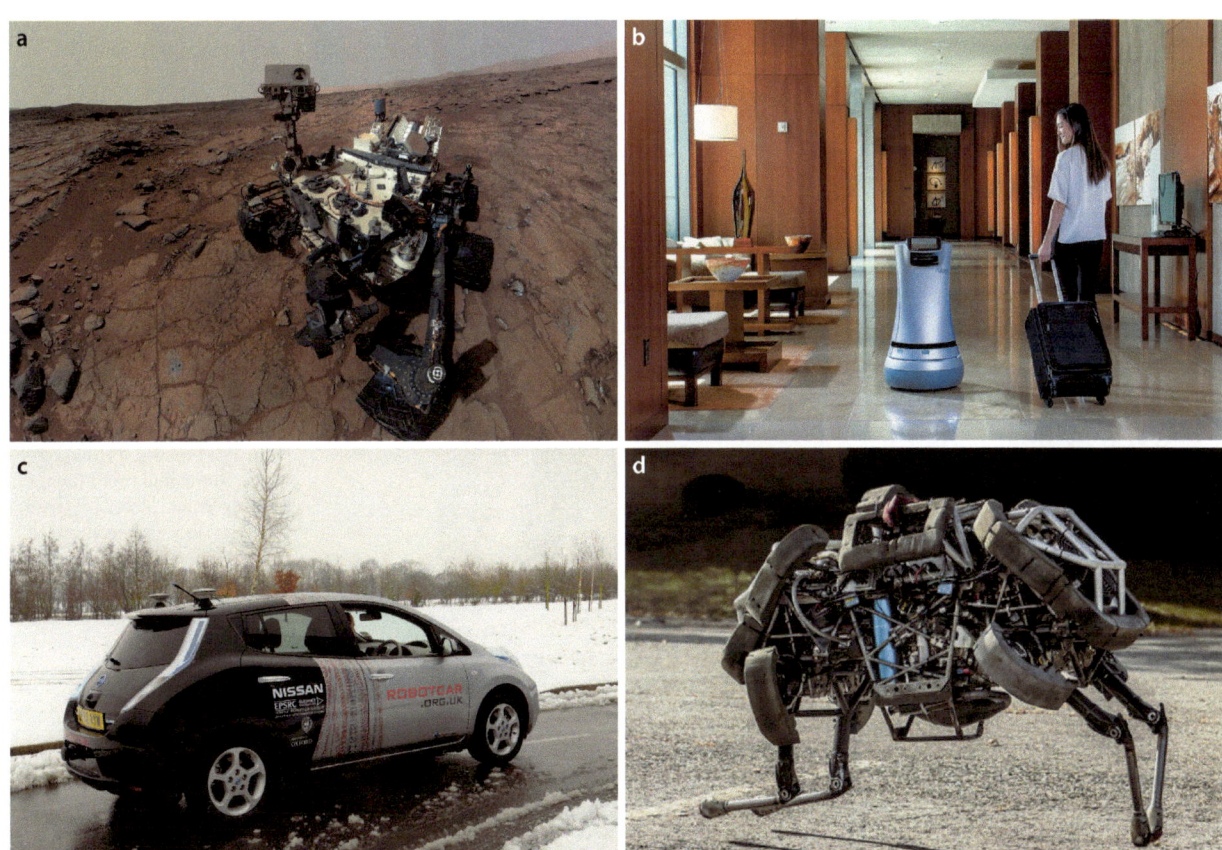

Fig. 1.5. Mobile robots. **a** Mars Science Lander, Curiosity, self portrait taken at "John Klein". The mast contains many cameras including two stereo camera pairs from which the robot can compute the 3-dimensional structure of its environment (image courtesy of NASA/JPL-Caltech/MSSS); **b** Savioke Relay delivery robot (image courtesy Savioke); **c** self driving car (image courtesy Dept. Information Engineering, Oxford Univ.); **d** Cheetah legged robot (image courtesy Boston Dynamics)

So what is a robot? There are many definitions and not all of them are particularly helpful. A definition that will serve us well in this book is

a goal oriented machine that can sense, plan and act.

A robot *senses* its environment and uses that information, together with a goal, to *plan* some *action*. The action might be to move the tool of an arm-robot to grasp an object or it might be to drive a mobile robot to some place.

Sensing is critical to robots. Proprioceptive sensors measure the state of the robot itself: the angle of the joints on a robot arm, the number of wheel revolutions on a mobile robot or the current drawn by an electric motor. Exteroceptive sensors measure the state of the world with respect to the robot. The sensor might be a simple bump sensor on a robot vacuum cleaner to detect collision. It might be a GPS receiver that measures distances to an orbiting satellite constellation, or a compass that measures the direction of the Earth's magnetic field vector relative to the robot. It might also be an active sensor that emits acoustic, optical (LIDAR) or radio (RADAR) pulses in order to measure the distance to points in the world based on the time taken for a reflection to return to the sensor. Or it might be a camera which is a passive sensor that captures patterns of ambient optical energy reflected from the scene. Our own experience is that eyes are a very effective sensor for recognition, navigation, obstacle avoidance and manipulation so vision has long been of interest to robotics researchers.

Telerobots are robot-like machines that are remotely controlled by a human operator. Perhaps the earliest was a radio controlled boat demonstrated by Nikola Tesla in 1898 and which he called a teleautomaton. According to the definition above these are not robots but they were an important precursor to robots and are still important

a

b
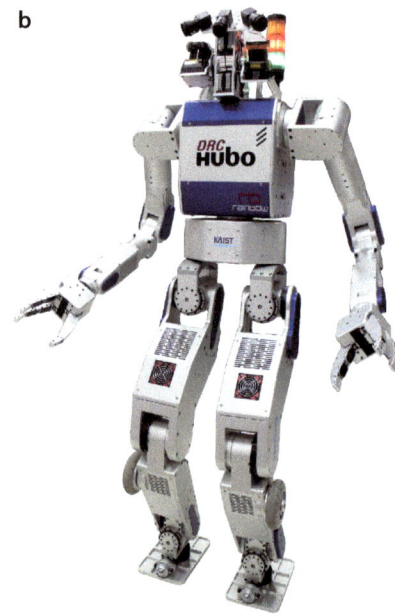

Fig. 1.6.
Humanoid robots. **a** Honda's
Asimo humanoid robot (image
courtesy Honda Motor Co. Japan);
b Hubo robot that won the
DARPA Robotics Challenge in
2015 (image courtesy KAIST,
Korea)

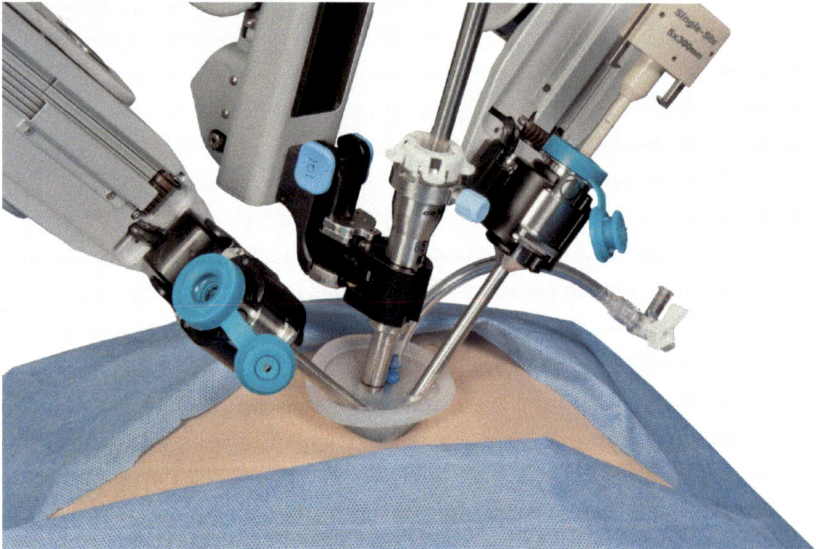

Fig. 1.7.
The working end of a surgical
robot, multiple tools working
through a single small inci-
sion (image © 2015 Intuitive
Surgical, Inc)

today for many tasks where people cannot work but which are too complex for a
machine to perform by itself. For example the underwater robots that surveyed the
wreck of the Titanic were technically telerobots or remotely operated vehicles (ROVs).
A modern surgical robot as shown in Fig. 1.7 is also teleoperated – the motion of the
small tools are remotely controlled by the surgeon and this makes it possible to use
much smaller incisions than the old-fashioned approach where the surgeon works
inside the body with their hands.

The various Mars rovers autonomously navigate the surface of Mars but human operators provide the high-level goals. That is, the operators tell the robot where to go and
the robot itself determines the details of the route. Local decision making on Mars is essential given that the communications delay is several minutes. Some robots are hybrids
and the control task is shared or traded with a human operator. In traded control, the
control function is passed back and forth between the human operator and the computer.
For example an aircraft pilot can pass control to an autopilot and take control back. In

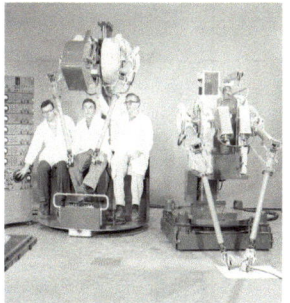
The Manhattan Project in World War 2 (WW II) developed the first nuclear weapons and this required handling of radioactive material. Remotely controlled arms were developed by Ray Goertz at Argonne National Laboratory to exploit the manual dexterity of human operators while keeping them away from the hazards of the material they were handling. The operators viewed the work space through thick lead-glass windows or via a television link and manipulated the master arm (on the left). The slave arm (on the right) followed the motion, and forces felt by the slave arm were reflected back to the master arm, allowing the operator to feel weight and interference force. Telerobotics is still important today for many tasks where people cannot work but which are too complex for a machine to perform by itself, for instance the underwater robots that surveyed the wreck of the Titanic. (Photo on the left: Courtesy Argonne National Laboratory)

shared control, the control function is performed by the human operator and the computer working together. For example an autonomous passenger car might have the computer keeping the car safely in the lane while the human driver just controls speed.

1.1   Robots, Jobs and Ethics

A number of ethical issues arise from the advent of robotics. Perhaps the greatest concern to the wider public is "robots taking jobs from people". This is a complex issue but we cannot shy away from the fact that many jobs now done by people will, in the future, be performed by robots. Clearly there are dangerous jobs which people should not do, for example handling hazardous substances or working in dangerous environments. There are many low-skilled jobs where human labor is increasingly hard to source, for instance in jobs like fruit picking or waste sorting. In many developed countries people no longer aspire to hard physical outdoor work in remote locations. What are the alternatives if people don't want to do the work? In areas like manufacturing, particularly car manufacturing, the adoption of robotic automation has been critical in raising productivity which has allowed that industry to be economically viable in high-wage countries like Europe, Japan and the USA. Without robots these industries could not exist; they would not employ any people, not pay any taxes, and not consume products and services from other parts of the economy. Automated industry might employ fewer people but it still makes an important contribution to society. Rather than taking jobs we could argue that robotics and automation has helped to keep manufacturing industries viable in high-labor cost countries. How do we balance the good of the society with the good of the individual?

There are other issues besides jobs. Consider self-driving cars. We are surprisingly accepting of manually driven cars even though they kill more than one million people every year, yet many are uncomfortable with the idea of self-driving cars even though they will dramatically reduce this loss of life. We worry about who to blame if a robotic car makes a mistake while the carnage caused by human drivers continues. Similar concerns are raised when talking about robotic healthcare and surgery – human surgeons are not perfect but robots are seemingly held to a much higher account. There is a lot of talk about using robots to look after elderly people, but does this detract from their quality of life by removing human contact, conversation and companionship? Should we use robots to look after our children, and even teach them? What do we think of armies of robots fighting and killing human beings?

Robotic cars, health care, elder care and child care might bring economic benefits to our society but is it the right thing to do? Is it a direction that we want our society to go? Once again how do we balance the good of the society with the good of the individual? These are deep ethical questions that cannot and should not be decided by roboticists alone. But neither should roboticists ignore them. This is a discussion for all of society and roboticists have a duty to be active participants in this debate.

1.2   About the Book

Robotics is a big topic and the coverage in this book is necessarily broad. The intent is not to be shallow but rather to give the reader a flavor of what robotics is about and how robots work – consider it a grand tasting menu.

The goals of the book are:

- to provide a broad and solid base of understanding through theory and examples;
- to make abstract concepts tangible
- to tackle more complex problems than other more specialized textbooks by virtue of the powerful numerical tools and software that underpins it;
- to provide instant gratification by solving complex problems with relatively little code;
- to complement the many excellent texts in robotics;
- to encourage intuition through hands on numerical experimentation; and
- to limit the number of equations presented to those cases where I think they add real value or clarity.

The approach used is to present background, theory and examples in an integrated fashion. Code and examples are first-class citizens in this book and are not relegated to the end of the chapter or an associated web site. The examples are woven into the discussion like this

```
>> p = transl(Ts);
>> plot(t, p);
```

where the MATLAB® code illuminates the topic being discussed and generally results in a crisp numerical result or a graph in a figure that is then discussed. The examples illustrate how to use the associated Toolbox and that knowledge can then be applied to other problems. Most of the figures in this book have been generated by the code examples provided and they are available from the book's website as described in Appendix A.

1.2.1   MATLAB Software and the Toolboxes

To do good work, one must first have good tools.
Chinese proverb

The computational foundation of this book is MATLAB®, a software package developed by The MathWorks Inc. MATLAB is an interactive mathematical software environment that makes linear algebra, data analysis and high-quality graphics a breeze. MATLAB is a popular package and one that is very likely to be familiar to engineering students as well as researchers. It also supports a programming language▸ which allows the creation of complex algorithms.

One of the 20 most common programming languages.

A strength of MATLAB is its support for Toolboxes which are collections of functions targeted at particular topics. Thousands of Toolboxes are available from MathWorks, third party companies and individuals. Some Toolboxes are products to be purchased while others are open-source and generally free to use.

This book is based on an open-source Toolbox written by the author: the Robotics Toolbox for MATLAB (RTB). It provides a diverse range of functions for simulating mobile and arm-type robots. The Toolbox supports a very general method of representing the structure of serial-link manipulators using MATLAB objects and provides functions for forward and inverse kinematics and dynamics. The Toolbox includes functions for manipulating and converting between datatypes such as vectors, homogeneous transformations, 3-angle representations, twists and unit-quaternions which are necessary to represent 3-dimensional position and orientation. The Toolbox also includes functionality for simulating mobile robots and includes models of wheeled

The MATLAB software we use today has a long history. It starts with the LINPACK and EISPACK projects run by the Argonne National Laboratory in the 1970s to produce high quality, tested and portable mathematical software. LINPACK is a collection of routines for linear algebra and EISPACK is a library of numerical algorithms for computing eigenvalues and eigenvectors of matrices. These packages were written in Fortran which was then the language of choice for large-scale numerical problems.

Cleve Moler, then at the University of New Mexico, contributed to both projects and wrote the first version of MATLAB in the late 1970s. It allowed interactive use of LINPACK and EISPACK for problem solving without having to write and compile Fortran code. MATLAB quickly spread to other universities and found a strong audience within the applied mathematics and engineering community. In 1984 Cleve Moler and Jack Little founded The MathWorks Inc. which exploited the newly released IBM PC – the first widely available desktop computer.

Cleve Moler received his bachelor's degree from Caltech in 1961, and a Ph.D. from Stanford University. He was a professor of mathematics and computer science at universities including University of Michigan, Stanford University, and the University of New Mexico. He has served as president of the Society for Industrial and Applied Mathematics (SIAM) and was elected to the National Academy of Engineering in 1997.

See also http://www.mathworks.com/company/aboutus/founders/clevemoler.html which includes a video of Cleve Moler and also http://history.siam.org/pdfs2/Moler_final.pdf.

vehicles and quadrotors and controllers for these vehicles. It also provides standard algorithms for robot path planning, localization, map making and SLAM. The Toolbox plus MATLAB, turns a personal computer into a powerful and convenient environment for investigating complex problems in robotics and control. The Toolbox is free to use and is distributed under the GNU Lesser General Public License (GNU LGPL).

The *Machine Vision Toolbox* (MVTB) provides a rich collection of functions for camera modeling, image processing, image feature extraction, multi-view geometry and vision-based control. The MVTB also contains functions for image acquisition and display; filtering; blob, point and line feature extraction; mathematical morphology; image warping; stereo vision; homography and fundamental matrix estimation; robust estimation; bundle adjustment; visual Jacobians; geometric camera models; camera calibration and color space operations. For modest image sizes on a modern computer the processing rate can be sufficiently "real-time" to allow for closed-loop control.

If you're starting out in robotics the Toolbox is a significant initial base of code on which to build your project, and is provided in source code form. The bulk of the code is written in the MATLAB M-language but a few functions are written in C◄ or Java for increased computational efficiency. In general the Toolbox code is written in a straightforward manner to facilitate understanding, perhaps at the expense of computational efficiency. Appendix A provides details of how to obtain the Toolboxes and pointers to online resources including discussion groups.

This book provides examples of how to use many Toolbox functions in the context of solving specific problems but it is not a reference manual. Comprehensive documentation of all Toolbox functions is available through the MATLAB builtin help mechanism or the PDF format manual that is distributed with the Toolbox.

These are implemented as MEX files, which are written in C in a very specific way that allows them to be invoked from MATLAB just like a function written in M-language.

1.2.2 Notation, Conventions and Organization

The mathematical notation used in the book is summarized in the Nomenclature section on page xxi. Since the coverage of the book is broad there are just not enough good symbols to go around, so it is unavoidable that some symbols have different meanings in different parts of the book.

There is a lot of MATLAB code in the book indicated in blue fixed-width font such as

```
>> a = 2 + 2
a =
    4
```

The MATLAB command prompt is >> and what follows on that line is the command issued to MATLAB by the user. Subsequent lines, without the prompt, are MATLAB's response. All functions, classes and methods mentioned in the text or in code segments are cross-referenced and have their own indexes at the end of the book allowing you to find different ways that particular functions can be used.

Colored boxes are used to indicate different types of material. Orange informational boxes highlight material that is particularly important while red and orange warning boxes highlight points that are often traps for those starting out. Blue boxes provide technical, historical or biographical information that augment the main text but they are not critical to its understanding.

As an author there is a tension between completeness, clarity and conciseness. For this reason a lot of detail has been pushed into notes▸ and blue boxes and on a first reading these can be skipped. Some chapters have an Advanced Topics section at the end that can also be skipped on a first reading. However if you are trying to understand a particular algorithm and apply it to your own problem then understanding the details and nuances can be important and the notes or advanced topics are for you.

They are placed as marginal notes near the corresponding marker.

Each chapter ends with a *Wrapping Up* section that summarizes the important lessons from the chapter, discusses some suggested further reading, and provides some exercises. For clarity, references are cited sparingly in the text of each chapter. The *Further Reading* subsection discusses prior work and references that provide more rigor or more complete description of the algorithms. *Resources* provides links to relevant online code and datasets. *MATLAB Notes* provides additional details about the author's toolbox and those with similar functionality from MathWorks. *Exercises* extend the concepts discussed within the chapter and are generally related to specific code examples discussed in the chapter. The exercises vary in difficulty from straightforward extension of the code examples to more challenging problems.

1.2.3   Audience and Prerequisites

The book is intended primarily for third or fourth year engineering undergraduate students, Masters students and first year Ph.D. students. For undergraduates the book will serve as a companion text for a robotics course or major project. Students should study Part I and the appendices for foundational concepts, and then the relevant part of the book: Part II (mobile robotics) or Part III (arm robots). The Toolbox provides a solid set of tools for problem solving, and the exercises at the end of each chapter provide additional problems beyond the worked examples in the book.

For students commencing graduate study in robotics, and who have previously studied engineering or computer science, the book will help fill the gaps between what you learned as an undergraduate and what will be required to underpin your deeper study of robotics. The book's working code base can help bootstrap your research, enabling you to get started quickly and working productively on your own problems and ideas. Since the source code is available you can reshape it to suit your need, and when the time comes (as it usually does) to code your algorithms in some other language then the Toolbox can be used to cross-check your implementation.

For those who are no longer students, the researcher or industry practitioner, the book will serve as a useful companion for your own reference to a wide range of topics in robotics, as well as a handbook and guide for the Toolbox.

The book assumes undergraduate-level knowledge of linear algebra (matrices, vectors, eigenvalues), complex numbers, basic group theory, basic graph theory, probability, dynamics (forces, torques, inertia) and control theory. Some of these topics will likely be more familiar to engineering students than computer science students. Computer science students may struggle with some concepts in Chap. 4 and 9 such as the Laplace transform, transfer functions, linear control (proportional control, proportional-derivative control, proportional-integral control) and block diagram notation. This material could be skimmed over on a first reading and Albertos and Mareels (2010) may be a useful introduction to some of these topics. The book also assumes the reader is familiar with using and programming in MATLAB and also familiar with object-oriented programming techniques (perhaps C++, Java or Python). Familiarity with Simulink®, the graphical block-diagram modeling tool integrated with MATLAB will be helpful but not essential.

1.2.4   Learning with the Book

The best way to learn is by doing. Although the book shows the MATLAB commands and the response there is something special about doing it for yourself. Consider the book as an invitation to tinker. By running the commands yourself you can look at the results in ways that you prefer, plot the results in a different way, or try the algorithm on different data or with different parameters. The book is especially designed to stay open which enables you to type in commands as you read. You can also look at the online documentation for the Toolbox functions, discover additional features and options, and experiment with those, or read the code to see how it really works and perhaps modify it.

Most of the commands are quite short so typing them into MATLAB is not too onerous. However the book's web site, see Appendix A, provides access to all the MATLAB commands shown in the book (more than 1 800 lines). This is provided by MATLAB LiveScripts, one per chapter, that you can run directly or cut and paste from.

The Robot Academy is a free and open learning resource that comprises over 200 short video lessons that cover many of the topics in this book, as well as simple online quizzes. It can be found at http://robotacademy.net.au.

1.2.5   Teaching with the Book

The book can be used in support of courses in robotics and mechatronics. All courses should include the introduction to coordinate frames and their composition which is discussed in Chap. 2. For a mobile robotics course it is sufficient to teach only the 2-dimensional case. For robotic manipulators the 2- and 3-dimensional cases should be taught.

Most figures (MATLAB-generated and line drawings) in this book are available as PDF format files from the book's web site and you are free to use them with attribution in any course material that you prepare. All the code in this book can be downloaded from the web site and used as the basis for demonstrations in lectures or tutorials. See Appendix A for details.

The exercises at the end of each chapter can be used as the basis of assignments, or as examples to be worked in class or in tutorials. Most of the questions are rather open ended in order to encourage exploration and discovery of the effects of parameters and the limits of performance of algorithms. This exploration should be supported by discussion and debate about performance measures and what *best* means. True understanding of algorithms involves an appreciation of the effects of parameters, how algorithms fail and under what circumstances.

The teaching approach could also be inverted, by diving headfirst into a particular problem and then teaching the appropriate prerequisite material. Suitable problems could be chosen from the Application sections of Chap. 7 or from any of the exercises. Particularly challenging exercises are so marked.

If you wanted to consider a flipped learning approach, then the Robot Academy mentioned on page 11 could be used in conjunction with your class. Students would watch the videos and undertake some formative assessment out of the classroom, and you could use classroom time to work through problem sets.

For graduate level teaching the papers and textbooks mentioned in the *Further Reading* could form the basis of a student's reading list. They could also serve as candidate papers for a reading group or journal club.

1.2.6   Outline

I promised a book with instant gratification but before we can get started in robotics there are some fundamental concepts that we absolutely need to understand, and understand well. Part I introduces the concepts of pose and coordinate frames – how we represent the position and orientation of a robot, a camera or the objects that the robot needs to work with. We discuss how motion between two poses can be *decomposed* into a sequence of elementary translations and rotations, and how elementary motions can be *composed* into more complex motions. Chapter 2 discusses how pose can be represented in a computer, and Chap. 3 discusses the relationship between velocity and the derivative of pose, estimating motion from sensors and generating a sequence of poses that smoothly follow some path in space and time.

With these formalities out of the way we move on to the main event – robots. There are two important classes of robot: mobile robots and manipulator arms and these are covered in Parts II and III respectively.▶

Part II begins, in Chap. 4, with motion models for several types of wheeled vehicles and a multi-rotor flying vehicle. Various control laws are discussed for wheeled vehicles such as moving to a point, following a path and moving to a specific pose. Chapter 5 is concerned with navigation, that is, how a robot finds a path between points A and B in the world. Two important cases, with and without a map, are discussed. Most navigation techniques require knowledge of the robot's position and Chap. 6 discusses various approaches to this problem based on dead-reckoning, or landmark observation and a map. We also show how a robot can make a map, and even determine its location while simultaneously mapping an unknown region.

Part III is concerned with arm-type robots, or more precisely serial-link manipulators. Manipulator arms are used for tasks such as assembly, welding, material handling and even surgery. Chapter 7 introduces the topic of kinematics which relates the angles of the robot's joints to the 3-dimensional pose of the robot's tool. Techniques to generate smooth paths for the tool are discussed and two examples show how an arm-robot can draw a letter on a surface and how multiple arms (acting as legs) can be used to create a model for a simple walking robot. Chapter 8 discusses the relationships between the rates of change of joint angles and tool pose. It introduces the manipulator Jacobian matrix and concepts such as singularities, manipulability, null-space motion, and resolved-rate motion control. It also discusses under- and over-actuated robots and the general numerical solution to inverse kinematics. Chapter 9 introduces the design of joint control systems, the dynamic equations of motion for a serial-link manipulator, and the relationship between joint forces and joint motion. It discusses important topics such as variation in inertia, the effect of payload, flexible transmissions and independent joint versus nonlinear control strategies.

This is a big book but any one of the parts can be read standalone, with more or less frequent visits to the required earlier material. Chapter 2 is the only mandatory material. Parts II or III could be used respectively for an introduction to mobile robots or arm-type robots class. An alternative approach, following the instant gratification theme, is to jump straight into any chapter and start exploring – visiting the earlier material as required.

Although robot arms came first chronologically, mobile robotics is mostly a 2-dimensional problem and easier to understand than the 3-dimensional arm-robot case.

Further Reading

The Handbook of Robotics (Siciliano and Khatib 2016) provides encyclopedic coverage of the field of robotics today, covering theory, technology and the different types of robot such as telerobots, service robots, field robots, flying robots, underwater robots and so on. The classic work by Sheridan (2003) discusses the spectrum of autonomy from remote control, through shared and traded control to full autonomy. A solid introduction to artificial intelligence is the text by Russell and Norvig (2009).

A number of recent books discuss the future impacts of robotics and artificial intelligence on society, for example Ford (2015), Brynjolfsson and McAfee (2014), Bostrom (2016) and Neilson (2011). The YouTube video Grey (2014) makes some powerful points about the future of work and is always a great discussion starter.

Resources for learning MATLAB are available at https://matlabacademy.mathworks.com. Some are free and open, while others require a MATLAB licence.

Part I Foundations

Chapter 2   **Representing Position and Orientation**

Chapter 3   **Time and Motion**

2

Representing Position and Orientation

Numbers are an important part of mathematics. We use numbers for counting: *there are 2 apples*. We use *denominate numbers*, a number plus a unit, to specify distance: *the object is 2* m *away*. We also call this single number a *scalar*. We use a vector, a denominate number plus a direction, to specify a location: *the object is 2* m *due north*. We may also want to know the orientation of the object: *the object is 2* m *due north and facing west*. The combination of position and orientation we call *pose*.

A point in space is a familiar concept from mathematics and can be described by a coordinate vector, as shown in Fig. 2.1a. The vector represents the displacement of the point with respect to some reference coordinate frame – we call this a bound vector since it cannot be freely moved. A coordinate frame, or Cartesian coordinate system, is a set of orthogonal axes which intersect at a point known as the origin. A vector can be described in terms of its components, a linear combination of unit vectors which are parallel to the axes of the coordinate frame. Note that points and vectors are different types of mathematical objects even though each can be described by a tuple of numbers. We can add vectors but adding points makes no sense. The difference of two points is a vector, and we can add a vector to a point to obtain another point.

A point is an interesting mathematical abstraction, but a real object comprises infinitely many points. An object, unlike a point, also has an orientation. If we attach a coordinate frame to an object, as shown in Fig. 2.1b, we can describe every point within the object as a constant vector with respect to that frame.◄ Now we can describe the position and orientation – the pose – of that coordinate frame with respect to the reference coordinate frame. To distinguish the different frames we label them and in this case the object coordinate frame is labeled $\{B\}$ and its axes are labeled x_B and y_B, adopting the frame's label as their subscript.

To completely describe the pose of a rigid object in a 3-dimensional world we need 6 not 3 dimensions: 3 to describe its position and 3 to describe its orientation. These dimensions behave quite differently. If we increase the value of one of the position dimensions the object will move continuously in a straight line, but if we increase the value of one of the orientation dimensions the object will rotate in some way and soon get back to its original orientation – this dimension is curved. We clearly need to treat the position and orientation dimensions quite differently.

We assume that the object is rigid, that is, the points do not move with respect to each other.

Fig. 2.1.
a The point **P** is described by a coordinate vector with respect to an absolute coordinate frame. **b** The points are described with respect to the object's coordinate frame $\{B\}$ which in turn is described by a relative pose ξ_B. Axes are denoted by thick lines with an open arrow, vectors by thin lines with a swept arrow head and a pose by a thick line with a solid head

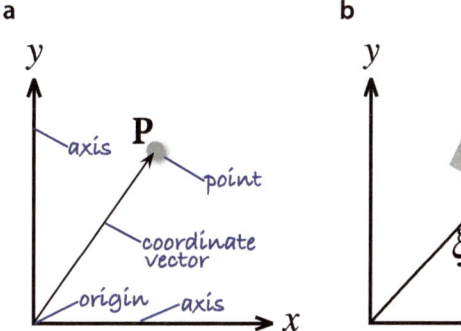

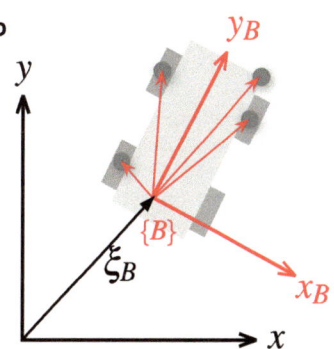

The pose of the coordinate frame is denoted by the symbol ξ – pronounced ksi. Figure 2.2 shows two frames $\{A\}$ and $\{B\}$ and the relative pose $^A\xi_B$ which describes $\{B\}$ with respect to $\{A\}$. The leading superscript denotes the reference coordinate frame and the subscript denotes the frame being described. We could also think of $^A\xi_B$ as describing some motion – imagine picking up $\{A\}$ and applying a displacement and a rotation so that it is transformed to $\{B\}$. If the initial superscript is missing we assume that the change in pose is relative to the world coordinate frame which is generally denoted $\{O\}$.

The point **P** in Fig. 2.2 can be described with respect to *either* coordinate frame by the vectors $^A\boldsymbol{p}$ or $^B\boldsymbol{p}$ respectively. Formally they are related by

$$^A\boldsymbol{p} = {^A\xi_B} \cdot {^B\boldsymbol{p}} \tag{2.1}$$

where the right-hand side expresses the motion from $\{A\}$ to $\{B\}$ and then to **P**. The operator $\cdot$ *transforms* the vector, resulting in a new vector that describes the same point but with respect to a different coordinate frame.

An important characteristic of relative poses is that they can be *composed* or *compounded*. Consider the case shown in Fig. 2.3. If one frame can be described in terms of another by a relative pose then they can be applied sequentially

$$^A\xi_C = {^A\xi_B} \oplus {^B\xi_C}$$

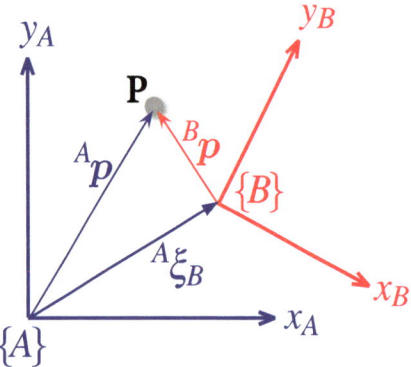

Fig. 2.2.
The point **P** can be described by coordinate vectors relative to either frame $\{A\}$ or $\{B\}$. The pose of $\{B\}$ relative to $\{A\}$ is $^A\xi_B$

In relative pose composition we can check that we have our reference frames correct by ensuring that the subscript and superscript on each side of the $\oplus$ operator are matched. We can then *cancel out* the intermediate subscripts and superscripts

$$^X\xi_Z = {^X\xi_X} \oplus {^X\xi_Z}$$

leaving just the end most subscript and superscript which are shown highlighted.

Euclid of Alexandria (ca. 325 BCE–265 BCE) was a Greek mathematician, who was born and lived in Alexandria Egypt, and is considered the "father of geometry". His great work *Elements* comprising 13 books, captured and systematized much early knowledge about geometry and numbers. It deduces the properties of planar and solid geometric shapes from a set of 5 axioms and 5 postulates.

Elements is probably the most successful book in the history of mathematics. It describes plane geometry and is the basis for most people's first introduction to geometry and formal proof, and is the basis of what we now call Euclidean geometry. Euclidean distance is simply the distance between two points on a plane. Euclid also wrote *Optics* which describes geometric vision and perspective.

which says, in words, that the pose of {C} relative to {A} can be obtained by compounding the relative poses from {A} to {B} and then {B} to {C}. We use the operator ⊕ to indicate *composition* of relative poses.

For this case the point **P** can be described by

$$^A\boldsymbol{p} = \left({}^A\xi_B \oplus {}^B\xi_C \right) \cdot {}^C\boldsymbol{p}$$

Later in this chapter we will convert these abstract notions of ξ, • and ⊕ into standard mathematical objects and operators that we can implement in MATLAB®.

In the examples so far we have shown 2-dimensional coordinate frames. This is appropriate for a large class of robotics problems, particularly for mobile robots which operate in a planar world. For other problems we require 3-dimensional coordinate frames to describe objects in our 3-dimensional world such as the pose of a flying or underwater robot or the end of a tool carried by a robot arm.

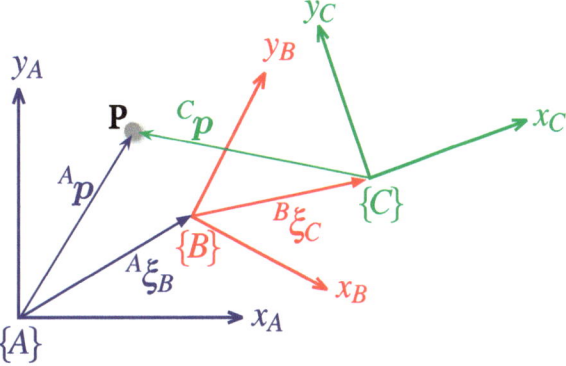

Fig. 2.3.
The point **P** can be described by coordinate vectors relative to either frame {A}, {B} or {C}. The frames are described by relative poses

Euclidean versus Cartesian geometry. Euclidean geometry is concerned with points and lines in the Euclidean plane (2D) or Euclidean space (3D). It is entirely based on a set of axioms and makes no use of arithmetic. Descartes added a coordinate system (2D or 3D) and was then able to describe points, lines and other curves in terms of algebraic equations. The study of such equations is called analytic geometry and is the basis of all modern geometry. The Cartesian plane (or space) is the Euclidean plane (or space) with all its axioms and postulates *plus* the extra facilities afforded by the added coordinate system. The term Euclidean geometry is often used to mean that Euclid's fifth postulate (parallel lines never intersect) holds, which is the case for a planar surface but not for a curved surface.

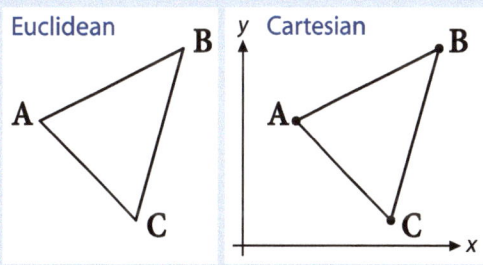

René Descartes (1596–1650) was a French philosopher, mathematician and part-time mercenary. He is famous for the philosophical statement "*Cogito, ergo sum*" or "*I am thinking, therefore I exist*" or "*I think, therefore I am*". He was a sickly child and developed a life-long habit of lying in bed and thinking until late morning. A possibly apocryphal story is that during one such morning he was watching a fly walk across the ceiling and realized that he could describe its position in terms of its distance from the two edges of the ceiling. This is the basis of the *Cartesian* coordinate system and modern (analytic) geometry, which he described in his 1637 book La Géométrie. For the first time mathematics and geometry were connected, and modern calculus was built on this foundation by Newton and Leibniz. In Sweden at the invitation of Queen Christina he was obliged to rise at 5 A.M., breaking his lifetime habit – he caught pneumonia and died. His remains were later moved to Paris, and are now lost apart from his skull which is in the Musée de l'Homme. After his death, the Roman Catholic Church placed his works on the Index of Prohibited Books.

Figure 2.4 shows a more complex 3-dimensional example in a graphical form where we have attached 3D coordinate frames to the various entities and indicated some relative poses. The fixed camera observes the object from its fixed viewpoint and estimates the object's pose $^F\xi_B$ relative to itself. The other camera is not fixed, it is attached to the robot at some constant relative pose and estimates the object's pose $^C\xi_B$ relative to itself.

An alternative representation of the spatial relationships is a directed graph (see Appendix I) which is shown in Fig. 2.5.▶ Each node in the graph represents a pose and each edge represents a *relative* pose. An arrow from node X to node Y denotes $^X\xi_Y$ the pose of Y relative to X. Recalling that we can compose relative poses using the $\oplus$ operator we can write some spatial relationships

It is quite possible that a pose graph can be inconsistent, that is, two paths through the graph give different results. In robotics these poses are only ever derived from noisy sensor data.

$$\xi_F \oplus {}^F\xi_B = \xi_R \oplus {}^R\xi_C \oplus {}^C\xi_B$$

$$\xi_F \oplus {}^F\xi_R = \xi_R$$

and each equation represents a loop in the graph with each side of the equation starting and ending at the same node. Each side of the first equation represents a path through the network from {0} to {B}, a sequence of edges (arrows) written in order.

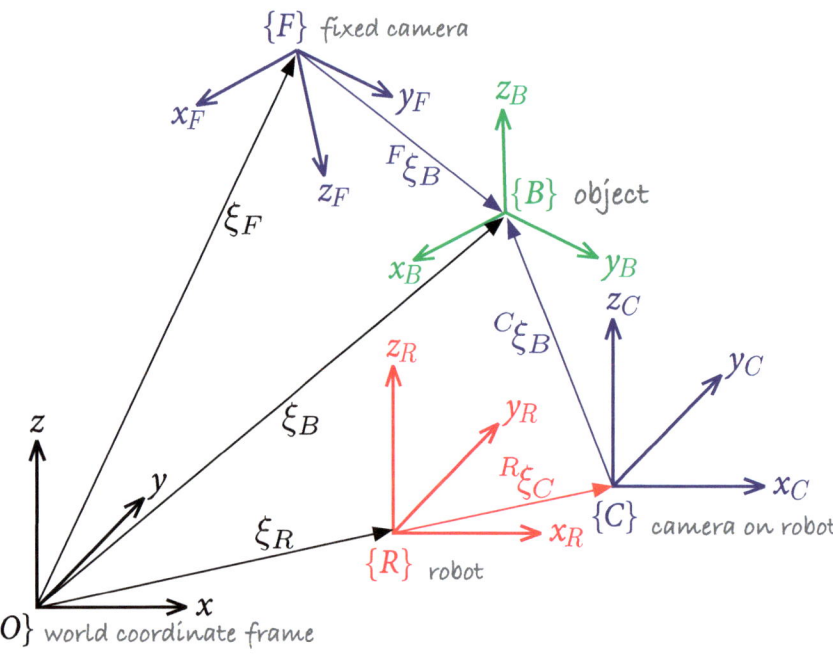

Fig. 2.4.
Multiple 3-dimensional coordinate frames and relative poses

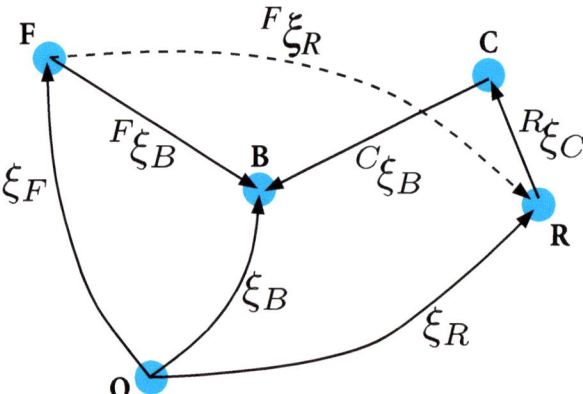

Fig. 2.5.
Spatial example of Fig. 2.4 expressed as a directed graph

In mathematical terms poses constitute a group – a set of objects that supports an associative binary operator (composition) whose result belongs to the group, an inverse operation and an identity element. In this case the group is the special Euclidean group in either 2 or 3 dimensions which are commonly referred to as **SE**(2) or **SE**(3) respectively.

There are just a few algebraic rules:◄

$$\xi \oplus 0 = \xi, \ \xi \ominus 0 = \xi$$
$$\xi \ominus \xi = 0, \ \ominus\xi \oplus \xi = 0$$

where 0 represents a zero relative pose. A pose has an inverse

$$\ominus {}^X\xi_Y = {}^Y\xi_X$$

which is represented graphically by an arrow from $\{Y\}$ to $\{X\}$. Relative poses can also be composed or compounded

$$ {}^X\xi_Y \oplus {}^Y\xi_Z = {}^X\xi_Z$$

It is important to note that the algebraic rules for poses are different to normal algebra and that composition is *not* commutative

$$\xi_1 \oplus \xi_2 \neq \xi_2 \oplus \xi_1$$

with the exception being the case where $\xi_1 \oplus \xi_2 = 0$. A relative pose can transform a point expressed as a vector relative to one frame to a vector relative to another

$$ {}^X\boldsymbol{p} = {}^X\xi_Y \cdot {}^Y\boldsymbol{p}$$

A very useful property of poses is the ability to perform algebra. The second loop equation says, in words, that the pose of the robot is the same as composing two relative poses: from the world frame to the fixed camera and from the fixed camera to the robot. We can subtract ξ_F from both sides of the equation◄ by adding the inverse of ξ_F which we denote as $\ominus\xi_F$ and this gives

Order is important here, and we add $\ominus\xi_F$ to the left on each side of the equation.

$$\ominus\xi_F \oplus \xi_F \oplus {}^F\xi_R = \ominus\xi_F \oplus \xi_R$$
$$ {}^F\xi_R = \ominus\xi_F \oplus \xi_R$$

which is the pose of the robot relative to the fixed camera, shown as a dashed line in Fig. 2.5.

We can write these expressions quickly by inspection. To find the pose of node X with respect to node Y:

- find a path from Y to X and write down the relative poses on the edges in a left to right order;
- if you traverse the edge in the direction of its arrow precede it with the $\oplus$ operator, otherwise use $\ominus$.

So what is ξ? It can be any mathematical object that supports the algebra described above and is suited to the problem at hand. It will depend on whether we are considering a 2- or 3-dimensional problem. Some of the objects that we will discuss in the rest of this chapter will be familiar to us, for example vectors, but others will be more exotic mathematical objects such as homogeneous transformations, orthonormal rotation matrices, twists and quaternions. Fortunately all these mathematical objects are well suited to the mathematical programming environment of MATLAB.

To recap:

1. A point is described by a bound coordinate vector that represents its displacement from the origin of a reference coordinate system.
2. Points and vectors are different things even though they are each described by a tuple of numbers. We can add vectors but not points. The difference between two points is a vector.
3. A set of points that represent a rigid object can be described by a single coordinate frame, and its constituent points are described by constant vectors relative to that coordinate frame.
4. The position and orientation of an object's coordinate frame is referred to as its pose.
5. A relative pose describes the pose of one coordinate frame with respect to another and is denoted by an algebraic variable ξ.
6. A coordinate vector describing a point can be represented with respect to a different coordinate frame by applying the relative pose to the vector using the • operator.
7. We can perform algebraic manipulation of expressions written in terms of relative poses and the operators $\oplus$ and $\ominus$.

The remainder of this chapter discusses concrete representations of ξ for various common cases that we will encounter in robotics and computer vision. We start by considering the two-dimensional case which is comparatively straightforward and then extend those concepts to three dimensions. In each case we consider rotation first, and then add translation to create a description of pose.

2.1   Working in Two Dimensions (2D)

A 2-dimensional world, or plane, is familiar to us from high-school Euclidean geometry. We use a right-handed▶ Cartesian coordinate system or coordinate frame with orthogonal axes denoted x and y and typically drawn with the x-axis horizontal and the y-axis vertical. The point of intersection is called the origin. Unit-vectors parallel to the axes are denoted $\hat{x}$ and $\hat{y}$. A point is represented by its x- and y-coordinates (x, y) or as a bound vector

The relative orientation of the x- and y-axes obey the right-hand rule as shown on page 29.

$$\boldsymbol{p} = x\hat{\boldsymbol{x}} + y\hat{\boldsymbol{y}} \qquad (2.2)$$

Figure 2.6 shows a red coordinate frame $\{B\}$ that we wish to describe with respect to the blue reference frame $\{A\}$. We can see clearly that the origin of $\{B\}$ has been displaced by the vector $\boldsymbol{t} = (x, y)$ and then rotated counter-clockwise by an angle θ.

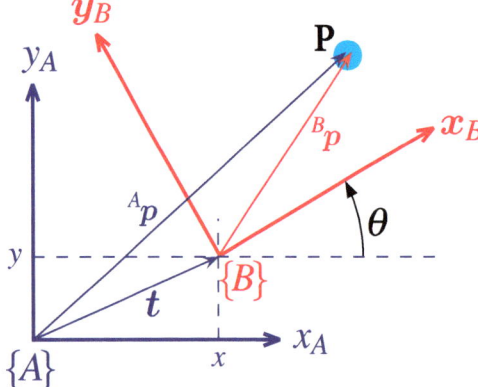

Fig. 2.6.
Two 2D coordinate frames $\{A\}$ and $\{B\}$ and a world point **P**. $\{B\}$ is rotated and translated with respect to $\{A\}$

A concrete representation of pose is therefore the 3-vector $^A\xi_B \sim (x, y, \theta)$, and we use the symbol $\sim$ to denote that the two representations are equivalent. Unfortunately this *representation* is not convenient for compounding since

$$\left(x_1, y_1, \theta_1\right) \oplus \left(x_2, y_2, \theta_2\right)$$

is a complex trigonometric function of both poses. Instead we will look for a different way to represent rotation and pose. We will consider the problem in two parts: rotation and then translation.

2.1.1   Orientation in 2-Dimensions

2.1.1.1   Orthonormal Rotation Matrix

Consider an arbitrary point **P** which we can express with respect to each of the coordinate frames shown in Fig. 2.6. We create a new frame $\{V\}$ whose axes are parallel to those of $\{A\}$ but whose origin is the same as $\{B\}$, see Fig. 2.7. According to Eq. 2.2 we can express the point **P** with respect to $\{V\}$ in terms of the unit-vectors that define the axes of the frame

$$
\begin{aligned}
^V\!p &= {}^V\!x\hat{\boldsymbol{x}}_V + {}^V\!y\hat{\boldsymbol{y}}_V \\
&= \begin{pmatrix} \hat{\boldsymbol{x}}_V & \hat{\boldsymbol{y}}_V \end{pmatrix} \begin{pmatrix} {}^V\!x \\ {}^V\!y \end{pmatrix}
\end{aligned}
\tag{2.3}
$$

which we have written as the product of a row and a column vector.

The coordinate frame $\{B\}$ is completely described by its two orthogonal axes which we represent by two unit vectors

$$
\begin{aligned}
\hat{\boldsymbol{x}}_B &= \cos\theta\hat{\boldsymbol{x}}_V + \sin\theta\hat{\boldsymbol{y}}_V \\
\hat{\boldsymbol{y}}_B &= -\sin\theta\hat{\boldsymbol{x}}_V + \cos\theta\hat{\boldsymbol{y}}_V
\end{aligned}
$$

which can be factorized into matrix form as

$$
\begin{pmatrix} \hat{\boldsymbol{x}}_B & \hat{\boldsymbol{y}}_B \end{pmatrix} = \begin{pmatrix} \hat{\boldsymbol{x}}_V & \hat{\boldsymbol{y}}_V \end{pmatrix} \begin{pmatrix} \cos\theta & -\sin\theta \\ \sin\theta & \cos\theta \end{pmatrix}
\tag{2.4}
$$

Using Eq. 2.2 we can represent the point **P** with respect to $\{B\}$ as

$$
\begin{aligned}
^B\!p &= {}^B\!x\hat{\boldsymbol{x}}_B + {}^B\!y\hat{\boldsymbol{y}}_B \\
&= \begin{pmatrix} \hat{\boldsymbol{x}}_B & \hat{\boldsymbol{y}}_B \end{pmatrix} \begin{pmatrix} {}^B\!x \\ {}^B\!y \end{pmatrix}
\end{aligned}
$$

Fig. 2.7.
Rotated coordinate frames in 2D. The point **P** can be considered with respect to the red or blue coordinate frame

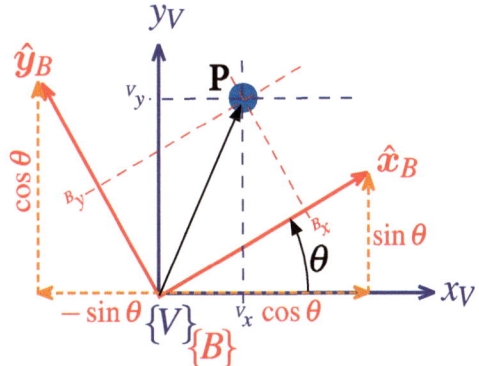

and substituting Eq. 2.4 we write

$$^{B}p = \begin{pmatrix} \hat{x}_V & \hat{y}_V \end{pmatrix} \begin{pmatrix} \cos\theta & -\sin\theta \\ \sin\theta & \cos\theta \end{pmatrix} \begin{pmatrix} ^{B}x \\ ^{B}y \end{pmatrix} \tag{2.5}$$

Now by equating the coefficients of the right-hand sides of Eq. 2.3 and Eq. 2.5 we write

$$\begin{pmatrix} ^{V}x \\ ^{V}y \end{pmatrix} = \begin{pmatrix} \cos\theta & -\sin\theta \\ \sin\theta & \cos\theta \end{pmatrix} \begin{pmatrix} ^{B}x \\ ^{B}y \end{pmatrix}$$

which describes how points are transformed from frame {B} to frame {V} when the frame is rotated. This type of matrix is known as a rotation matrix since it transforms a point from frame {V} to {B} and is denoted $^{V}R_B$

$$\begin{pmatrix} ^{V}x \\ ^{V}y \end{pmatrix} = {}^{V}R_B \begin{pmatrix} ^{B}x \\ ^{B}y \end{pmatrix} \tag{2.6}$$

$$^{X}R_Y(\theta) = \begin{pmatrix} \cos\theta & -\sin\theta \\ \sin\theta & \cos\theta \end{pmatrix}$$

is a 2-dimensional rotation matrix with some special properties:

- it is *orthonormal* (also called *orthogonal*) since each of its columns is a unit vector and the columns are orthogonal.▶
- the columns are the unit vectors that define the axes of the rotated frame Y with respect to X and are by definition both unit-length and orthogonal.
- it belongs to the special orthogonal group of dimension 2 or $R \in \mathbf{SO}(2) \subset \mathbb{R}^{2\times2}$. This means that the product of any two matrices belongs to the group, as does its inverse.
- its *determinant* is $+1$, which means that the length of a vector is unchanged after transformation, that is, $\|^{Y}p\| = \|^{X}p\|, \forall\theta$.
- the inverse is the same as the transpose, that is, $R^{-1} = R^{T}$.

See Appendix B which provides a refresher on vectors, matrices and linear algebra.

We can rearrange Eq. 2.6 as

$$\begin{pmatrix} ^{B}x \\ ^{B}y \end{pmatrix} = \left(^{V}R_B\right)^{-1} \begin{pmatrix} ^{V}x \\ ^{V}y \end{pmatrix} = \left(^{V}R_B\right)^{T} \begin{pmatrix} ^{V}x \\ ^{V}y \end{pmatrix} = {}^{B}R_V \begin{pmatrix} ^{V}x \\ ^{V}y \end{pmatrix}$$

Note that inverting the matrix is the same as swapping the superscript and subscript, which leads to the identity $R(-\theta) = R(\theta)^{T}$.

It is interesting to observe that instead of representing an angle, which is a scalar, we have used a 2×2 matrix that comprises four elements, however these elements are not independent. Each column has a unit magnitude which provides two constraints. The columns are orthogonal which provides another constraint. Four elements and three constraints are effectively one independent value. The rotation matrix is an example of a nonminimum representation and the disadvantages such as the increased memory it requires are outweighed, as we shall see, by its advantages such as composability.

The Toolbox allows easy creation of these rotation matrices

```
>> R = rot2(0.2)
R =
    0.9801   -0.1987
    0.1987    0.9801
```

where the angle is specified in radians. We can observe some of the properties such as

```
>> det(R)
ans =
     1
```

and the product of two rotation matrices is also a rotation matrix

```
>> det(R*R)
ans =
     1
```

You will need to have the MATLAB Symbolic Math Toolbox™ installed.

The Toolbox also supports symbolic mathematics◄ for example

```
>> syms theta
>> R = rot2(theta)
R =
[ cos(theta), -sin(theta)]
[ sin(theta),  cos(theta)]
>> simplify(R*R)
ans =
[ cos(2*theta), -sin(2*theta)]
[ sin(2*theta),  cos(2*theta)]
>> simplify(det(R))
ans =
1
```

2.1.1.2 Matrix Exponential

Consider a pure rotation of 0.3 radians expressed as a rotation matrix

```
>> R = rot2(0.3)
ans =
    0.9553   -0.2955
    0.2955    0.9553
```

We can compute the logarithm of this matrix using the MATLAB builtin function `logm`◄

logm is different to the builtin function log which computes the logarithm of each element of the matrix. A logarithm can be computed using a power series, with a matrix rather than scalar argument. For a matrix the logarithm is not unique and logm computes the principal logarithm of the matrix.

```
>> S = logm(R)
S =
    0.0000   -0.3000
    0.3000    0.0000
```

and the result is a simple matrix with two elements having a magnitude of 0.3, which intriguingly is the original rotation angle. There is something deep and interesting going on here – we are on the fringes of Lie group theory which we will encounter throughout this chapter.

In 2 dimensions the skew-symmetric matrix is

$$[\omega]_\times = \begin{pmatrix} 0 & -\omega \\ \omega & 0 \end{pmatrix} \tag{2.7}$$

which has clear structure and only one unique element $\omega \in \mathbb{R}$. A simple example of Toolbox support for skew-symmetric matrices is

```
>> skew(2)
ans =
     0    -2
     2     0
```

and the inverse operation is performed using the Toolbox function vex

```
>> vex(ans)
ans =
     2
```

This matrix has a zero diagonal and is an example of a 2×2 skew-symmetric matrix. The matrix has only one unique element and we can unpack it using the Toolbox function `vex`

```
>> vex(S)
ans =
    0.3000
```

to recover the rotation angle.

The inverse of a logarithm is exponentiation and using the builtin MATLAB matrix exponential function `expm`▶

```
>> expm(S)
ans =
    0.9553    -0.2955
    0.2955     0.9553
```

`expm` is different to the builtin function `exp` which computes the exponential of each element of the matrix.
$expm(A) = I + A + A^2/2! + A^3/3! + \cdots$

the result is, as expected, our original rotation matrix. In fact the command

```
>> R = rot2(0.3);
```

is equivalent to

```
>> R = expm(  skew(0.3)  );
```

Formally we can write

$$R = e^{[\theta]_\times} \in SO(2)$$

where θ is the rotation angle, and the notation $[\cdot]_\times : \mathbb{R} \mapsto \mathbb{R}^{2\times2}$ indicates a mapping from a scalar to a skew-symmetric matrix.

2.1.2   Pose in 2-Dimensions

2.1.2.1   Homogeneous Transformation Matrix

Now we need to account for the translation between the origins of the frames shown in Fig. 2.6. Since the axes $\{V\}$ and $\{A\}$ are parallel, as shown in Figs. 2.6 and 2.7, this is simply vectorial addition

$$\begin{pmatrix} {}^A x \\ {}^A y \end{pmatrix} = \begin{pmatrix} {}^V x \\ {}^V y \end{pmatrix} + \begin{pmatrix} x \\ y \end{pmatrix} \tag{2.8}$$

$$= \begin{pmatrix} \cos\theta & -\sin\theta \\ \sin\theta & \cos\theta \end{pmatrix} \begin{pmatrix} {}^B x \\ {}^B y \end{pmatrix} + \begin{pmatrix} x \\ y \end{pmatrix} \tag{2.9}$$

$$= \begin{pmatrix} \cos\theta & -\sin\theta & x \\ \sin\theta & \cos\theta & y \end{pmatrix} \begin{pmatrix} {}^B x \\ {}^B y \\ 1 \end{pmatrix} \tag{2.10}$$

or more compactly as

$$\begin{pmatrix} {}^A x \\ {}^A y \\ 1 \end{pmatrix} = \begin{pmatrix} {}^A R_B & t \\ 0_{1\times2} & 1 \end{pmatrix} \begin{pmatrix} {}^B x \\ {}^B y \\ 1 \end{pmatrix} \tag{2.11}$$

where $t = (x, y)$ is the translation of the frame and the orientation is ${}^A R_B$. Note that ${}^A R_B = {}^V R_B$ since the axes of frames $\{A\}$ and $\{V\}$ are parallel. The coordinate vectors for point **P** are now expressed in homogeneous form and we write

A vector $p = (x, y)$ is written in homogeneous form as $\tilde{p} \in \mathbb{P}^2$, $\tilde{p} = (x_1, x_2, x_3)$ where $x = x_1/x_3$, $y = x_2/x_3$ and $x_3 \neq 0$. The dimension has been increased by one and a point on a plane is now represented by a 3-vector. To convert a point to homogeneous form we typically append an element equal to one $\tilde{p} = (x, y, 1)$. The tilde indicates the vector is homogeneous.

Homogeneous vectors have the important property that $\tilde{p}$ is equivalent to $\lambda \tilde{p}$ for all $\lambda \neq 0$ which we write as $\tilde{p} \simeq \lambda \tilde{p}$. That is $\tilde{p}$ represents the same point in the plane irrespective of the overall scaling factor. Additional details are provided in Sect. C.2.

$$
{}^A\tilde{p} = \begin{pmatrix} {}^A\!R_B & t \\ \mathbf{0}_{1\times 2} & 1 \end{pmatrix} {}^B\tilde{p}
$$
$$
= {}^A\!T_B \, {}^B\tilde{p}
$$

and ${}^A\!T_B$ is referred to as a homogeneous transformation. The matrix has a very specific structure and belongs to the special Euclidean group of dimension 2 or $T \in \mathbf{SE}(2) \subset \mathbb{R}^{3\times 3}$.

By comparison with Eq. 2.1 it is clear that ${}^A\!T_B$ represents translation and orientation or relative pose. This is often referred to as a *rigid-body motion*.

$$
T = \begin{pmatrix} \cos\theta & -\sin\theta & x \\ \sin\theta & \cos\theta & y \\ 0 & 0 & 1 \end{pmatrix}
$$

A concrete representation of relative pose ξ is $\xi \sim T \in \mathbf{SE}(2)$ and $T_1 \oplus T_2 \mapsto T_1 T_2$ which is standard matrix multiplication

$$
T_1 T_2 = \begin{pmatrix} R_1 & t_1 \\ \mathbf{0}_{1\times 2} & 1 \end{pmatrix}\begin{pmatrix} R_2 & t_2 \\ \mathbf{0}_{1\times 2} & 1 \end{pmatrix} = \begin{pmatrix} R_1 R_2 & t_1 + R_1 t_2 \\ \mathbf{0}_{1\times 2} & 1 \end{pmatrix}
$$

One of the algebraic rules from page 19 is $\xi \oplus 0 = \xi$. For matrices we know that $TI = T$, where I is the identify matrix, so for pose $0 \mapsto I$ the identity matrix. Another rule was that $\xi \ominus \xi = 0$. We know for matrices that $TT^{-1} = I$ which implies that $\ominus T \mapsto T^{-1}$.

$$
T^{-1} = \begin{pmatrix} R & t \\ \mathbf{0}_{1\times 2} & 1 \end{pmatrix}^{-1} = \begin{pmatrix} R^T & -R^T t \\ \mathbf{0}_{1\times 2} & 1 \end{pmatrix}
$$

For a point described by $\tilde{p} \in \mathbb{P}^2$ then $T \cdot \tilde{p} \mapsto T\tilde{p}$ which is a standard matrix-vector product.

To make this more tangible we will show some numerical examples using MATLAB and the Toolbox. We create a homogeneous transformation which represents a translation of $(1, 2)$ followed by a rotation of 30°

```
>> T1 = transl2(1, 2) * trot2(30, 'deg')
T1 =
    0.8660   -0.5000    1.0000
    0.5000    0.8660    2.0000
         0         0    1.0000
```

The function `transl2` creates a relative pose with a finite translation but zero rotation, while `trot2` creates a relative pose with a finite rotation but zero translation. ◀ We can plot this, relative to the world coordinate frame, by

Many Toolbox functions have variants that return orthonormal rotation matrices or homogeneous transformations, for example, `rot2` and `trot2`.

```
>> plotvol([0 5 0 5]);
>> trplot2(T1, 'frame', '1', 'color', 'b')
```

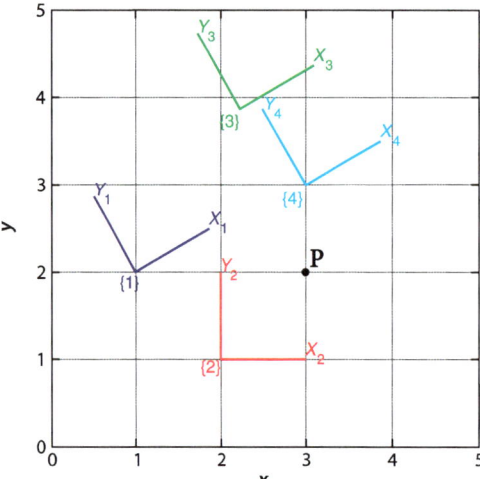

Fig. 2.8.
Coordinate frames drawn using
the Toolbox function `trplot2`

The options specify that the label for the frame is {1} and it is colored blue and this is shown in Fig. 2.8. We create another relative pose which is a displacement of (2, 1) and zero rotation

```
>> T2 = transl2(2, 1)
T2 =
     1     0     2
     0     1     1
     0     0     1
```

which we plot in red

```
>> trplot2(T2, 'frame', '2', 'color', 'r');
```

Now we can compose the two relative poses

```
>> T3 = T1*T2
T3 =
    0.8660   -0.5000    2.2321
    0.5000    0.8660    3.8660
         0         0    1.0000
```

and plot it, in green, as

```
>> trplot2(T3, 'frame', '3', 'color', 'g');
```

We see that the displacement of (2, 1) has been applied with respect to frame {1}. It is important to note that our final displacement is not (3, 3) because the displacement is with respect to the rotated coordinate frame. The noncommutativity of composition is clearly demonstrated by

```
>> T4 = T2*T1;
>> trplot2(T4, 'frame', '4', 'color', 'c');
```

and we see that frame {4} is different to frame {3}.

Now we define a point (3, 2) relative to the world frame

```
>> P = [3 ; 2 ];
```

which is a column vector and add it to the plot

```
>> plot_point(P, 'label', 'P', 'solid', 'ko');
```

To determine the coordinate of the point with respect to {1} we use Eq. 2.1 and write down

$$^{0}\boldsymbol{p} = {}^{0}\xi_1 \cdot {}^{1}\boldsymbol{p}$$

and then rearrange as

$$^1\boldsymbol{p} = {}^1\xi_0 \bullet {}^0\boldsymbol{p}$$

$$= \left({}^0\xi_1\right)^{-1} \bullet {}^0\boldsymbol{p}$$

Substituting numerical values

```
>> P1 = inv(T1) * [P; 1]
P1 =
    1.7321
   -1.0000
    1.0000
```

where we first converted the Euclidean point coordinates to *homogeneous form* by appending a one. The result is also in homogeneous form and has a negative *y*-coordinate in frame {1}. Using the Toolbox we could also have expressed this as

```
>> h2e( inv(T1) * e2h(P) )
ans =
    1.7321
   -1.0000
```

where the result is in Euclidean coordinates. The helper function `e2h` converts Euclidean coordinates to homogeneous and `h2e` performs the inverse conversion.

2.1.2.2 Centers of Rotation

We will explore the noncommutativity property in more depth and illustrate with the example of a pure rotation. First we create and plot a reference coordinate frame {0} and a target frame {*X*}

```
>> plotvol([-5 4 -1 5]);
>> T0 = eye(3,3);
>> trplot2(T0, 'frame', '0');
>> X = transl2(2, 3);
>> trplot2(X, 'frame', 'X');
```

and create a rotation of 2 radians (approximately 115°)

```
>> R = trot2(2);
```

and plot the effect of the two possible orders of composition

```
>> trplot2(R*X, 'framelabel', 'RX', 'color', 'r');
>> trplot2(X*R, 'framelabel', 'XR', 'color', 'r');
```

The results are shown as red coordinate frames in Fig. 2.9. We see that the frame {*RX*} has been rotated about the origin, while frame {*XR*} has been rotated about the origin of {*X*}.

What if we wished to rotate a coordinate frame about an arbitrary point? First of all we will establish a new point **C** and display it

```
>> C = [1 2]';
>> plot_point(C, 'label', ' C', 'solid', 'ko')
```

and then compute a transform to rotate about point **C**

```
>> RC = transl2(C) * R * transl2(-C)
RC =
   -0.4161   -0.9093    3.2347
    0.9093   -0.4161    1.9230
         0         0    1.0000
```

and applying this

```
>> trplot2(RC*X, 'framelabel', 'XC', 'color', 'r');
```

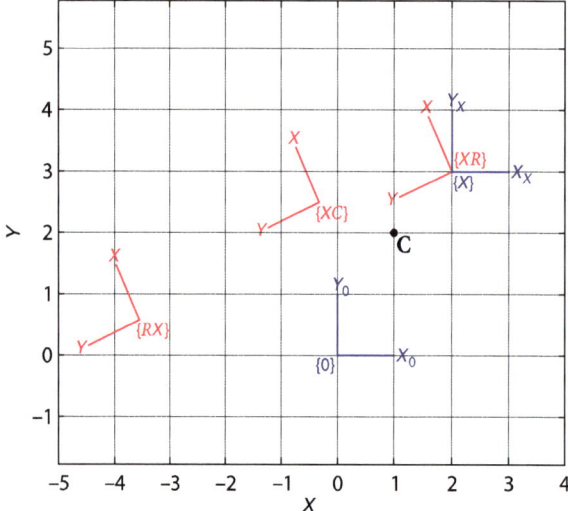

Fig. 2.9.
The frame {X} is rotated by
2 radians about {0} to give
frame {RX}, about {X} to
give {XR}, and about point C
to give frame {XC}

we see that the frame has indeed been rotated about point **C**. Creating the required transform was somewhat cumbersome and not immediately obvious. Reading from right to left▸ we first apply an origin shift, a translation from **C** to the origin of the reference frame, apply the rotation about that origin, and then apply the inverse origin shift, a translation from the reference frame origin back to **C**. A more descriptive way to achieve this is using twists.

RC left multiplies X, therefore the first transform applied to X is transl(-C), then R, then transl(C).

2.1.2.3   Twists in 2D

The corollary to what we showed in the last section is that, given any two frames we can find a rotational center that will *rotate* the first frame into the second. For the case of pure translational motion the rotational center will be at infinity. This is the key concept behind what is called a twist.

We can create a rotational twist about the point specified by the coordinate vector C

```
>> tw = Twist('R', C)
tw =
( 2 -1; 1 )
```

and the result is a `Twist` object that encodes a twist vector with two components: a 2-vector *moment* and a 1-vector *rotation*. The first argument `'R'` indicates a rotational twist is to be computed. This particular twist is a *unit twist* since the magnitude of the rotation, the last element of the twist, is equal to one.

To create an **SE**(2) transformation for a rotation about this unit twist by 2 radians we use the `T` method

```
>> tw.T(2)
ans =
   -0.4161   -0.9093    3.2347
    0.9093   -0.4161    1.9230
         0         0    1.0000
```

which is the same as that computed in the previous section, but more concisely specified in terms of the center of rotation. The center is also called the pole of the transformation and is encoded in the twist

```
>> tw.pole'
ans =
     1     2
```

If we wish to perform translational motion in the direction (1, 1) the relevant unit twist is◀

For a unit-translational twist the rotation is zero and the moment is a unit vector.

```
>> tw = Twist('T', [1 1])
tw =
( 0.70711 0.70711; 0 )
```

and for a displacement of $\sqrt{2}$ in the direction defined by this twist the **SE**(2) transformation is

```
>> tw.T(sqrt(2))
ans =
      1      0      1
      0      1      1
      0      0      1
```

which we see has a null rotation and a translation of 1 in the *x*- and *y*-directions.

For an arbitrary planar transform such as

```
>> T = transl2(2, 3) * trot2(0.5)
T =
    0.8776   -0.4794    2.0000
    0.4794    0.8776    3.0000
         0         0    1.0000
```

we can compute the twist vector

```
>> tw = Twist(T)
tw =
( 2.7082 2.4372; 0.5 )
```

and we note that the last element, the rotation, is not equal to one but is the required rotation angle of 0.5 radians. This is a nonunit twist. Therefore when we convert this to an **SE**(2) transform we don't need to provide a second argument since it is implicit in the twist

```
>> tw.T
ans =
    0.8776   -0.4794    2.0000
    0.4794    0.8776    3.0000
         0         0    1.0000
```

and we have regenerated our original homogeneous transformation.

2.2   Working in Three Dimensions (3D)

The 3-dimensional case is an extension of the 2-dimensional case discussed in the previous section. We add an extra coordinate axis, typically denoted by *z*, that is orthogonal to both the *x*- and *y*-axes. The direction of the *z*-axis obeys the *right-hand rule* and forms a *right-handed coordinate frame*. Unit vectors parallel to the axes are denoted $\hat{x}$, $\hat{y}$ and $\hat{z}$ such that◀

In all these identities, the symbols from left to right (across the equals sign) are a cyclic rotation of the sequence xyz.

$$\hat{z} = \hat{x} \times \hat{y}, \ \ \hat{x} = \hat{y} \times \hat{z}; \ \ \hat{y} = \hat{z} \times \hat{x} \tag{2.12}$$

A point **P** is represented by its *x*-, *y*- and *z*-coordinates (x, y, z) or as a bound vector

$$p = x\hat{x} + y\hat{y} + z\hat{z}$$

Figure 2.10 shows a red coordinate frame {*B*} that we wish to describe with respect to the blue reference frame {*A*}. We can see clearly that the origin of {*B*} has been

> **Right-hand rule.** A right-handed coordinate frame is defined by the first three fingers of your right hand which indicate the relative directions of the *x*-, *y*- and *z*-axes respectively.

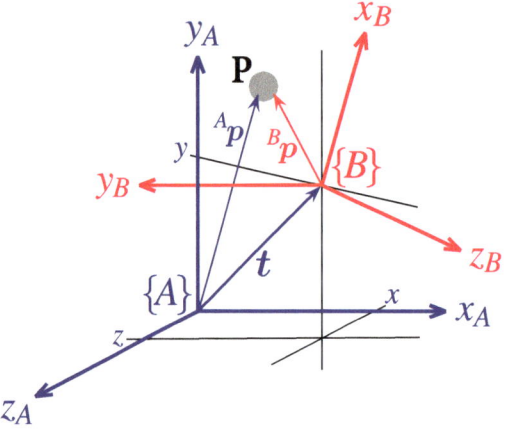

Fig. 2.10.
Two 3D coordinate frames {A}
and {B}. {B} is rotated and translated with respect to {A}

displaced by the vector $t = (x, y, z)$ and then rotated in some complex fashion. Just as for the 2-dimensional case the way we represent orientation is very important.

Our approach is to again consider an arbitrary point **P** with respect to each of the coordinate frames and to determine the relationship between $^A\!p$ and $^B\!p$. We will again consider the problem in two parts: rotation and then translation. Rotation is surprisingly complex for the 3-dimensional case and we devote all of the next section to it.

2.2.1 Orientation in 3-Dimensions

> *Any two independent orthonormal coordinate frames*
> *can be related by a sequence of rotations (not more than **three**)*
> *about coordinate axes, where no two successive rotations may be about the same axis.*
> Euler's rotation theorem (Kuipers 1999).

Figure 2.10 shows a pair of right-handed coordinate frames with very different orientations, and we would like some way to describe the orientation of one with respect to the other. We can imagine picking up frame {A} in our hand and rotating it until it looked just like frame {B}. *Euler's rotation theorem* states that any rotation can be considered as a sequence of rotations about different coordinate axes.

We start by considering rotation about a single coordinate axis. Figure 2.11 shows a right-handed coordinate frame, and that same frame after it has been rotated by various angles about different coordinate axes.

The issue of rotation has some subtleties which are illustrated in Fig. 2.12. This shows a sequence of two rotations applied in different orders. We see that the final orientation depends on the order in which the rotations are applied. This is a deep and confounding characteristic of the 3-dimensional world which has intrigued mathematicians for a long time. There are implication for the pose algebra we have used in this chapter:

> In 3-dimensions rotation is not commutative – the order in which rotations are applied makes a difference to the result.

Mathematicians have developed many ways to represent rotation and we will discuss several of them in the remainder of this section: orthonormal rotation matrices, Euler and Cardan angles, rotation axis and angle, exponential coordinates, and unit quaternions. All can be represented as vectors or matrices, the natural datatypes of MATLAB or as a Toolbox defined class. The Toolbox provides many function to convert between these representations and these are shown in Tables 2.1 and 2.2 (pages 55, 56).

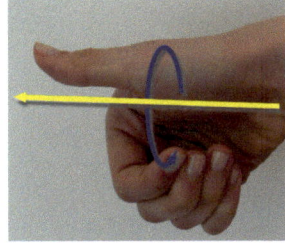

Rotation about a vector. Wrap your right hand around the vector with your thumb (your *x*-finger) in the direction of the arrow. The curl of your fingers indicates the direction of increasing angle.

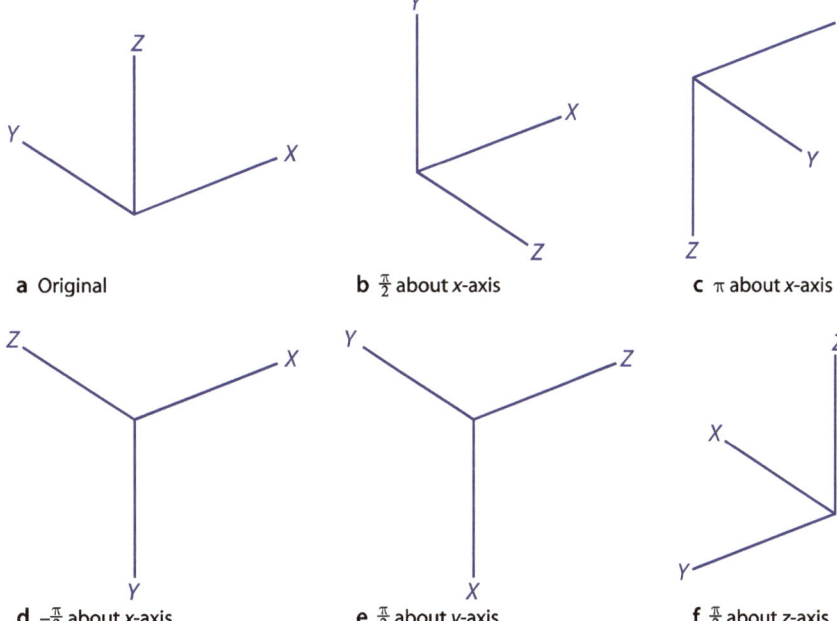

a Original   **b** $\frac{\pi}{2}$ about *x*-axis   **c** π about *x*-axis

Fig. 2.11.
Rotation of a 3D coordinate frame.
a The original coordinate frame,
b–f frame **a** after various rotations as indicated

d $-\frac{\pi}{2}$ about *x*-axis   **e** $\frac{\pi}{2}$ about *y*-axis   **f** $\frac{\pi}{2}$ about *z*-axis

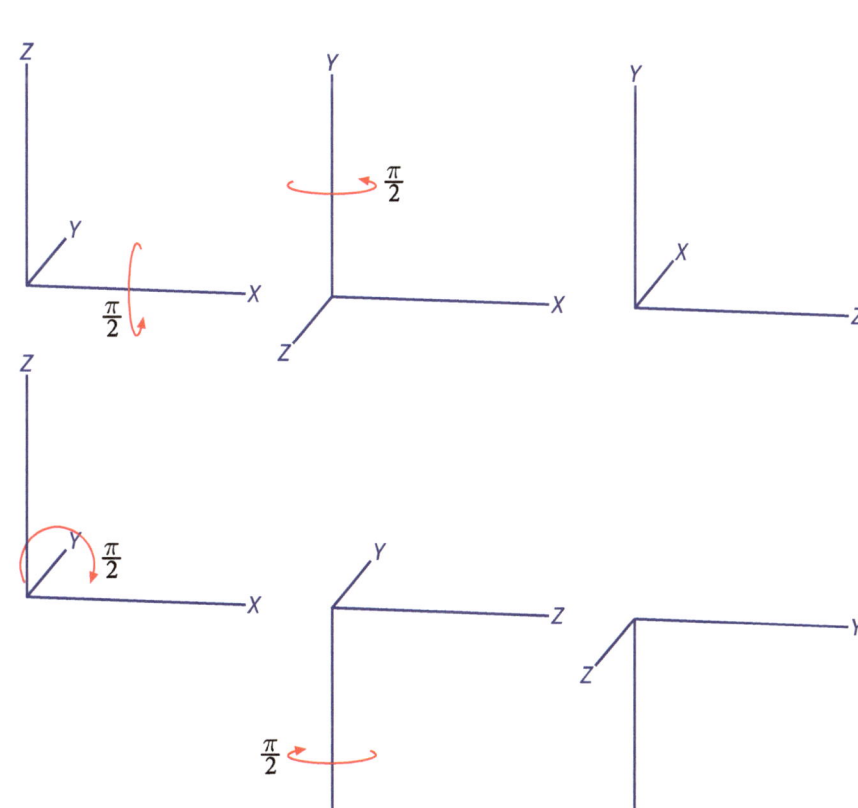

Fig. 2.12.
Example showing the noncommutativity of rotation. In the top row the coordinate frame is rotated by $\frac{\pi}{2}$ about the *x*-axis and then $\frac{\pi}{2}$ about the *y*-axis. In the bottom row the order of rotations has been reversed. The results are clearly different

2.2.1.1   Orthonormal Rotation Matrix

Just as for the 2-dimensional case we can represent the orientation of a coordinate frame by its unit vectors expressed in terms of the reference coordinate frame. Each unit vector has three elements and they form the columns of a 3×3 *orthonormal matrix* $^A R_B$

$$\begin{pmatrix} ^A x \\ ^A y \\ ^A z \end{pmatrix} = {}^A R_B \begin{pmatrix} ^B x \\ ^B y \\ ^B z \end{pmatrix} \tag{2.13}$$

which transforms the description of a vector defined with respect to frame $\{B\}$ to a vector with respect to $\{A\}$.

> A 3-dimensional rotation matrix $^X R_Y$ has some special properties:
>
> - it is *orthonormal* (also called *orthogonal*) since each of its columns is a unit vector and the columns are orthogonal.▸
> - the columns are the unit vectors that define the axes of the rotated frame Y with respect to X and are by definition both unit-length and orthogonal.
> - it belongs to the special orthogonal group of dimension 3 or $R \in \mathbf{SO}(3) \subset \mathbb{R}^{3\times3}$. This means that the product of any two matrices within the group also belongs to the group, as does its inverse.
> - its determinant is $+1$, which means that the length of a vector is unchanged after transformation, that is, $\|^Y p\| = \|^X p\|, \forall\theta$.
> - the inverse is the same as the transpose, that is, $R^{-1} = R^T$.

See Appendix B which provides a refresher on vectors, matrices and linear algebra.

The orthonormal rotation matrices for rotation of θ about the x-, y- and z-axes are

$$R_x(\theta) = \begin{pmatrix} 1 & 0 & 0 \\ 0 & \cos\theta & -\sin\theta \\ 0 & \sin\theta & \cos\theta \end{pmatrix}$$

$$R_y(\theta) = \begin{pmatrix} \cos\theta & 0 & \sin\theta \\ 0 & 1 & 0 \\ -\sin\theta & 0 & \cos\theta \end{pmatrix}$$

$$R_z(\theta) = \begin{pmatrix} \cos\theta & -\sin\theta & 0 \\ \sin\theta & \cos\theta & 0 \\ 0 & 0 & 1 \end{pmatrix}$$

The Toolbox provides functions to compute these elementary rotation matrices, for example $R_x(\theta)$ is

```
>> R = rotx(pi/2)
R =
    1.0000         0         0
         0    0.0000   -1.0000
         0    1.0000    0.0000
```

and its effect on a reference coordinate frame is shown graphically in Fig. 2.11b. The functions `roty` and `rotz` compute $R_y(\theta)$ and $R_z(\theta)$ respectively.

If we consider that the rotation matrix represents a pose then the corresponding coordinate frame can be displayed graphically

```
>> trplot(R)
```

which is shown in Fig. 2.13a. We can visualize a rotation more powerfully using the Toolbox function `tranimate` which animates a rotation

```
>> tranimate(R)
```

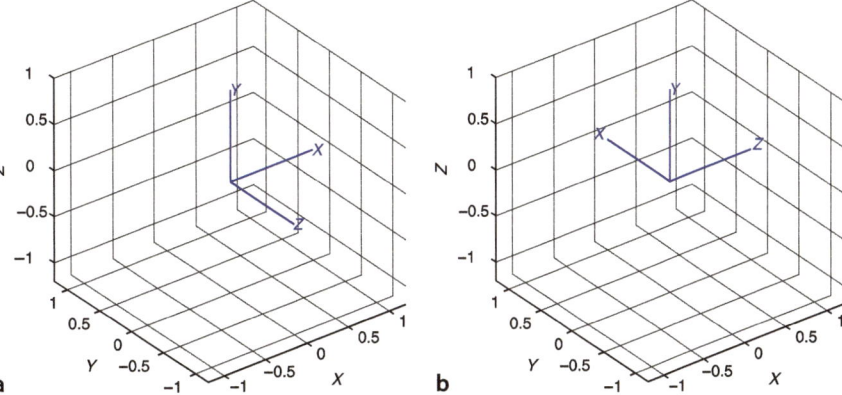

Fig. 2.13.
Coordinate frames displayed using `trplot`. **a** Reference frame rotated by $\frac{\pi}{2}$ about the *x*-axis, **b** frame **a** rotated by $\frac{\pi}{2}$ about the *y*-axis

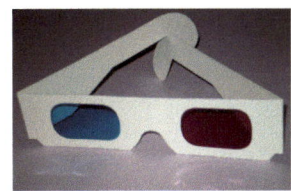

showing the world frame rotating into the specified coordinate frame. If you have a pair of anaglyph stereo glasses◄ you can see this in more realistic 3D by

```
>> tranimate(R, '3d')
```

To illustrate compounding of rotations we will rotate the frame of Fig. 2.13a again, this time around its *y*-axis

```
>> R = rotx(pi/2) * roty(pi/2)
R =
    0.0000        0    1.0000
    1.0000   0.0000   -0.0000
   -0.0000   1.0000    0.0000
>> trplot(R)
```

to give the frame shown in Fig. 2.13b. In this frame the *x*-axis now points in the direction of the world *y*-axis.

The noncommutativity of rotation can be shown by reversing the order of the rotations above

```
>> roty(pi/2)*rotx(pi/2)
ans =
    0.0000   1.0000    0.0000
         0   0.0000   -1.0000
   -1.0000   0.0000    0.0000
```

which has a very different value.

We recall that Euler's rotation theorem states that *any* rotation can be represented by *not more than three* rotations about coordinate axes. This means that in general an arbitrary rotation between frames can be decomposed into a sequence of three rotation angles and associated rotation axes – this is discussed in the next section.

The orthonormal rotation matrix has nine elements but they are not independent. The columns have unit magnitude which provides three constraints. The columns are orthogonal to each other which provides another three constraints.◄ Nine elements and six constraints is effectively three independent values.

If the column vectors are $c_i, i \in 1 \cdots 3$ then $c_1 \cdot c_2 = c_2 \cdot c_3 = c_3 \cdot c_1 = 0$ and $\|c_i\| = 1$.

> **Reading an orthonormal *rotation matrix*,** the columns from left to right tell us the directions of the new frame's axes in terms of the current coordinate frame. For example if
>
> ```
> R =
>     1.0000        0        0
>          0   0.0000  -1.0000
>          0   1.0000   0.0000
> ```
>
> the new frame has its *x*-axis in the old *x*-direction $(1, 0, 0)$, its *y*-axis in the old *z*-direction $(0, 0, 1)$, and the new *z*-axis in the old negative *y*-direction $(0, -1, 0)$. In this case the *x*-axis was unchanged since this is the axis around which the rotation occurred. The rows are the converse – the current frame axes in terms of the new frame axes.

2.2.1.2   Three-Angle Representations

Euler's rotation theorem requires successive rotation about three axes such that no two successive rotations are about the same axis. There are two classes of rotation sequence: Eulerian and Cardanian, named after Euler and Cardano respectively.

The Eulerian type involves repetition, but not successive, of rotations about one particular axis: XYX, XZX, YXY, YZY, ZXZ, or ZYZ. The Cardanian type is characterized by rotations about all three axes: XYZ, XZY, YZX, YXZ, ZXY, or ZYX.

It is common practice to refer to all 3-angle representations as Euler angles but this is underspecified since there are twelve different types to choose from. The particular angle sequence is often a convention within a particular technological field.

The ZYZ sequence

$$R = R_z(\phi)R_y(\theta)R_z(\psi) \qquad (2.14)$$

is commonly used in aeronautics and mechanical dynamics, and is used in the Toolbox. The Euler angles are the 3-vector $\boldsymbol{\Gamma} = (\phi, \theta, \psi)$.

For example, to compute the equivalent rotation matrix for $\boldsymbol{\Gamma} = (0.1, 0.2, 0.3)$ we write

```
>> R = rotz(0.1) * roty(0.2) * rotz(0.3);
```

or more conveniently

```
>> R = eul2r(0.1, 0.2, 0.3)
R =
    0.9021   -0.3836    0.1977
    0.3875    0.9216    0.0198
   -0.1898    0.0587    0.9801
```

The inverse problem is finding the Euler angles that correspond to a given rotation matrix

```
>> gamma = tr2eul(R)
gamma =
    0.1000    0.2000    0.3000
```

However if θ is negative

```
>> R = eul2r(0.1 , -0.2, 0.3)
R =
    0.9021   -0.3836   -0.1977
    0.3875    0.9216   -0.0198
    0.1898   -0.0587    0.9801
```

the inverse function

```
>> tr2eul(R)
ans =
   -3.0416    0.2000   -2.8416
```

returns a positive value for θ and quite different values for ϕ and ψ. However the corresponding rotation matrix

Leonhard Euler (1707–1783) was a Swiss mathematician and physicist who dominated eighteenth century mathematics. He was a student of Johann Bernoulli and applied new mathematical techniques such as calculus to many problems in mechanics and optics. He also developed the functional notation, $y = f(x)$, that we use today. In robotics we use his rotation theorem and his equations of motion in rotational dynamics.

He was prolific and his collected works fill 75 volumes. Almost half of this was produced during the last seventeen years of his life when he was completely blind.

```
>> eul2r(ans)
ans =
    0.9021   -0.3836   -0.1977
    0.3875    0.9216   -0.0198
    0.1898   -0.0587    0.9801
```

is the same – the two different sets of Euler angles correspond to the one rotation matrix. The mapping from a rotation matrix to Euler angles is not unique and the Toolbox *always* returns a positive angle for θ.

For the case where $\theta = 0$

```
>> R = eul2r(0.1, 0, 0.3)
R =
    0.9211   -0.3894        0
    0.3894    0.9211        0
         0         0   1.0000
```

the inverse function returns

```
>> tr2eul(R)
ans =
         0         0    0.4000
```

which is clearly quite different but the result is the same rotation matrix. The explanation is that if $\theta = 0$ then $R_y = I$ and Eq. 2.14 becomes

$$R = R_z(\phi) R_z(\psi) = R_z(\phi + \psi)$$

which is a function of the sum $\phi + \psi$. Therefore the inverse operation can do no more than determine this sum, and by convention we choose $\phi = 0$. The case $\theta = 0$ is a singularity and will be discussed in more detail in the next section.

Another widely used convention are the Cardan angles: roll, pitch and yaw. Confusingly there are two different versions in common use. Text books seem to define the roll-pitch-yaw sequence as ZYX or XYZ depending on whether they have a mobile robot or robot arm focus.◄ When describing the attitude of vehicles such as ships, aircraft and cars the convention is that the *x*-axis points in the forward direction and the *z*-axis points either up or down. It is intuitive to apply the rotations in the sequence: yaw (direction of travel), pitch (elevation of the front with respect to horizontal) and then finally roll (rotation about the forward axis of the vehicle). This leads to the ZYX angle sequence

> Well known texts such as Siciliano et al. (2008), Spong et al. (2006) and Paul (1981) use the XYZ sequence. The Toolbox supports both formats by means of the `'xyz'` and `'zyx'` options. The ZYX order is default for Release 10, but for Release 9 the default was XYZ.

$$R = R_z(\theta_y) R_y(\theta_p) R_x(\theta_r) \tag{2.15}$$

Roll-pitch-yaw angles are also known as Tait-Bryan angles◄ or nautical angles, and for aeronautical applications they can be called bank, attitude and heading angles respectively.

> Named after Peter Tait a Scottish physicist and quaternion supporter, and George Bryan an early Welsh aerodynamicist.

Gerolamo Cardano (1501–1576) was an Italian Renaissance mathematician, physician, astrologer, and gambler. He was born in Pavia, Italy, the illegitimate child of a mathematically gifted lawyer. He studied medicine at the University of Padua and later was the first to describe typhoid fever. He partly supported himself through gambling and his book about games of chance *Liber de ludo aleae* contains the first systematic treatment of probability as well as effective cheating methods. His family life was problematic: his eldest son was executed for poisoning his wife, and his daughter was a prostitute who died from syphilis (about which he wrote a treatise). He computed and published the horoscope of Jesus, was accused of heresy, and spent time in prison until he abjured and gave up his professorship.

He published the solutions to the cubic and quartic equations in his book *Ars magna* in 1545, and also invented the combination lock, the gimbal consisting of three concentric rings allowing a compass or gyroscope to rotate freely (see Fig. 2.15), and the Cardan shaft with universal joints – the drive shaft used in motor vehicles today.

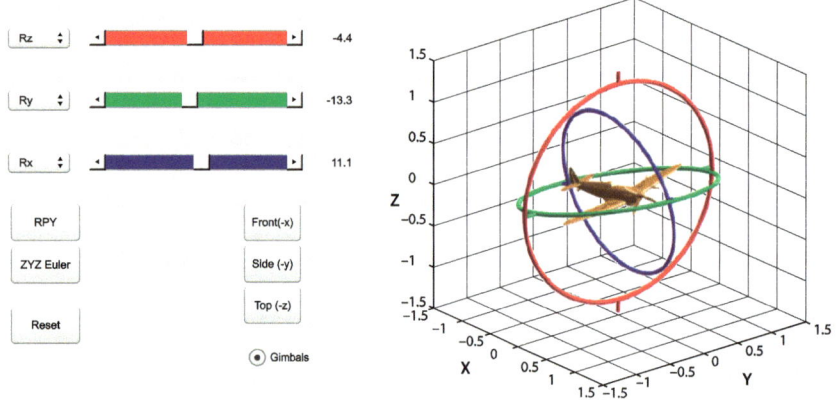

Fig. 2.14.
The Toolbox application
`tripleangle` allows you to
experiment with Euler angles
and roll-pitch-yaw angles and
see how the attitude of a body
changes

When describing the attitude of a robot gripper, as shown in Fig. 2.16, the convention is that the z-axis points forward and the x-axis is either up or down. This leads to the XYZ angle sequence

$$R = R_x\left(\theta_y\right)R_y\left(\theta_p\right)R_z\left(\theta_r\right) \tag{2.16}$$

The Toolbox defaults to the ZYX sequence but can be overridden using the `'xyz'` option. For example

```
>> R = rpy2r(0.1, 0.2, 0.3)
R =
    0.9363   -0.2751    0.2184
    0.2896    0.9564   -0.0370
   -0.1987    0.0978    0.9752
```

and the inverse is

```
>> gamma = tr2rpy(R)
gamma =
    0.1000    0.2000    0.3000
```

The roll-pitch-yaw sequence allows all angles to have arbitrary sign and it has a singularity when $\theta_p = \pm\frac{\pi}{2}$ which is fortunately outside the range of feasible attitudes for most vehicles.

The Toolbox includes an interactive graphical tool

```
>> tripleangle
```

that allows you to experiment with Euler angles or roll-pitch-yaw angles and see their effect on the orientation of a body as shown in Fig. 2.14.

2.2.1.3   Singularities and Gimbal Lock

A fundamental problem with all the three-angle representations just described is singularity. This is also known as gimbal lock, a term made famous in the movie Apollo 13. This occurs when the rotational axis of the middle term in the sequence becomes parallel to the rotation axis of the first or third term.

A mechanical gyroscope used for spacecraft navigation is shown in Fig. 2.15. The innermost assembly is the *stable member* which has three orthogonal gyroscopes that hold it at a constant orientation with respect to the universe. It is mechanically connected to the spacecraft via a gimbal mechanism which allows the spacecraft to move around the stable platform without exerting any torque on it. The attitude of the spacecraft is determined directly by measuring the angles of the gimbal axes with respect to the stable platform – giving a direct indication of roll-pitch-yaw angles which in this design are a Cardanian YZX sequence.▸

"The LM Body coordinate system is right-handed, with the +X axis pointing up through the thrust axis, the +Y axis pointing right when facing forward which is along the +Z axis. The rotational transformation matrix is constructed by a 2-3-1 Euler sequence, that is: Pitch about Y, then Roll about Z and, finally, Yaw about X. Positive rotations are pitch up, roll right, yaw left."(Hoag 1963).

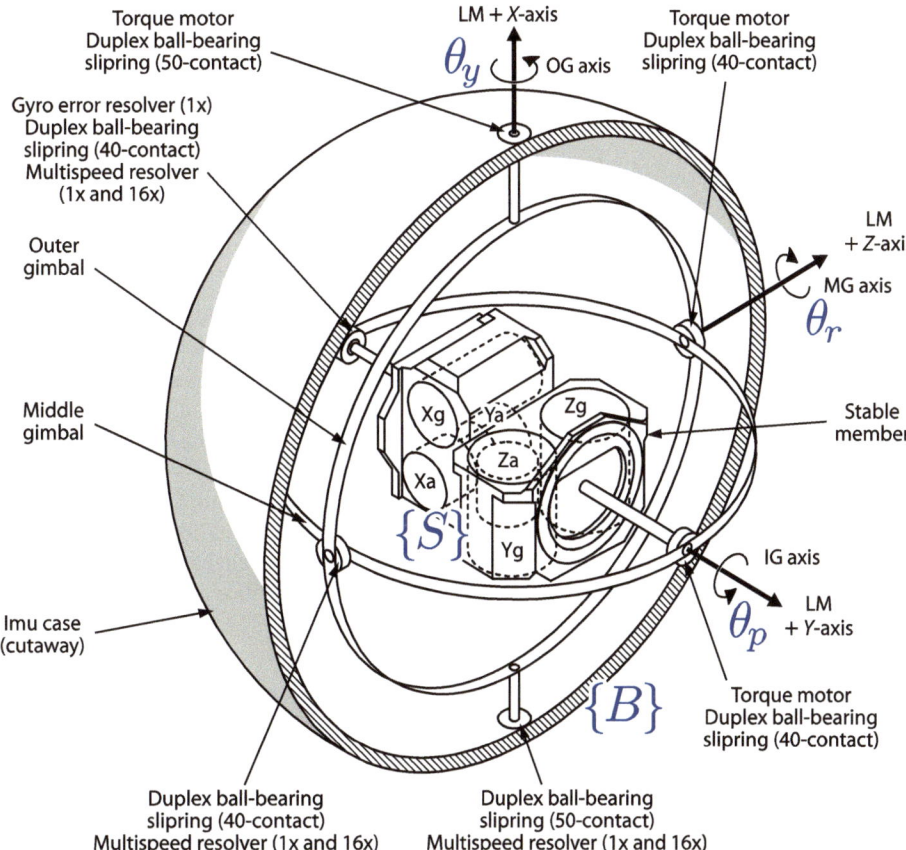

Fig. 2.15.
Schematic of Apollo Lunar Module (LM) inertial measurement unit (IMU). The vehicle's coordinate system has the x-axis pointing up through the thrust axis, the z-axis forward, and the y-axis pointing right. Starting at the stable platform {S} and working outwards toward the spacecraft's body frame {B} the rotation angle sequence is YZX. The components labeled X_g, Y_g and Z_g are the x-, y- and z-axis gyroscopes and those labeled X_a, Y_a and Z_a are the x-, y- and z-axis accelerometers (redrawn after Apollo Operations Handbook, LMA790-3-LM)

Consider the situation when the rotation angle of the middle gimbal (rotation about the spacecraft's z-axis) is 90° – the axes of the inner and outer gimbals are aligned and they share the *same* rotation axis. Instead of the original three rotational axes, since two are parallel, there are now only two effective rotational axes – we say that one degree of freedom has been lost.◄

In mathematical, rather than mechanical, terms this problem can be seen using the definition of the Lunar module's coordinate system where the rotation of the spacecraft's body-fixed frame {B} with respect to the stable platform frame {S} is

$$^{S}\!R_B = R_y\big(\theta_p\big)R_z\big(\theta_r\big)R_x\big(\theta_y\big)$$

For the case when $\theta_r = \frac{\pi}{2}$ we can apply the identity◄

$$R_y(\theta)R_z\big(\tfrac{\pi}{2}\big) \equiv R_z\big(\tfrac{\pi}{2}\big)R_x(\theta)$$

leading to

$$^{S}\!R_B = R_z\big(\tfrac{\pi}{2}\big)R_x\big(\theta_p\big)R_x\big(\theta_y\big) = R_z\big(\tfrac{\pi}{2}\big)R_x\big(\theta_p + \theta_y\big)$$

which is unable to represent any rotation about the y-axis. This is not a good thing because spacecraft rotation about the y-axis would rotate the stable element and thus ruin its precise alignment with the stars: hence the anxiety on Apollo 13.

The loss of a degree of freedom means that mathematically we cannot invert the transformation, we can only establish a linear relationship between two of the angles. In this case the best we can do is determine the sum of the pitch and yaw angles. We observed a similar phenomena with the Euler angle singularity earlier.

Operationally this was a significant limiting factor with this particular gyroscope (Hoag 1963) and could have been alleviated by adding a fourth gimbal, as was used on other spacecraft. It was omitted on the Lunar Module for reasons of weight and space.

Rotations obey the cyclic rotation rules
$Rx(\tfrac{\pi}{2})\,Ry(\theta)\,Rx(\tfrac{\pi}{2})^T \equiv Rz(\theta)$
$Ry(\tfrac{\pi}{2})\,Rz(\theta)\,Ry(\tfrac{\pi}{2})^T \equiv Rx(\theta)$
$Rz(\tfrac{\pi}{2})\,Rx(\theta)\,Rz(\tfrac{\pi}{2})^T \equiv Ry(\theta)$
and anti-cyclic rotation rules
$Ry(\tfrac{\pi}{2})^T\,Rx(\theta)\,Ry(\tfrac{\pi}{2}) \equiv Rz(\theta)$
$Rz(\tfrac{\pi}{2})^T\,Ry(\theta)\,Rz(\tfrac{\pi}{2}) \equiv Rx(\theta)$.

Apollo 13 mission clock: 02 08 12 47

- **Flight:** "Go, Guidance."
- **Guido:** "He's getting close to <u>gimbal lock</u> there."
- **Flight:** "Roger. CapCom, recommend he bring up C3, C4, B3, B4, C1 and C2 thrusters, and advise he's getting close to <u>gimbal lock</u>."
- **CapCom:** "Roger."

Apollo 13, mission control communications loop (1970) (Lovell and Kluger 1994, p 131; NASA 1970).

All three-angle representations of attitude, whether Eulerian or Cardanian, suffer this problem of *gimbal lock* when two consecutive axes become aligned. For ZYZ-Euler angles this occurs when $\theta = k\pi$, $k \in \mathbb{Z}$ and for roll-pitch-yaw angles when pitch $\theta_p = \pm(2k + 1)\frac{\pi}{2}$. The best that can be hoped for is that the singularity occurs for an attitude which does not occur during normal operation of the vehicle – it requires judicious choice of angle sequence and coordinate system.

Singularities are an unfortunate consequence of using a minimal representation. To eliminate this problem we need to adopt different representations of orientation. Many in the Apollo LM team would have preferred a four gimbal system and the clue to success, as we shall see shortly in Sect. 2.2.1.7, is to introduce a fourth parameter.

2.2.1.4   Two Vector Representation

For arm-type robots it is useful to consider a coordinate frame {E} attached to the end-effector as shown in Fig. 2.16. By convention the axis of the tool is associated with the z-axis and is called the *approach vector* and denoted $\hat{a} = (a_x, a_y, a_z)$. For some applications it is more convenient to specify the approach vector than to specify Euler or roll-pitch-yaw angles.

However specifying the direction of the z-axis is insufficient to describe the coordinate frame – we also need to specify the direction of the x- and y-axes. An orthogonal vector that provides orientation, perhaps between the two fingers of the robot's gripper is called the *orientation vector*, $\hat{o} = (o_x, o_y, o_z)$. These two unit vectors are sufficient to completely define the rotation matrix

$$R = \begin{pmatrix} n_x & o_x & a_x \\ n_y & o_y & a_y \\ n_z & o_z & a_z \end{pmatrix} \tag{2.17}$$

since the remaining column, the normal vector, can be computed using Eq. 2.12 as $\hat{n} = \hat{o} \times \hat{a}$. Consider an example where the gripper's approach and orientation vectors are parallel to the world x- and y-directions respectively. Using the Toolbox this is implemented by

```
>> a = [1 0 0]';
>> o = [0 1 0]';
>> R = oa2r(o, a)
R =
     0     0     1
     0     1     0
    -1     0     0
```

Any two nonparallel vectors are sufficient to define a coordinate frame. Even if the two vectors $\hat{a}$ and $\hat{o}$ are not orthogonal they still define a plane and the computed $\hat{n}$ is normal to that plane. In this case we need to compute a new value for $\hat{o}' = \hat{a} \times \hat{n}$ which lies in the plane but is orthogonal to each of $\hat{a}$ and $\hat{n}$.

For a camera we might use the optical axis, by convention the z-axis, and the left side of the camera which is by convention the x-axis. For a mobile robot we might use

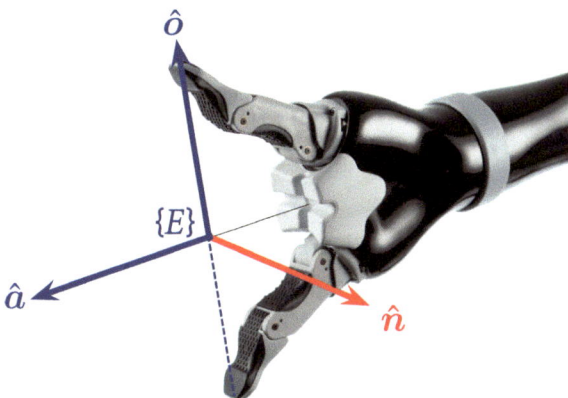

Fig. 2.16.
Robot end-effector coordinate system defines the pose in terms of an *approach* vector $\hat{a}$ and an *orientation* vector $\hat{o}$, from which $\hat{n}$ can be computed. $\hat{n}$, $\hat{o}$ and $\hat{a}$ vectors correspond to the x-, y- and z-axes respectively of the end-effector coordinate frame. (courtesy of Kinova Robotics)

the gravitational acceleration vector (measured with accelerometers) which is by convention the z-axis and the heading direction (measured with an electronic compass) which is by convention the x-axis.

2.2.1.5   Rotation about an Arbitrary Vector

Two coordinate frames of arbitrary orientation are related by a *single* rotation about some axis in space. For the example rotation used earlier

```
>> R = rpy2r(0.1 , 0.2, 0.3);
```

we can determine such an angle and vector by

```
>> [theta, v] = tr2angvec(R)
th =
    0.3655
v =
    0.1886    0.5834    0.7900
```

This is not unique. A rotation of $-\texttt{theta}$ about the vector $-\texttt{v}$ results in the same orientation.

where `theta` is the angle of rotation and `v` is the vector◄ around which the rotation occurs.

This information is encoded in the eigenvalues and eigenvectors of R. Using the built-in MATLAB function `eig`

```
>> [x,e] = eig(R)
x =
   -0.6944 + 0.0000i   -0.6944 + 0.0000i    0.1886 + 0.0000i
    0.0792 + 0.5688i    0.0792 - 0.5688i    0.5834 + 0.0000i
    0.1073 - 0.4200i    0.1073 + 0.4200i    0.7900 + 0.0000i
e =
    0.9339 + 0.3574i    0.0000 + 0.0000i    0.0000 + 0.0000i
    0.0000 + 0.0000i    0.9339 - 0.3574i    0.0000 + 0.0000i
    0.0000 + 0.0000i    0.0000 + 0.0000i    1.0000 + 0.0000i
```

Both matrices are complex, but some elements are real (zero imaginary part).

the eigenvalues are returned on the diagonal of the matrix `e` and the corresponding eigenvectors are the corresponding columns of `x`.◄

From the definition of eigenvalues and eigenvectors we recall that

$$Rv = \lambda v$$

where v is the eigenvector corresponding to the eigenvalue λ. For the case $\lambda = 1$

$$Rv = v$$

which implies that the corresponding eigenvector v is *unchanged* by the rotation. There is only one such vector and that is the one *about which* the rotation occurs. In the example the third eigenvalue is equal to one, so the rotation axis is the third column of `x`.

Olinde Rodrigues (1795–1850) was a French banker and mathematician who wrote extensively on politics, social reform and banking. He received his doctorate in mathematics in 1816 from the University of Paris, for work on his first well known formula which is related to Legendre polynomials. His eponymous rotation formula was published in 1840 and is perhaps the first time the representation of a rotation as a scalar and a vector was articulated. His formula is sometimes, and inappropriately, referred to as the Euler-Rodrigues formula. He is buried in the Pere-Lachaise cemetery in Paris.

An orthonormal rotation matrix will always have one real eigenvalue at $\lambda = 1$ and in general a complex pair $\lambda = \cos\theta \pm i\sin\theta$ where θ is the rotation angle. The angle of rotation◀ in this case is

```
>> theta = angle(e(1,1))
theta =
    0.3655
```

The inverse problem, converting from angle and vector to a rotation matrix, is achieved using Rodrigues' rotation formula

$$R = I_{3\times3} + \sin\theta[\hat{v}]_\times + (1-\cos\theta)[\hat{v}]_\times^2 \qquad (2.18)$$

where $[\hat{v}]_\times$ is a skew-symmetric matrix. We can use this formula to determine the rotation of $\frac{\pi}{2}$ about the x-axis

It can also be shown that the trace of a rotation matrix $tr(R) = 1 + 2\cos\theta$ from which we can compute the magnitude of θ but not its sign.

```
>> R = angvec2r(pi/2, [1 0 0])
R =
    1.0000        0        0
         0   0.0000  -1.0000
         0   1.0000   0.0000
```

It is interesting to note that this representation of an arbitrary rotation is parameterized by four numbers: three for the rotation axis, and one for the angle of rotation. This is far fewer than the nine numbers required by a rotation matrix. However the direction can be represented by a unit vector which has only two parameters▶ and the angle can be encoded in the length to give a 3-parameter representation such as $\hat{v}\theta$, $\hat{v}\sin(\theta/2)$, $\hat{v}\tan(\theta)$ or the Rodrigues' vector $\hat{v}\tan(\theta/2)$. While these forms are minimal and efficient in terms of data storage they are analytically problematic and ill-defined when $\theta = 0$.

Imagine a unit-sphere. All possible unit vectors from the center can be described by the latitude and longitude of the point at which they touch the surface of the sphere.

2.2.1.6 Matrix Exponentials

Consider an x-axis rotation expressed as a rotation matrix

```
>> R = rotx(0.3)
R =
    1.0000        0        0
         0   0.9553  -0.2955
         0   0.2955   0.9553
```

As we did for the 2-dimensional case we can compute the logarithm of this matrix using the MATLAB builtin function logm▶

logm is different to the builtin function log which computes the logarithm of each element of the matrix. A logarithm can be computed using a power series, with a matrix rather than scalar argument. For a matrix the logarithm is not unique and logm computes the principal logarithm of the matrix.

```
>> S = logm(R)
S =
         0        0        0
         0   0.0000  -0.3000
         0   0.3000   0.0000
```

and the result is a sparse matrix with two elements that have a magnitude of 0.3, which is the original rotation angle. This matrix has a zero diagonal and is another example of a skew-symmetric matrix, in this case 3×3.

Applying vex to the skew-symmetric matrix gives

```
>> vex(S)'
ans =
    0.3000        0        0
```

and we find the original rotation angle is in the first element, corresponding to the *x*-axis about which the rotation occurred. For the 3-dimensional case the Toolbox function `trlog` is equivalent◀

`trlog` uses a more efficient closed-form solution as well as being able to return the angle and axis information separately.

```
>> [th,w] = trlog(R)
th =
    0.3000
w =
    1.0000
         0
         0
```

The inverse of a logarithm is exponentiation and applying the builtin MATLAB matrix exponential function `expm`◀

`expm` is different to the builtin function `exp` which computes the exponential of each element of the matrix:
$expm(A) = I + A + A^2/2! + A^3/3! + \cdots$

```
>> expm(S)
ans =
    1.0000         0         0
         0    0.9553   -0.2955
         0    0.2955    0.9553
```

we have regenerated our original rotation matrix. In fact the command

```
>> R = rotx(0.3);
```

is equivalent to

```
>> R = expm( skew([1 0 0]) * 0.3 );
```

where we have specified the rotation in terms of a rotation angle and a rotation axis (as a unit-vector). This generalizes to rotation about *any* axis and formally we can write

$$\boldsymbol{R} = e^{[\hat{\omega}]_\times \theta} \in \mathbf{SO}(3)$$

where θ is the rotation angle, $\hat{\omega}$ is a unit-vector parallel to the rotation axis, and the notation $[\cdot]_\times : \mathbb{R}^3 \mapsto \mathbb{R}^{3\times3}$ indicates a mapping from a vector to a skew-symmetric matrix. Since $[\omega]_\times \theta = [\omega\theta]_\times$ we can treat $\omega\theta \in \mathbb{R}^3$ as a rotational parameter called exponential coordinates. For the 3-dimensional case, Rodrigues' rotation formula (Eq. 2.18) is a computationally efficient means of computing the matrix exponential for the special case where the argument is a skew-symmetric matrix, and this is used by the Toolbox function `trexp` which is equivalent to `expm`.

In 3-dimensions the skew-symmetric matrix has the form

$$[\omega]_\times = \begin{pmatrix} 0 & -\omega_z & \omega_y \\ \omega_z & 0 & -\omega_x \\ -\omega_y & \omega_x & 0 \end{pmatrix} \tag{2.19}$$

which has clear structure and only three unique elements $\omega \in \mathbb{R}^3$. The matrix can be used to implement the vector cross product $v_1 \times v_2 = [v_1]_\times v_2$. A simple example of Toolbox support for skew-symmetric matrices is

```
>> skew([1 2 3])
ans =
     0    -3     2
     3     0    -1
    -2     1     0
```

and the inverse operation is performed using the Toolbox function `vex`

```
>> vex(ans)'
ans =
     1     2     3
```

Both functions work for the 3D case, shown here, and the 2D case where the vector is a 1-vector.

2.2.1.7   Unit Quaternions

> *Quaternions came from Hamilton after his really good work had been done;*
> *and, though beautifully ingenious, have been an unmixed evil to those*
> *who have touched them in any way, including Clark Maxwell.*
>
> Lord Kelvin, 1892

Quaternions were discovered by Sir William Hamilton over 150 years ago and, while initially controversial, have great utility for robotics. The quaternion is an extension of the complex number – a hypercomplex number – and is written as a scalar plus a vector

$$\begin{aligned} \boldsymbol{q} &= s + \boldsymbol{v} \\ &= s + v_1 i + v_2 j + v_3 k \end{aligned} \tag{2.20}$$

where $s \in \mathbb{R}$, $\boldsymbol{v} \in \mathbb{R}^3$ and the orthogonal complex numbers i, j and k are defined such that

$$i^2 = j^2 = k^2 = ijk = -1 \tag{2.21}$$

and we denote a quaternion as

$$\boldsymbol{q} = s < v_1, \ v_2, \ v_3 >$$

In the Toolbox quaternions are implemented by the `Quaternion` class. Quaternions support addition and subtraction, performed element-wise, multiplication by a scalar and multiplication

$$\boldsymbol{q}_1 \circ \boldsymbol{q}_2 = s_1 s_2 - v_1 \cdot v_2 < s_1 v_2 + s_2 v_1 + v_1 \times v_2 >$$

which is known as the quaternion or Hamilton product.▶

One early objection to quaternions was that multiplication was not commutative but as we have seen above this is exactly what we require for rotations. Despite the initial controversy quaternions are elegant, powerful and computationally straightforward and they are *widely* used for robotics, computer vision, computer graphics and aerospace navigation systems.

To represent rotations we use unit-quaternions denoted by $\mathring{q}$. These are quaternions of unit magnitude; that is, those for which $\|q\| = s^2 + v_1^2 + v_2^2 + v_3^2 = 1$. They can be considered as a rotation of θ about the unit vector $\hat{\boldsymbol{v}}$ which are related to the quaternion components by▶

If we write the quaternion as a 4-vector (s, v_1, v_2, v_2) then multiplication can be expressed as a matrix-vector product where

$$\boldsymbol{q} \circ \boldsymbol{q}' = \begin{pmatrix} s & -v_1 & -v_2 & -v_3 \\ v_1 & s & -v_3 & v_2 \\ v_2 & v_3 & s & -v_1 \\ v_3 & -v_2 & v_1 & s \end{pmatrix} \begin{pmatrix} s' \\ v_1' \\ v_2' \\ v_3' \end{pmatrix}$$

As for the angle-vector representation this is not unique. A rotation of θ about the vector $-v$ results in the same orientation. This is referred to as a double mapping or double cover.

Sir William Rowan Hamilton (1805–1865) was an Irish mathematician, physicist, and astronomer. He was a child prodigy with a gift for languages and by age thirteen knew classical and modern European languages as well as Persian, Arabic, Hindustani, Sanskrit, and Malay. Hamilton taught himself mathematics at age 17, and discovered an error in Laplace's Celestial Mechanics. He spent his life at Trinity College, Dublin, and was appointed Professor of Astronomy and Royal Astronomer of Ireland while still an undergraduate. In addition to quaternions he contributed to the development of optics, dynamics, and algebra. He also wrote poetry and corresponded with Wordsworth who advised him to devote his energy to mathematics.

According to legend the key quaternion equation, Eq. 2.21, occured to Hamilton in 1843 while walking along the Royal Canal in Dublin with his wife, and this is commemorated by a plaque on Broome bridge:

Here as he walked by on the 16th of October 1843 Sir William Rowan Hamilton in a flash of genius discovered the fundamental formula for quaternion multiplication $i^2 = j^2 = k^2 = ijk = -1$ & cut it on a stone of this bridge.

His original carving is no longer visible, but the bridge is a pilgrimage site for mathematicians and physicists.

For the case of unit quaternions our generalized pose is a rotation $\xi \sim \mathring{q} \in \mathbb{S}^3$ and

$$\mathring{q}_1 \oplus \mathring{q}_2 \mapsto \mathring{q}_1 \circ \mathring{q}_2$$

and

$$\ominus \mathring{q} \mapsto \mathring{q}^{-1} = s <-v>$$

which is the quaternion conjugate. The zero rotation $0 \mapsto 1 <0, 0, 0>$ which is the identity quaternion. A vector $v \in \mathbb{R}^3$ is rotated by

$$\mathring{q} \cdot v \mapsto \mathring{q} \circ \mathring{v} \circ \mathring{q}^{-1}$$

where $\mathring{v} = 0 <v>$ is known as a pure quaternion.

$$\mathring{q} = \cos\tfrac{\theta}{2} < \hat{v}\sin\tfrac{\theta}{2}> \tag{2.22}$$

and has similarities to the angle-axis representation of Sect. 2.2.1.5.

In the Toolbox these are implemented by the `UnitQuaternion` class and the constructor converts a passed argument such as a rotation matrix to a unit quaternion, for example

```
>> q = UnitQuaternion( rpy2tr(0.1, 0.2, 0.3)  )
q =
0.98335 < 0.034271, 0.10602, 0.14357 >
```

This class overloads a number of standard methods and functions. Quaternion multiplication◄ is invoked through the overloaded multiplication operator

```
>> q = q * q;
```

and inversion, the conjugate of a unit quaternion, is

```
>> inv(q)
ans =
0.93394 < -0.0674, -0.20851, -0.28236 >
```

Multiplying a quaternion by its inverse yields the identity quaternion

```
>> q*inv(q)
ans =
1 < 0, 0, 0 >
```

which represents a null rotation, or more succinctly

```
>> q/q
ans =
1 < 0, 0, 0 >
```

The quaternion can be converted to an orthonormal rotation matrix by

```
>> q.R
ans =
    0.7536   -0.4993    0.4275
    0.5555    0.8315   -0.0081
   -0.3514    0.2436    0.9040
```

and we can also plot the orientation represented by a quaternion

```
>> q.plot()
```

which produces a result similar in style to that shown in Fig. 2.13. A vector is rotated by a quaternion using the overloaded multiplication operator

```
>> q*[1 0 0]'
ans =
    0.7536
    0.5555
   -0.3514
```

Compounding two orthonormal rotation matrices requires 27 multiplications and 18 additions. The quaternion form requires 16 multiplications and 12 additions. This saving can be particularly important for embedded systems.

The Toolbox implementation is quite complete and the `UnitQuaternion` class has many methods and properties which are described fully in the online documentation.

2.2.2   Pose in 3-Dimensions

We return now to representing relative pose in three dimensions – the position and orientation change between the two coordinate frames as shown in Fig. 2.10. This is often referred to as a rigid-body displacement or rigid-body motion.

We have discussed several different representations of orientation, and we need to combine one of these with translation, to create a tangible representation of relative pose.

2.2.2.1   Homogeneous Transformation Matrix

The derivation for the homogeneous transformation matrix is similar to the 2D case of Eq. 2.11 but extended to account for the z-dimension. $^At_B \in \mathbb{R}^3$ is a vector defining the origin of frame $\{B\}$ with respect to frame $\{A\}$, and ^AR_B is the 3×3 orthonormal matrix which describes the orientation of the axes of frame $\{B\}$ with respect to frame $\{A\}$.

$$\begin{pmatrix} ^Ax \\ ^Ay \\ ^Az \\ 1 \end{pmatrix} = \begin{pmatrix} ^AR_B & ^At_B \\ 0_{1\times3} & 1 \end{pmatrix} \begin{pmatrix} ^Bx \\ ^By \\ ^Bz \\ 1 \end{pmatrix}$$

If points are represented by homogeneous coordinate vectors then

$$\begin{aligned} ^A\tilde{p} &= \begin{pmatrix} ^AR_B & ^At_B \\ 0_{1\times3} & 1 \end{pmatrix} {}^B\tilde{p} \\ &= {}^AT_B \, {}^B\tilde{p} \end{aligned} \tag{2.23}$$

and ^AT_B is a 4×4 homogeneous transformation matrix. This matrix has a very specific structure and belongs to the special Euclidean group of dimension 3 or $T \in \mathrm{SE}(3) \subset \mathbb{R}^{4\times4}$.

A concrete representation of relative pose is $\xi \sim T \in \mathrm{SE}(3)$ and $T_1 \oplus T_2 \mapsto T_1 T_2$ which is standard matrix multiplication.

$$T_1 T_2 = \begin{pmatrix} R_1 & t_1 \\ 0_{1\times3} & 1 \end{pmatrix} \begin{pmatrix} R_2 & t_2 \\ 0_{1\times3} & 1 \end{pmatrix} = \begin{pmatrix} R_1 R_2 & t_1 + R_1 t_2 \\ 0_{1\times3} & 1 \end{pmatrix} \tag{2.24}$$

One of the rules of pose algebra from page 19 is $\xi \oplus 0 = \xi$. For matrices we know that $TI = T$, where I is the identify matrix, so for pose $0 \mapsto I$ the identity matrix. Another rule of pose algebra was that $\xi \ominus \xi = 0$. We know for matrices that $TT^{-1} = I$ which implies that $\ominus T \mapsto T^{-1}$

$$T^{-1} = \begin{pmatrix} R & t \\ 0_{1\times3} & 1 \end{pmatrix}^{-1} = \begin{pmatrix} R^T & -R^T t \\ 0_{1\times3} & 1 \end{pmatrix} \tag{2.25}$$

The 4×4 homogeneous transformation is very commonly used in robotics, computer graphics and computer vision. It is supported by the Toolbox and will be used throughout this book as a concrete representation of 3-dimensional pose.

The Toolbox has many functions to create homogeneous transformations. For example we can demonstrate composition of transforms by

```
>> T = transl(1, 0, 0) * trotx(pi/2) * transl(0, 1, 0)
T =
    1.0000         0         0    1.0000
         0    0.0000   -1.0000    0.0000
         0    1.0000    0.0000    1.0000
         0         0         0    1.0000
```

The function `transl` creates a relative pose with a finite translation but no rotation, while `trotx` creates a relative pose corresponding to a rotation of $\frac{\pi}{2}$ about the x-axis with zero translation.◄ We can think of this expression as representing a walk along the x-axis for 1 unit, then a rotation by 90° about the x-axis and then a walk of 1 unit along the new y-axis which was the previous z-axis. The result, as shown in the last column of the resulting matrix is a translation of 1 unit along the original x-axis and 1 unit along the original z-axis. The orientation of the final pose shows the effect of the rotation about the x-axis. We can plot the corresponding coordinate frame by

> Many Toolbox functions have variants that return orthonormal rotation matrices or homogeneous transformations, for example, `rotx` and `trotx`, `rpy2r` and `rpy2tr` etc. Some Toolbox functions accept an orthonormal rotation matrix or a homogeneous transformation and ignore the translational component, for example, `tr2rpy`.

```
>> trplot(T)
```

The rotation matrix component of `T` is

```
>> t2r(T)
ans =
    1.0000         0         0
         0    0.0000   -1.0000
         0    1.0000    0.0000
```

and the translation component is a column vector

```
>> transl(T)'
ans =
    1.0000    0.0000    1.0000
```

2.2.2.2   Vector-Quaternion Pair

A compact and practical representation is the vector and unit quaternion pair. It represents pose using just 7 numbers, is easy to compound, and singularity free.◄

> This representation is not implemented in the Toolbox.

> For the vector-quaternion case $\xi \sim (\boldsymbol{t}, \mathring{q})$ where $\boldsymbol{t} \in \mathbb{R}^3$ is a vector defining the frame's origin with respect to the reference coordinate frame, and $\mathring{q} \in \mathbb{S}^3$ is the frame's orientation with respect to the reference frame.
> Composition is defined by
> $$\xi_1 \oplus \xi_2 = (\boldsymbol{t}_1 + \mathring{q}_1 \cdot \boldsymbol{t}_2, \mathring{q}_1 \circ \mathring{q}_2)$$
> and negation is
> $$\ominus \xi = \left(-\mathring{q}^{-1} \cdot \boldsymbol{t}, \mathring{q}^{-1}\right)$$
> and a point coordinate vector is transformed to a coordinate frame by
> $$^X\boldsymbol{p} = {}^X\xi_Y \cdot {}^Y\boldsymbol{p} = \mathring{q} \cdot {}^Y\boldsymbol{p} + \boldsymbol{t}$$

2.2.2.3   Twists

In Sect. 2.1.2.3 we introduced twists for the 2D case. Any rigid-body motion in 3D space is equivalent to a screw motion – motion about and along some line in space.◄ We represent a screw as a pair of 3-vectors $\boldsymbol{s} = (\boldsymbol{v}, \boldsymbol{\omega}) \in \mathbb{R}^6$.

> Pure translation can be considered as rotation about a point at infinity.

The $\boldsymbol{\omega}$ component of the twist vector is the direction of the screw axis. The $\boldsymbol{v}$ component is called the moment and encodes the position of the line of the twist axis in space and also the pitch of the screw. The pitch is the ratio of the distance along the screw axis to the rotation about the screw axis.

Consider the example of a rotation of 0.3 radians about the *x*-axis. We first specify a unit twist► with an axis that is parallel to the *x*-axis and passes through the origin

A rotational unit twist has $\|\omega\| = 1$.

```
>> tw = Twist('R', [1 0 0], [0 0 0])
tw =
( -0 -0 -0; 1 0 0 )
```

which we convert, for the required rotation angle, to an **SE**(3)-homogeneous transformation

```
>>  tw.T(0.3)
ans =
    1.0000        0         0         0
         0    0.9553   -0.2955         0
         0    0.2955    0.9553         0
         0         0         0    1.0000
```

and has the same value we would obtain using `trotx(0.3)`.

For pure translation in the *y*-direction the unit twist► would be

A translational unit twist has $\|v\| = 1$ and $\omega = 0$.

```
>> tw = Twist('T', [0 1 0])
tw =
( 0 1 0; 0 0 0 )
```

which we convert, for the required translation distance, to an **SE**(3)-homogeneous transformation.

```
>> tw.T(2)
ans =
     1     0     0     0
     0     1     0     2
     0     0     1     0
     0     0     0     1
```

which is, as expected, an identity matrix rotational component (no rotation) and a translational component of 2 in the *y*-direction.

To illustrate the underlying screw model we define a coordinate frame {*X*}

```
>> X = transl(3, 4, -4);
```

which we will rotate by a range of angles

```
>> angles = [0:0.3:15];
```

around a screw axis parallel to the *z*-axis, direction $(0, 0, 1)$, through the point $(2, 3, 2)$ and with a pitch of 0.5

```
>> tw = Twist('R', [0 0 1], [2 3 2], 0.5);
```

The next line packs a lot of functionality. For values of θ drawn successively from the vector `angles` we use an anonymous function to evaluate the twist for each value of θ and apply it to the frame {*X*}. This sequence is animated and each frame in the sequence is retained

```
>> tranimate( @(theta) tw.T(theta) * X, angles, ...
       'length', 0.5, 'retain', 'rgb', 'notext');
```

and the result is shown in Fig. 2.17. We can clearly see the screw motion in the successive poses of the displaced reference frame as it is rotated about the screw axis.

The screw axis is the line

```
>> L = tw.line
L =
{ 3  -2   0; 0   0   1 }
```

which is described in terms of its Plücker coordinates which we can plot

```
>> L.plot('k:', 'LineWidth', 2)
```

Finally we can convert an arbitrary homogeneous transformation to a nonunit twist

```
>> T = transl(1, 2, 3) * eul2tr(0.3, 0.4, 0.5);
>> tw = Twist(T)
tw =
( 1.1204 1.6446 3.1778; 0.041006 0.4087 0.78907 )
```

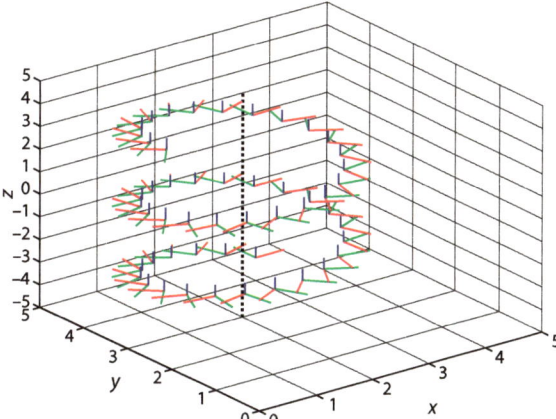

Fig. 2.17.
A coordinate frame {X} displayed for different values of θ about a screw parallel to the z-axis and passing through the point (2, 3, 2). The x-, y- and z-axes are indicated by red, green and blue lines respectively

which has a pitch of

```
>> tw.pitch
ans =
    3.2256
```

and the rotation about the axis is

```
>> tw.theta
ans =
    0.8896
```

and a point lying on the twist axis is

```
>> tw.pole'
ans =
    0.0011    0.8473    -0.4389
```

2.3   Advanced Topics

2.3.1   Normalization

The IEEE standard for double precision floating point, the standard MATLAB numeric format, has around 16 decimal digits of precision.

Floating-point arithmetic has finite precision◄ and consecutive operations will accumulate error. A rotation matrix has by definition, a determinant of one

```
>> R = eye(3,3);
>> det(R) - 1
ans =
    0
```

but if we repeatedly multiply by a valid rotation matrix the result

```
>> for i=1:100
     R = R * rpy2r(0.2, 0.3, 0.4);
   end
>> det(R) - 1
ans =
  4.4409e-15
```

indicates a small error – the determinant is no longer equal to one and the matrix is no longer a proper orthonormal rotation matrix. To fix this we need to normalize the matrix, a process which enforces the constraints on the columns c_i of an orthonormal matrix $R = [c_1, c_2, c_3]$. We need to assume that one column has the correct direction

$$c_3' = c_3$$

then the first column is made orthogonal to the last two

$$c_1' = c_2 \times c_3'$$

However the last two columns may not have been orthogonal so

$$c_2' = c_1' \times c_3'$$

Finally the columns are all normalized to unit magnitude

$$c_i'' = \frac{c_i'}{\|c_i'\|}, \quad i = 1 \cdots 3$$

In the Toolbox normalization is implemented by

```
>> R = trnorm(R);
```

and the determinant is now much closer to one▶

```
>> det(R) - 1
ans =
   -2.2204e-16
```

This error is now at the limit of double precision arithmetic which is 2.2204×10^{-16} and given by the MATLAB function `eps`.

A similar issue arises for unit quaternions when the norm, or magnitude, of the unit quaternion is no longer equal to one. However this is much easier to fix since normalizing the quaternion simply involves dividing all elements by the norm

$$\mathring{q}' = \frac{\mathring{q}}{\|\mathring{q}\|}$$

which is implemented by the `unit` method

```
>> q = q.unit();
```

The `UnitQuaternion` class also supports a variant of multiplication

```
>> q = q .* q2;
```

which performs an explicit normalization after the multiplication.

Normalization does not need to be done after every multiplication since it is an expensive operation. However for situations like the example above where one transform is being repeatedly updated it is advisable.

2.3.2   Understanding the Exponential Mapping

In this chapter we have glimpsed some connection between rotation matrices, skew-symmetric matrices and matrix exponentiation. The basis for this lies in the mathematics of Lie groups which are covered in text books on algebraic geometry and algebraic topology. These require substantial knowledge of advanced mathematics and many people starting out in robotics will find their content quite inaccessible. An introduction to the essentials of this topic is given in Appendix D. In this section we will use an intuitive approach, based on undergraduate engineering mathematics, to shed some light on these relationships.

Consider a point **P**, defined by a coordinate vector p, being rotated with an angular velocity ω which is a vector whose direction defines the axis of rotation and whose magnitude $\|\omega\|$ specifies the rate of rotation about the axis which we assume passes through the origin.▶ We wish to rotate the point by an angle θ about this axis and the velocity of the point is known from mechanics to be

Angular velocity will be properly introduced in the next chapter.

$$\dot{p} = \omega \times p$$

and we replace the cross product with a skew-symmetric matrix giving a matrix-vector product

$$\dot{p} = [\omega]_\times p \tag{2.26}$$

We can find the solution to this first-order differential equation by analogy to the simple scalar case

$$\dot{x} = ax$$

whose solution is

$$x(t) = e^{at} x(0)$$

This implies that the solution to Eq. 2.26 is

$$p(t) = e^{[\omega]_\times t} p(0)$$

If $\|\omega\| = 1$ then after t seconds the vector will have rotated by t radians. We require a rotation by θ so we can set $t = \theta$ to give

$$p(\theta) = e^{[\hat{\omega}]_\times \theta} p(0)$$

which describes the vector $p(0)$ being rotated to $p(\theta)$. A matrix that rotates a vector is a rotation matrix, and this implies that our matrix exponential is a rotation matrix

$$R(\theta, \hat{\omega}) = e^{[\hat{\omega}]_\times \theta} \in \mathbf{SO}(3)$$

Now consider the more general case of rotational and translational motion. We can write

$$\dot{p} = [\omega]_\times p + v$$

and rearranging into matrix form

$$\begin{pmatrix} \dot{p} \\ 0 \end{pmatrix} = \begin{pmatrix} [\omega]_\times & v \\ 0 & 0 \end{pmatrix} \begin{pmatrix} p \\ 1 \end{pmatrix}$$

and introducing homogeneous coordinates this becomes

$$\begin{aligned} \dot{\tilde{p}} &= \begin{pmatrix} [\omega]_\times & v \\ 0 & 0 \end{pmatrix} \tilde{p} \\ &= \Sigma \tilde{p} \end{aligned}$$

where Σ is a 4×4 augmented skew-symmetric matrix. Again, by analogy with the scalar case we can write the solution as

$$\tilde{p}(\theta) = e^{\Sigma \theta} \tilde{p}(0)$$

A matrix that rotates and translates a point in homogeneous coordinates is a homogeneous transformation matrix, and this implies that our matrix exponential is a homogeneous transformation matrix

$$T(\theta, \hat{\omega}, v) = e^{\begin{pmatrix} [\hat{\omega}]_\times & v \\ 0 & 0 \end{pmatrix} \theta} \in \mathbf{SE}(3)$$

where $[\hat{\omega}]_\times \theta$ defines the magnitude and axis of rotation and $v\theta$ is the translation.

The exponential of a scalar can be computed using a power series, and the matrix case is analogous and relatively straightforward to compute. The MATLAB function `expm` uses a polynomial approximation for the general matrix case. If A is skew-symmetric or augmented-skew-symmetric then an efficient closed-form solution for a rotation matrix – the Rodrigues' rotation formula (Eq. 2.18) – can be used and this is implemented by the Toolbox function `trexp`.

2.3.3   More About Twists

In this chapter we introduced and applied twists and here we will more formally define them. We also highlight the very close relationship between twists and homogeneous transformation matrices via the exponential mapping.

The key concept comes from Chasle's theorem: "*any displacement of a body in space can be accomplished by means of a rotation of the body about a unique line in space accompanied by a translation of the body parallel to that line*". Such a line is called a screw axis and is illustrated in Fig. 2.18. The mathematics of screw theory was developed by Sir Robert Ball in the late 19th century for the analysis of mechanisms. At the core of screw theory are pairs of vectors: angular and linear velocity; forces and moments; and Plücker coordinates (see Sect. C.1.2.2).

The general displacement of a rigid body in 3D can be represented by a twist vector

$$S = (v, \omega) \in \mathbb{R}^6$$

where $v \in \mathbb{R}^3$ is referred to as the moment and encodes the position of the action line in space and the pitch of the screw and $\omega \in \mathbb{R}^3$ is the direction of the screw axis.

For rotational motion where the screw axis is parallel to the vector $\hat{a}$, passes through a point **Q** defined by its coordinate vector q, and the screw pitch p is the ratio of the distance along the screw axis to the rotation about the axis, the twist elements are

$$S = (q \times \hat{a} + p\hat{a}, \hat{a})$$

and the pitch can be recovered by

$$p = \hat{w}^T v$$

For the case of pure rotation the pitch of the screw is zero and the unit twist is

$$S = (q \times \hat{a}, \hat{a})$$

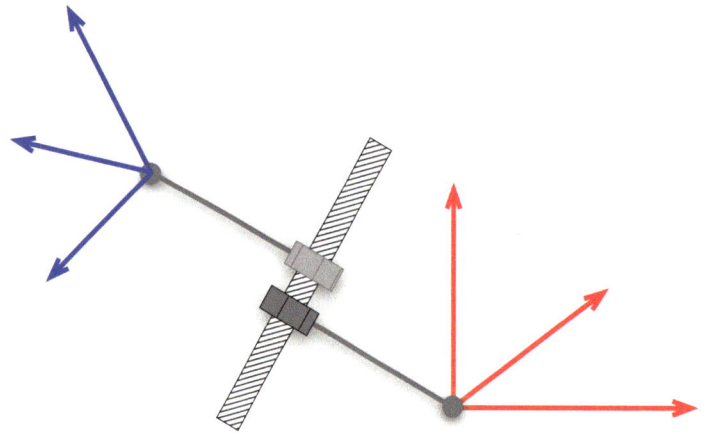

Fig. 2.18.
Conceptual depiction of a screw. A coordinate frame is attached to a nut by a rigid rod and rotated around the screw thread. The pose changes from the red frame to the blue frame. The corollary is that given any two frames we can determine a screw axis to rotate one into the other

Michel Chasles (1793–1880) was a French mathematician born at Épernon. He studied at the École Polytechnique in Paris under Poisson and in 1814 was drafted to defend Paris in the War of the Sixth Coalition. In 1837 he published a work on the origin and development of methods in geometry, which gained him considerable fame and he was appointed as professor at the École Polytechnique in 1841, and at the Sorbonne in 1846.

He was an avid collector and purchased over 27 000 forged letters purporting to be from Newton, Pascal and other historical figures – all written in French! One from Pascal claimed he had discovered the laws of gravity before Newton, and in 1867 Chasles took this to the French Academy of Science but scholars recognized the fraud. Eventually Chasles admitted he had been deceived and revealed he had spent nearly 150 000 francs on the letters. He is buried in Cimetière du Père Lachaise in Paris.

For purely translational motion in the direction parallel to the vector $\boldsymbol{a}$, the pitch is infinite which leads to a zero rotational component and the unit twist is

$$\boldsymbol{S} = (\hat{\boldsymbol{a}}, 0)$$

A twist is related to the rigid-body displacement in $\mathbf{SE}(3)$ by the exponential mapping already discussed.

$$\boldsymbol{T}(\theta, \boldsymbol{S}) = e^{[S]\theta} \in \mathbf{SE}(3)$$

where the augmented skew-symmetric matrix

$$[\boldsymbol{S}] = \left(\begin{array}{ccc|c} 0 & -\omega_3 & \omega_2 & v_1 \\ \omega_3 & 0 & -\omega_1 & v_2 \\ -\omega_2 & \omega_1 & 0 & v_3 \\ \hline 0 & 0 & 0 & 0 \end{array}\right) \in \mathbf{se}(3)$$

belongs to the Lie algebra $\mathbf{se}(3)$ and is the *generator* of the rigid-body displacement. The matrix exponential has an efficient closed-form

$$\boldsymbol{T}(\theta, \boldsymbol{S}) = \begin{pmatrix} \boldsymbol{R}(\theta, \hat{\boldsymbol{\omega}}) & \left(\boldsymbol{I}_{3\times3}\theta + (1-\cos\theta)[\hat{\boldsymbol{\omega}}]_\times + (\theta - \sin\theta)[\hat{\boldsymbol{\omega}}]_\times^2\right)\boldsymbol{v} \\ 0 & 1 \end{pmatrix}$$

where $\boldsymbol{R}(\theta, \hat{\omega})$ is computed using Rodrigues' rotation formula (Eq. 2.18). For a nonunit rotational twist, that is $\|\boldsymbol{\omega}\| \neq 1$, then $\theta = \|\boldsymbol{\omega}\|$.

For real numbers, if $x = \log X$ and $y = \log Y$ then

$$Z = XY = e^x e^y = e^{x+y}$$

but for the matrix case this is *only true* if the matrices commute, and rotation matrices do not, therefore

$$Z = XY = e^x e^y \neq e^{x+y} \quad \text{if } x, y \in \mathbf{so}(n) \text{ or } \mathbf{se}(n)$$

The bottom line is that there is no shortcut to compounding rotations, we must compute $z = \log(e^x e^y)$ not $z = x + y$.

The Toolbox provides many ways to create twists and to convert them to rigid-body displacements expressed as homogeneous transformations. Now that we understand more about the exponential mapping we will revisit the example from page 46

```
>> tw = Twist('R', [1 0 0], [0 0 0])
tw =
( -0 -0 -0; 1 0 0 )
```

A unit twist describes a *family* of motions that have a single parameter, either a rotation and translation about and along some screw axis, or a pure translation in some direction. We can visualize it as a mechanical screw in space, or represent it as a 6-vector $S = (v, \omega)$ where $\|\omega\| = 1$ for a rotational twist and $\|v\| = 1, \omega = 0$ for a translational twist.

A *particular* rigid-body motion is described by a unit-twist s and a motion parameter θ which is a scalar specifying the amount of rotation or translation. The motion is described by the twist $S\theta$ which is in general not a unit-twist. The exponential of this in 4×4 matrix format is the 4×4 homogeneous transformation matrix describing that particular rigid-body motion in **SE(3)**.

which is a unit twist that describes rotation about the *x*-axis in **SE(3)**. The `Twist` has a number of properties

```
>> tw.S'
ans =
     0     0     0     1     0     0
>> tw.v'
ans =
     0     0     0
>> tw.w'
ans =
     1     0     0
```

as well as various methods. We can create the **se**(3) Lie algebra using the `se` method of this class

```
>> tw.se
ans =
     0     0     0     0
     0     0    -1     0
     0     1     0     0
     0     0     0     0
```

which is the augmented skew-symmetric version of S. The method `T` performs the exponentiation▶ of this to create an **SE(3)** homogeneous transformation for the specified rotation about the unit twist

The `expm` method is synonomous and both invoke the Toolbox function `trexp`.

```
>>  tw.T(0.3)
ans =
     1.0000          0          0          0
          0     0.9553    -0.2955          0
          0     0.2955     0.9553          0
          0          0          0     1.0000
```

The Toolbox functions `trexp` and `trlog` are respectively closed-form alternatives to `expm` and `logm` when the arguments are in **so**(3)/**se**(3) or **SO**(3)/**SE**(3).

The `line` method returns a `Plucker` object that represents the line of the screw in Plücker coordinates

```
>> tw.line
ans =
{ 0  0  0; 1  0  0 }
```

Finally, the overloaded multiplication operator for the `Twist` class will compound two twists.

```
>> t2 = tw * tw
t2 =
( -0  -0  -0; 2  0  0 )
>> tr2angvec(t2.T)
Rotation: 2.000000 rad x [1.000000 0.000000 0.000000]
```

and the result in this case is a nonunit twist of two units, or 2 rad, about the *x*-axis.

2.3.4   Dual Quaternions

1845–1879, an English mathematician and geometer.

Quaternions were developed by William Hamilton in 1843 and we have already seen their utility for representing orientation, but using them to represent pose proved more difficult. One early approach was Hamilton's bi-quaternion where the quaternion coefficients were complex numbers. Somewhat later William Clifford◀ developed the dual number, defined as an ordered pair $d = (x, y)$ which can be written as $d = x + y\varepsilon$ where $\varepsilon^2 = 0$ and for which specific addition and multiplication rules exist. Clifford created a quaternion dual number with $x, y \in \mathbb{H}$ which he also called a bi-quaternion but is today called a dual quaternion

$$\mathring{\underline{q}}_\xi = r + \tfrac{1}{2}\varepsilon \mathring{t} \circ \mathring{r}$$

where $\mathring{r} \in \mathbb{H}$ is a unit quaternion representing the rotational part of the pose and $\mathring{t} \in \mathbb{H}$ is a pure quaternion representing translation. This type of mathematical object has been largely eclipsed by modern matrix and vector approaches, but there seems to be a recent resurgence of interest in alternative approaches. The dual quaternion is quite compact, requiring just 8 numbers; it is easy to compound using a special multiplication table; and it is easy to renormalize to eliminate the effect of imprecise arithmetic. However it has no real useful computational advantage over matrix methods.

2.3.5   Configuration Space

We have so far considered the pose of objects in terms of the position and orientation of a coordinate frame affixed to them. For an arm-type robot we might affix a coordinate frame to its end-effector, while for a mobile robot we might affix a frame to its body – its body-fixed frame. This is sufficient to describe the state of the robot in the familiar 2D or 3D Euclidean space which is referred to as the task space or operational space since it is where the robot performs tasks or operates.

An alternative way of thinking about this comes from classical mechanics and is referred to as the *configuration* of a system. The configuration is the smallest set of parameters, called generalized coordinates, that are required to fully describe the position of *every* particle in the system. This is not as daunting as it may appear since in general a robot comprises one or more rigid elements, and in each of these the particles maintain a constant relative offset to each other.

If the *system* is a train moving along a track then all the particles comprising the train move together and we need only a single generalized coordinate q, the distance along the track from some datum, to describe their location. A robot arm with a fixed base and two rigid links, connected by two rotational joints has a configuration that is completely described by two generalized coordinates – the two joint angles (q_1, q_2). The generalized coordinates can, as their name implies, represent displacements or rotations.

Sir Robert Ball (1840–1913) was an Irish astronomer born in Dublin. He became Professor of Applied Mathematics at the Royal College of Science in Dublin in 1867, and in 1874 became Royal Astronomer of Ireland and Andrews Professor of Astronomy at the University of Dublin. In 1892 he was appointed Lowndean Professor of Astronomy and Geometry at Cambridge University and became director of the Cambridge Observatory. He was a Fellow of the Royal Society and in 1900 became the first president of the Quaternion Society.

He is best known for his contributions to the science of kinematics described in his treatise "The Theory of Screws" (1876), but he also published "A Treatise on Spherical Astronomy" (1908) and a number of popular articles on astronomy. He is buried at the Parish of the Ascension Burial Ground in Cambridge.

The number of independent▸ generalized coordinates N is known as the number of degrees of freedom of the system. Any configuration of the system is represented by a point in its N-dimensional configuration space, or C-space, denoted by $\mathcal{C}$ and $q \in \mathcal{C}$. We can also say that $\dim \mathcal{C} = N$. For the train example $\mathcal{C} \subset \mathbb{R}$ which says that the displacement is a bounded real number. For the 2-joint robot the generalized coordinates are both angles so $\mathcal{C} \subset \mathbb{S}^1 \times \mathbb{S}^1$.

That is, there are no holonomic constraints on the system.

Any point in the configuration space can be mapped to a point in the task space $q \in \mathcal{C} \mapsto \tau \in \mathcal{T}$ but the inverse is not necessarily true. This mapping depends on the task space that we choose and this, as its name suggests, is task specific.

Consider again the train moving along its rail. We might be interested to describe the train in terms of its position on a plane in which case the task space would be $\mathcal{T} \subset \mathbb{R}^2$, or in terms of its latitude and longitude, in which case the task space would be $\mathcal{T} \subset \mathbb{S}^1 \times \mathbb{S}^1$. We might choose a 3-dimensional task space $\mathcal{T} \subset \mathbf{SE}(3)$ to account for height changes as the train moves up and down hills and its orientation changes as it moves around curves. However in all these case the dimension of the task space exceeds the dimension of the configuration space $\dim \mathcal{T} > \dim \mathcal{C}$ and this means that the train cannot *access* all points in the task space. While every point along the rail line can be mapped to the task space, most points in the task space will not map to a point on the rail line. The train is constrained by its fixed rails to move in a subset of the task space.

The simple 2-joint robot arm can access a subset of points in a plane so a useful task space might be $\mathcal{T} \subset \mathbb{R}^2$. The dimension of the task space equals the dimension of the configuration space $\dim \mathcal{T} = \dim \mathcal{C}$ and this means that the mapping between task and configuration spaces is bi-directional but it is not necessarily unique – for this type of robot, in general, two different configurations map to a single point in task space. Points in the task space beyond the physical reach of the robot are not mapped to the configuration space. If we chose a task space with more dimensions such as $\mathbf{SE}(2)$ or $\mathbf{SE}(3)$ then $\dim \mathcal{T} > \dim \mathcal{C}$ and the robot would only be able to access points within a subset of that space.

Now consider a snake-robot arm, such as shown in Fig. 8.9, with 20 joints and $\mathcal{C} \subset \mathbb{S}^1 \times \cdots \times \mathbb{S}^1$ and $\dim \mathcal{T} < \dim \mathcal{C}$. In this case an infinite number of configurations in a $20 - 6 = 14$-dimensional subspace of the 20-dimensional configuration space will map to the same point in task space. This means that in addition to the task of positioning the robot's end-effector we can *simultaneously* perform motion in the configuration subspace to control the shape of the arm to avoid obstacles in the environment. Such a robot is referred to as over-actuated or redundant and this topic is covered in Sect. 8.4.2.

The body of a quadrotor, such as shown in Fig. 4.19d, is a single rigid-body whose configuration is completely described by six generalized coordinates, its position and orientation in 3D space $\mathcal{C} \subset \mathbb{R}^3 \times \mathbb{S}^1 \times \mathbb{S}^1 \times \mathbb{S}^1$ where the orientation is expressed in some three-angle representation. For such a robot the most logical task space would be $\mathbf{SE}(3)$ which is equivalent to the configuration space and $\dim \mathcal{T} = \dim \mathcal{C}$. However the quadrotor has only four actuators which means it cannot *directly* access all the points in its configuration space and hence its task space. Such a robot is referred to as under-actuated and we will revisit this in Sect. 4.2.

2.4   Using the Toolbox

The Toolbox supports all the different representations discussed in this chapter as well as conversions between many of them. The representations and possible conversions are shown in tabular form in Tables 2.1 and 2.2 for the 2D and 3D cases respectively.

In this chapter we have mostly used native MATLAB matrices to represent rotations and homogeneous transformations▸ and historically this has been what the Toolbox supported – the Toolbox *classic* functions. From Toolbox release 10 there are classes that

Quaternions and twists are implemented as classes not native types, but in very old versions of the Toolbox quaternions were 1×4 vectors.

represent rotations and homogeneous transformations, named respectively SO2 and SE2 for 2 dimensions and SO3 and SE3 for 3 dimensions. These provide real advantages in terms of code readability and type safety and can be used in an almost identical fashion to the native matrix types. They are also polymorphic meaning they support many of the same operations which makes it very easy to switch between using say rotation matrices and quaternions or lifting a solution from 2- to 3-dimensions. A quick illustration of the new functionality is the example from page 25 which becomes

```
>> T1 = SE2(1, 2, 30, 'deg');
>> about T1
T1 [SE2] : 1x1 (176 bytes)
```

The size of the object in bytes, shown in parentheses, will vary between MATLAB. versions and computer types.

which results in an SE2 class object not a 3 × 3 matrix. If we display it however it does look like a 3 × 3 matrix

If you have the cprintf package from MATLAB File Exchange installed then the rotation submatrix will be colored red.

```
>> T1
 T1 =
    0.8660   -0.5000         1
    0.5000    0.8660         2
         0         0         1
```

The matrix is encapsulated within the object and we can extract it readily if required

```
>> T1.T
ans =
    0.8660   -0.5000    1.0000
    0.5000    0.8660    2.0000
         0         0    1.0000
>> about ans
ans [double] : 3x3 (72 bytes)
```

Returning to that earlier example we can quite simply transform the vector

```
>> inv(T1) * P
ans =
    1.7321
   -1.0000
```

and the class handles the details of converting the vector between Euclidean and homogeneous forms.

This new functionality is also covered in Tables 2.1 and 2.2, and Table 2.3 is a map between the classic and new functionality to assist you in using the Toolbox. From here on the book will use a mixture of classic functions and the newer classes.

Table 2.1. Toolbox supported data types for representing 2D pose: constructors and conversions

	Output type							
Input type	*t*	*θ*	*R*	*T*	**Twist vector**	**Twist**	**SO2**	**SE2**
t (2-vector)				transl2		**Twist**('T')		**SE2**()
θ (scalar)			rot2	trot2		**Twist**('R')	**SO2**()	**SE2**()
R (2 × 2 matrix)				r2t			**SO2**()	**SE2**()
T (3 × 3 matrix)	transl2		t2r			**Twist**()		**SE2**()
Twist vector (1- or 3-vector)			trexp2	trexp2		**Twist**()	SO2.exp()	**SE3**.exp()
Twist				.T	.S			.SE
SO2		.theta	.R	.T	.log			.SE2
SE2	.t	.theta	.R	.T	.log	.Twist	.SO2	

Dark grey boxes are not possible conversions. Light grey boxes are possible conversions but the Toolbox has no direct conversion, you need to convert via an intermediate type. Red text indicates classical Robotics Toolbox functions that work with native MATLAB® vectors and matrices. **Bold text** indicates a Toolbox class. Class.type() indicates a static factory method that constructs a Class object from input of that type. Functions shown starting with a dot are a method on the class corresponding to that row.

Input type	t	Euler	RPY	θ, v	R	T	Twist vector	Twist	Unit-Quaternion	SO3	SE3
t (3-vector)						transl		Twist('T')			SE3()
Euler (3-vector)					eul2r	eul2tr			UnitQuaternion.eul()	SO3.eul()	SE3.eul()
RPY (3-vector)					rpy2r	rpy2tr			UnitQuaternion.rpy()	SO3.rpy()	SE3.rpy()
θ, v (scalar + 3-vector)					angvec2r	angvec2tr			UnitQuaternion.angvec()	SO3.angvec()	SE3.angvec()
R (3×3 matrix)		tr2eul	tr2rpy	tr2angvec		r2t	trlog		UnitQuaternion()	SO3()	SE3()
T (4×4 matrix)	transl	tr2eul	tr2rpy	tr2angvec	t2r		trlog	Twist()	UnitQuaternion()	SO3()	SE3()
Twist vector (3- or 6-vector)					trexp	trexp		Twist()		SO3.exp()	SE3.exp()
Twist						.T	.S				.SE
Unit-Quaternion		.toeul	.torpy	.toangvec	.R	.T				.SO3	.SE3
SO3		.toeul	.torpy	.toangvec	.R	.T	.log		.UnitQuaternion		.SE3
SE3	.t	.toeul	.torpy	.toangvec	.R	.T	.log	.Twist	.UnitQuaternion	.SO3	

Output type

Dark grey boxes are not possible conversions. Light grey boxes are possible conversions but the Toolbox has no direct conversion, you need to convert via an intermediate type. Red text indicates classical Robotics Toolbox functions that work with native MATLAB® vectors and matrices. Class.type() indicates a static factory method that constructs a Class object from input of that type. Functions shown starting with a dot are a method on the class corresponding to that row.

Table 2.2. Toolbox supported data types for representing 3D pose: constructors and conversions

2.5   Wrapping Up

In this chapter we learned how to represent points and poses in 2- and 3-dimensional worlds. Points are represented by coordinate vectors relative to a coordinate frame. A set of points that belong to a rigid object can be described by a coordinate frame, and its constituent points are described by constant vectors in the object's coordinate frame. The position and orientation of any coordinate frame can be described relative to another coordinate frame by its relative pose ξ. We can think of a relative pose as a motion – a rigid-body motion – and these motions can be applied sequentially (composed or compounded). It is important to remember that composition is noncommutative – the order in which relative poses are applied is important.

We have shown how relative poses can be expressed as a pose graph or manipulated algebraically. We can also use a relative pose to transform a vector from one coordinate frame to another. A simple graphical summary of key concepts is given in Fig. 2.19.

We have discussed a variety of mathematical objects to tangibly represent pose. We have used orthonormal rotation matrices for the 2- and 3-dimensional case to represent orientation and shown how it can rotate a points' coordinate vector from one coordinate frame to another. Its extension, the homogeneous transformation matrix, can be used to represent both orientation and translation and we have shown how it can rotate and translate a point expressed in homogeneous coordinates from one frame

Orientation		Pose	
Classic	**New**	**Classic**	**New**
rot2	SO2	trot2	SE2
		transl2	SE2
trplot2	.plot	trplot2	.plot
rotx, roty, rotz	SO3.Rx, SO3.Ry, SO3.Rz	trotx, troty, trotz	SE3.Rx, SE3.Ry, SE3.Rz
		T = transl(v)	SE3(v)
eul2r, rpy2r	SO3.eul, SO3.rpy	eul2tr, rpy2tr	SE3.eul, SE3.rpy
angvec2r	SO3.angvec	angvec2tr	SE3.angvec
oa2r	SO3.oa	oa2tr	SE3.oa
		v = transl(T)	.t, .transl
tr2eul, tr2rpy	.toeul, .torpy	tr2eul, tr2rpy	.toeul, .torpy
tr2angvec	.toangvec	tr2angvec	.toangvec
trexp	SO3.exp	trexp	SE3.exp
trlog	.log	trlog	.log
trplot	.plot	trplot	.plot

Functions starting with dot are methods on the new objects. You can use them in functional form `toeul(R)` or in dot form `R.toeul()` or `R.toeul`. It's a personal preference. The trailing parentheses are not required if no arguments are passed, but it is a useful convention and reminder that you that you are invoking a method not reading a property. The old function `transl` appears twice since it maps a vector to a matrix as well as the inverse.

Table 2.3. Table of subsitutions from classic Toolbox functions that operate on and return a matrix, to the corresponding new classes and methods

to another. Rotation in 3-dimensions has subtlety and complexity and we have looked at various parameterizations such as Euler angles, roll-pitch-yaw angles and unit quaternions. Using Lie group theory we showed that rotation matrices, from the group **SO**(2) or **SO**(3), are the result of exponentiating skew-symmetric generator matrices. Similarly, homogeneous transformation matrices, from the group **SE**(2) or **SE**(3), are the result of exponentiating augmented skew-symmetric generator matrices. We have also introduced twists as a concise way of describing relative pose in terms of rotation around a screw axis, a notion that comes to us from screw theory and these twists are the unique elements of the generator matrices.

There are two important lessons from this chapter. The first is that there are *many* mathematical objects that can be used to represent pose and these are summarized in Table 2.4. There is no right or wrong – each has strengths and weaknesses and we typically choose the representation to suit the problem at hand. Sometimes we wish for a vectorial representation, perhaps for interpolation, in which case (x, y, θ) or (x, y, z, Γ) might be appropriate, but this representation cannot be easily compounded. Sometime we may only need to describe 3D rotation in which case Γ or $\mathring{q}$ is appropriate. Converting between representations is easy as shown in Tables 2.1 and 2.2.

The second lesson is that coordinate frames are your friend. The essential first step in many vision and robotics problems is to assign coordinate frames to all objects of interest, indicate the relative poses as a directed graph, and write down equations for the loops. Figure 2.20 shows you how to build a coordinate frame out of paper that you can pick up and rotate – making these ideas more tangible. Don't be shy, embrace the coordinate frame.

We now have solid foundations for moving forward. The notation has been defined and illustrated, and we have started our hands-on work with MATLAB. The next chapter discusses motion and coordinate frames that change with time, and after that we are ready to move on and discuss robots.

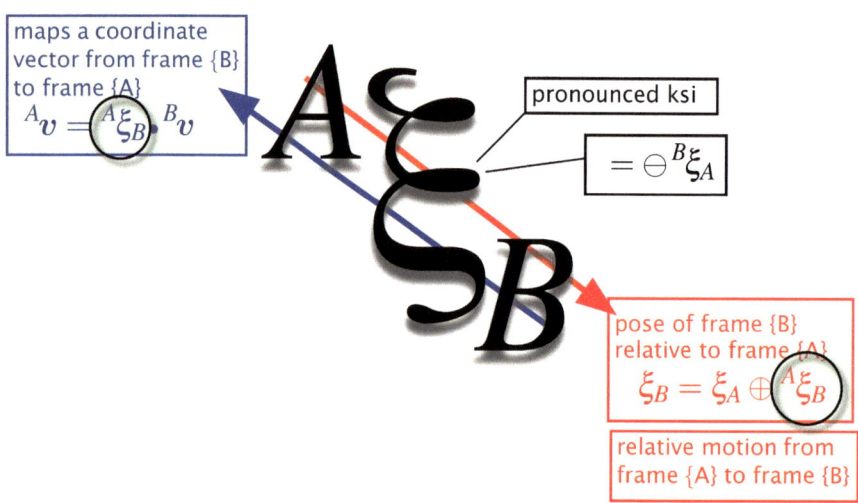

Fig. 2.19.
Everything you need to know about pose

	2D	Composition	3D	Composition
Position	2-vector	+	3-vector	+
Orientation	Angle	+	3 angles $\boldsymbol{\Gamma}$: Euler, RPY, etc.	⊗
	2 × 2 rotation matrix	*	2 vectors: OA	⊗
			(angle, vector)	⊗
			UnitQuaternion $\mathring{q}$	*
			3 × 3 rotation matrix	*
Pose	(angle, 2-vector)	⊗	(3 angles, 3-vector)	⊗
	3 × 3 transformation matrix	*	(3-vector, UnitQuaternion)	☺
			4 × 4 transformation matrix	*

Toolbox composition operators are shown in blue. Composition operators shown in red are ⊗ difficult to implement, ☺ less difficult to implement.

Table 2.4. Summary of the various concrete representations of pose ξ introduced in this chapter

Further Reading

The treatment in this chapter is a hybrid mathematical and graphical approach that covers the 2D and 3D cases by means of abstract representations and operators which are later made tangible. The standard robotics textbooks such as Kelly (2013), Siciliano et al. (2009), Spong et al. (2006), Craig (2005), and Paul (1981) all introduce homogeneous transformation matrices for the 3-dimensional case but differ in their approach. These books also provide good discussion of the other representations such as angle-vector and 3-angle representations. Spong et al. (2006, sect. 2.5.1) have a good discussion of singularities. The book Lynch and Park (2017) covers the standard matrix approaches but also introduces twists and screws. Siegwart et al. (2011) explicitly cover the 2D case in the context of mobile robotics.

Quaternions are discussed in Kelly (2013) and briefly in Siciliano et al. (2009). The book by Kuipers (1999) is a very readable and comprehensive introduction to quaternions. Quaternion interpolation is widely used in computer graphics and animation and the classic paper by Shoemake (1985) is very readable introduction to this topic. The first publication about quaternions for robotics is probably Taylor (1979), and followed up in subsequent work by Funda et al. (1990).

You will encounter a wide variety of different notation for rotations and transformations in textbooks and research articles. This book uses $^{A}T_{B}$ to denote a transform giving the pose of frame {B} with respect to frame {A}. A common alternative notation is T_{B}^{A} or even $_{B}^{A}T$. To denote points this book uses $^{A}p_{B}$ to denote a vector from the origin of frame {A} to the point **B** whereas others use p_{B}^{A}, or even $^{C}p_{B}^{A}$ to denote a vector

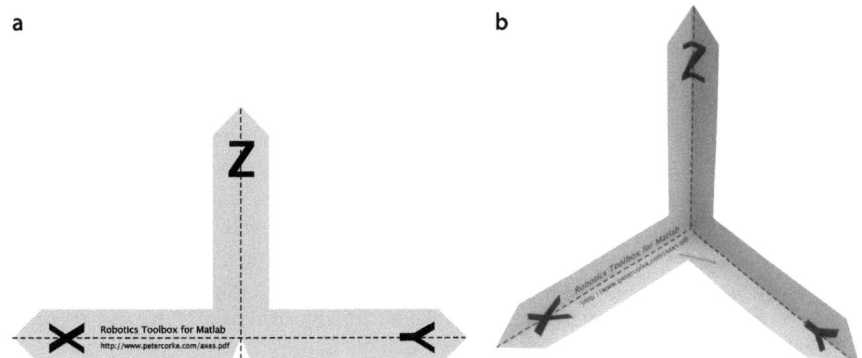

Fig. 2.20.
Build your own coordinate frame. **a** Get the PDF file from http://www.petercorke.com/axes.pdf; **b** cut it out, fold along the dotted lines and add a staple. Voila!

from the origin of frame {*A*} to the point **B** but with respect to coordinate frame {*C*}. Twists can be written as either (v, ω) as in this book, or as (ω, v).

Historical and general. Hamilton and his supporters, including Peter Tait, were vigorous in defending Hamilton's precedence in inventing quaternions, and for opposing the concept of vectors which were then beginning to be understood and used. Rodrigues developed his eponymous formula in 1840 although Gauss discovered it in 1819 but, as usual, did not publish it. It was published in 1900. Quaternions had a tempestuous beginning. The paper by Altmann (1989) is an interesting description on this tussle of ideas, and quaternions have even been woven into fiction (Pynchon 2006).

Exercises

1. Explore the many options associated with `trplot`.
2. Animate a rotating cube
   a) Write a function to plot the edges of a cube centered at the origin.
   b) Modify the function to accept an argument which is a homogeneous transformation which is applied to the cube vertices before plotting.
   c) Animate rotation about the *x*-axis.
   d) Animate rotation about all axes.
3. Create a vector-quaternion class to describe pose and which supports composition, inverse and point transformation.
4. Create a 2D rotation matrix. Visualize the rotation using `trplot2`. Use it to transform a vector. Invert it and multiply it by the original matrix; what is the result? Reverse the order of multiplication; what is the result? What is the determinant of the matrix and its inverse?
5. Create a 3D rotation matrix. Visualize the rotation using `trplot` or `tranimate`. Use it to transform a vector. Invert it and multiply it by the original matrix; what is the result? Reverse the order of multiplication; what is the result? What is the determinant of the matrix and its inverse?
6. Compute the matrix exponential using the power series. How many terms are required to match the result shown to standard MATLAB precision?
7. Generate the sequence of plots shown in Fig. 2.12.
8. For the 3-dimensional rotation about the vector [2, 3, 4] by 0.5 rad compute an **SO**(3) rotation matrix using: the matrix exponential functions `expm` and `trexp`, Rodrigues' rotation formula (code this yourself), and the Toolbox function `angvec2tr`. Compute the equivalent unit quaternion.
9. Create two different rotation matrices, in 2D or 3D, representing frames {*A*} and {*B*}. Determine the rotation matrix $^A\boldsymbol{R}_B$ and $^B\boldsymbol{R}_A$. Express these as a rotation axis and angle, and compare the results. Express these as a twist.

10. Create a 2D or 3D homogeneous transformation matrix. Visualize the rigid-body displacement using `tranimate`. Use it to transform a vector. Invert it and multiply it by the original matrix, what is the result? Reverse the order of multiplication; what happens?

11. Create two different rotation matrices, in 2D or 3D, representing frames $\{A\}$ and $\{B\}$. Determine the rotation matrix $^A\!R_B$ and $^B\!R_A$. Express these as a rotation axis and angle and compare the results. Express these as a twist.

12. Create three symbolic variables to represent roll, pitch and yaw angles, then use these to compute a rotation matrix using `rpy2r`. You may want to use the `simplify` function on the result. Use this to transform a unit vector in the z-direction. Looking at the elements of the rotation matrix devise an algorithm to determine the roll, pitch and yaw angles. Hint – find the pitch angle first.

13. Experiment with the `tripleangle` application in the Toolbox. Explore roll, pitch and yaw motions about the nominal attitude and at singularities.

14. If you have an iPhone or iPad download from the App Store the free "Euler Angles" app by École de Technologie Supérieure and experiment with it.

15. Using Eq. 2.24 show that $TT^{-1} = I$.

16. Is the inverse of a homogeneous transformation matrix equal to its transpose?

17. In Sect. 2.1.2.2 we rotated a frame about an arbitrary point. Derive the expression for computing RC that was given.

18. Explore the effect of negative roll, pitch or yaw angles. Does transforming from RPY angles to a rotation matrix then back to RPY angles give a different result to the starting value as it does for Euler angles?

19. From page 51 show that $e^x e^y \neq e^{x+y}$ for the case of matrices. Hint – expand the first few terms of the exponential series.

20. A camera has its z-axis parallel to the vector $[0, 1, 0]$ in the world frame, and its y-axis parallel to the vector $[0, 0, -1]$. What is the attitude of the camera with respect to the world frame expressed as a rotation matrix and as a unit quaternion?

3

Time and Motion

Fig. 149

In the previous chapter we learned how to describe the pose of objects in 2- or 3-dimensional space. This chapter extends those concepts to poses that change as a function of time. Section 3.1 introduces the derivative of time-varying position, orientation and pose and relates that to concepts from mechanics such as velocity and angular velocity. Discrete-time approximations to the derivatives are covered which are useful for computer implementation of algorithms such as inertial navigation. Section 3.2 is a brief introduction to the dynamics of objects moving under the influence of forces and torques and discusses the important difference between inertial and noninertial reference frames.

Section 3.3 discusses how to generate a temporal sequence of poses, a trajectory, that smoothly changes from an initial pose to a final pose. For robots this could be the path of a robot gripper moving to grasp an object or the flight path of a flying robot. Section 3.4 brings many of these topics together for the important application of inertial navigation. We introduce three common types of inertial sensor and learn how to how to use their measurements to update the estimate of pose for a moving object such as a robot.

3.1 Time-Varying Pose

In this section we discuss how to describe the rate of change of pose which has both a translational and rotational velocity component. The translational velocity is straightforward: it is the rate of change of the position of the origin of the coordinate frame. Rotational velocity is a little more complex.

3.1.1 Derivative of Pose

There are many ways to represent the orientation of a coordinate frame but most convenient for present purposes is the exponential form

$$^{A}\boldsymbol{R}_{B}(t) = e^{\left[^{A}\hat{\boldsymbol{\omega}}(t)\right]_{\times}\theta(t)} \in \mathbf{SO}(3)$$

where the rotation is described by a rotational axis $^{A}\hat{\omega}(t)$ defined with respect to frame $\{A\}$ and a rotational angle $\theta(t)$, and where $[\cdot]_{\times}$ is a skew-symmetric matrix.

At an instant in time t we will assume that the axis has a fixed direction and the frame is rotating around the axis. The derivative with respect to time is

$$^{A}\dot{\boldsymbol{R}}_{B}(t) = \left[^{A}\hat{\boldsymbol{\omega}}(t)\right]_{\times}\dot{\theta}\, e^{\left[^{A}\hat{\boldsymbol{\omega}}(t)\right]_{\times}\theta(t)} \in \mathbb{R}^{3\times 3}$$

$$= \left[^{A}\hat{\boldsymbol{\omega}}(t)\right]_{\times}\dot{\theta}\, ^{A}\boldsymbol{R}_{B}(t)$$

© Springer Nature Switzerland AG 2022
P. Corke, *Robotics and Control*, Springer Tracts in Advanced Robotics 141,
https://doi.org/10.1007/978-3-030-79179-7_3

which we write succinctly as

$$
{}^A\dot{\boldsymbol{R}}_B = \left[{}^A\boldsymbol{\omega}\right]_\times {}^A\boldsymbol{R}_B \in \mathbb{R}^{3\times3} \tag{3.1}
$$

where ${}^A\boldsymbol{\omega} = {}^A\hat{\boldsymbol{\omega}}\dot{\theta}$ is the angular velocity in frame $\{A\}$. This is a vector quantity ${}^A\boldsymbol{\omega} = (\omega_x, \omega_y, \omega_z)$ that defines the *instantaneous* axis and rate of rotation. The direction of ${}^A\boldsymbol{\omega}$ is parallel to the axis about which the coordinate frame is rotating at a particular instant of time, and the magnitude $\|{}^A\boldsymbol{\omega}\|$ is the rate of rotation about that axis.▶ Note that the derivative of a rotation matrix is not a rotation matrix, it is a general 3×3 matrix.

For a tumbling object the axis of rotation changes with time.

Consider now that angular velocity is expressed in frame $\{B\}$ and we know that

$$
{}^A\boldsymbol{\omega} = {}^A\boldsymbol{R}_B\,{}^B\boldsymbol{\omega}
$$

and using the identity $[\boldsymbol{Av}]_\times = \boldsymbol{A}[\boldsymbol{v}]_\times \boldsymbol{A}^T$ it follows that

$$
{}^A\dot{\boldsymbol{R}}_B = {}^A\boldsymbol{R}_B\left[{}^B\boldsymbol{\omega}\right]_\times \in \mathbb{R}^{3\times3} \tag{3.2}
$$

The derivative of a unit quaternion, the quaternion equivalent of Eq. 3.1, is defined as

$$
{}^A\dot{\boldsymbol{q}}_B = \tfrac{1}{2}{}^A\mathring{\boldsymbol{\omega}} \circ {}^A\mathring{\boldsymbol{q}}_B = \tfrac{1}{2}{}^A\mathring{\boldsymbol{q}}_B \circ {}^B\mathring{\boldsymbol{\omega}} \in \mathbb{H} \tag{3.3}
$$

where $\mathring{\boldsymbol{\omega}}$ is a pure quaternion formed from the angular velocity vector. These are implemented by the Toolbox methods `dot` and `dotb` respectively. The derivative of a unit-quaternion is not a unit-quaternion, it is a regular quaternion which can also be considered as a 4-vector.

The derivative of pose can be determined by expressing pose as a homogeneous transformation matrix

$$
\xi \sim {}^A\boldsymbol{T}_B = \begin{pmatrix} {}^A\boldsymbol{R}_B & {}^A\boldsymbol{t}_B \\ 0_{1\times3} & 1 \end{pmatrix}
$$

and taking the derivative with respect to time and substituting Eq. 3.1 gives

$$
\dot{\xi} \sim {}^A\dot{\boldsymbol{T}}_B = \begin{pmatrix} {}^A\dot{\boldsymbol{R}}_B & {}^A\dot{\boldsymbol{t}}_B \\ 0_{1\times3} & 0 \end{pmatrix} = \begin{pmatrix} \left[{}^A\boldsymbol{\omega}\right]_\times {}^A\boldsymbol{R}_B & {}^A\dot{\boldsymbol{t}}_B \\ 0_{1\times3} & 0 \end{pmatrix}
$$

The rate of change can be described in terms of the current orientation ${}^A\boldsymbol{R}_B$ and *two* velocities. The linear or translational velocity $\boldsymbol{v} = {}^A\dot{\boldsymbol{t}}_B$ is the velocity of the origin of $\{B\}$ with respect to $\{A\}$. The angular velocity ${}^A\boldsymbol{\omega}_B$ we have already introduced. We can combine these two velocity vectors to create the spatial velocity vector

$$
{}^A\boldsymbol{\nu}_B = \left({}^A\boldsymbol{v}_B,\, {}^A\boldsymbol{\omega}_B\right) \in \mathbb{R}^6
$$

which is the instantaneous velocity of frame $\{B\}$ with respect to $\{A\}$.

Every point in the body has the same angular velocity. Knowing that, plus the translational velocity vector of any point is enough to fully describe the instantaneous motion of a rigid body. It is common to place $\{B\}$ at the body's center of mass.

3.1.2   Transforming Spatial Velocities

The velocity of a moving body can be expressed with respect to a world reference frame $\{A\}$ or the moving body frame $\{B\}$ as shown in Fig. 3.1. The spatial velocities are linearly related by

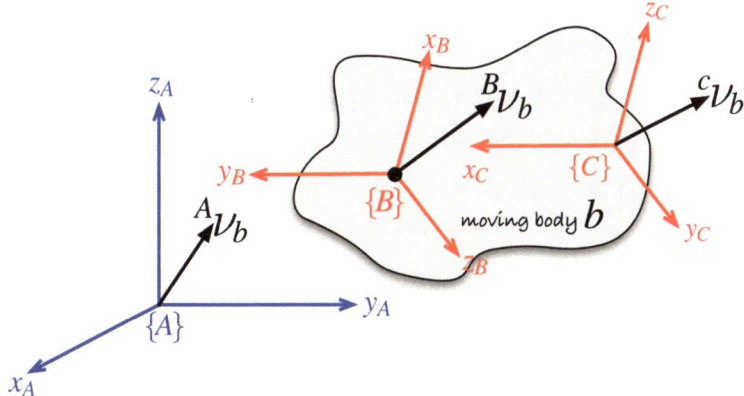

Fig. 3.1.
Representing the spatial velocity of a moving body b with respect to various coordinate frames. Note that ν is a 6-dimensional vector

$$^{A}\nu = \begin{pmatrix} ^{A}\boldsymbol{R}_B & \boldsymbol{0}_{3\times3} \\ \boldsymbol{0}_{3\times3} & ^{A}\boldsymbol{R}_B \end{pmatrix} {}^{B}\nu = {}^{A}\boldsymbol{J}_B\left(^{A}\xi_B\right){}^{B}\nu \qquad (3.4)$$

where $^{A}\xi_B \sim (^{A}\boldsymbol{R}_B, {}^{A}\boldsymbol{t}_B)$ and $^{A}\boldsymbol{J}_B(\cdot)$ is a Jacobian or interaction matrix. For example, we can define a body-fixed frame and a spatial velocity in that frame

```
>> TB = SE3(1, 2, 0) * SE3.Rx(pi/2);
>> vb = [0.2 0.3 0 0 0 0.5]';
```

and the spatial velocity in the world frame is

```
>> va = TB.velxform * vb;
>> va'
ans =
    0.2000    0.0000    0.3000         0   -0.5000    0.0000
```

For the case where frame {C} is also on the moving body the transformation becomes

$$^{C}\nu = \begin{pmatrix} ^{C}\boldsymbol{R}_B & \left[^{C}\boldsymbol{t}_B\right]_{\times}{}^{C}\boldsymbol{R}_B \\ \boldsymbol{0}_{3\times3} & ^{C}\boldsymbol{R}_B \end{pmatrix} {}^{B}\nu = \mathrm{Ad}\left(^{C}\xi_B\right){}^{B}\nu$$

and involves the adjoint matrix of the relative pose which is discussed in Appendix D. Continuing the example above we will define an additional frame {C} relative to frame {B}

```
>> TBC = SE3(0, 0.4, 0);
```

To determine velocity at the origin of this frame we first compute $^{C}\xi_B$

```
>> TCB = inv(TBC);
```

and the velocity in frame {C} is

```
>> vc = TCB.Ad * vb;
>> vc'
ans =
         0    0.3000         0         0         0    0.5000
```

which has zero velocity in the x_C-direction since the rotational and translational velocity components cancel out.

Lynch and Park (2017) use the term velocity twist while Murray et al. 1994 call this a spatial velocity.

The scalar product of a velocity twist and a wrench represents power.

Some texts introduce a *velocity twist* $\boldsymbol{V}$ which is different to the spatial velocity introduced above.◄ The velocity twist of a body-fixed frame {B} is $^{B}\boldsymbol{V} = (^{B}\boldsymbol{v}, {}^{B}\boldsymbol{\omega})$ which has a translational and rotational velocity component but $^{B}\boldsymbol{v}$ is the body-frame velocity of an imaginary point rigidly attached to the body and located at the world frame origin. The body- and world-frame velocity twists are related by the adjoint matrix rather than Eq. 3.4. The velocity twist is the dual of the wrench described in Sect. 3.2.2.◄

3.1.3   Incremental Rotation

The physical meaning of $\dot{R}$ is not intuitively obvious – it is simply the way that the elements of R change with time. To gain some insight we consider a first-order approximation to the derivative▶

$$\dot{R} \approx \frac{R\langle t+\delta_t \rangle - R\langle t \rangle}{\delta_t} \in \mathbb{R}^{3\times 3} \tag{3.5}$$

The only valid operator for the group SO(n) is composition ⊕, so the result of subtraction cannot belong to the group. The result is a 3×3 matrix of element-wise differences. Groups are introduced in Appendix D.

Consider an object whose body frames {B} at two consecutive timesteps are related by a small rotation $^B R_\Delta$ expressed in the body frame

$$R_B \langle t+\delta_t \rangle = R_B \langle t \rangle {}^B R_\Delta$$

We substitute Eq. 3.2 into 3.5 and rearrange to obtain

$$^B R_\Delta \approx \delta_t \left[{}^B \omega \right]_\times + I_{3\times 3} \tag{3.6}$$

which says that an infinitesimally small rotation can be approximated by the sum of a skew-symmetric matrix and an identity matrix.▶ For example

```
>> rotx(0.001)
ans =
    1.0000         0         0
         0    1.0000   -0.0010
         0    0.0010    1.0000
```

This is the first two terms of the Rodrigues' rotation formula on, Eq. 2.18, when $\theta = \delta_t \omega$.

Equation 3.6 directly relates rotation between timesteps to the angular velocity. Rearranging it allows us to compute the approximate angular velocity vector

$$\omega \approx \frac{1}{\delta_t} \vee_\times \left(R_B \langle t \rangle^T R_B \langle t+\delta_t \rangle - I_{3\times 3} \right)$$

from two consecutive rotation matrices where $\vee_\times(\cdot)$ is the inverse skew-symmetric matrix operator such that if $S = [v]_\times$ then $v = \vee_\times(S)$. Alternatively, if the angular velocity in the body frame is known we can approximately update the rotation matrix

$$R_B \langle t+\delta_t \rangle \approx R_B \langle t \rangle + \delta_t R_B \langle t \rangle [\omega]_\times \tag{3.7}$$

which is cheap to compute, involves no trigonometric operations, and is key to inertial navigation systems which we discuss in Sect. 3.4.

The only valid operator for the group SO(n) is composition ⊕, so the result of addition cannot be within the group. The result is a general 3×3 matrix.

Adding any nonzero matrix to a rotation matrix results in a matrix that is *not* a rotation matrix.◀ However if the increment is sufficiently small, that is the angular velocity and/or sample time is small,▶ the result will be close to orthonormal. We can *straighten it up*, that is make it a proper rotation matrix, by normalization as discussed in Sect. 2.3.1. This is a common approach when implementing inertial navigation systems on low-end computing hardware.

Which is why inertial navigation systems operate at a high sample rate and δ_t is small.

We can also approximate the quaternion derivative by a first-order difference▶

$$\dot{\mathring{q}} \approx \frac{\mathring{q}\langle k+1 \rangle - \mathring{q}\langle k \rangle}{\delta_t} \in \mathbb{H}$$

which combined with Eq. 3.3 gives us the approximation

Similar to the case for SO(n), addition and subtraction are not operators for the unit-quaternion group $\mathbb{S}^3$ so the result will be a quaternion $q \in \mathbb{H}$ for which addition and substraction are permitted. The Toolbox supports this with overloaded operators + and − and appropriate object class conversions.

$$\hat{q}\langle k+1 \rangle \approx \hat{q}\langle k \rangle + \frac{\delta_t}{2}\hat{\omega} \circ \hat{q}\langle k \rangle \qquad (3.8)$$

which is even cheaper to compute than the rotation matrix approach. Adding a non-zero vector to a unit-quaternion results in a nonunit quaternion but if the angular velocity and/or sample time is small then the approximation is reasonable. Normalizing the result to create a unit-quaternion is computationally cheaper than normalizing a rotation matrix, as discussed in Sect. 2.3.1.

3.1.4   Incremental Rigid-Body Motion

Consider two poses ξ_1 and ξ_2 which differ infinitesimally and are related by

$$\xi_2 = \xi_1 \oplus \xi_\Delta$$

where $\xi_\Delta = \ominus \xi_1 \oplus \xi_2$. In homogeneous transformation matrix form

$$\xi_\Delta \sim \boldsymbol{T}_\Delta = \begin{pmatrix} \boldsymbol{R}_\Delta & \boldsymbol{t}_\Delta \\ 0_{1\times3} & 1 \end{pmatrix}$$

where $\boldsymbol{t}_\Delta$ is an incremental displacement and $\boldsymbol{R}_\Delta$ is an incremental rotation matrix which will be skew symmetric with only three unique elements $\vee_x(\boldsymbol{R}_\Delta - \boldsymbol{I}_{3\times3})$ plus an identity matrix. The incremental rigid-body motion can therefore be described by just six parameters

$$\Delta\big(\xi_1, \xi_2\big) \mapsto \boldsymbol{\Delta}_\xi \in \mathbb{R}^6$$

where $\boldsymbol{\Delta}_\xi = (\boldsymbol{\Delta}_t, \boldsymbol{\Delta}_R)$ can be considered as a spatial displacement.◀ A body with constant spatial velocity ν for δ_t seconds undergoes a spatial displacement of $\boldsymbol{\Delta}_\xi = \delta_t \nu$.

The inverse operator

$$\Delta^{-1}\big(\boldsymbol{\Delta}_\xi\big) \mapsto \xi_\Delta \in \mathbf{SE}(3)$$

is given by

This is useful in optimization procedures that seek to minimize the error between two poses: we can choose the cost function $e = \|\Delta(\xi_1, \xi_2)\|$ which is equal to zero when $\xi_1 \equiv \xi_2$. This is very approximate when the poses are significantly different, but becomes ever more accurate as $\xi_1 \rightarrow \xi_2$.

$$\xi_\Delta \sim \boldsymbol{T}_\Delta = \begin{pmatrix} \big[\boldsymbol{\Delta}_R\big]_\times + \boldsymbol{I}_{3\times3} & \boldsymbol{\Delta}_t \\ \boldsymbol{0}_{1\times3} & 1 \end{pmatrix}$$

The spatial displacement operator and its inverse are implemented by the Toolbox functions `tr2delta` and `delta2tr` respectively. These functions assume that the displacements are infinitesimal and become increasingly approximate with displacement magnitude.

Sir Isaac Newton (1642–1727) was an English mathematician and alchemist. He was Lucasian professor of mathematics at Cambridge, Master of the Royal Mint, and the thirteenth president of the Royal Society. His achievements include the three laws of motion, the mathematics of gravitational attraction, the motion of celestial objects and the theory of light and color, and building the first reflecting telescope.

Many of these results were published in 1687 in his great 3-volume work "The Philosophiae Naturalis Principia Mathematica" (Mathematical principles of natural philosophy). In 1704 he published "Opticks" which was a study of the nature of light and color and the phenomena of diffraction. The SI unit of force is named in his honor. He is buried in Westminster Abbey, London.

3.2   Accelerating Bodies and Reference Frames

So far we have considered only the first derivative, the velocity of coordinate frames. However all motion is ultimately caused by a force or a torque which leads to acceleration and the consideration of dynamics.

3.2.1   Dynamics of Moving Bodies

For translational motion Newton's second law describes, in the inertial frame, the acceleration of a particle with position x and mass m

$$m\,{}^0\ddot{x} = {}^0f \qquad (3.9)$$

due to the applied force f.

Rotational motion in $\mathbf{SO}(3)$ is described by Euler's equations of motion which relates the angular acceleration of the body in the body frame

$$ {}^B\!J\,{}^B\dot{\omega} + {}^B\omega \times \left({}^B\!J\,{}^B\omega\right) = {}^B\tau \qquad (3.10)$$

to the applied torque or moment τ and a positive-definite rotational inertia matrix ${}^B\!J \in \mathbb{R}^{3\times 3}$.▶ Nonzero angular acceleration implies that angular velocity, the axis and/or angle of rotation, evolves over time.◀

Consider the motion of a tumbling object which we can easily simulate. We define an inertia matrix◀

```
>> J = [2 -1 0;-1 4 0;0 0 3];
```

and initial conditions for orientation and angular velocity

```
>> attitude = UnitQuaternion();
>> w = 0.2*[1 2 2]';
```

The simulation loop computes angular acceleration with Eq. 3.10, uses rectangular integration to obtain angular velocity and attitude, and then updates a graphical coordinate frame

```
>> dt = 0.05;
>> h = attitude.plot();
>> for t=0:dt:10
       wd = -inv(J) * (cross(w, J*w));
       w = w + wd*dt; attitude = attitude .* UnitQuaternion.omega(wd*dt);
       attitude.plot('handle', h); pause(dt)
   end
```

Notice that inertia has an associated reference frame, it is a matrix and its elements depend on the choice of the coordinate frame.

In the absence of torque a body generally rotates with a time-varying angular velocity – this is quite different to the linear velocity case. It is angular momentum $h = J\omega$ in the inertial frame that is constant.

The matrix must be positive definite, that is symmetric and all its eigenvalues are positive.

The rotational inertia of a body that moves in $\mathbf{SE}(3)$ is represented by the 3×3 symmetric matrix

$$J = \begin{pmatrix} J_{xx} & J_{xy} & J_{xz} \\ J_{xy} & J_{yy} & J_{yz} \\ J_{xz} & J_{yz} & J_{zz} \end{pmatrix}$$

The diagonal elements are the positive moments of inertia, and the off-diagonal elements are products of inertia. Only six of these nine elements are unique: three moments and three products of inertia. The products of inertia are all zero if the object's mass distribution is symmetrical with respect to the coordinate frame.

3.2.2   Transforming Forces and Torques

The spatial velocity is a vector quantity that represents translational and rotational velocity. In a similar fashion we can combine translational force and rotational torque into a 6-vector that is called a wrench $W = (f_x, f_y, f_z, m_x, m_y, m_z) \in \mathbb{R}^6$. A wrench $^B W$ is defined with respect to the coordinate frame $\{B\}$ and applied at the origin of that frame.

The wrench $^C W$ is equivalent if it causes the same motion of the body when applied to the origin of coordinate frame $\{C\}$ and defined with respect to $\{C\}$. The wrenches are related by

$$^C W = \begin{pmatrix} ^B R_C & \left[^B t_C \right]_\times {}^B R_C \\ 0_{3\times3} & ^B R_C \end{pmatrix}^T {}^B W = \mathrm{Ad}\left(^B \xi_C \right)^T {}^B W \tag{3.11}$$

which is similar to the spatial velocity transform of Eq. 3.4 but uses the transpose of the adjoint of the *inverse* relative pose.

Continuing the MATLAB example from page 63 we define a wrench with respect to frame $\{B\}$ with forces of 3 and 4 Nm in the x- and y-directions respectively

```
>> WB = [3 4 0 0 0 0]';
```

The equivalent wrench in frame $\{C\}$ would be

```
>> WC = TBC.Ad' * WB;
>> WC'
ans =
    3.0000    4.0000         0         0         0    1.2000
```

which is the same forces as applied at $\{B\}$ *plus* a torque of 1.2 Nm about the z-axis to counter the moment due to the application of the x-axis force along a different line of action.

3.2.3   Inertial Reference Frame

The term *inertial reference frame* is frequently used in robotics and it is crisply defined as "*a reference frame that is not accelerating or rotating*".

Consider a particle P at rest with respect to a stationary reference frame $\{0\}$. Frame $\{B\}$ is moving with constant velocity $^0 v_B$ relative to frame $\{0\}$. From the perspective of $\{B\}$ the particle would be moving at constant velocity, in fact $^B v_P = -^0 v_B$. The particle is not accelerating and obeys Newton's first law "*that in the absence of an applied force a particle moves at a constant velocity*". Frame $\{B\}$ is therefore also an inertial reference frame.

Now imagine that frame $\{B\}$ is accelerating at a constant acceleration $^0 a_B$ with respect to $\{0\}$. From the perspective of $\{B\}$ the particle appear to be accelerating, in fact $^B a_P = -^0 a_B$ and this violates Newton's first law. An observer in frame $\{B\}$ who was aware of Newton's theories might invoke some magical force to explain what they observe. We call such a force a fictitious, apparent, pseudo, inertial or d'Alembert force – they only exist in an accelerating or noninertial reference frame. This accelerating

Gaspard-Gustave de Coriolis (1792–1843) was a French mathematician, mechanical engineer and scientist. Born in Paris, in 1816 he became a tutor at the École Polytechnique where he carried out experiments on friction and hydraulics and later became a professor at the École des Ponts and Chaussées (School of Bridges and Roads). He extended ideas about kinetic energy and work to rotating systems and in 1835 wrote the famous paper *Sur les équations du mouvement relatif des systèmes de corps* (On the equations of relative motion of a system of bodies) which dealt with the transfer of energy in rotating systems such as waterwheels. In the late 19th century his ideas were picked up by the meteorological community to incorporate effects due to the Earth's rotation. He is buried in Paris's Montparnasse Cemetery.

frame {*B*} is *not* an inertial reference frame. In Newtonian mechanics, gravity is considered a real body force *mg* – a free object will accelerate relative to the inertial frame.▶

An everyday example of a noninertial reference frame is an accelerating car or airplane. Inside an accelerating vehicle we observe fictitious forces pushing objects around in a way that is not explained by Newton's law in an inertial reference frame. We also experience real forces acting on our body which, in this case, are provided by the seat and the restraint.

For a rotating reference frame things are more complex still. Imagine that you and a friend are standing on a large rotating turntable, and throwing a ball back and forth. You will observe that the ball follows a curved path in space.▶ As a Newton-aware observer in this noninertial reference frame you would have to resort to invoking some magical force that explains why flying objects follow curved paths.

If the reference frame {*B*} is rotating with angular velocity ω about its origin then Newton's second law Eq. 3.9 becomes

$$m\left({}^{B}\dot{\boldsymbol{v}} + \underbrace{\boldsymbol{\omega}\times\left(\boldsymbol{\omega}\times{}^{B}\boldsymbol{p}\right)}_{\text{centripetal}} + \underbrace{2\boldsymbol{\omega}\times{}^{B}\boldsymbol{v}}_{\text{Coriolis}} + \underbrace{\frac{\mathrm{d}\boldsymbol{\omega}}{\mathrm{d}t}\times{}^{B}\boldsymbol{p}}_{\text{Euler}} \right) = {}^{0}\boldsymbol{f}$$

with three *new* acceleration terms. Centripetal acceleration always acts inward toward the origin. If the point is moving then Coriolis acceleration will be normal to its velocity. If rotational velocity is time varying then Euler acceleration will be normal to the position vector. Frequently the centripetal term is moved to the right-hand side in which case it becomes a fictitious outward centrifugal force. This complexity is symptomatic of being in a noninertial reference frame, and another definition of an inertial frame is one in which the "*physical laws hold good in their simplest form*".▶

In robotics the term inertial frame and world coordinate frame tend to be used loosely and interchangeably to indicate a frame fixed to some point on the Earth. This is to distinguish it from the body-frame attached to the robot or vehicle. The surface of the Earth is an approximation of an inertial reference frame – the effect of the Earth's rotation is a finite acceleration less than 0.04 m s^{-2} due to centripetal acceleration. From the perspective of an Earth-bound observer a moving body will experience Coriolis acceleration. Both effects are small,▶ dependent on latitude, and typically ignored.

3.3   Creating Time-Varying Pose

In robotics we often need to generate a time-varying pose that moves smoothly in translation and rotation. A path is a spatial construct – a locus in space that leads from an initial pose to a final pose. A trajectory is a path with specified timing. For example there is a path from A to B, but there is a trajectory from A to B in 10 s or at 2 m s^{-1}.

An important characteristic of a trajectory is that it is *smooth* – position and orientation vary smoothly with time. We start by discussing how to generate smooth trajectories in one dimension. We then extend that to the multi-dimensional case and then to piecewise-linear trajectories that visit a number of intermediate points without stopping.

3.3.1   Smooth One-Dimensional Trajectories

We start our discussion with a scalar function of time. Important characteristics of this function are that its initial and final value are specified and that it is *smooth*. Smoothness in this context means that its first few temporal derivatives are continuous. Typically velocity and acceleration are required to be continuous and sometimes also the derivative of acceleration or jerk.

Albert Einstein's equivalence principle is that "*we assume the complete physical equivalence of a gravitational field and a corresponding acceleration of the reference system*" – we are unable to distinguish between gravity and being on a rocket accelerating at 1 *g* far from the gravitational influence of any celestial object.

Of course if we look down onto the turntable from an inertial reference frame the ball is moving in a straight line.

Einstein, "*The foundation of the general theory of relativity*".

Coriolis acceleration is significant for weather systems and meteorological prediction but below the sensitivity of low-cost sensors.

An obvious candidate for such a function is a polynomial function of time. Polynomials are simple to compute and can easily provide the required smoothness and boundary conditions. A quintic (fifth-order) polynomial is often used

$$s(t) = At^5 + Bt^4 + Ct^3 + Dt^2 + Et + F \qquad (3.12)$$

where time $t \in [0, T]$. The first- and second-derivatives are also smooth polynomials

$$\dot{s}(t) = 5At^4 + 4Bt^3 + 3Ct^2 + 2Dt + E \qquad (3.13)$$

$$\ddot{s}(t) = 20At^3 + 12Bt^2 + 6Ct + 2D \qquad (3.14)$$

Time	s	$\dot{s}$	$\ddot{s}$
$t = 0$	s_0	$\dot{s}_0$	$\ddot{s}_0$
$t = T$	s_T	$\dot{s}_T$	$\ddot{s}_T$

The trajectory has defined boundary conditions for position, velocity and acceleration◄ and frequently the velocity and acceleration boundary conditions are all zero.

Writing Eq. 3.12 to Eq. 3.14 for the boundary conditions $t = 0$ and $t = T$ gives six equations which we can write in matrix form as

$$\begin{pmatrix} s_0 \\ s_T \\ \dot{s}_0 \\ \dot{s}_T \\ \ddot{s}_0 \\ \ddot{s}_T \end{pmatrix} = \begin{pmatrix} 0 & 0 & 0 & 0 & 0 & 1 \\ T^5 & T^4 & T^3 & T^2 & T & 1 \\ 0 & 0 & 0 & 0 & 1 & 0 \\ 5T^4 & 4T^3 & 3T^2 & 2T & 1 & 0 \\ 0 & 0 & 0 & 2 & 0 & 0 \\ 20T^3 & 12T^2 & 6T & 2 & 0 & 0 \end{pmatrix} \begin{pmatrix} A \\ B \\ C \\ D \\ E \\ F \end{pmatrix}$$

This is the reason for choice of quintic polynomial. It has six coefficients that enable it to meet the six boundary conditions on initial and final position, velocity and acceleration.

Since the matrix is square◄ we can solve for the coefficient vector (A, B, C, D, E, F) using standard linear algebra methods such as the MATLAB \-operator. For a quintic polynomial acceleration will be a smooth cubic polynomial, and jerk will be a parabola.

The Toolbox function `tpoly` generates a quintic polynomial trajectory as described by Eq. 3.12. For example

```
>> tpoly(0, 1, 50);
```

generates a polynomial trajectory and plots it, along with the corresponding velocity and acceleration, as shown in Fig. 3.2a. We can get these values into the workspace by providing output arguments

```
>> [s,sd,sdd] = tpoly(0, 1, 50);
```

where `s`, `sd` and `sdd` are respectively the trajectory, velocity and acceleration – each a 50×1 column vector. We observe that the initial and final velocity and acceleration

Fig. 3.2. Quintic polynomial trajectory. From top to bottom is position, velocity and acceleration versus time step. **a** With zero-velocity boundary conditions, **b** initial velocity of 0.5 and a final velocity of 0. Note that velocity and acceleration are in units of timestep not seconds

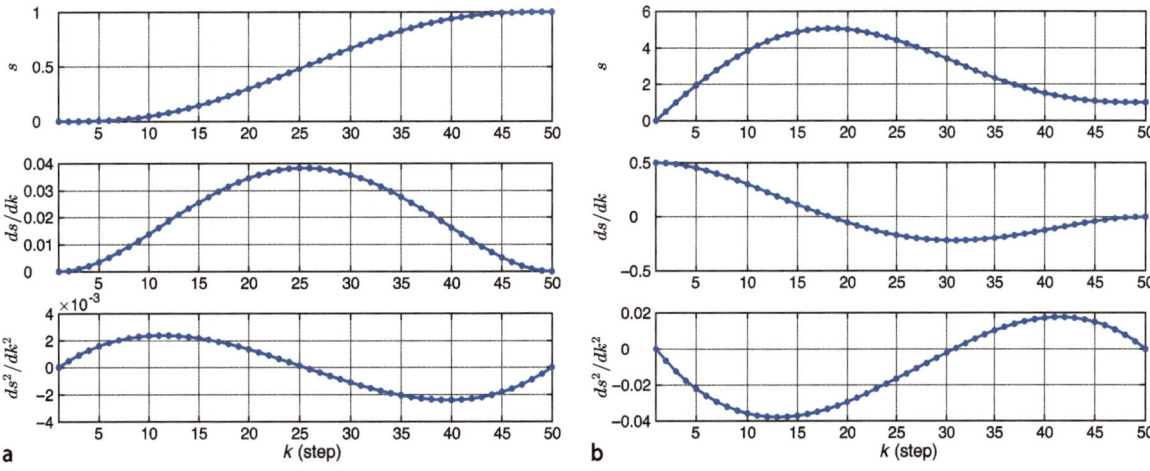

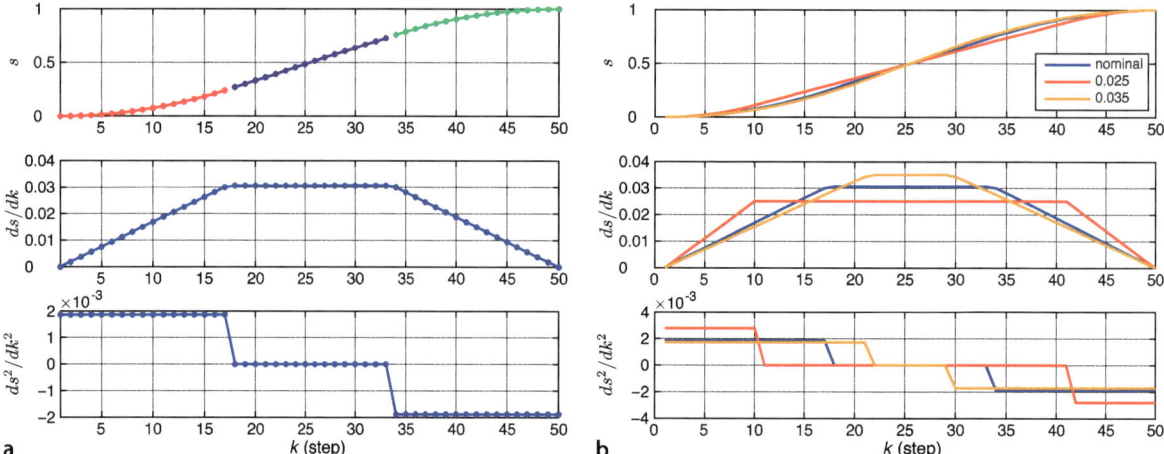

a   **b**

are all zero – the default value. The initial and final velocities can be set to nonzero values

```
>> tpoly(0, 1, 50, 0.5, 0);
```

Fig. 3.3. Linear segment with parabolic blend (LSPB) trajectory: **a** default velocity for linear segment; **b** specified linear segment velocity values

in this case, an initial velocity of 0.5 and a final velocity of 0. The results shown in Fig. 3.2b illustrate an important problem with polynomials. The nonzero initial velocity causes the polynomial to overshoot the terminal value – it peaks at 5 on a trajectory from 0 to 1.

Another problem with polynomials, a very practical one, can be seen in the middle graph of Fig. 3.2a. The velocity peaks at $k = 25$ which means that for most of the time the velocity is far less than the maximum. The mean velocity

```
>> mean(sd) / max(sd)
ans =
    0.5231
```

is only 52% of the peak so we are not using the motor as fully as we could. A real robot joint has a well defined maximum velocity and for minimum-time motion we want to be operating at that maximum for as much of the time as possible. We would like the velocity curve to be *flatter* on top.

A well known alternative is a hybrid trajectory which has a constant velocity segment with polynomial segments for acceleration and deceleration. Revisiting our first example the hybrid trajectory is

```
>> lspb(0, 1, 50);
```

where the arguments have the same meaning as for tpoly and the trajectory is shown in Fig. 3.3a. The trajectory comprises a linear segment (constant velocity) with parabolic blends, hence the name lspb. The term blend is commonly used to refer to a trajectory segment that smoothly joins linear segments. As with tpoly we can also return the trajectory and its velocity and acceleration

```
>> [s,sd,sdd] = lspb(0, 1, 50);
```

This type of trajectory is also referred to as trapezoidal due to the shape of the velocity curve versus time, and is commonly used in industrial motor drives.▶

The trapezoidal trajectory is smooth in velocity, but not in acceleration.

The function lspb has *chosen* the velocity of the linear segment to be

```
>> max(sd)
ans =
    0.0306
```

but this can be overridden by specifying it as a fourth input argument

```
>> s = lspb(0, 1, 50, 0.025);
>> s = lspb(0, 1, 50, 0.035);
```

The trajectories for these different cases are overlaid in Fig. 3.3b. We see that as the velocity of the linear segment increases its duration decreases and ultimately its duration would be zero. In fact the velocity cannot be chosen arbitrarily◄, too high or too low a value for the maximum velocity will result in an infeasible trajectory and the function returns an error.

The system has one design degree of freedom. There are six degrees of freedom (blend time, three parabolic coefficients and two linear coefficients) and five constraints (total time, initial and final position and velocity).

3.3.2 Multi-Dimensional Trajectories

Most useful robots have more than one axis of motion and it is quite straightforward to extend the smooth scalar trajectory to the vector case. In terms of configuration space (Sect. 2.3.5), these axes of motion correspond to the dimensions of the robot's configuration space – to its degrees of freedom. We represent the robot's configuration as a vector $q \in \mathbb{R}^N$ where N is the number of degrees of freedom. The configuration of a 3-joint robot would be its joint angles $q = (q_1, q_2, q_3)$. The configuration vector of wheeled mobile robot might be its position $q = (x, y)$ or its position and heading angle $q = (x, y, \theta)$. For a 3-dimensional body that had an orientation in $\mathbf{SO}(3)$ we would use a configuration vector $q = (\theta_r, \theta_p, \theta_y)$ or for a pose in $\mathbf{SE}(3)$ we would use $q = (x, y, z, \theta_r, \theta_p, \theta_y)$◄. In all these cases we would require smooth multi-dimensional motion from an initial configuration vector to a final configuration vector.

Or an equivalent 3-angle representation.

In the Toolbox this is achieved using the function `mtraj` and to move from configuration $(0, 2)$ to $(1, -1)$ in 50 steps we write

```
>> q = mtraj(@lspb, [0 2], [1 -1], 50);
```

which results in a 50 × 2 matrix `q` with one row per time step and one column per axis. The first argument is a handle to a function that generates a *scalar* trajectory, `@lspb` as in this case or `@tpoly`. The trajectory for the `@lspb` case

```
>> plot(q)
```

is shown in Fig. 3.4.

If we wished to create a trajectory for 3-dimensional pose we might consider converting a pose T to a 6-vector by a command like

```
q = [T1.t' T1.torpy]
```

though as we shall see later interpolation of 3-angle representations has some limitations.

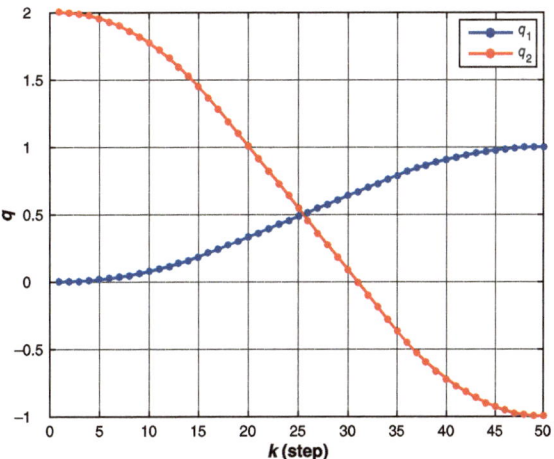

Fig. 3.4.
Multi-dimensional motion. q_1 varies from $0 \rightarrow 1$ and q_2 varies from $2 \rightarrow -1$

3.3.3   Multi-Segment Trajectories

In robotics applications there is often a need to move smoothly along a path through one or more intermediate or *via* points without stopping. This might be to avoid obstacles in the workplace, or to perform a task that involves following a piecewise continuous trajectory such as welding a seam or applying a bead of sealant in a manufacturing application.

To formalize the problem consider that the trajectory is defined by M configurations q_k, $k \in [1, M]$ and there are $M-1$ motion segments. As in the previous section $q_k \in \mathbb{R}^N$ is a *vector* representation of configuration.

The robot starts from q_1 at rest and finishes at q_M at rest, but moves through (or close to) the intermediate configurations without stopping. The problem is over constrained and in order to attain continuous velocity we surrender the ability to reach each intermediate configuration. This is easiest to understand for the 1-dimensional case shown in Fig. 3.5. The motion comprises linear motion segments with polynomial blends, like `lspb`, but here we choose quintic polynomials because they are able to match boundary conditions on position, velocity and acceleration at their start and end points.

The first segment of the trajectory accelerates from the initial configuration q_1 and zero velocity, and joins the line heading toward the second configuration q_2. The blend time is set to be a constant t_{acc} and $t_{acc}/2$ before reaching q_2 the trajectory executes a polynomial blend, of duration t_{acc}, onto the line from q_2 to q_3, and the process repeats. The constant velocity $\dot{q}_k$ can be specified for each segment. The average acceleration during the blend is

$$\ddot{q} = \frac{\dot{q}_{k+1} - \dot{q}_k}{t_{acc}}$$

If the maximum acceleration capability of the axis is known then the minimum blend time can be computed.▶

On a particular motion segment each axis will have a different distance to travel and traveling at its maximum speed there will be a minimum time before it can reach its goal. The first step in planning a segment is to determine which axis will be the slowest to complete the segment, based on the distance that each axis needs to travel for the segment and its maximum achievable velocity. From this the duration of the segment can be computed and then the required velocity of each axis. This ensures that all axes reach the next target q_k at the *same time*.

The Toolbox function `mstraj` generates a multi-segment multi-axis trajectory based on a matrix of via points. For example 2-axis motion via the corners of a rotated square can be generated by

```
>> via = SO2(30, 'deg') * [-1 1; 1 1; 1 -1; -1 -1]';
>> q0 = mstraj(via(:,[2 3 4 1])', [2,1], [], via(:,1)', 0.2, 0);
```

The first argument is the matrix of via points, each row is the coordinates of a point. The remaining arguments are respectively: a vector of maximum speeds per axis, a vector of

The real limit of the axis will be its peak, rather than average, acceleration. The peak acceleration for the blend can be determined from Eq. 3.14 once the quintic coefficients are known.

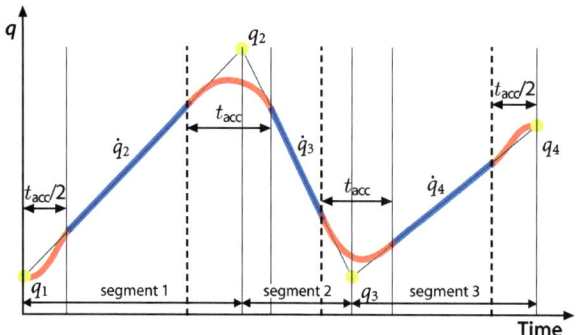

Fig. 3.5.
Notation for multi-segment trajectory showing four points and three motion segments. Blue indicates constant velocity motion, red indicates regions of acceleration

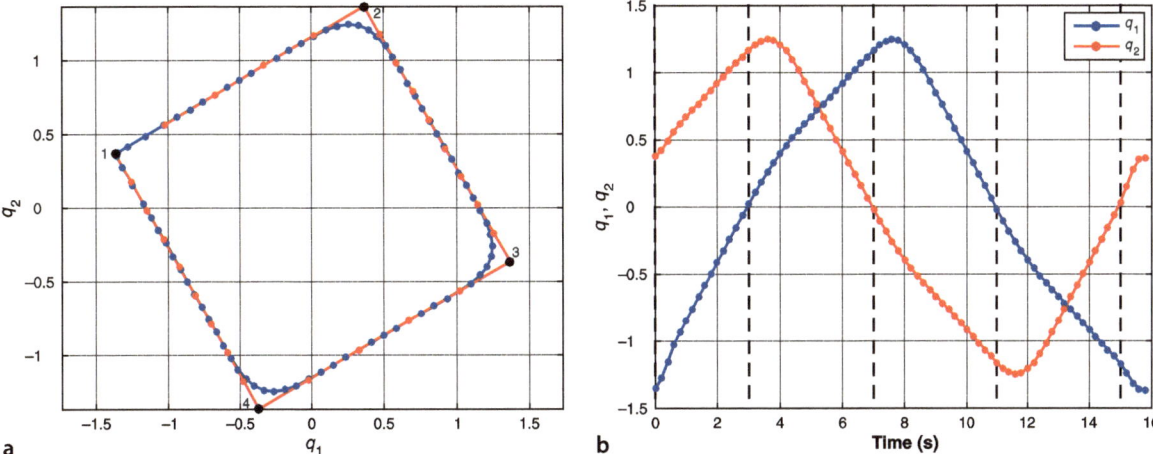

Fig. 3.6. Multi-segment multi-axis trajectories: **a** configuration of robot (tool position) for acceleration time of $t_{acc} = 0$ s (red) and $t_{acc} = 2$ s (blue), the via points are indicated by solid black markers; **b** configuration versus time with segment transitions ($t_{acc} = 2$ s) indicated by dashed black lines. The discrete-time points are indicated by dots

Only one of the maximum axis speed or time per segment can be specified, the other is set to MATLAB's empty matrix [ ].

Acceleration time if given is rounded up internally to a multiple of the time step.

durations for each segment,◤ the initial configuration, the sample time step, and the acceleration time.◤ The function `mstraj` returns a matrix with one row per time step and the columns correspond to the axes. We can plot q_2 against q_1 to see the path of the robot

```
>> plot(q0(:,1), q0(:,2))
```

and is shown by the red path in Fig. 3.6a. If we increase the acceleration time

```
>> q2 = mstraj(via(:,[2 3 4 1])', [2,1], [], via(:,1)', 0.2, 2);
```

the trajectory becomes more rounded (blue path) as the polynomial blending functions do their work. The smoother trajectory also takes more time to complete.

```
>> [numrows(q0) numrows(q2)]
ans =
   28    80
```

The configuration variables as a function of time are shown in Fig. 3.6b. This function also accepts optional initial and final velocity arguments and t_{acc} can be a vector specifying different acceleration times for each of the N blends.

Keep in mind that this function simply interpolates pose represented as a vector. In this example the vector was assumed to be Cartesian coordinates, but this function could also be applied to Euler or roll-pitch-yaw angles but this is not an ideal way to interpolate rotation. This leads us nicely to the next section where we discuss interpolation of orientation.

3.3.4 Interpolation of Orientation in 3D

In robotics we often need to interpolate orientation, for example, we require the end-effector of a robot to smoothly change from orientation $\boldsymbol{\xi}_0$ to $\boldsymbol{\xi}_1$ in SO(3). We require some function $\boldsymbol{\xi}(s) = \sigma(\boldsymbol{\xi}_0, \boldsymbol{\xi}_1, s)$ where $s \in [0, 1]$ which has the boundary conditions $\sigma(\boldsymbol{\xi}_0, \boldsymbol{\xi}_1, 0) = \boldsymbol{\xi}_0$ and $\sigma(\boldsymbol{\xi}_0, \boldsymbol{\xi}_1, 1) = \boldsymbol{\xi}_1$ and where $\sigma(\boldsymbol{\xi}_0, \boldsymbol{\xi}_1, s)$ varies *smoothly* for intermediate values of s. How we implement this depends very much on our concrete representation of $\boldsymbol{\xi}$.

If pose is represented by an orthonormal rotation matrix, $\boldsymbol{\xi} \sim \boldsymbol{R} \in$ SO(3), we might consider a simple linear interpolation $\sigma(\boldsymbol{R}_0, \boldsymbol{R}_1, s) = (1 - s)\boldsymbol{R}_0 + s\boldsymbol{R}_1$ but this would not, in general, be a valid orthonormal matrix which has strict column norm and inter-column orthogonality constraints.

A workable and commonly used approach is to consider a 3-angle representation such as Euler or roll-pitch-yaw angles, $\boldsymbol{\xi} \sim \boldsymbol{\Gamma} \in \mathbb{S}^1 \times \mathbb{S}^1 \times \mathbb{S}^1$ and use linear interpolation

$$\sigma(\boldsymbol{\Gamma}_0, \boldsymbol{\Gamma}_1, s) = (1 - s)\boldsymbol{\Gamma}_0 + s\boldsymbol{\Gamma}_1$$

and converting the interpolated angles back to a rotation matrix always results in a valid form. For example we define two orientations

```
>> R0 = SO3.Rz(-1) * SO3.Ry(-1);
>> R1 = SO3.Rz(1) * SO3.Ry(1);
```

and find the equivalent roll-pitch-yaw angles

```
>> rpy0 = R0.torpy();  rpy1 = R1.torpy();
```

and create a trajectory between them over 50 time steps

```
>> rpy = mtraj(@tpoly, rpy0, rpy1, 50);
```

which is most easily visualized as an animation▸

```
>> SO3.rpy( rpy ).animate;
```

rpy is a 50 × 3 matrix and the result of SO3.rpy is a 1 × 50 vector of SO3 objects, and their animate method is then called.

For large orientation changes we see that the axis around which the coordinate frame rotates changes along the trajectory. The motion, while smooth, sometimes looks uncoordinated. There will also be problems if either ξ_0 or ξ_1 is close to a singularity in the particular 3-angle system being used. This particular trajectory passes very close to the singularity, at around steps 24 and 25, and a symptom of this is the very rapid rate of change of roll-pitch-yaw angles at this point. The frame is not rotating faster at this point – you can verify that in the animation – the rotational parameters are changing very quickly and this is consequence of the particular representation.

Interpolation of unit-quaternions is only a little more complex than for 3-angle vectors and produces a change in orientation that is a rotation around a *fixed* axis in space. Using the Toolbox we first find the two equivalent quaternions

```
>> q0 = R0.UnitQuaternion;  q1 = R1.UnitQuaternion;
```

and then interpolate them

```
>> q = interp(q0, q1, 50);
>> about(q)
q [UnitQuaternion] : 1x50 (1.7 kB)
```

which results in a vector of 50 UnitQuaternion objects which we can animate by

```
>> q.animate
```

Quaternion interpolation is achieved using spherical linear interpolation (*slerp*) in which the unit quaternions follow a great circle path on a 4-dimensional hypersphere. The result in 3-dimensions is rotation about a fixed axis in space.

3.3.4.1   Direction of Rotation

When traveling on a circle we can move clockwise or counter-clockwise to reach the goal – the result is the same but the distance traveled may be different. On a sphere or hypersphere the principle is the same but now we are traveling on a great circle▸. In this example we animate a rotation about the *z*-axis, from an angle of -2 radians to $+2$ radians

```
>> q0 = UnitQuaternion.Rz(-2);  q1 = UnitQuaternion.Rz(2);
>> q = interp(q0, q1, 50);
>> q.animate()
```

A great circle on a sphere is the intersection of the sphere and a plane that passes through the center. On Earth the equator and all lines of longitude are great circles. Ships and aircraft prefer to follow great circles because they represent the shortest path between two points on the surface of a sphere.

but this is taking the long way around the circle, moving 4 radians when we could travel $2\pi - 4 \approx 2.28$ radians in the opposite direction. The 'shortest' option requests the rotational interpolation to select the shortest path

```
>> q = interp(q0, q1, 50, 'shortest');
>> q.animate()
```

and the animation clearly shows the difference.

3.3.5   Cartesian Motion in 3D

Another common requirement is a smooth path between two poses in **SE**(3) which involves change in position as well as in orientation. In robotics this is often referred to as Cartesian motion.

We represent the initial and final poses as homogeneous transformations

```
>> T0 = SE3([0.4, 0.2, 0]) * SE3.rpy(0, 0, 3);
>> T1 = SE3([-0.4, -0.2, 0.3]) * SE3.rpy(-pi/4, pi/4, -pi/2);
```

The `SE3` object has a method `interp` that interpolates between two poses for normalized distance $s \in [0, 1]$ along the path, for example the midway pose between `T0` and `T1` is

```
>> interp(T0, T1, 0.5)
ans =
    0.0975   -0.7020    0.7055         0
    0.7020    0.5510    0.4512         0
   -0.7055    0.4512    0.5465    0.15
         0         0         0
```

where the translational component is linearly interpolated and the rotation is spherically interpolated using the unit-quaternion interpolation method `interp`.

A trajectory between the two poses in 50 steps is created by

```
>> Ts = interp(T0, T1, 50);
```

where the arguments are the initial and final pose and the trajectory length.◄ The resulting trajectory `Ts` is a vector of `SE3` objects

This could also be written as `T0.interp(T1, 50)`.

```
>> about(Ts)
Ts [SE3] : 1x50 (6.5 kB)
```

representing the pose at each time step. The homogeneous transformation for the first point on the path is

```
>> Ts(1)
ans =
   -0.9900   -0.1411         0       0.4
    0.1411   -0.9900         0       0.2
         0         0         1         0
         0         0         0         1
```

and once again the easiest way to visualize this is by animation

```
>> Ts.animate
```

which shows the coordinate frame moving and rotating from pose `T0` to pose `T1`.

The translational part of this trajectory is obtained by◄

```
>> P = Ts.transl;
```

The `.t` property applied to a vector of `SE3` objects returns a MATLAB comma-separated list of translation vectors. The `.transl` method returns the translations in a more useful matrix form.

which returns the Cartesian position for the trajectory in matrix form

```
>> about(P)
P [double] : 50x3 (1.2 kB)
```

which has one row per time step that is the corresponding position vector. This is plotted

```
>> plot(P);
```

in Fig. 3.7 along with the orientation in roll-pitch-yaw format

```
>> rpy = Ts.torpy;
>> plot(rpy);
```

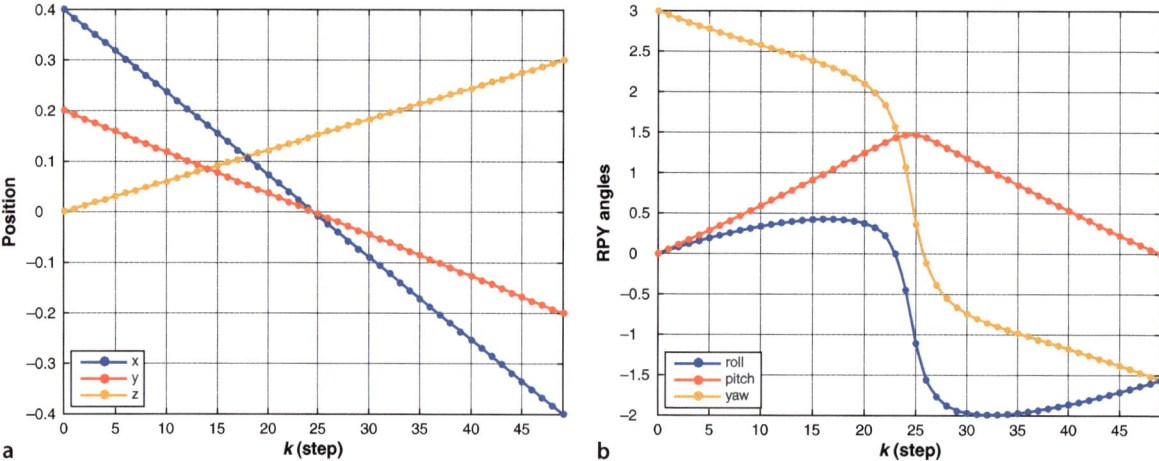

a   b

Fig. 3.7. Cartesian motion. **a** Cartesian position versus time, **b** roll-pitch-yaw angles versus time

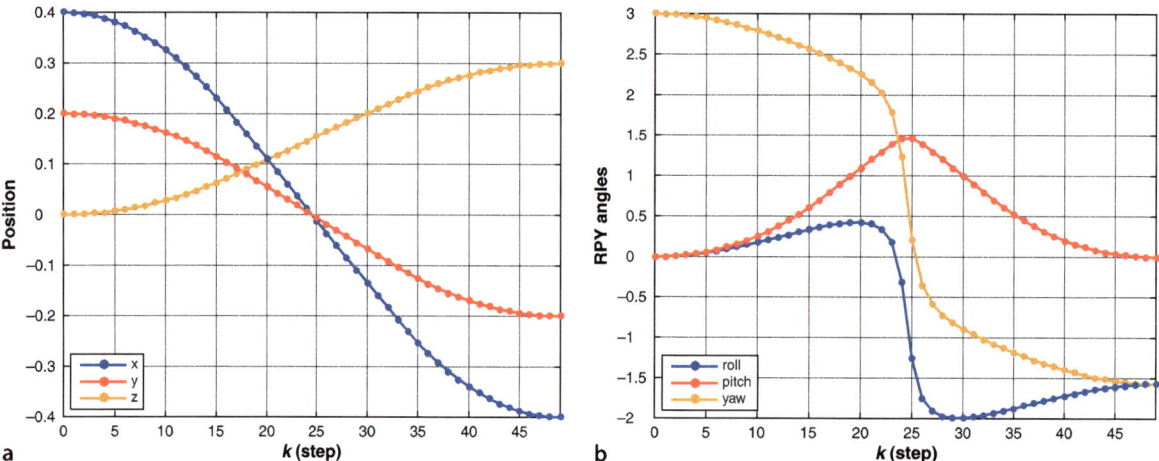

a   b

We see that the position coordinates vary smoothly and linearly with time and that orientation varies smoothly with time. ◀

However the motion has a velocity and acceleration *discontinuity* at the first and last points. While the path is smooth in space the distance *s* along the path is not smooth in time. Speed along the path jumps from zero to some finite value and then drops to zero at the end – there is no initial acceleration or final deceleration. The scalar functions `tpoly` and `lspb` discussed earlier can be used to generate *s* so that motion *along* the path is smooth. We can pass a vector of normalized distances along the path as the second argument to `interp`

```
>> Ts = T0.interp(T1, lspb(0, 1, 50) );
```

The trajectory is unchanged but the coordinate frame now accelerates to a constant speed along the path and then decelerates and this is reflected in smoother curves for the trajectory shown in Fig. 3.8. The Toolbox provides a convenient shorthand `ctraj` for the above

```
>> Ts = ctraj(T0, T1, 50);
```

where the arguments are the initial and final pose and the number of time steps.

Fig. 3.8. Cartesian motion with LSPB path distance profile. **a** Cartesian position versus time, **b** roll-pitch-yaw angles versus time

The roll-pitch-yaw angles do not vary linearly with time because they represent a nonlinear transformation of the linearly varying quaternion.

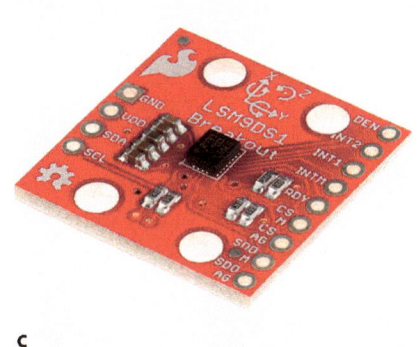

Fig. 3.9. a SPIRE (Space Inertial Reference Equipment) from 1953 was 1.5 m in diameter and weighed 1 200 kg. **b** A modern inertial navigation system the LORD Micro-Strain 3DM-GX4-25 has triaxial gyroscopes, accelerometers and magnetometer, a pressure altimeter, is only $36 \times 24 \times 11$ mm and weighs 16 g (image courtesy of LORD MicroStrain); **c** 9 Degrees of Freedom IMU Breakout (LSM9DS1-SEN-13284 from Spark-Fun Electronics), the chip itself is only 3.5×3 mm

3.4   Application: Inertial Navigation

An inertial navigation system or INS is a "black box" that estimates its velocity, orientation and position by measuring accelerations and angular velocities and integrating them over time. Importantly it has no external inputs such as radio signals from satellites. This makes it well suited to applications such as submarine, spacecraft and missile guidance where it is not possible to communicate with radio navigation aids or which must be immune to radio jamming. These particular applications drove development of the technology during the cold war and space race of the 1950s and 1960s. Those early systems were large, see Fig. 3.9a, extremely expensive and the technical details were national secrets. Today INSs are considerably cheaper and smaller as shown in Fig. 3.9b; the sensor chips shown in Fig. 3.9c can cost as little as a few dollars and they are built into every smart phone.

An INS estimates its pose with respect to an inertial reference frame which is typically denoted {0} and fixed to some point on the Earth's surface – the world coordinate frame.◄ The frame typically has its z-axis upward or downward and the x- and y-axes establish a local tangent plane. Two common conventions have the x-, y- and z-axes respectively parallel to north-east-down (NED) or east-north-up (ENU) directions. The coordinate frame {B} is attached to the moving vehicle or robot and is known as the body- or body-fixed frame.

As discussed in Sect. 3.2.3 the Earth's surface is not an inertial reference frame, but for most robots with nonmilitary grade sensors this is a valid assumption.

3.4.1   Gyroscopes

Any sensor that measures the rate of change of orientation is known, for historical reasons, as a gyroscope.

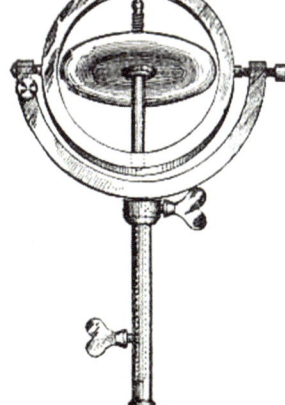

3.4.1.1   How Gyroscopes Work

The term gyroscope conjures up an image of a childhood toy – a spinning disk in a round frame that can balance on the end of a pencil. Gyroscopes are confounding devices – you try to turn them one way but they resist and turn (precess) in a different direction. This unruly behavior is described by a simplified version of Eq. 3.10

$$\tau = \boldsymbol{\omega} \times \boldsymbol{h} \tag{3.15}$$

where $\boldsymbol{h}$ is the angular momentum of the gyroscope, a vector parallel to the rotor's axis of spin and with magnitude $\|\boldsymbol{h}\| = J\varpi$, where J is the rotor's inertia and ϖ its rotational speed. It is the cross product in Eq. 3.15 that makes the gyroscope move in a contrary way.

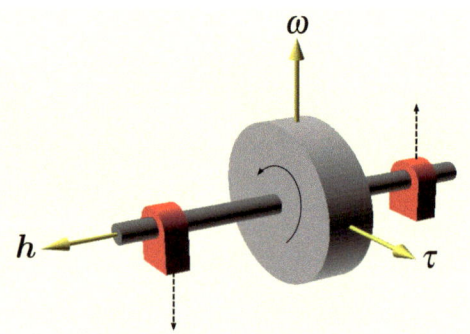

Fig. 3.10.
Gyroscope in strapdown configuration. Angular velocity ω induces a torque τ which can be sensed as forces at the bearings shown in red

If no torque is applied to the gyroscope its angular momentum remains constant in the inertial reference frame which implies that the axis will maintain a *constant direction* in that frame. Two gyroscopes with orthogonal axes form a stable platform that will maintain a *constant orientation* with respect to the inertial reference frame – fixed with respect to the universe. This was the principle of many early spacecraft navigation systems such as that shown in Fig. 2.15 – the vehicle was able to rotate about the stable platform and the spacecraft's orientation could be measured with respect to the platform.▶

The challenge was to create a mechanism that allowed the vehicle to rotate around the stable platform without exerting any torque on the gyroscopes. This required exquisitely engineered low-friction gimbals and bearing systems.

Alternatively we can fix the gyroscope to the vehicle in the strapdown configuration as shown in Fig. 3.10. If the vehicle rotates with an angular velocity ω the attached gyroscope will *resist* and exert an orthogonal torque τ which can be measured.▶ If the magnitude of h is high then this kind of sensor is very sensitive – a very small angular velocity leads to an easily measurable torque.

Typically by strain gauges attached to the bearings of the rotor shaft.

Over the last few decades this rotating disk technology has been eclipsed by sensors based on optical principles such as the ring-laser gyroscope (RLG) and the fiber-optic gyroscope (FOG). These are high quality sensors but expensive and bulky. The low-cost sensors used in mobile phones and drones are based on micro-electro-mechanical systems (MEMS) fabricated on silicon chips. Details of the designs vary but all contain a mass vibrating at high frequency▶ in a plane, and rotation about an axis normal to the plane causes an orthogonal displacement within the plane that is measured capacitively.

Typically over 10 kHz.

Gyroscopic angular velocity sensors measure rotation about a single axis. Typically three gyroscopes are packaged together and arranged so that their sensitive axes are orthogonal. The three outputs of such a triaxial gyroscope are the components of the angular velocity vector $^B\omega^\#$ measured in the body frame $\{B\}$, and we introduce the # superscript to explicitly indicate a sensor measurement.

Interestingly, nature has invented gyroscopic sensors. All vertebrates have angular velocity sensors as part of their vestibular system. In each inner ear we have three semi-circular canals – fluid filled organs that measure angular velocity. They are arranged orthogonally, just like a triaxial gyroscope, with two measurement axes in a vertical plane and one diagonally across the head.

3.4.1.2 Estimating Orientation

If we assume that $^B\omega$ is constant over a time interval δ_t the equivalent rotation at the timestep k is

$$^B\xi_\Delta\langle k\rangle \sim e^{\left[^B\omega^\#\right]_\times \delta_t} \tag{3.16}$$

If the orientation of the sensor frame is initially ξ_B then the evolution of estimated pose can be written in discrete-time form as

$$\hat{\xi}_B\langle k+1\rangle \leftarrow \hat{\xi}_B\langle k\rangle \oplus {}^B\xi_\Delta\langle k\rangle \tag{3.17}$$

Much important development was undertaken by the MIT Instrumentation Laboratory under the leadership of Charles Stark Draper. In 1953 the feasibility of inertial navigation for aircraft was demonstrated in a series of flight tests with a system called SPIRE (Space Inertial Reference Equipment) shown in Fig. 3.9a. It was 1.5 m in diameter and weighed 1 200 kg. SPIRE guided a B-29 bomber on a 12 hour trip from Massachusetts to Los Angeles without the aid of a pilot and with Draper aboard. In 1954 the first self-contained submarine navigation system (SINS) was introduced to service. The Instrumentation Lab also developed the Apollo Guidance Computer, a one-cubic-foot computer that guided the Apollo Lunar Module to the surface of the Moon in 1969.

Today high-performance inertial navigation systems based on fiber-optic gyroscopes are widely available and weigh around one 1 kg while low-cost systems based on MEMS technology can weigh just a few grams and cost a few dollars.

where we use the hat notation to explicitly indicate an estimate of pose and $k \in \mathbb{Z}^+$ is the index of the time step. In concrete terms we can compute this *update* using $\mathbf{SO}(3)$ rotation matrices or unit-quaternions as discussed in Sect. 3.1.3 and taking care to normalize the rotation after each step.

We will demonstrate this integration using unit quaternions and simulated angular velocity data for a tumbling body. The script

```
>> ex_tumble
```

creates a matrix w whose columns represent consecutive body-frame angular velocity measurements with corresponding times given by elements of the vector t. We choose the initial pose to be the null rotation

```
>> attitude(1) = UnitQuaternion();
```

and then for each time step we update the orientation and keep the orientation history in a vector of quaternions

```
>> for k=1:numcols(w)-1
       attitude(k+1) = attitude(k) .* UnitQuaternion.omega( w(:,k)*dt );
   end
```

The omega method creates a unit-quaternion corresponding to a rotation angle and axis given by the magnitude and direction of its argument. The .* operator performs quaternion multiplication and normalizes the product, ensuring the result has a unit norm.◄ We can animate the changing orientation of the body frame

The .increment method of the UnitQuaternion class does this in a single call.

```
>> attitude.animate('time', t)
```

or view the roll-pitch-yaw angles as a function of time

```
>> mplot(t, attitude.torpy() )
```

3.4.2   Accelerometers

Accelerometers are sensors that measure acceleration. Even when not moving they sense the acceleration due to gravity which defines the direction we know as *downward*. Gravitational acceleration is a function of the material in the Earth beneath us and our distance from the Earth's center. The Earth is not a perfect sphere◄ and points in the equatorial region are further from the center. Gravitational acceleration can be approximated by

The technical term is an oblate spheroid, it bulges out at the equator because of centrifugal acceleration due to the Earth's rotation. The equatorial diameter is around 40 km greater than the polar diameter.

$$g \approx 9.780327\left(1 + 0.0053024\sin^2\theta - 0.0000058\sin^2 2\theta\right) - 0.000003086h$$

where θ is the angle of latitude and h is height above sea level. A map of gravity showing the effect of latitude and topography is shown in Fig. 3.11.

Charles Stark (Doc) Draper (1901–1987) was an American scientist and engineer, often referred to as "the father of inertial navigation." Born in Windsor, Missouri, he studied at the University of Missouri then Stanford where he earned a B.A. in psychology in 1922, then at MIT an S.B. in electro-chemical engineering and an S.M. and Sc.D. in physics in 1928 and 1938 respectively. He started teaching while at MIT and became a full professor in aeronautical engineering in 1939. He was the founder and director of the MIT Instrumentation Laboratory which made important contributions to the theory and practice of inertial navigation to meet the needs of the cold war and the space program.

Draper was named one of Time magazine's Men of the Year in 1961 and inducted to the National Inventors Hall of Fame in 1981. The Instrumentation lab was renamed Charles Stark Draper Laboratory (CSDL) in his honor. (Photo courtesy of The Charles Stark Draper Laboratory Inc.)

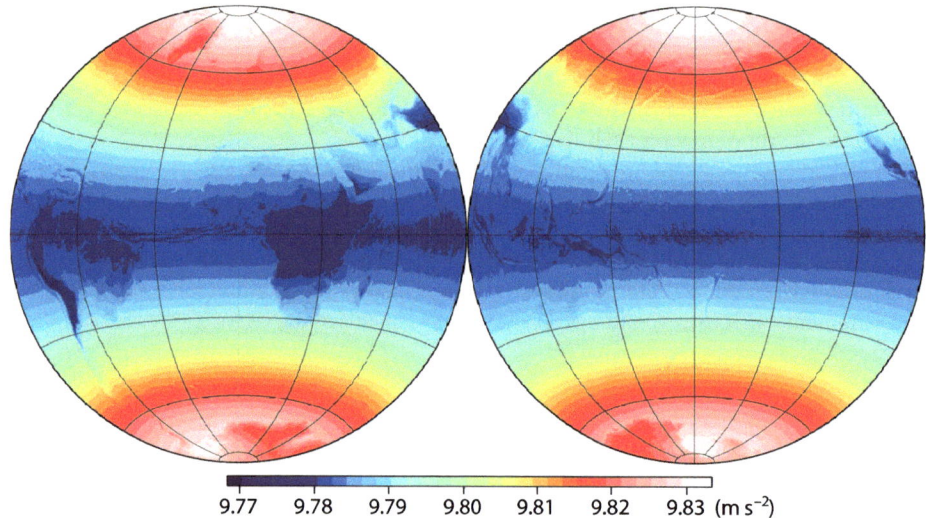

9.77   9.78   9.79   9.80   9.81   9.82   9.83 (m s^{-2})

Fig. 3.11.
Variation in Earth's gravitational acceleration, continents and mountain ranges are visible. The hemispheres shown are centerd on the prime (left) and anti (right) meridian respectively (from Hirt et al. 2013)

3.4.2.1   How Accelerometers Work

An accelerometer is conceptually a very simple device comprising a mass, known as the proof mass, supported by a spring as shown in Fig. 3.12. In the inertial reference frame Newton's second law for the proof mass is

$$m\ddot{x}_m = F_s - mg \tag{3.18}$$

and for a spring with natural length l_0 the relationship between force and extension d is

$$F_s = kd$$

The various displacements are related

$$x_b - (l_0 + d) = x_m$$

and taking the double derivative then substituting Eq. 3.18 gives

$$\ddot{x}_b - \ddot{d} = \tfrac{1}{m}(kd - mg)$$

The quantity we wish to measure is the acceleration of the accelerometer $a = \ddot{x}_b$ ▶ and the relative displacement of the proof mass

$$d = \frac{m}{k}(a + g)$$

We assume that $\ddot{d} = 0$ in steady state. Typically there would be a damping element to increase friction and stop the proof mass oscillating. This adds a term $-B\dot{x}_m$ to the right-hand side of Eq. 3.18.

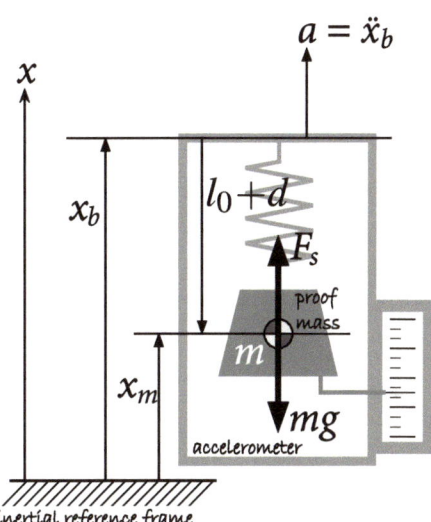

Fig. 3.12.
The essential elements of an
accelerometer and notation

is linearly related to that acceleration. In an accelerometer the displacement is measured and scaled by k/m so that the output of the sensor is

$$a^{\#} = a + g \ \mathrm{m \, s^{-2}}$$

If this accelerometer is stationary then $a = 0$ yet the measured acceleration would be $a^{\#} = 0 + g = g$ in the upward direction. This is because our model has included the Newtonian gravity force mg, as discussed in Sect. 3.2.3. Accelerometer output is sometimes referred to as specific, inertial or proper acceleration.

> The fact that a stationary accelerometer indicates an upward acceleration of 1 g is unintuitive since the accelerometer is clearly stationary and not accelerating. Intuition would suggest, that if anything, the acceleration should be in the downward direction where the device would accelerate if dropped. However the reality is that an accelerometer at rest in a gravity field reports upward acceleration. ◄

A number of iPhone sensor apps incorrectly report acceleration in the *downward* direction when the phone is stationary.

Accelerometers measure acceleration along a single axis. Typically three accelerometers are packaged together and arranged so that their sensitive axes are orthogonal. The three outputs of such a triaxial accelerometer are the components of the acceleration vector $^{B}a^{\#}$ measured in the body frame $\{B\}$.

Nature has also invented the accelerometer. All vertebrates have acceleration sensors called ampullae as part of their vestibular system. We have two in each inner ear: the saccule which measures vertical acceleration, and the utricle which measures front-to-back acceleration, and they help us maintain balance. ◄ The proof mass in the am-

Inconsistency between motion sensed in our ears and motion perceived by our eyes is the root cause of motion sickness.

pullae is a collection of calcium carbonate crystals called otoliths, literally ear stones, on a gelatinous substrate which serves as the spring and damper. Hair cells embedded in the substrate measure the displacement of the otoliths due to acceleration.

3.4.2.2   Estimating Pose and Body Acceleration

In frame $\{0\}$ with its z-axis vertically upward, the gravitational acceleration vector is

$$^{0}a = \begin{pmatrix} 0 \\ 0 \\ g \end{pmatrix}$$

where g is the local gravitational acceleration from Fig. 3.11. In a body-fixed frame $\{B\}$ at an arbitrary orientation expressed in terms of ZYX roll-pitch-yaw angles▶

We could use any 3-angle sequence.

$$^0\xi_B = \mathscr{R}_z\big(\theta_y\big) \oplus \mathscr{R}_y\big(\theta_p\big) \oplus \mathscr{R}_x\big(\theta_r\big)$$

the gravitational acceleration will be

$$^B\boldsymbol{a} = \big(\ominus\,^0\xi_B\big)\boldsymbol{\cdot}\,^0\boldsymbol{a} = \begin{pmatrix} -g\sin\theta_p \\ g\cos\theta_p\sin\theta_r \\ g\cos\theta_p\cos\theta_r \end{pmatrix} \tag{3.19}$$

The *measured* acceleration vector from the sensor in frame $\{B\}$ is

$$^B\boldsymbol{a}^{\#} = \begin{pmatrix} a_x \\ a_y \\ a_z \end{pmatrix}$$

and equating this with Eq. 3.19 we can solve for the roll and pitch angles

$$\sin\hat{\theta}_p = \frac{-a_x}{g} \tag{3.20}$$

$$\tan\hat{\theta}_r = \frac{a_y}{a_z}, \quad \theta_p \neq \pm\frac{\pi}{2} \tag{3.21}$$

and we use the hat notation to indicate that these are estimates of the angles.▶ Notice that there is no solution for the yaw angle and in fact θ_y does not even appear in Eq. 3.19. The gravity vector is parallel to the vertical axis and rotating around that axis, yaw rotation, will not change the measured value at all.▶

These angles are sufficient to determine whether a phone, tablet or camera is in portrait or landscape orientation.

> We have made a very strong assumption that the measured acceleration $^B\boldsymbol{a}^{\#}$ is only due to gravity. On a robot the sensor will experience additional acceleration as the vehicle moves and this will introduce an error in the estimated orientation.

Another way to consider this is that we are essentially measuring the direction of the gravity vector with respect to the frame $\{B\}$ and a vector provides only two unique *pieces* of directional information, since one component of a unit vector can be written in terms of the other two.

Frequently we want to estimate the motion of the vehicle in the inertial frame, and the total measured acceleration in $\{0\}$ is due to gravity *and* motion

$$^0\boldsymbol{a}^{\#} = \,^0\boldsymbol{g} + \,^0\boldsymbol{a}_v$$

We observe acceleration in the body frame so the vehicle acceleration in the world frame is

$$^0\hat{\boldsymbol{a}}_v = \,^0\hat{\boldsymbol{R}}_B\,^B\boldsymbol{a}^{\#} - \,^0\boldsymbol{g} \tag{3.22}$$

and we assume that $^0\hat{\boldsymbol{R}}_B$ and g are both known.▶ Integrating that with respect to time

The first assumption is a strong one and problematic in practice. Any error in the rotation matrix results in incorrect cancellation of the gravity component of $a^{\#}$ which leads to an error in the estimated body acceleration.

$$^0\hat{\boldsymbol{v}}_v(t) = \int\,^0\hat{\boldsymbol{a}}_v(t)\,\mathrm{d}t \tag{3.23}$$

gives the velocity of the vehicle, and integrating again

$$^0\hat{\boldsymbol{p}}_v(t) = \int\,^0\hat{\boldsymbol{v}}_v(t)\,\mathrm{d}t \tag{3.24}$$

gives its position. Note that we can assume vehicle acceleration is zero and estimate attitude, or assume attitude and estimate vehicle acceleration. We cannot estimate both since there are more unknowns than measurements.

3.4.3   Magnetometers

The Earth is a massive but weak magnet. The poles of this geomagnet are the Earth's north and south magnetic poles which are constantly moving and located quite some distance from the planet's rotational axis.

At any point on the planet the magnetic flux lines can be considered a vector m whose magnitude and direction can be accurately predicted and mapped as shown in Fig. 3.13. We describe the vector's direction in terms of two angles: declination and inclination. A horizontal projection of the vector m points in the direction of magnetic north and the declination angle D is measured from true north◀ clockwise to that projection. The inclination angle I of the vector is measured in a vertical plane downward◀ from horizontal to m. The length of the vector, the magnetic field intensity, is measured by a magnetometer in units of Tesla (T) and for the Earth this varies from $25-65\ \mu T^\blacktriangleright$ as shown in Fig. 3.13a.

The direction of the Earth's north rotational pole, where the rotational axis intersects the surface of the northern hemisphere.

In the Northern hemisphere inclination is positive, that is, the vector points into the ground.

By comparison a modern MRI machine has a magnetic field strength of 4-8 T.

3.4.3.1   How Magnetometers Work

The key element of most modern magnetometers is a Hall-effect sensor, a semiconductor device which produces a voltage proportional to the magnetic field intensity in a direction normal to the current flow. Typically three Hall-effect sensors are packaged together and arranged so that their sensitive axes are orthogonal. The three outputs of such a triaxial magnetometer are the components of the Earth's magnetic field intensity vector $^B m^{\#}$ measured in the body frame {B}.

Yet again nature leads, and creatures from bacteria to turtles and birds are known to sense magnetic fields. The effect is particularly well known in pigeons and there is debate about whether or not humans have this sense. The actual biological sensing mechanism has not yet been discovered.

3.4.3.2   Estimating Heading

Consider an inertial coordinate frame {0} with its z-axis vertically upward and its x-axis pointing toward *magnetic* north. The magnetic field intensity vector therefore lies in the xz-plane

$$
^0 m = B\begin{pmatrix} \cos I \\ 0 \\ \sin I \end{pmatrix}
$$

where B is the magnetic field intensity and I the inclination angle which are both known from Fig. 3.13. In a body-fixed frame {B} at an arbitrary orientation expressed in terms of roll-pitch-yaw angles◀

We could use any 3-angle sequence.

$$
^0\xi_B = \mathscr{R}_z\left(\theta_y\right) \oplus \mathscr{R}_y\left(\theta_p\right) \oplus \mathscr{R}_x\left(\theta_r\right)
$$

Edwin Hall (1855–1938) was an American physicist born in Maine. His Ph.D. research in physics at the Johns Hopkins University in 1880 discovered that a magnetic field exerts a force on a current in a conductor. He passed current through thin gold leaf and in the presence of a magnetic field normal to the leaf was able to measure a very small potential difference between the sides of the leaf. This is now known as the Hall effect. While it was then known that a magnetic field exerted a force on a current carrying conductor it was believed the force acted on the conductor not the current itself – electrons were yet to be discovered. He was appointed as professor of physics at Harvard in 1895 where he worked on thermoelectric effects.

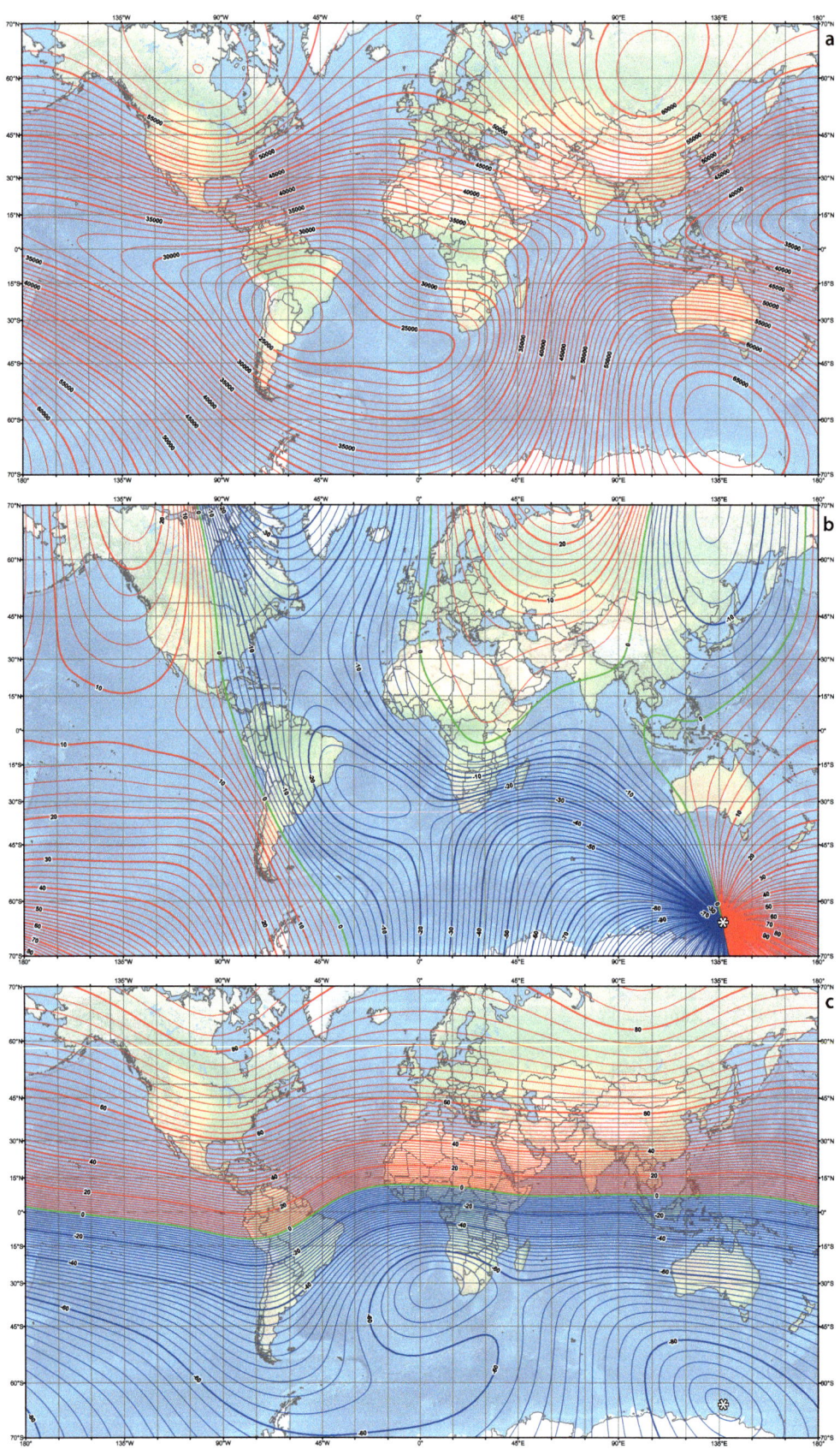

Fig. 3.13.
A predicted model of the Earth magnetic field parameters for 2015. **a** Magnetic field intensity (nT); **b** magnetic declination (degrees); **c** magnetic inclination (degrees). Magnetic poles indicated by *asterisk* (maps by NOAA/NGDC and CIRES http://ngdc. noaa.gov/geomag/WMM, published Dec 2014)

the magnetic field intensity will be

$$^{B}\boldsymbol{m} = \left(\ominus^{0}\xi_{B}\right) \bullet {}^{0}\boldsymbol{m} \tag{3.25}$$

The measured magnetic field intensity vector from the sensor in frame {B} is

$$^{B}\boldsymbol{m}^{\#} = \begin{pmatrix} m_{x} \\ m_{y} \\ m_{z} \end{pmatrix}$$

and equating this with Eq. 3.25 we can solve for the yaw angle

$$\hat{\theta}_{y} = \tan^{-1} \frac{\cos\theta_{p}\left(m_{z}\sin\theta_{r} - m_{y}\cos\theta_{r}\right)}{m_{x} + B\sin I \sin\theta_{p}}$$

Many triaxial Hall-effect sensor chips also include a triaxial accelerometer for just this purpose.

assuming that the roll and pitch angles have been determined, perhaps using measured acceleration and Eq. 3.21.◄

We defined yaw angle as the orientation of the frame {B} x-axis◄ with respect to *magnetic* north. To obtain the heading angle with respect to true-north we subtract the local declination angle

Typically in vehicle navigation the x-axis points forward and the yaw angle is also called the heading angle.

$$^{\text{tn}}\hat{\theta}_{y} = \hat{\theta}_{y} - D$$

Magnetometers are great in theory but problematic in practice. Firstly, our modern world is full of magnets and electromagnets. Buildings contain electrical wiring and robots themselves are full of electric motors, batteries and electronics. These all add to, or overwhelm, the local geomagnetic field. Secondly, many objects in our world contain ferromagnetic materials such as the reinforcing steel in buildings or the steel bodies of cars or ships. These distort the geomagnetic field leading to local changes in its direction. These effects are referred to respectively as hard- and soft-iron distortion of the magnetic field.◄

These can be calibrated out but the process requires that the sensor is rotated by 360 degrees.

3.4.4   Sensor Fusion

An inertial navigation system uses the devices we have just discussed to determine the pose of a vehicle – its position and its orientation. Early inertial navigation systems, such as shown in Fig. 2.15, used mechanical gimbals to keep the accelerometers at a constant attitude with respect to the stars using a gyro-stabilized platform. The acceleration measured on this platform is by definition referred to the inertial frame and simply needs to be integrated to obtain the velocity of the platform, and integrated again to obtain its position. In order to achieve accurate position estimates over periods of hours or days the gimbals and gyroscopes had to be of extremely high quality so that the stable platform did not drift, and the acceleration sensors needed to be extremely accurate.

The modern strapdown inertial measurement configuration uses no gimbals. The angular velocity, acceleration and magnetic field sensors are rigidly attached to the vehicle. The collection of inertial sensors is referred to as an inertial measurement unit or IMU. A 6-DOF IMU comprises triaxial gyroscopes and accelerometers while a 9-DOF IMU comprises triaxial gyroscopes, accelerometers and magnetometers.◄ A system that only determines attitude is called an attitude and heading reference system or AHRS.

Increasingly these sensor packages also include a barometric pressure sensor to measure changes in altitude.

The sensors we use, particularly the low-cost ones in phones and drones, are far from perfect. Consider any sensor value – gyroscope, accelerometer or magnetometer – the measured signal

$$x^{\#} = sx + b + \varepsilon$$

is related▶ to the unknown true value x by a scale factor s, offset or bias b and random noise ε. s is usually specified by the manufacturer to some tolerance, perhaps $\pm 1\%$, and for a particular sensor this can be determined by some calibration procedure. Bias b is ideally equal to zero but will vary from device to device. Bias that varies over time is often called sensor drift. Scale factor and bias are typically both a function of temperature.▶

In practice bias is the biggest problem because it varies with time and temperature and has a very deleterious effect on the estimated pose and position. Consider a positive bias on the output of a gyroscopic sensor – the output is higher than it should be. At each time step in Eq. 3.17 the incremental rotation will be bigger than it should be, which means that the orientation error will grow linearly with time.▶

If we use Eq. 3.22 to estimate the vehicle's acceleration then the error in attitude means that the measured gravitation acceleration is incorrectly canceled out and will be indistinguishable from *actual* vehicle acceleration. This offset in acceleration becomes a linear time error in velocity and a quadratic time error in position. Given that the pose error is already linear in time we end up with a cubic time error in position, and this is ignoring the effects of accelerometer bias. Sensor bias is problematic! A rule of thumb is that gyroscopes with bias stability of 0.01 deg h^{-1} will lead to position error growing at a rate of 1 nmi h^{-1} (1.85 km h^{-1}). Military grade systems have very impressive stability, for missiles <0.00002 deg h^{-1} which is in stark contrast to consumer grade devices which are in the range $0.01-0.2$ deg per *second*.

A simple approach to this problem is to estimate bias by leaving the IMU stationary for a few seconds and computing the average value of all the sensors.▶ This value is then subtracted from future sensor readings. This is really only valid over a short time period because the bias is not constant.

A more sophisticated approach is to estimate the bias online▶, but to do this we need to combine information from different sensors – an approach known as sensor fusion. We rely on the fact that different sensors have complementary characteristics. Bias on angular rate sensors causes the attitude estimate error to grow with time, but for accelerometers it will only cause an attitude offset. However accelerometers respond to motion of the vehicle while good gyroscopes do not. Magnetometers provide partial information about roll, pitch and yaw, are immune to acceleration, but do respond to stray magnetic fields and other distortions. There are many ways to achieve this kind of fusion. A common approach is to use an estimation tool called an extended Kalman filter described in Appendix H. Given a full nonlinear mathematical model that relates the sensor signals and their biases to the vehicle pose and knowledge about the noise (uncertainty) on the sensor signals, the filter gives an optimal estimate of the pose and bias that best explain the sensor signals.

Here we will consider a simple but still very effective alternative called the explicit complementary filter. The rotation update step is performed using Eq. 3.17 but compared to Eq. 3.16 the incremental rotation is more complex

$$^{B}\xi_{\Delta}\langle k\rangle = e^{\left[^{B}\omega^{\#}\langle k\rangle - \hat{b}\langle k\rangle + k_{P}\sigma_{R}\langle k\rangle\right]_{\times}\delta_{t}} \tag{3.26}$$

The key differences are that the estimated bias $\hat{b}$ is subtracted from the sensor measurement and a term based on the orientation error σ_{R} is added. The estimated bias changes with time according to

$$\hat{b}\langle k+1\rangle \leftarrow \hat{b}\langle k\rangle - k_{I}\sigma_{R}\langle k\rangle \tag{3.27}$$

and also depends on the orientation error σ_{R}. $k_{P} > 0$ and $k_{I} > 0$ are both well chosen constants.

The orientation error is derived from N *vector measurements* $^{0}v_{i}^{\#}$

$$\sigma_{R}\langle k+1\rangle = \sum_{i=1}^{N} k_{i} {}^{0}v_{i} \times {}^{0}v_{i}^{\#}\langle k\rangle$$

We assume a linear relationship but check the fine print in a datasheet to understand what a sensor really does.

Some sensors also exhibit cross-sensitivity. They may give a weak response to a signal in an orthogonal direction or from a different mode, quite commonly low-cost gyroscopes respond to vibration and acceleration as well as rotation.

The effect of an attitude error is dangerous on something like a quadrotor. For example, if the estimated pitch angle is too high then the vehicle control system will pitch down by the same amount to keep the craft "level", and this will cause it to accelerate forward.

A lot of hobby drones do this just before they take off.

Our brain has an online mechanism to cancel out the bias in our vestibular gyroscopes. It uses the recent average rotation as the bias, based on the reasonable assumption that we do not undergo prolonged rotation. If we do, then that angular rate becomes the new normal so that when we stop rotating we perceive rotation in the opposite direction. We call this dizziness.

where ^0v_i is the known value of a vector signal in the inertial frame (for example gravitational acceleration) and

$$^0v_i^\#\langle k\rangle = {}^0\hat{\xi}_B\langle k\rangle \bullet {}^Bv^\#\langle k\rangle$$

is the value measured in the body-fixed frame and rotated into the inertial frame by the estimated orientation $^0\hat{\xi}_B$. Any error in direction between these vectors will yield a nonzero cross-product which is the axis around which to rotate one vector into the other. The filter uses this difference – the innovation – to improve the orientation estimate by feeding it back into Eq. 3.26. This filter allows an unlimited number of vectorial measurements ^0v_i to be fused together; for example we could add magnetic field or any other kind of direction data such as the altitude and azimuth of visual landmarks, stars or planets.

The script

```
>> ex_tumble
```

provides simulated "measured" gyroscope, accelerometer and magnetometer data organized as columns of the matrices `wm`, `gm` and `mm` respectively and all include a fixed bias. Corresponding times are given by elements of the vector `t`. Firstly we will repeat the example from page 79 but now with sensor bias

```
>> attitude(1) = UnitQuaternion();
>> for k=1:numcols(wm)-1
       attitude(k+1) = attitude(k) .* UnitQuaternion.omega( wm(:,k)*dt );
   end
```

To see the effect of bias on the estimated attitude we will compare it to the true attitude `truth` that was also computed by the script. As a measure of error we plot the angle between the corresponding unit quaternions in the sequence

```
>> plot(t, angle(attitude, truth), 'r' );
```

which is shown as the red line in Fig. 3.14a. We can clearly see growth in angular error over time. Now we implement the explicit complementary filter with just a few extra lines of code

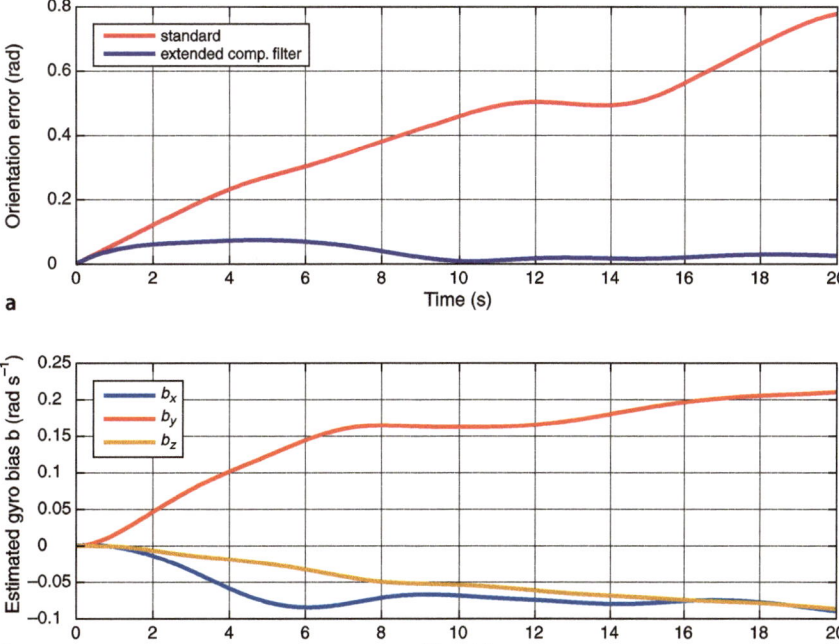

Fig. 3.14.
a Effect of gyroscope bias on naïve INS (red) and explicit complementary filter (blue); b estimated gyroscope bias from the explicit complementary filter

```
>> kI = 0.2; kP = 1;
>> b = zeros(3, numcols(w));
>> attitude_ecf(1) = UnitQuaternion(); b = [0 0 0]';
>> for k=1:numcols(wm)-1
       invq = inv( attitude_ecf(k) );
       sigmaR = cross(gm(:,k), invq*g0) + cross(mm(:,k), invq*m0);
       wp = wm(:,k) - b(:,k) + kP*sigmaR;
       attitude_ecf(k+1) = attitude_ecf(k) .* UnitQuaternion.omega( wp*dt );
       b(:,k+1) = b(:,k) - kI*sigmaR*dt;
   end
```

and plot the angular difference between the estimated and the attitude as a blue line

```
>> plot(t, angle(attitude_ecf, truth), 'b' );
```

Bringing together information from multiple sensors has checked the growth in attitude error, despite all the sensors having a bias. The estimated gyroscope bias is shown in Fig. 3.14b and we can see the bias estimates converging on their true value.

3.5   Wrapping Up

In this chapter we have considered pose that varies as a function of time from several perspectives.

Firstly we took a calculus perspective and showed that the temporal derivative of an orthonormal rotation matrix or a quaternion is a function of the angular velocity of the body – a concept from mechanics. The skew-symmetric matrix appears in the rotation matrix case and we should no longer be surprised about this given its intimate connection to rotation via Lie group theory. We then looked at finite time differences as an approximation to the derivative and showed how these lead to computationally cheap methods to update rotation matrices and quaternions given knowledge of angular velocity. We also discussed the dynamics of moving bodies that translate and rotate under the influence of forces and torques, inertial and noninertial reference frames and the notion of fictitious forces.

The second perspective was to create motion – a sequence of poses, a trajectory, that a robot can follow. An important characteristic of a trajectory is that it is *smooth* – the position and orientation changes smoothly with time. We started by discussing how to generate smooth trajectories in one dimension and then extended that to the multi-dimensional case and then to piecewise-linear trajectories that visit a number of intermediate points. Smoothly varying rotation was achieved by interpolating roll-pitch-yaw angles and quaternions.

With all this under our belt we were then able to tackle an application, the important problem of inertial navigation. Given imperfect measurements from sensors on a moving body we are able to estimate the pose of that moving body.

Further Reading

The earliest work on manipulator Cartesian trajectory generation was by Paul (1972, 1979) and Taylor (1979). The multi-segment trajectory is discussed by Paul (1979, 1981) and the concept of segment transitions or blends is discussed by Lloyd and Hayward (1991). These early papers, and others, are included in the compilation on Robot Motion (Brady et al. 1982). Polynomial and LSPB trajectories are described in detail by Spong et al. (2006) and multi-segment trajectories are covered at length in Siciliano et al. (2009) and Craig (2005).

The book *Digital Apollo* (Mindell 2008) is a very readable story of the development of the inertial navigation system for the Apollo Moon landings. The article by Corke et al. (2007) describes the principles of inertial sensors and the functionally equivalent sensors located in the inner ear of mammals that play a key role in maintaining balance.

There is a lot of literature related to the theory and practice of inertial navigation systems. The thesis of Achtelik (2014) describes a sophisticated extended Kalman filter for estimating the pose, velocity and sensor bias for a small flying robot. The explicit complementary filter used in this chapter is described by Hua et al. (2014). The recently revised book Groves (2013) covers inertial and terrestrial radio and satellite navigation and has a good coverage of Kalman filter state estimation techniques. Titterton and Weston (2005) provides a clear and concise description of the principles underlying inertial navigation with a focus on the sensors themselves but is perhaps a little dated with respect to modern low-cost sensors. Data sheets on many low-cost inertial and magnetic field sensing chips can be found at https://www.sparkfun.com in the Sensors category.

Exercises

1. Express the incremental rotation ${}^B R_\Delta$ as an exponential series and verify Eq. 3.7.
2. Derive the unit-quaternion update equation Eq. 3.8.
3. Make a simulation with a particle moving at constant velocity and a rotating reference frame. Plot the position of the particle in the inertial and the rotating reference frame and observe how the motion changes as a function of the inertial frame velocity.
4. Redo the quaternion-based angular velocity integration on page 79 using rotation matrices.
5. Derive the expression for fictitious forces in a rotating reference frame from Sect. 3.2.3.
6. At your location determine the magnitude and direction of centripetal acceleration you would experience. If you drove at 100 km h^{-1} due east what is the magnitude and direction of the Coriolis acceleration you would experience? What about at 100 km h^{-1} due north? The vertical component is called the Eötvös effect, how much lighter does it make you?
7. For a `tpoly` trajectory from 0 to 1 in 50 steps explore the effects of different initial and final velocities, both positive and negative. Under what circumstances does the quintic polynomial overshoot and why?
8. For a `lspb` trajectory from 0 to 1 in 50 steps explore the effects of specifying the velocity for the constant velocity segment. What are the minimum and maximum bounds possible?
9. For a trajectory from 0 to 1 and given a maximum possible velocity of 0.025 compare how many time steps are required for each of the `tpoly` and `lspb` trajectories?
10. Use `animate` to compare rotational interpolation using quaternions, Euler angles and roll-pitch-yaw angles. Hint: use the quaternion `interp` method and `mtraj`.
11. Repeat the example of Fig. 3.7 for the case where:
    a) the interpolation does *not* pass through a singularity. Hint – change the start or goal pitch angle. What happens?
    b) the final orientation is at a singularity. What happens?
12. Develop a method to quantitatively compare the performance of the different orientation interpolation methods. Hint: plot the locus followed by $\hat{z}$ on a unit sphere.
13. For the `mstraj` example (page 73)
    a) Repeat with different initial and final velocity.
    b) Investigate the effect of increasing the acceleration time. Plot total time as a function of acceleration time.
14. Modify `mstraj` so that acceleration limits are taken into account when determining the segment time.
15. There are a number of iOS and Android apps that display sensor data from gyros, accelerometers and magnetometers. You could also use MATLAB, see http://mathworks.com/hardware-support/iphone-sensor.html. Run one of these and explore how the sensor signals change with orientation and movement. What happens when you throw the phone into the air?

16. Consider a gyroscope with a 20 mm diameter steel rotor that is 4 mm thick and rotating at 10 000 rpm. What is the magnitude of h? For an angular velocity of 5 deg s^{-1}, what is the generated torque?

17. Using Eq. 3.15 can you explain how a toy gyroscope is able to balance on a single point with its spin axis horizontal? What holds it up?

18. A triaxial accelerometer has fallen off the table, ignoring air resistance what value does it return as it falls?

19. Implement the algorithm to determine roll and pitch angles from accelerometer measurements.

    a) Devise an algorithm to determine if you are in portrait or landscape orientation?

    b) Create a trajectory for the accelerometer using `tpoly` to generate motion in either the x- or y-direction. What effect does the acceleration along the path have on the estimated angle?

    c) Calculate the orientation using quaternions rather than roll-pitch-yaw angles.

20. You are in an aircraft flying at 30 000 feet over your current location. How much lighter are you?

21. Determine the Euler angles as a function of the measured acceleration. If you have the Symbolic Math Toolbox™ you might like to use that.

22. Determine the magnetic field strength, declination and inclination at your location. Visit the website http://www.ngdc.noaa.gov/geomag-web.

23. Using the sensor reading app from above, orient the phone so that the magnetic field vector has only a z-axis component, where is the magnetic field vector with respect to your phone?

24. Using the sensor reading app from above log some inertial sensor data from a phone while moving it around. Use that data to estimate the changing attitude or full pose of the phone. Can you do this in real time?

25. Experiment with varying the parameters of the explicit complementary filter on page 88. Change the bias or add Gaussian noise to the simulated sensor readings.

Part II Mobile Robots

Chapter 4 **Mobile Robot Vehicles**

Chapter 5 **Navigation**

Chapter 6 **Localization**

Mobile Robots

In this part we discuss mobile robots, a class of robots that are able to move through the environment. The figures show an assortment of mobile robots that can move over the ground, over the water, through the air, or through the water. This highlights the diversity of what is referred to as the *robotic platform* – the robot's physical embodiment and means of locomotion as shown in Figs. II.2 through II.4.

However these mobile robots are very similar in terms of what they do and how they do it. One of the most important functions of a mobile robot is to move to some place. That place might be specified in terms of some feature in the environment, for instance move to the light, or in terms of some geometric coordinate or map reference. In either case the robot will take some path to reach its destination and it faces challenges such as obstacles that might block its way or having an incomplete map, or no map at all.

One strategy is to have very simple sensing of the world and to react to what is sensed. For example *Elsie* the robotic tortoise, shown in Fig. II.1a, was built in the 1940s and *reacted* to her environment to seek out a light source without having any explicit plan or knowledge of the position of the light. An alternative to the reactive approach was embodied in the 1960s robot Shakey, shown in Fig. II.1b, which was capable of 3D perception and created a map of its environment and then reasoned about the map to plan a path to its destination.

These two approaches exemplify opposite ends of the spectrum for mobile robot navigation. Reactive systems can be fast and simple since sensation is connected directly to action – there is no need for resources to hold and maintain a representation of the world nor any capability to reason about that representation. In nature such

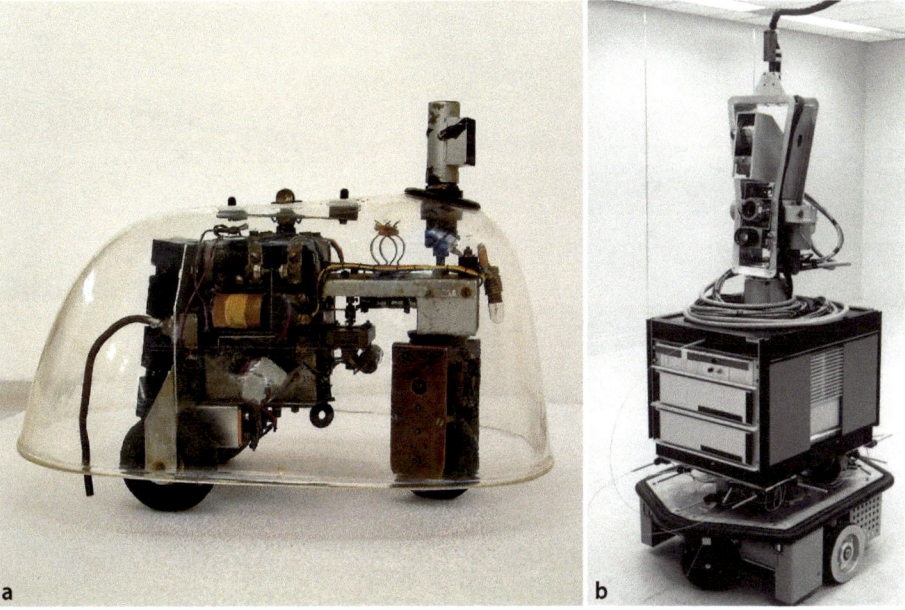

Fig. II.1.
a Elsie the tortoise. Burden Institute Bristol (1948). Now in the collection of the Smithsonian Institution but not on display (photo courtesy Reuben Hoggett collection). **b** Shakey. SRI International (1968). Now in the Computer Museum in Mountain View (photo courtesy SRI International)

strategies are used by simple organisms such as insects. Systems that make maps and reason about them require more resources but are capable of performing more complex tasks. In nature such strategies are used by more complex creatures such as mammals.

The first commercial applications of mobile robots came in the 1980s when automated guided vehicles (AGVs) were developed for transporting material around factories and these have since become a mature technology. Those early free-ranging mobile wheeled vehicles typically use fixed infrastructure for guidance, for example, a painted line on the floor, a buried cable that emits a radio-frequency signal, or wall-mounted bar codes. The last decade has seen significant achievements in mobile robotics that can operate without navigational infrastructure. Figure II.2a shows a robot vacuum cleaner which use reactive strategies to clean the floor, after the fashion of *Elsie*. Figure II.2b shows an early self-driving vehicle developed for the DARPA series of grand challenges for autonomous cars (Buehler et al. 2007, 2010). We see a multitude of sensors that provide the vehicle with awareness of its surroundings. Other examples are shown in Figs. 1.4 to 1.6. Mobile robots are not just limited to operations on the ground. Figure II.3 shows examples of unmanned aerial vehicles (UAVs), autonomous underwater vehicles (AUVs), and robotic boats which are known as autonomous surface vehicles (ASVs). Field robotic systems such as trucks in mines, container transport vehicles in shipping ports, and self-driving tractors for broad-acre agriculture are now commercially available for various applications are shown in Fig. II.4.

Fig. II.2.
Some mobile ground robots: **a** The Roomba robotic vacuum cleaner, 2008 (photo courtesy iRobot Corporation). **b** *Boss*, Tartan racing team's autonomous car that won the Darpa Urban Grand Challenge, 2007 (Carnegie-Mellon University)

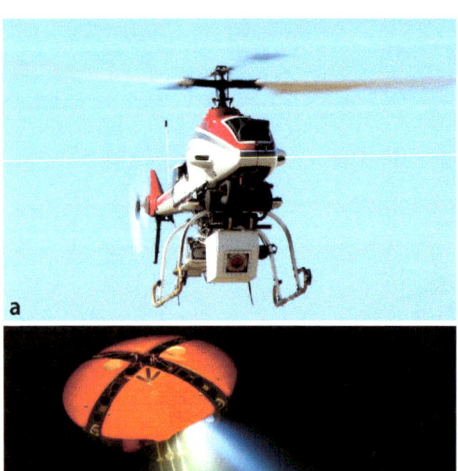 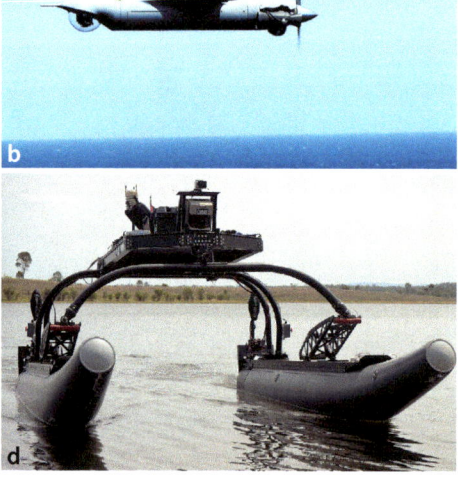

Fig. II.3.
Some mobile air and water robots: **a** Yamaha RMAX helicopter with 3 m blade diameter (photo by Sanjiv Singh). **b** Fixed-wing robotic aircraft (photo of ScanEagle courtesy of Insitu). **c** DEPTHX: Deep Phreatic Thermal Explorer, a 6-thruster under-water robot. Stone Aerospace/CMU (2007) (photo by David Wettergreen, © Carnegie-Mellon University). **d** Autonomous Surface Vehicle (photo by Matthew Dunbabin)

Fig. II.4.
a Exploration: Mars Science Laboratory (MSL) rover, known as Curiosity, undergoing testing (image courtesy NASA/Frankie Martin). **b** Logistics: an automated straddle carrier that moves containers; Port of Brisbane, 2006 (photo courtesy of Port of Brisbane Pty Ltd). **c** Mining: autonomous haul truck (Copyright © 2015 Rio Tinto). **d** Agriculture: broad-acre weeding robot (image courtesy Owen Bawden)

The chapters in this part of the book cover the fundamentals of mobile robotics. Chapter 4 discusses the motion and control of two exemplar robot platforms: wheeled vehicles that operate on a planar surface, and flying robots that move in 3-dimensional space – specifically quadrotor flying robots. Chapter 5 is concerned with navigation. We will cover in some detail the reactive and plan-based approaches to guiding a robot through an environment that contains obstacles. Most navigation strategies require knowledge of the robot's position and this is the topic of Chap. 6 which examines techniques such dead reckoning and the use of maps along with observations of landmarks. We also show how a robot can make a map, and even determine its location while simultaneously mapping an unknown region.

4

Mobile Robot Vehicles

This chapter discusses how a robot platform moves, that is, how its pose changes with time as a function of its control inputs. There are many different types of robot platform as shown on pages 93–95 but in this chapter we will consider only four important exemplars. Section 4.1 covers three different types of wheeled vehicle that operate in a 2-dimensional world. They can be propelled forwards or backwards and their heading direction controlled by some steering mechanism. Section 4.2 describes a quadrotor, a flying vehicle, which is an example of a robot that moves in 3-dimensional space. Quadrotors are becoming increasing popular as a robot platform since they are low cost and can be easily modeled and controlled.

Section 4.3 revisits the concept of configuration space and dives more deeply into important issues of under-actuation and nonholonomy.

4.1 Wheeled Mobile Robots

Wheeled locomotion is one of humanity's great innovations. The wheel was invented around 3000 BCE and the two-wheeled cart around 2000 BCE. Today four-wheeled vehicles are ubiquitous and the total automobile population of the planet is over one billion. The effectiveness of cars, and our familiarity with them, makes them a natural choice for robot platforms that move across the ground.

We know from our everyday experience with cars that there are limitations on how they move. It is not possible to drive sideways, but with some practice we can learn to follow a path that results in the vehicle being to one side of its initial position – this is parallel parking. Neither can a car rotate on the spot, but we can follow a path that results in the vehicle being at the same position but rotated by 180° – a three-point turn. The necessity to perform such maneuvers is the hall mark of a system that is nonholonomic – an important concept which is discussed further in Sect. 4.3. Despite these minor limitations the car is the simplest and most effective means of moving in a planar world that we have yet found. The car's motion model and the challenges it raises for control will be discussed in Sect. 4.1.1.

In Sect. 4.1.2 we will introduce differentially-steered vehicles which are mechanically simpler than cars and do not have steered wheels. This is a common configuration for small mobile robots and also for larger machines like bulldozers. Section 4.1.3 introduces novel types of wheels that *are* capable of omnidirectional motion and then models a vehicle based on these wheels.

4.1.1 Car-Like Mobile Robots

Cars with steerable wheels are a very effective class of vehicle and the archetype for most ground robots such as those shown in Fig. II.4a–c. In this section we will create a model for a car-like vehicle and develop controllers that can drive the car to a point, along a line, follow an arbitrary trajectory, and finally, drive to a specific pose.

© Springer Nature Switzerland AG 2022
P. Corke, *Robotics and Control*, Springer Tracts in Advanced Robotics 141,
https://doi.org/10.1007/978-3-030-79179-7_4

A commonly used model for the low-speed behavior of a four-wheeled car-like vehicle is the kinematic bicycle model▶ shown in Fig. 4.1. The bicycle has a rear wheel fixed to the body and the plane of the front wheel rotates about the vertical axis to steer the vehicle. We assume that the velocity of each wheel is in the plane of the wheel, and that the wheel rolls without slipping sideways

Often incorrectly called the Ackermann model.

$$^B\!v = (v, 0)$$

The pose of the vehicle is represented by its body coordinate frame $\{B\}$ shown in Fig. 4.1, with its x-axis in the vehicle's forward direction and its origin at the center of the rear axle. The *configuration* of the vehicle is represented by the generalized coordinates $q = (x, y, \theta) \in \mathcal{C}$ where $\mathcal{C} \subset \mathbb{R}^2 \times \mathbb{S}^1$.

The dashed lines show the direction along which the wheels cannot move, the lines of no motion, and these intersect at a point known as the Instantaneous Center of Rotation (ICR). The reference point of the vehicle thus follows a circular path and its angular velocity is

$$\dot{\theta} = \frac{v}{R_B} \qquad\qquad (4.1)$$

and by simple geometry the turning radius is $R_B = L / \tan\gamma$ where L is the length of the vehicle or *wheel base*. As we would expect the turning circle increases with vehicle length. The steering angle γ is typically limited mechanically and its maximum value dictates the minimum value of R_B.

> **Vehicle coordinate system.** The coordinate system that we will use, and a common one for vehicles of all sorts is that the x-axis is forward (longitudinal motion), the y-axis is to the left side (lateral motion) which implies that the z-axis is upward. For aerospace and underwater applications the z-axis is often downward and the x-axis is forward.

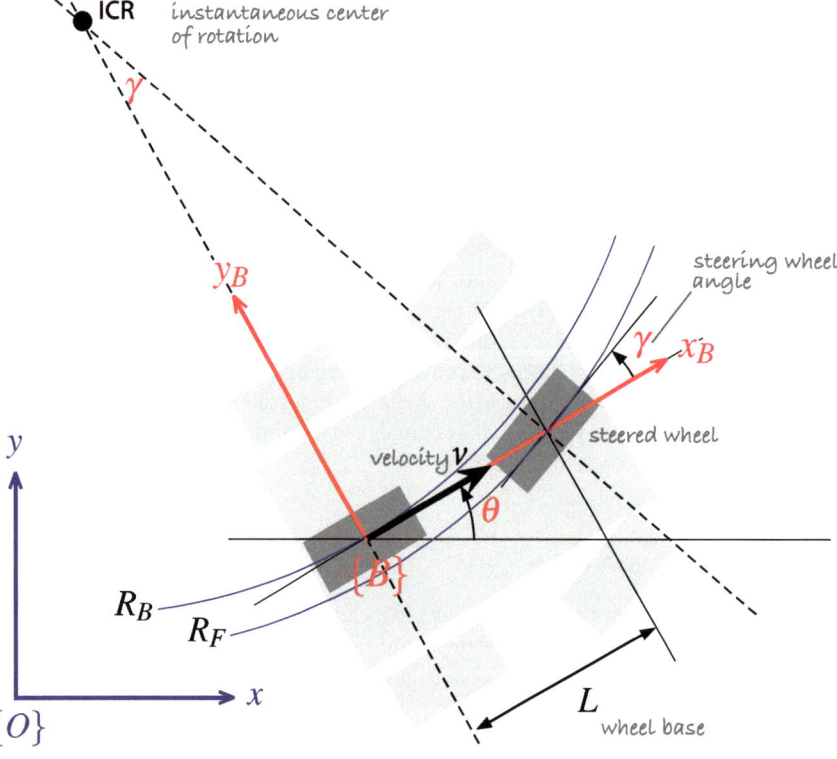

Fig. 4.1.
Bicycle model of a car. The car is shown in *light grey*, and the bicycle approximation is *dark grey*. The vehicle's body frame is shown in *red*, and the world coordinate frame in *blue*. The steering wheel angle is γ and the velocity of the back wheel, in the x-direction, is v. The two wheel axes are extended as dashed lines and intersect at the Instantaneous Center of Rotation (ICR) and the distance from the ICR to the back and front wheels is R_B and R_F respectively

Arcs with smoothly varying radius. Dubbins and Reeds-Shepp paths comprises constant radius circular arcs and straight line segments.

For a fixed steering wheel angle the car moves along a circular arc. For this reason curves on roads are circular arcs or clothoids◄ which makes life easier for the driver since constant or smoothly varying steering wheel angle allow the car to follow the road. Note that $R_F > R_B$ which means the front wheel must follow a longer path and therefore rotate more quickly than the back wheel. When a four-wheeled vehicle goes around a corner the two steered wheels follow circular paths of different radii and therefore the angles of the steered wheels γ_L and γ_R should be very slightly different. This is achieved by the commonly used Ackermann steering mechanism which results in lower wear and tear on the tyres. The driven wheels must rotate at different speeds on corners which is why a differential gearbox is required between the motor and the driven wheels.

The velocity of the robot in the world frame is $(v\cos\theta, v\sin\theta)$ and combined with Eq. 4.1 we write the equations of motion as

$$\dot{x} = v\cos\theta$$
$$\dot{y} = v\sin\theta$$
$$\dot{\theta} = \frac{v}{L}\tan\gamma$$

(4.2)

Fig. 124.

From Sharp 1896

This model is referred to as a kinematic model since it describes the velocities of the vehicle but not the forces or torques that cause the velocity. The rate of change of heading $\dot{\theta}$ is referred to as turn rate, heading rate or yaw rate and can be measured by a gyroscope. It can also be deduced from the angular velocity of the nondriven wheels on the left- and right-hand sides of the vehicle which follow arcs of different radius, and therefore rotate at different speeds.

Equation 4.2 captures some other important characteristics of a car-like vehicle. When $v = 0$ then $\dot{\theta} = 0$; that is, it is not possible to change the vehicle's orientation when it is not moving. As we know from driving, we must be moving in order to turn. When the steering angle $\gamma = \frac{\pi}{2}$ the front wheel is orthogonal to the back wheel, the vehicle cannot move forward and the model enters an undefined region.

In the world coordinate frame we can write an expression for velocity in the vehicle's y-direction

$$\dot{y}\cos\theta - \dot{x}\sin\theta \equiv 0$$

(4.3)

which is the called a nonholonomic constraint and will be discussed further in Sect. 4.3.1. This equation cannot be integrated to form a relationship between x, y and θ.

The Simulink® system

The model also includes a maximum velocity limit, a velocity rate limiter to model finite acceleration, and a limiter on the steering angle to model the finite range of the steered wheel. These can be accessed by double clicking the Bicycle block in Simulink.

```
>> sl_lanechange
```

shown in Fig. 4.2 uses the Toolbox `Bicycle` block which implements Eq. 4.2◄. The velocity input is a constant, and the steering wheel angle is a finite positive pulse followed by a negative pulse. Running the model simulates the motion of the vehicle and adds a new variable `out` to the workspace

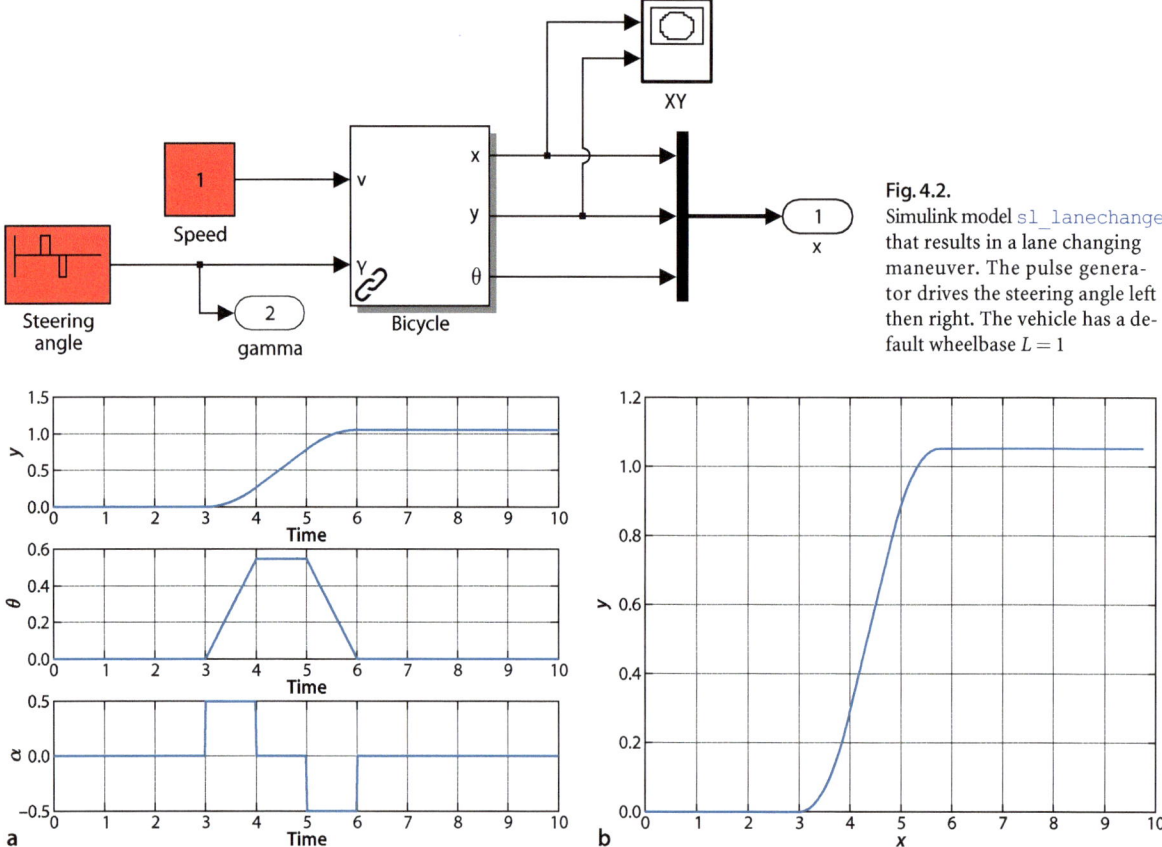

Fig. 4.2.
Simulink model `sl_lanechange` that results in a lane changing maneuver. The pulse generator drives the steering angle left then right. The vehicle has a default wheelbase $L = 1$

Fig. 4.3. Simple lane changing maneuver. **a** Vehicle response as a function of time, **b** motion in the xy-plane, the vehicle moves in the positive x-direction

```
>> out
Simulink.SimulationOutput:
    t: [504x1 double]
    y: [504x4 double]
```

from which we can retrieve the simulation time and other variables

```
>> t = out.get('t'); q = out.get('y');
```

Configuration is plotted against time

```
>> mplot(t, q)
```

in Fig. 4.3a and the result in the xy-plane

```
>> plot(q(:,1), q(:,2))
```

shown in Fig. 4.3b demonstrates a simple *lane-changing* trajectory.

4.1.1.1   Moving to a Point

Consider the problem of moving toward a goal point (x^*, y^*) in the plane. We will control the robot's velocity to be proportional to its distance from the goal

$$v^* = K_v \sqrt{\left(x^* - x\right)^2 + \left(y^* - y\right)^2}$$

and to steer toward the goal which is at the vehicle-relative angle▸ in the world frame of

$$\theta^* = \tan^{-1}\frac{y^* - y}{x^* - x}$$

This angle can be anywhere in the interval $[-\pi, \pi)$ and is computed using the `atan2` function.

using a proportional controller

$$\gamma = K_h\big(\theta^* \ominus \theta\big), \ \ K_h > 0$$

which turns the steering wheel toward the target. Note the use of the operator $\ominus$ since θ^* and θ are angles $\in \mathbb{S}^1$ not real numbers◄. A Simulink model

```
>> sl_drivepoint
```

The Toolbox function `angdiff` computes the difference between two angles and returns a difference in the interval $[-\pi, \pi)$. This is also the shortest distance around the circle, as discussed in Sect. 3.3.4.1. Also available in the Toolbox Simulink blockset `roblocks`.

is shown in Fig. 4.4. We specify a goal coordinate

```
>> xg = [5 5];
```

and an initial pose

```
>> x0 = [8 5 pi/2];
```

and then simulate the motion

```
>> r = sim('sl_drivepoint');
```

The variable `r` is an object that contains the simulation results from which we extract the configuration as a function of time

```
>> q = r.find('y');
```

The vehicle's path in the plane is

```
>> plot(q(:,1), q(:,2));
```

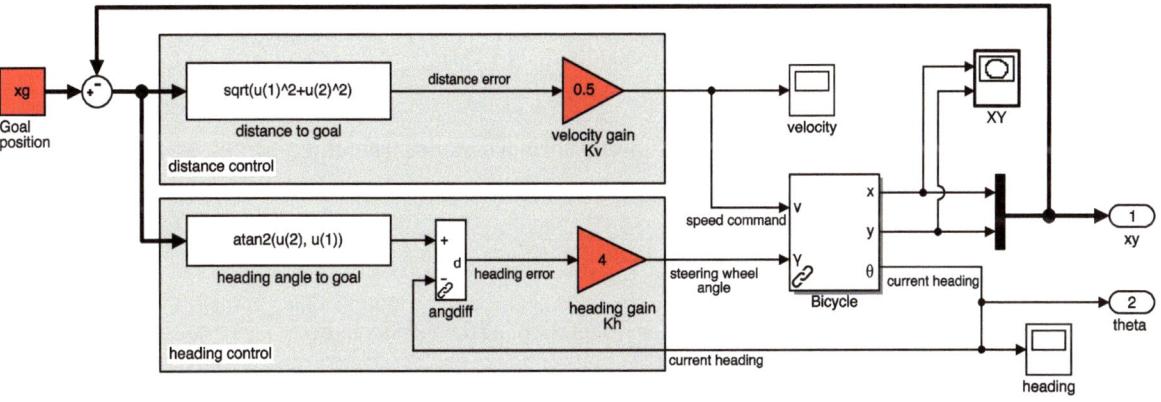

Fig. 4.4. `sl_drivepoint`, the Simulink model that drives the vehicle to a point. Red blocks have parameters that you can adjust to investigate the effect on performance

To run the Simulink model called `model` we first load it

```
>> model
```

and a new window is popped up that displays the model in block-diagram form. The simulation can be started by pressing the play button on the toolbar of the model's window. The model can also be run directly from the MATLAB command line

```
>> sim('model')
```

Many Toolbox models create additional figures to display robot animations or graphs as they run.

All models in this chapter have the simulation data export option set to create a MATLAB `SimulationOutput` object. All the unconnected output signals are concatenated, in port number order, to form a row vector and these are stacked to form a matrix `y` with one row per timestep. The corresponding time values form a vector `t`. These variables are packaged in a `SimulationOutput` object which is written to the workspace variable `out` or returned if the simulation is invoked from MATLAB

```
>> r = sim('model')
```

Displaying `r` or `out` lists the variables that it contains and their value is obtained using the `find` method, for example

```
>> t = r.find('t');
```

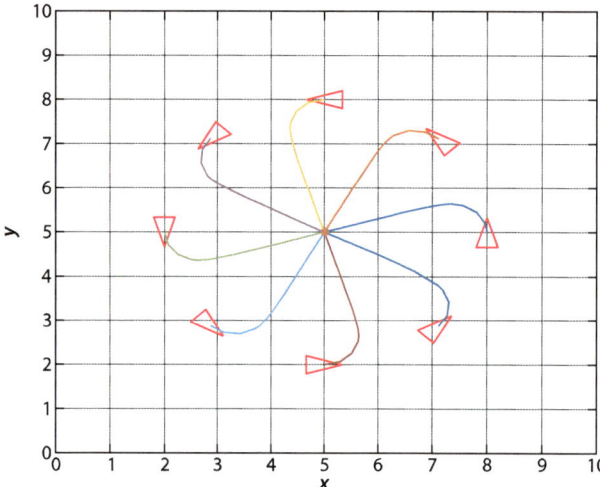

Fig. 4.5.
Simulation results for
`sl_drivepoint` for different
initial poses. The goal is (5, 5)

which is shown in Fig. 4.5 for a number of starting poses. In each case the vehicle has moved forward and turned onto a path toward the goal point. The final part of each path is a straight line and the final orientation therefore depends on the starting point.

4.1.1.2   Following a Line

Another useful task for a mobile robot is to follow a line on the plane▸ defined by $ax + by + c = 0$. This requires two controllers to adjust steering. One controller

2-dimensional lines in homogeneous form are discussed in Sect. C.2.1.

$$\alpha_d = -K_d d, \; K_d > 0$$

turns the robot toward the line to minimize the robot's normal distance from the line

$$d = \frac{(a, b, c) \cdot (x, y, 1)}{\sqrt{a^2 + b^2}}$$

The second controller adjusts the heading angle, or orientation, of the vehicle to be parallel to the line

$$\theta^* = \tan^{-1} \frac{-a}{b}$$

using the proportional controller

$$\alpha_h = K_h \left(\theta^* \ominus \theta \right), \; K_h > 0$$

The combined control law

$$\gamma = -K_d d + K_h \left(\theta^* \ominus \theta \right)$$

turns the steering wheel so as to drive the robot toward the line and move along it.
The Simulink model

```
>> sl_driveline
```

is shown in Fig. 4.6. We specify the target line as a 3-vector (a, b, c)

```
>> L = [1 -2 4];
```

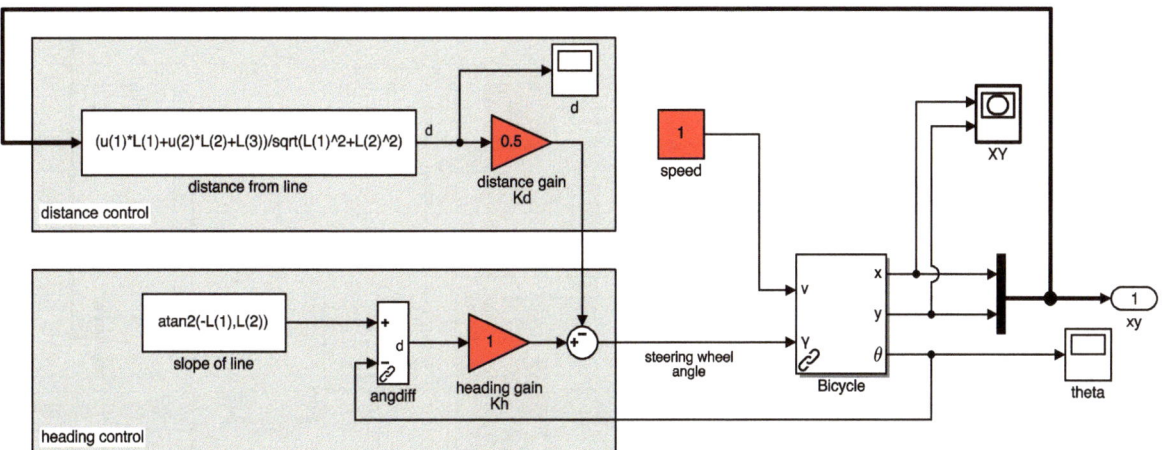

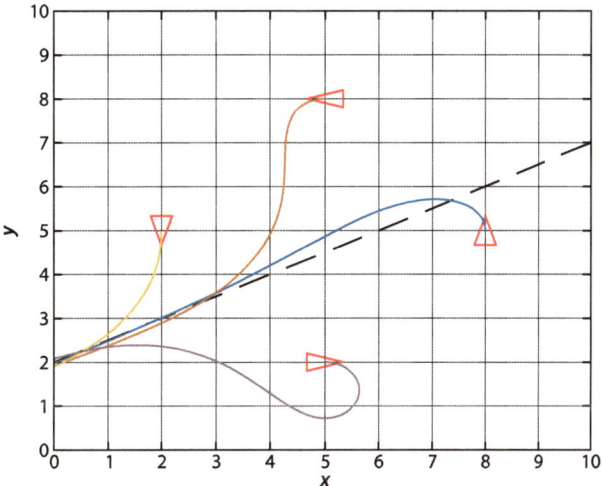

Fig. 4.6. The Simulink model `sl_driveline` drives the vehicle along a line. The line parameters (a, b, c) are set in the workspace variable `L`. Red blocks have parameters that you can adjust to investigate the effect on performance

Fig. 4.7. Simulation results from different initial poses for the line $(1, -2, 4)$

and an initial pose

```
>> x0 = [8 5 pi/2];
```

and then simulate the motion

```
>> r = sim('sl_driveline');
```

The vehicle's path for a number of different starting poses is shown in Fig. 4.7.

4.1.1.3 Following a Trajectory

Instead of a straight line we might wish to follow a trajectory that is a timed sequence of points on the xy-plane. This might come from a motion planner, such as discussed in Sect. 3.3 or 5.2, or in real-time based on the robot's sensors.

A simple and effective algorithm for trajectory following is pure pursuit in which the goal point $(x^*\langle t\rangle, y^*\langle t\rangle)$ moves along the trajectory, in its simplest form at constant speed. The vehicle always heads toward the goal – think carrot and donkey.

This problem is very similar to the control problem we tackled in Sect. 4.1.1.1, moving to a point, except this time the point is moving. The robot maintains a distance d^* behind the pursuit point and we formulate an error

$$e = \sqrt{\left(x^* - x\right)^2 + \left(y^* - y\right)^2} - d^*$$

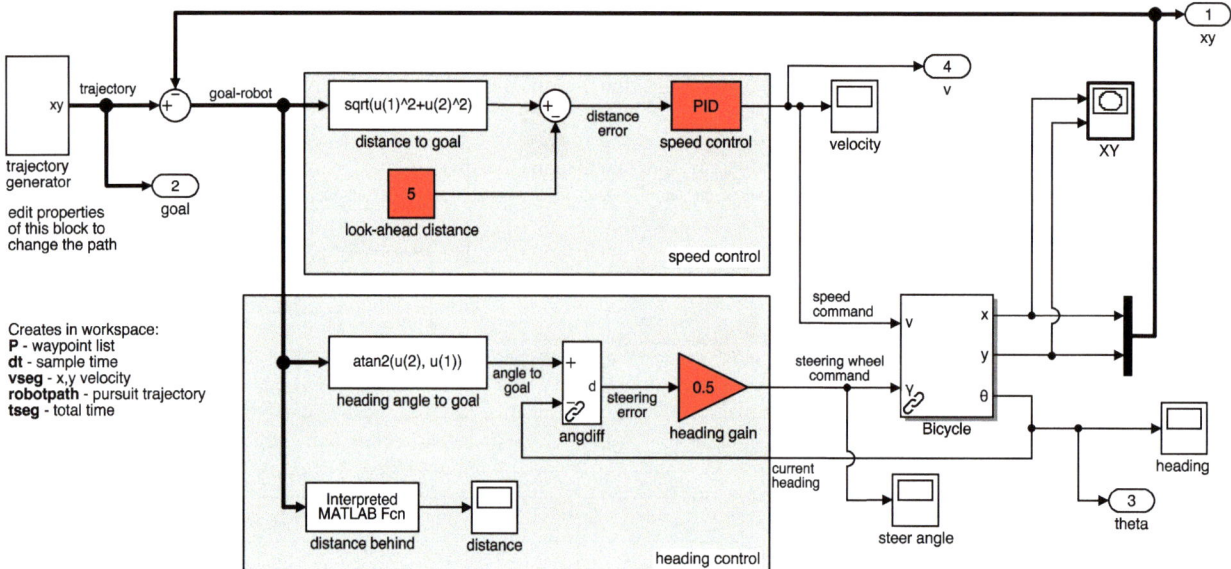

that we regulate to zero by controlling the robot's velocity using a proportional-integral (PI) controller

$$v^* = K_v e + K_i \int e\, \mathrm{d}t$$

The integral term is required to provide a nonzero velocity demand v^* when the following error is zero. The second controller steers the robot toward the target which is at the relative angle

$$\theta^* = \tan^{-1} \frac{y^* - y}{x^* - x}$$

and a simple proportional controller

$$\gamma = K_h\!\left(\theta^* \ominus \theta\right), \quad K_h > 0$$

turns the steering wheel so as to drive the robot toward the target.
   The Simulink model

```
>> sl_pursuit
```

shown in Fig. 4.8 includes a target that moves at constant velocity along a piecewise linear path defined by a number of waypoints. It can be simulated

```
>> r = sim('sl_pursuit')
```

and the results are shown in Fig. 4.9a. The robot starts at the origin but catches up to, and follows, the moving goal. Figure 4.9b shows how the speed converges on a steady state value when following at the desired distance. Note the slow down at the end of each segment as the robot *short cuts* across the corner.

Fig. 4.8. The Simulink model `sl_pursuit` drives the vehicle along a piecewise linear trajectory. Red blocks have parameters that you can adjust to investigate the effect on performance

4.1.1.4 Moving to a Pose

The final control problem we discuss is driving to a specific pose (x^*, y^*, θ^*). The controller of Fig. 4.4 could drive the robot to a goal position but the final orientation depended on the starting position.

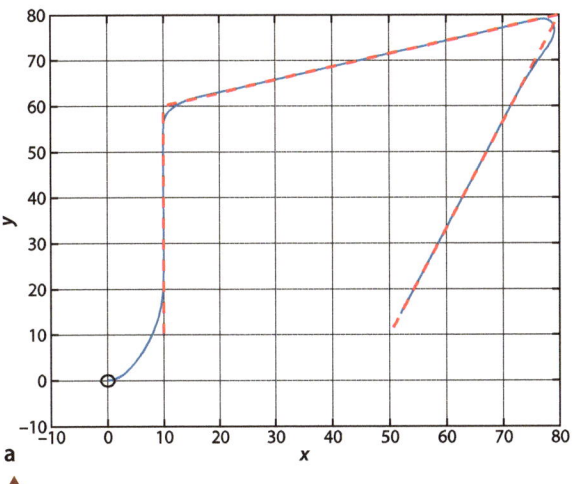

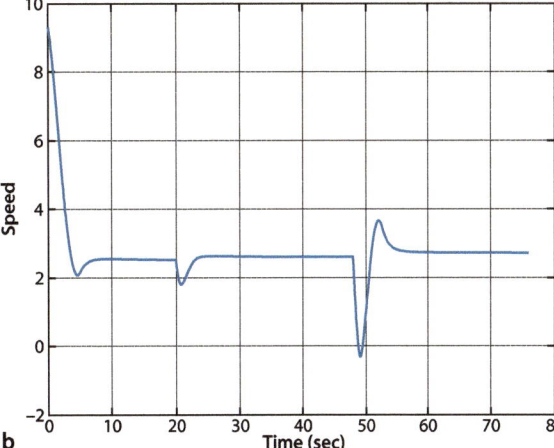

a   b

Fig. 4.9. Simulation results from pure pursuit. **a** Path of the robot in the *xy*-plane. The red dashed line is the path to be followed and the blue line in the path followed by the robot, which starts at the origin. **b** The speed of the robot versus time

Fig. 4.10.
Polar coordinate notation for the bicycle model vehicle moving toward a goal pose: ρ is the distance to the goal, β is the angle of the goal vector with respect to the world frame, and α is the angle of the goal vector with respect to the vehicle frame

In order to control the final orientation we first rewrite Eq. 4.2 in matrix form

$$\begin{pmatrix} \dot{x} \\ \dot{y} \\ \dot{\theta} \end{pmatrix} = \begin{pmatrix} \cos\theta & 0 \\ \sin\theta & 0 \\ 0 & 1 \end{pmatrix} \begin{pmatrix} v \\ \omega \end{pmatrix}$$

where the inputs to the vehicle model are the speed v and the turning rate ω which can be achieved by applying the steering angle◄

We have effectively converted the Bicycle kinematic model to a Unicycle model which we discuss in Sect. 4.1.2.

$$\gamma = \tan^{-1}\frac{\omega L}{v}$$

We then transform the equations into polar coordinate form using the notation shown in Fig. 4.10 and apply a change of variables

$$\rho = \sqrt{\Delta_x^2 + \Delta_y^2}$$

$$\alpha = \tan^{-1}\frac{\Delta_y}{\Delta_x} - \theta$$

$$\beta = -\theta - \alpha$$

which results in

$$
\begin{pmatrix} \dot{\rho} \\ \dot{\alpha} \\ \dot{\beta} \end{pmatrix} = \begin{pmatrix} -\cos\alpha & 0 \\ \dfrac{\sin\alpha}{\rho} & -1 \\ -\dfrac{\sin\alpha}{\rho} & 0 \end{pmatrix} \begin{pmatrix} v \\ \omega \end{pmatrix}, \ \text{if } \alpha \in \left(-\dfrac{\pi}{2}, \dfrac{\pi}{2} \right]
$$

and assumes the goal frame $\{G\}$ is in front of the vehicle. The linear control law

$$
v = k_\rho \rho
$$
$$
\omega = k_\alpha \alpha + k_\beta \beta
$$

drives the robot to a unique equilibrium▸ at $(\rho, \alpha, \beta) = (0, 0, 0)$. The intuition behind this controller is that the terms $k_\rho \rho$ and $k_\alpha \alpha$ drive the robot along a line toward $\{G\}$ while the term $k_\beta \beta$ rotates the line so that $\beta \to 0$. The closed-loop system

The control law introduces a discontinuity at $\rho = 0$ which satisfies Brockett's theorem.

$$
\begin{pmatrix} \dot{\rho} \\ \dot{\alpha} \\ \dot{\beta} \end{pmatrix} = \begin{pmatrix} -k_\rho \rho \cos\alpha \\ k_\rho \sin\alpha - k_\alpha \alpha - k_\beta \beta \\ -k_\rho \sin\alpha \end{pmatrix}
$$

is stable so long as

$$
k_\rho > 0, \ k_\beta < 0, \ k_\alpha - k_\rho > 0
$$

The distance and bearing to the goal (ρ, α) could be measured by a camera or laser range finder, and the angle β could be derived from α and vehicle orientation θ as measured by a compass.

For the case where the goal is behind the robot, that is $\alpha \notin (-\frac{\pi}{2}, \frac{\pi}{2}]$, we reverse the vehicle by negating v and γ in the control law. The velocity v always has a constant sign which depends on the initial value of α.

So far we have described a *regulator* that drives the vehicle to the pose $(0, 0, 0)$. To move the robot to an arbitrary pose (x^*, y^*, θ^*) we perform a change of coordinates

$$
x' = x - x^*, \ y' = y - y^*, \ \theta' = \theta, \ \beta' = \beta + \theta^*
$$

This pose controller is implemented by the Simulink model

```
>> sl_drivepose
```

shown in Fig. 4.11 and the transformation from Bicycle to Unicycle kinematics is clearly shown, mapping angular velocity ω to steering wheel angle γ. We specify a goal pose

Fig. 4.11. The Simulink model `sl_drivepose` drives the vehicle to a pose. The initial and final poses are set by the workspace variable `x0` and `xf` respectively. Red blocks have parameters that you can adjust to investigate the effect on performance

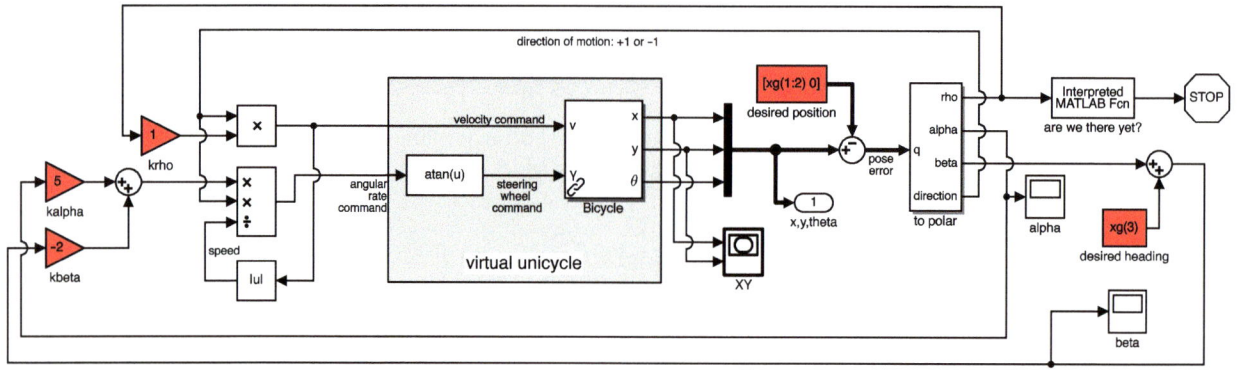

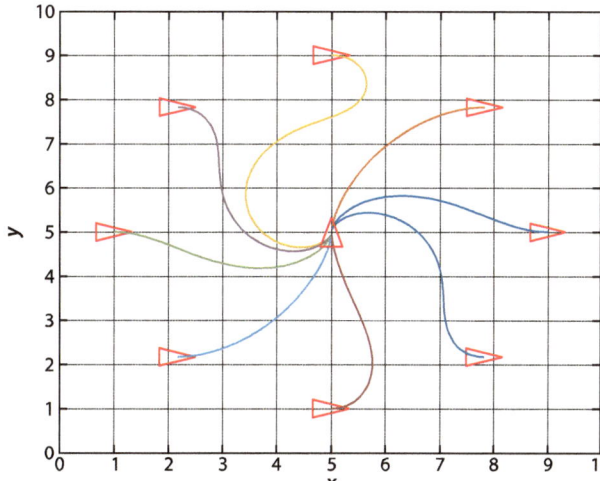

Fig. 4.12.
Simulation results from different initial poses to the final pose $(5, 5, \frac{\pi}{2})$. Note that in some cases the robot has *backed into* the final pose

```
>> xg = [5 5 pi/2];
```

and an initial pose

```
>> x0 = [9 5 0];
```

and then simulate the motion

```
>> r = sim('sl_drivepose');
```

As before, the simulation results are stored in `r` and can be plotted

```
>> q = r.find('y');
>> plot(q(:,1), q(:,2));
```

The controller is based on the bicycle model but the Simulink model `Bicycle` has additional hard nonlinearities including steering angle limits and velocity rate limiting. If those limits are violated the pose controller may fail.

to show the vehicle's path in the plane. The vehicle's path for a number of starting poses is shown in Fig. 4.12. The vehicle moves forwards or backward and takes a smooth path to the goal pose. ◄

4.1.2 Differentially-Steered Vehicle

Having steerable wheels as in a car-like vehicle is mechanically complex. Differential steering does away with this and steers by independently controlling the speed of the wheels on each side of the vehicle – if the speeds are not equal the vehicle will turn. Very simple differential steer robots have two driven wheels and a front and back castor to provide stability. Larger differential steer vehicles such as the one shown in Fig. 4.13 employ a pair of wheels on each side, with each pair sharing a drive motor via some mechanical transmission. Very large differential steer vehicles such as bulldozers and tanks sometimes employ caterpillar tracks instead of wheels. The vehicle's velocity is by definition v in the vehicle's x-direction, and zero in the y-direction since the wheels cannot slip sideways. In the vehicle frame $\{B\}$ this is

$$^B\boldsymbol{v} = (v, 0)$$

The pose of the vehicle is represented by the body coordinate frame $\{B\}$ shown in Fig. 4.14, with its x-axis in the vehicle's forward direction and its origin at the centroid of the four wheels. The configuration of the vehicle is represented by the generalized coordinates $\boldsymbol{q} = (x, y, \theta) \in \mathcal{C}$ where $\mathcal{C} \subset \mathbb{R}^2 \times \mathbb{S}^1$.

The vehicle follows a curved path centered on the Instantaneous Center of Rotation (ICR). The left-hand wheels move at a speed of v_L along an arc with a radius of R_L

Fig. 4.13.
Clearpath Husky robot with differential drive steering (photo by Tim Barfoot)

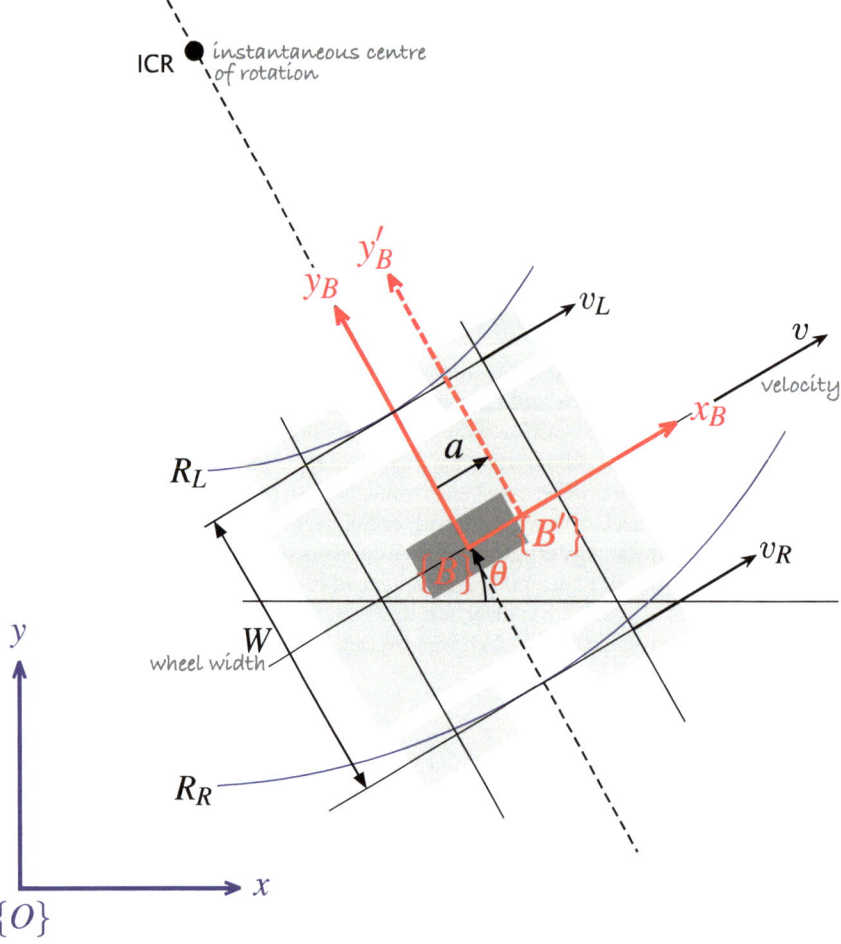

Fig. 4.14.
Differential drive robot is shown in *light grey*, and the unicycle approximation is *dark grey*. The vehicle's body coordinate frame is shown in *red*, and the world coordinate frame in *blue*. The vehicle follows a path around the Instantaneous Center of Rotation (ICR) and the distance from the ICR to the left and right wheels is R_L and R_R respectively. We can use the alternative body frame $\{B'\}$ for trajectory tracking control

while the right-hand wheels move at a speed of v_R along an arc with a radius of R_R. The angular velocity of $\{B\}$ is

$$\dot{\theta} = \frac{v_L}{R_L} = \frac{v_R}{R_R}$$

and since $R_R = R_L + W$ we can write the turn rate

$$\dot{\theta} = \frac{v_R - v_L}{W} \tag{4.4}$$

in terms of the differential velocity and wheel separation W. The equations of motion are therefore

$$\dot{x} = v\cos\theta$$
$$\dot{y} = v\sin\theta$$
$$\dot{\theta} = \frac{v_\Delta}{W} \tag{4.5}$$

where $v = \frac{1}{2}(v_R + v_L)$ and $v_\Delta = v_R - v_L$ are the average and differential velocities respectively. For a desired speed v and turn rate $\dot{\theta}$ we can solve for v_R and v_L. This kinematic model is often called the *unicycle model*.

There are similarities and differences to the bicycle model of Eq. 4.2. The turn rate for this vehicle is directly proportional to v_Δ and is independent of speed – the vehicle can turn even when not moving forward. For the 4-wheel case shown in Fig. 4.14 the axes of the wheels do not intersect the ICR, so when the vehicle is turning the wheel velocity vectors v_L and v_R are not tangential to the path – there is a component in the lateral direction which violates the no-slip constraint. This causes skidding or scuffing◀ which is extreme when the vehicle is turning on the spot – hence differential steering is also called skid steering. Similar to the car-like vehicle we can write an expression for velocity in the vehicle's y-direction expressed in the world coordinate frame

$$\dot{y}\cos\theta - \dot{x}\sin\theta \equiv 0 \tag{4.6}$$

which is the nonholonomic constraint. It is important to note that the ability to turn on the spot does not make the vehicle holonomic and is fundamentally different to the ability to move in an arbitrary direction which we will discuss next.

If we move the vehicle's reference frame to $\{B'\}$ and ignore orientation we can rewrite Eq. 4.5 in matrix form as

$$\begin{pmatrix} \dot{x} \\ \dot{y} \end{pmatrix} = \begin{pmatrix} \cos\theta & -a\sin\theta \\ \sin\theta & a\cos\theta \end{pmatrix}\begin{pmatrix} v \\ \omega \end{pmatrix}$$

and if $a \neq 0$ this can be be inverted

$$\begin{pmatrix} v \\ \omega \end{pmatrix} = \begin{pmatrix} \cos\theta & \sin\theta \\ -\frac{1}{a}\sin\theta & \frac{1}{a}\cos\theta \end{pmatrix}\begin{pmatrix} \dot{x} \\ \dot{y} \end{pmatrix} \tag{4.7}$$

to give the required forward speed and turn rate to achieve an arbitrary velocity $(\dot{x}, \dot{y})$ for the origin of frame $\{B'\}$.

The Toolbox Simulink block library `roblocks` contains a block called `Unicycle` to implement this model and the coordinate frame shift a is one of its parameters. It has the same outputs as the `Bicycle` model used in the last section. Equation 4.7 is implemented in the block called `Tracking Controller`.

Fɪɢ 171.

From Sharp 1896

For indoor applications this can destroy carpet.

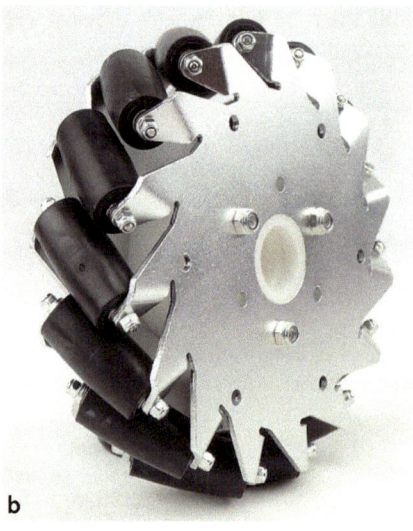

a   **b**

Fig. 4.15.
Two types of omnidirectional wheel, note the different roller orientation. **a** Allows the wheel to *roll* sideways (courtesy Vex Robotics); **b** allows the wheel to *drive* sideways (courtesy of Nexus Robotics)

4.1.3   Omnidirectional Vehicle

The vehicles we have discussed so far have a constraint on lateral motion, the non-holonomic constraint, which necessitates complex maneuvers in order to achieve some goal poses. Alternative wheel designs such as shown in Fig. 4.15 remove this constraint and allow omnidirectional motion. Even more radical is the spherical wheel shown in Fig. 4.16.

In this section we will discuss the mecanum or "Swedish" wheel▶ shown in Fig. 4.15b and schematically in Fig. 4.17. It comprises a number of rollers set around the circumference of the wheel with their axes at an angle of α relative to the axle of the wheel. The dark roller is the one on the bottom of the wheel and currently in contact with the ground. The rollers have a barrel shape so only one point on the roller is in contact with the ground at any time.

As shown in Fig. 4.17 we establish a wheel coordinate frame $\{W\}$ with its x-axis pointing in the direction of wheel motion. Rotation of the wheel will cause forward velocity of $R\varpi\hat{\boldsymbol{x}}_w$ where R is the wheel radius and ϖ is the wheel rotational rate. However because the roller is free to roll in the direction indicated by the green line, normal to the roller's axis, there is potentially arbitrary velocity in that direction. A desired velocity $\boldsymbol{v}$ can be resolved into two components, one parallel to the direction of wheel motion $\hat{\boldsymbol{x}}_w$ and one parallel to the rolling direction

▶ Mecanum was a Swedish company where the wheel was invented by Bengt Ilon in 1973. It is described in US patent 3876255.

$$
\begin{aligned}
\boldsymbol{v} &= \underbrace{v_w\hat{\boldsymbol{x}}_w}_{\text{driven}} + \underbrace{v_r\big(\cos\alpha\hat{\boldsymbol{x}}_w + \sin\alpha\hat{\boldsymbol{y}}_w\big)}_{\text{rolling}} \\
&= \big(v_w + v_r\cos\alpha\big)\hat{\boldsymbol{x}}_w + v_r\sin\alpha\hat{\boldsymbol{y}}_w
\end{aligned} \tag{4.8}
$$

where v_w is the speed due to wheel rotation and v_r is the rolling speed. Expressing $\boldsymbol{v} = v_x\hat{\boldsymbol{x}}_w + v_y\hat{\boldsymbol{y}}_w$ in component form allows us to solve for the rolling speed $v_r = \boldsymbol{v}_y / \sin\alpha$ and substituting this into the first term we can solve for the required wheel velocity

$$
v_w = \boldsymbol{v}_x - \boldsymbol{v}_y\cot\alpha \tag{4.9}
$$

The required wheel rotation rate is then $\varpi = v_w / R$. If $\alpha = 0$ then v_w is undefined since the roller axes are parallel to the wheel axis and the wheel can provide no traction. If $\alpha = \frac{\pi}{2}$ as in Fig. 4.15a, the wheel allows sideways rolling but not sideways driving since there is zero coupling from v_w to $\boldsymbol{v}_y$.

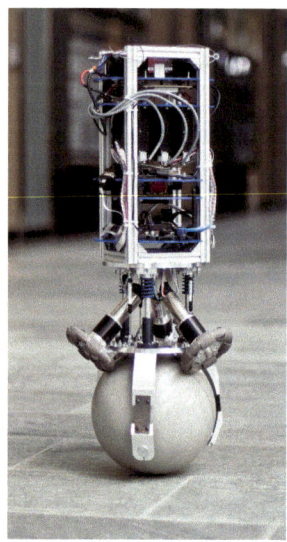

Fig. 4.16. The Rezero ballbot developed at ETH Zurich (photo by Péter Fankhauser)

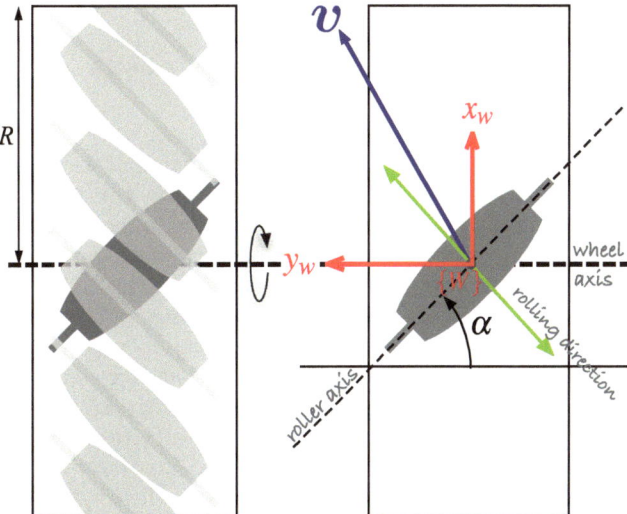

Fig. 4.17.
Schematic of a mecanum wheel in plan view. The *light rollers* are on top of the wheel, the *dark roller* is in contact with the ground. The *green arrow* indicates the rolling direction

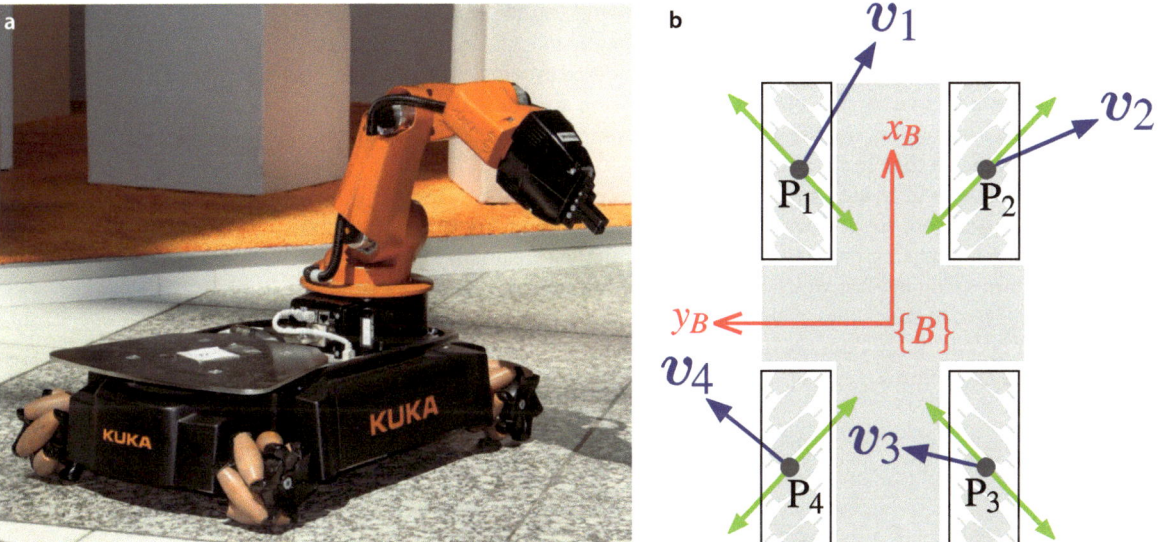

Fig. 4.18. a Kuka youBot, which has four mecanum wheels (image courtesy youBot Store); **b** schematic of a vehicle with four mecanum wheels in the youBot configuration

A single mecanum wheel does not allow any control in the rolling direction but for three or more mecanum wheels, suitably arranged, the motion in the rolling direction of any one wheel will be driven by the other wheels. A vehicle with four mecanum wheels is shown in Fig. 4.18. Its pose is represented by the body frame {B} with its x-axis in the vehicle's forward direction and its origin at the centroid of the four wheels. The configuration of the vehicle is represented by the generalized coordinates $q = (x, y, \theta) \in \mathcal{C}$ where $\mathcal{C} \subset \mathbb{R}^2 \times \mathbb{S}^1$. The rolling axes of the wheels are orthogonal which means that when the wheels are not rotating the vehicle cannot roll in any direction or rotate.

The four wheel contact points indicated by *grey dots* have coordinate vectors ${}^B\!p_i$. For a desired body velocity ${}^B\!v_B$ and angular rate ${}^B\!\omega$ the velocity at each wheel contact point is

$$ {}^B\!v_i = {}^B\!v_B + {}^B\!\omega \hat{z}_B \times {}^B\!p_i $$

and we then apply Eq. 4.8 and 4.9 to determine wheel rotational rates ϖ_i, while noting that α has the opposite sign for wheels 2 and 4 in Eq. 4.8.

4.2   Flying Robots

In order to fly, all one must do is simply miss the ground.
Douglas Adams

Flying robots or unmanned aerial vehicles (UAV) are becoming increasingly common and span a huge range of size and shape as shown in shown in Fig. 4.19. Applications include military operations, surveillance, meteorological observation, robotics research, commercial photography and increasingly hobbyist and personal use. A growing class of flying machines are known as micro air vehicles or MAVs which are smaller than 15 cm in all dimensions. Fixed wing UAVs are similar in principle to passenger aircraft with wings to provide lift, a propeller or jet to provide forward thrust and control surface for maneuvering. Rotorcraft UAVs have a variety of configurations that include conventional *helicopter* design with a main and tail rotor, a *coax* with counter-rotating coaxial rotors and *quadrotors*. Rotorcraft UAVs have the advantage of being able to take off vertically and to hover.

Flying robots differ from ground robots in some important ways. Firstly they have 6 degrees of freedom and their configuration $q \in \mathcal{C}$ where $\mathcal{C} \subset \mathbb{R}^3 \times \mathbb{S}^1 \times \mathbb{S}^1 \times \mathbb{S}^1$. Secondly they are actuated by forces; that is their motion model is expressed in terms of forces, torques and accelerations rather than velocities as was the case for the ground vehicle models – we use a dynamic rather than a kinematic model. Underwater robots have many similarities to flying robots and can be considered as vehicles that *fly through water* and there are underwater equivalents to fixed wing aircraft and rotorcraft. The principal differences underwater are an upward buoyancy force, drag forces that are much more significant than in air, and added mass.

In this section we will create a model for a quadrotor flying vehicle such as shown in Fig. 4.19d. Quadrotors are now widely available, both as commercial products and as open-source projects. Compared to fixed wing aircraft they are highly maneuverable and can be flown safely indoors which makes them well suited for laboratory or hobbyist use. Compared to conventional helicopters, with a large main rotor and tail rotor, the quadrotor is easier to fly, does not have the complex swash plate mechanism and is easier to model and control.

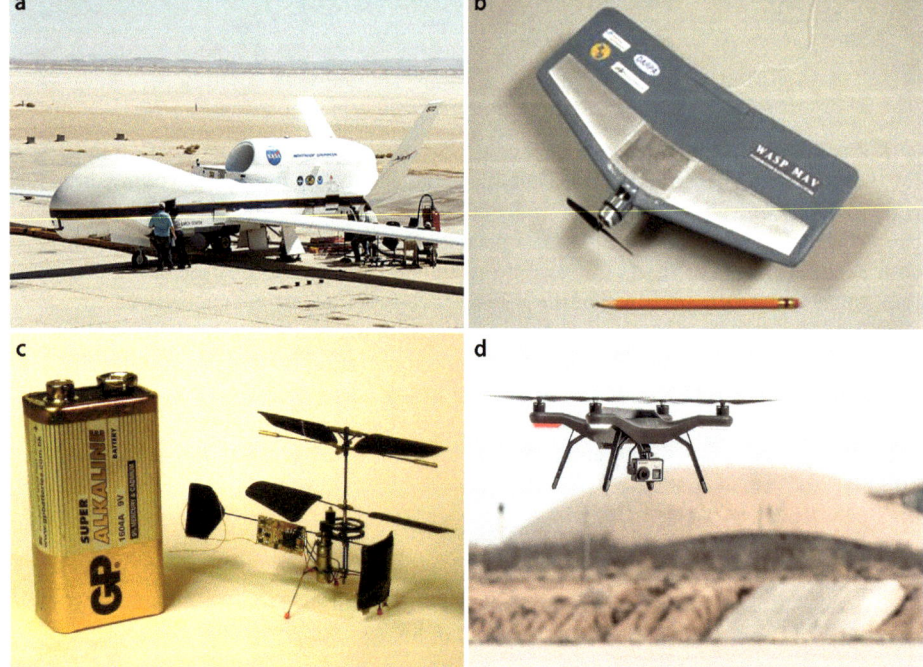

Fig. 4.19.
Flying robots. **a** Global Hawk unmanned aerial vehicle (UAV) (photo courtesy of NASA), **b** a micro air vehicle (MAV) (photo courtesy of AeroVironment, Inc.), **c** a 1 gram co-axial helicopter with 70 mm rotor diameter (photo courtesy of Petter Muren and Proxflyer AS), **d** a quadrotor which has four rotors and a block of sensing and control electronics in the middle (photo courtesy of 3DRobotics)

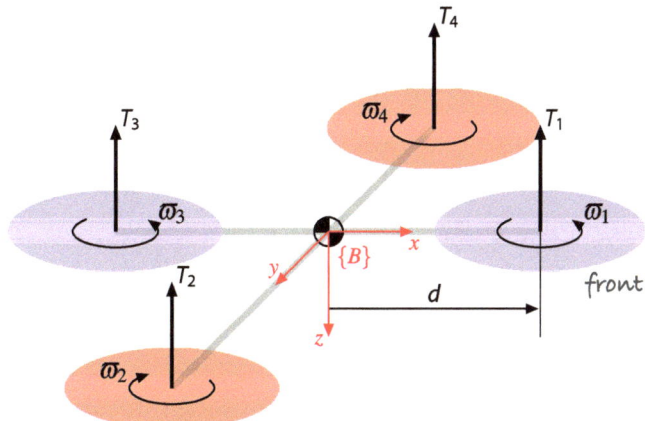

Fig. 4.20.
Quadrotor notation showing the
four rotors, their thrust vectors
and directions of rotation. The
body frame {*B*} is attached to the
vehicle and has its origin at the
vehicle's center of mass. Rotors
1 and 3 (*blue*) rotate counter-
clockwise (viewed from above)
while rotors 2 and 4 (*red*) rotate
clockwise

The notation for the quadrotor model is shown in Fig. 4.20. The body coordinate frame {*B*} has its *z*-axis downward following the aerospace convention. The quadrotor has four rotors, labeled 1 to 4, mounted at the end of each cross arm. Hex- and octo-rotors are also popular, with the extra rotors providing greater payload lift capability. The approach described here can be generalized to N rotors, where N is even.

The rotors are driven by electric motors powered by electronic speed controllers. Some low-cost quadrotors use small motors and reduction gearing to achieve sufficient torque. The rotor speed is ϖ_i and the thrust is an upward vector

$$T_i = b\varpi_i^2, \quad i = 1, 2, 3, 4 \tag{4.10}$$

in the vehicle's negative *z*-direction, where $b > 0$ is the lift constant that depends on the air density, the cube of the rotor blade radius, the number of blades, and the chord length of the blade. ◀

Close to the ground, height <2*d*, the vehicle experiences increased lift due to a cushion of air beneath it – this is ground effect.

The translational dynamics of the vehicle in world coordinates is given by Newton's second law

$$m\dot{\boldsymbol{v}} = \begin{pmatrix} 0 \\ 0 \\ mg \end{pmatrix} - {}^0\!\boldsymbol{R}_B \begin{pmatrix} 0 \\ 0 \\ T \end{pmatrix} - B\boldsymbol{v} \tag{4.11}$$

where $\boldsymbol{v}$ is the velocity of the vehicle's center of mass in the world frame, g is gravitational acceleration, m is the total mass of the vehicle, B is aerodynamic friction and $T = \Sigma T_i$ is the total upward thrust. The first term is the force of gravity which acts downward in the world frame, the second term is the total thrust in the vehicle frame rotated into the world coordinate frame and the third term is aerodynamic drag.

Pairwise differences in rotor thrusts cause the vehicle to rotate. The torque about the vehicle's *x*-axis, the *rolling* torque, is generated by the moments

$$\tau_x = dT_4 - dT_2$$

> The propeller blades on a rotor craft have fascinating dynamics. When flying into the wind the blade tip coming forward experiences greater lift while the receding blade has less lift. This is equivalent to a torque about an axis pointing into the wind and the rotor blades behave like a gyroscope (see Sect. 3.4.1.1) so the net effect is that the rotor blade plane pitches up by an amount proportional to the apparent or nett wind speed, countered by the blade's bending stiffness and the change in lift as a function of blade bending. The pitched blade plane causes a component of the thrust vector to retard the vehicle's forward motion and this velocity dependent force acts like a friction force. This is known as blade flapping and is an important characteristic of blades on all types of rotorcraft.

where d is the distance from the rotor axis to the center of mass. We can write this in terms of rotor speeds by substituting Eq. 4.10

$$\tau_x = db\left(\varpi_4^2 - \varpi_2^2\right) \tag{4.12}$$

and similarly for the y-axis, the *pitching* torque is

$$\tau_y = db\left(\varpi_1^2 - \varpi_3^2\right) \tag{4.13}$$

The torque applied to each propeller by the motor is opposed by aerodynamic drag

$$Q_i = k\varpi_i^2$$

where k depends on the same factors as b. This torque exerts a reaction torque on the airframe which acts to rotate the airframe about the propeller shaft in the opposite direction to its rotation. The total reaction torque about the z-axis is

$$\begin{aligned}
\tau_z &= Q_1 - Q_2 + Q_3 - Q_4 \\
&= k\left(\varpi_1^2 + \varpi_3^2 - \varpi_2^2 - \varpi_4^2\right)
\end{aligned} \tag{4.14}$$

where the different signs are due to the different rotation directions of the rotors. A yaw torque can be created simply by appropriate coordinated control of all four rotor speeds.

The total torque applied to the airframe according to Eq. 4.12 to 4.14 is $\boldsymbol{\tau} = (\tau_x, \tau_y, \tau_z)^T$ and the rotational acceleration is given by Euler's equation of motion from Eq. 3.10

$$\boldsymbol{J}\dot{\boldsymbol{\omega}} = -\boldsymbol{\omega} \times \boldsymbol{J}\boldsymbol{\omega} + \boldsymbol{\tau} \tag{4.15}$$

where $\boldsymbol{J}$ is the 3×3 inertia matrix of the vehicle and $\boldsymbol{\omega}$ is the angular velocity vector.

The motion of the quadrotor is obtained by integrating the forward dynamics equations Eq. 4.11 and Eq. 4.15 where the forces and moments on the airframe

$$\begin{pmatrix} T \\ \boldsymbol{\tau} \end{pmatrix} = \begin{pmatrix} -b & -b & -b & -b \\ 0 & -db & 0 & db \\ db & 0 & -db & 0 \\ k & -k & k & -k \end{pmatrix} \begin{pmatrix} \varpi_1^2 \\ \varpi_2^2 \\ \varpi_3^2 \\ \varpi_4^2 \end{pmatrix} = \boldsymbol{A} \begin{pmatrix} \varpi_1^2 \\ \varpi_2^2 \\ \varpi_3^2 \\ \varpi_4^2 \end{pmatrix} \tag{4.16}$$

are functions of the rotor speeds. The matrix $\boldsymbol{A}$ is constant, and full rank if $b, k, d > 0$ and can be inverted

$$\begin{pmatrix} \varpi_1^2 \\ \varpi_2^2 \\ \varpi_3^2 \\ \varpi_4^2 \end{pmatrix} = \boldsymbol{A}^{-1} \begin{pmatrix} T \\ \tau_x \\ \tau_y \\ \tau_z \end{pmatrix} \tag{4.17}$$

to solve for the rotor speeds► required to apply a specified thrust T and moment $\boldsymbol{\tau}$ to the airframe.

To control the vehicle we will employ a nested control structure which we describe for pitch and x-translational motion. The innermost loop uses a proportional and derivative controller► to compute the required pitching torque on the airframe

$$\tau_y^* = K_{\tau,p}\left(\theta_p^* - \theta_p^\#\right) + K_{\tau,d}\left(\dot{\theta}_p^* - \dot{\theta}_p^\#\right) \tag{4.18}$$

based on the error between desired and actual pitch angle.► The gains $K_{\tau,p}$ and $K_{\tau,d}$ are determined by classical control design approaches based on an approximate dy-

The direction of rotation is as shown in Fig. 4.20. Control of motor velocity is discussed in Sect. 9.1.6.

The rotational dynamics has a second-order transfer function of $\Theta_y(s) / \tau_y(s) = 1 / (Js^2 + Bs)$ where J is rotational inertia and B is aerodynamic damping which is generally quite small. To regulate a second-order system requires a proportional-derivative controller.

The term $\dot{\theta}_p^*$ is commonly ignored.

namic model and then tuned to achieve good performance. The actual vehicle pitch angle $\theta_p^\#$ would be estimated by an inertial navigation system as discussed in Sect. 3.4 and $\dot{\theta}_p^\#$ would be derived from gyroscopic sensors. The required rotor speeds are then determined using Eq. 4.17.

Consider a coordinate frame $\{B'\}$ attached to the vehicle and with the same origin as $\{B\}$ but with its x- and y-axes in the horizontal plane and parallel to the ground. The thrust vector is parallel to the z-axis of frame $\{B\}$ and pitching the nose down, rotating about the y-axis by θ_p, generates a force

$$
{}^{B'}\!f = \mathcal{R}_y\!\left(\theta_p\right) \cdot \begin{pmatrix} 0 \\ 0 \\ T \end{pmatrix} = \begin{pmatrix} T\sin\theta_p \\ 0 \\ T\cos\theta_p \end{pmatrix}
$$

which has a component

$$
{}^{B'}\!f_x = T\sin\theta_p \approx T\theta_p
$$

that accelerates the vehicle in the ${}^{B'}x$-direction, and we have assumed that θ_p is small. We can control the velocity in this direction with a proportional control law

$$
{}^{B'}\!f_x^* = mK_f\!\left({}^{B'}\!v_x^* - {}^{B'}\!v_x^\#\right)
$$

where $K_f > 0$ is a gain. Combining these two equations we obtain the desired pitch angle

$$
\theta_p^* \approx \frac{m}{T}K_f\!\left({}^{B'}\!v_x^* - {}^{B'}\!v_x^\#\right) \tag{4.19}
$$

required to achieve the desired forward velocity. Using Eq. 4.18 we compute the required pitching torque, and then using Eq. 4.17 the required rotor speeds. For a vehicle in vertical equilibrium the total thrust equals the weight force so $m/T \approx 1/g$.

The actual vehicle velocity ${}^{B}v_x$ would be estimated by an inertial navigation system as discussed in Sect. 3.4 or a GPS receiver. If the position of the vehicle in the xy-plane of the world frame is $p \in \mathbb{R}^2$ then the desired velocity is given by the proportional control law

$$
{}^{0}\!v^* = K_p\!\left({}^{0}\!p^* - {}^{0}\!p^\#\right) \tag{4.20}
$$

based on the error between the desired and actual position. The desired velocity in the xy-plane of frame$\{B'\}$ is

$$
{}^{B'}\!v = \ominus{}^{0}\mathcal{R}_{B'}\!\left(\theta_y\right) \cdot {}^{0}\!v, \; \mathcal{R} \in \mathbf{SO}(2)
$$

which is a function of the yaw angle θ_y

$$
\begin{pmatrix} {}^{B'}\!v_x \\ {}^{B'}\!v_y \end{pmatrix} = \begin{pmatrix} \cos\theta_y & -\sin\theta_y \\ \sin\theta_y & \cos\theta_y \end{pmatrix}^T \begin{pmatrix} v_x \\ v_y \end{pmatrix}
$$

This model is hierarchical and organized in terms of subsystems. Click the down arrow on a subsystem (can be seen onscreen but not in the figure) to reveal the detail. Double-click on the subsystem box to modify its parameters.

Figure 4.21 shows a Simulink model of the complete control system for a quadrotor◀ which can be loaded and displayed by

```
>> sl_quadrotor
```

Working our way left to right and starting at the top we have the desired position of the quadrotor in world coordinates. The position error is rotated from the world frame to the body frame and becomes the desired velocity. The velocity controller implements Eq. 4.19 and its equivalent for the roll axis and outputs the desired pitch and roll angles of the quadrotor. The attitude controller is a proportional-derivative controller that determines the appropriate pitch and roll torques to achieve these

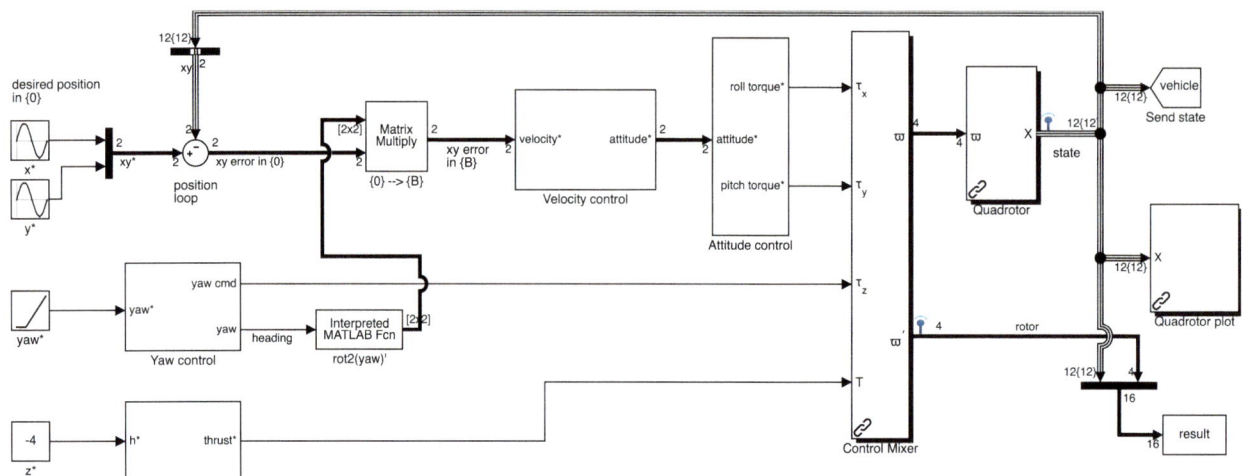

angles based on feedback of current attitude and attitude rate.◀ The yaw control block determines the error in heading angle and implements a proportional-derivative controller to compute the required yaw torque which is achieved by speeding up one pair of rotors and slowing the other pair.

Altitude is controlled by a proportional-derivative controller

$$T = K_p\left(z^* - z^{\#}\right) + K_d\left(\dot{z}^* - \dot{z}^{\#}\right) + T_0$$

which determines the average rotor speed. $T_0 = mg$ is the weight of the vehicle and this is an example of feedforward control – used here to counter the effect of gravity which otherwise is a constant disturbance to the altitude control loop. The alternatives to feedforward control would be to have very high gain for the altitude loop which often leads to actuator saturation and instability, or a proportional-integral (PI) controller which might require a long time for the integral term to increase to a useful value and then lead to overshoot. We will revisit gravity compensation in Chap. 9 applied to arm-type robots.

The control mixer block combines the three torque demands and the vertical thrust demand and implements Eq. 4.17 to determine the appropriate rotor speeds. Rotor speed limits are applied here. These are input to the quadrotor block▶ which implements the forward dynamics integrating Eq. 4.16 to give the position, velocity, orientation and orientation rate. The output of this block is the state vector $x = ({}^0p, {}^0\mathbf{\Gamma}, {}^B\dot{p}, {}^B\dot{\mathbf{\Gamma}}) \in \mathbb{R}^{12}$. As is common in aerospace applications we represent orientation $\mathbf{\Gamma}$ and orientation rate $\dot{\mathbf{\Gamma}}$ in terms of roll-pitch-yaw angles. Note that position and attitude are in the world frame while the rates are expressed in the body frame.

The parameters of a specific quadrotor can be loaded

```
>> mdl_quadrotor
```

which creates a structure called `quadrotor` in the workspace, and its elements are the various dynamic properties of the quadrotor. The simulation can be run using the Simulink menu or from the MATLAB command line

```
>> sim('sl_quadrotor');
```

and it displays an animation in a separate window.▶ The vehicle lifts off and flies around a circle while spinning slowly about its own z-axis. A snapshot is shown in Fig. 4.22. The simulation writes the results from each timestep into a matrix in the workspace

```
>> about result
result [double] : 2412x16 (308.7 kB)
```

Fig. 4.21. The Simulink® model `sl_quadrotor` which is a closed-loop simulation of the quadrotor. The vehicle takes off and flies in a circle at constant altitude. A Simulink bus is used for the 12-element state vector X output by the `Quadrotor` block. To reduce the number of lines in the diagram we have used `Goto` and `From` blocks to transmit and receive the state vector

Note that according to the coordinate conventions shown in Fig. 4.20 x-direction motion requires a negative rotation about the y-axis (pitch angle) and y-direction motion requires a positive rotation about the x-axis (roll angle) so the gains have different signs for the roll and pitch loops.

The Simulink library `roblocks` also includes a block for an N-rotor vehicle.

Loading and displaying the model using `>> sl_quadrotor` automatically loads the default quadrotor model. This is done by the PreLoadFcn callback set from model's properties File+Model Properties+Model Properties+Call-backs+PreLoadFcn.

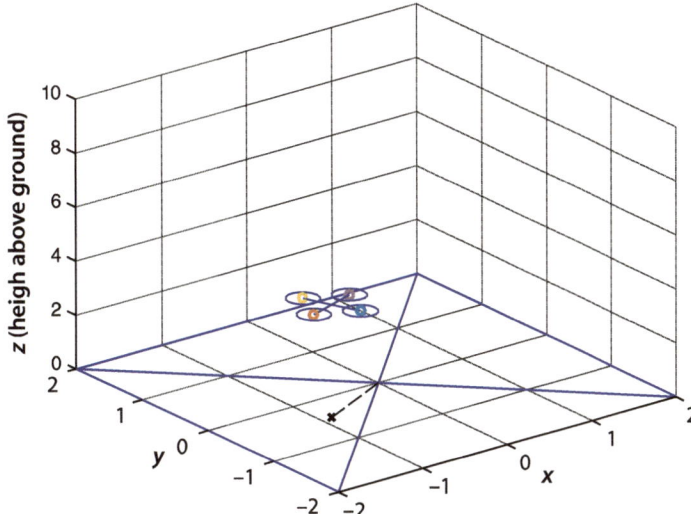

Fig. 4.22.
One frame from the quadrotor
simulation. The marker on the
ground plane is a projection of
the vehicle's centroid

which has one row per timestep, and each row contains the time followed by the state
vector (elements 2–13) and the commanded rotor speeds ω_i (elements 14–17). To
plot x and y versus time is

```
>> plot(result(:,1), result(:,2:3));
```

To recap on control of the quadrotor. A position error results in a required translational velocity. To achieve this requires appropriate pitch and roll angles so that a
component of the vehicle's thrust acts in the horizontal plane and generates a force to
accelerate the vehicle. ◀ As it approaches its goal the airframe must be rotated in the
opposite direction so that a component of thrust decelerates the motion. To achieve
the pitch and roll angles requires differential propeller thrust to create a moment that
rotationally accelerates the airframe.

*The total thrust must be increased so
that the vertical thrust component still
balances gravity.*

This indirection from translational motion to rotational motion is a consequence
of the vehicle being under-actuated – we have just four rotor speeds to adjust but the
vehicle's configuration space is 6-dimensional. In the configuration space we cannot
move in the x- or y-direction, but we can move in the pitch- or roll-direction which
results in motion in the x- or y-direction. The cost of under actuation is once again a
maneuver. The pitch and roll angles are a means to achieve translation control and
cannot be independently set.

4.3   Advanced Topics

4.3.1   Nonholonomic and Under-Actuated Systems

We introduced the notion of configuration space in Sect. 2.3.5 and it is useful to revisit it now that we have discussed several different types of mobile robot platform.
Common vehicles – as diverse as cars, hovercrafts, ships and aircraft – are all able to
move forward effectively but are unable to instantaneously move sideways. This is a
very sensible tradeoff that simplifies design and caters to the motion we most commonly require of the vehicle. Sideways motion for occasional tasks such as parking a
car, docking a ship or landing an aircraft are possible, albeit with some complex maneuvering but humans can learn this skill.

Consider a hovercraft which moves over a planar surface. To fully describe all its constituent particles we need to specify three generalized coordinates: its position in the
xy-plane and its rotation angle. It has three degrees of freedom and its configuration
space is $\mathcal{C} \subset \mathbb{R}^2 \times \mathbb{S}^1$. This hovercraft has two propellers whose axes are parallel but not

collinear. The sum of their thrusts provide a forward force and the difference in thrusts generates a yawing torque for steering. The number of actuators, two, is less than its degrees of freedom dim $\mathcal{C} = 3$ and we call this an under-actuated system. This imposes significant limitations on the way in which it can move. At any point in time we can control the forward (parallel to the thrust vectors) acceleration and the rotational acceleration of the hovercraft but there is zero sideways (or lateral) acceleration since it cannot generate any lateral thrust. Nevertheless with some clever maneuvering, like with a car, the hovercraft can follow a path that will take it to a place to one side of where it started. In the hovercraft's 3-dimensional configuration space this means that at any point there are certain directions in which *acceleration* is not possible. We can reach points in those direction but not directly, only by following some circuitous path.

All flying and underwater vehicles have a configuration that is completely described by six generalized coordinates – their position and orientation in 3D space. $\mathcal{C} \subset \mathbb{R}^3 \times \mathbb{S}^1 \times \mathbb{S}^1 \times \mathbb{S}^1$ where the orientation is expressed in some three-angle representation – since dim $\mathcal{C} = 6$ the vehicles have six degrees of freedom. A quadrotor has four actuators, four thrust-generating propellers, and this is fewer than its degrees of freedom making it under-actuated. Controlling the four propellers causes motion in the up/down, roll, pitch and yaw directions of the configuration space but not in the forward/backward or left/right directions. To access those degrees of freedom it is necessary to perform a *maneuver*: pitch down so that the thrust vector provides a horizontal force component, accelerate forward, pitch up so that the thrust vector provides a horizontal force component to decelerate, and then level out.

For a helicopter only four of the six degrees of freedom are practically useful: up/down, forward/backward, left/right and yaw. Therefore a helicopter requires a minimum of four actuators: the main rotor generates a thrust vector whose magnitude is controlled by the collective pitch and whose direction is controlled by the lateral and longitudinal cyclic pitch. The tail rotor provides a yawing moment. This leaves two degrees of freedom unactuated, roll and pitch angles, but clever design ensures that gravity actuates them and keeps them close to zero – without gravity a helicopter cannot work. A fixed-wing aircraft moves forward very efficiently and also has four actuators: engine thrust provides acceleration in the forward direction and the ailerons, elevator and rudder exert respectively roll, pitch and yaw moments on the aircraft.▸ To access the missing degrees of freedom such as up/down and left/right translation, the aircraft must pitch or yaw while moving forward.

Some low-cost hobby aircraft have no rudder and rely only on ailerons to bank and turn the aircraft. Even cheaper hobby aircraft have no elevator and rely on engine speed to control height.

The advantage of under-actuation is having fewer actuators. In practice this means real savings in terms of cost, complexity and weight. The consequence is that at any point in its configuration space there are certain directions in which the vehicle cannot move. Full actuation is possible but not common, for example the DEPTHX underwater robot shown on page 94 has six degrees of freedom and six actuators. These can exert an arbitrary force and torque on the vehicle, allowing it to accelerate in any direction or about any axis.

A 4-wheeled car has many similarities to the hovercraft discussed above. It moves over a planar surface and its configuration can be fully described by its generalized coordinates: its position in the *xy*-plane and a rotation angle. It has three degrees of freedom and its configuration space is $\mathcal{C} \subset \mathbb{R}^2 \times \mathbb{S}^1$. A car has two actuators, one to move forwards or backwards and one to change the heading direction. A car, like a hovercraft, is under-actuated.

We know from our experience with cars that we cannot move directly in certain directions and sometimes needs to perform a maneuver to reach our goal. A differential- or skid-steered vehicle, such as a tank, is also under-actuated – it has only two actuators, one for each track. While this type of vehicle can turn on the spot it cannot move sideways. To do that it has to turn, proceed, stop then turn – this need to maneuver is the clear signature of an under-actuated system.

We might often wish for an ability to drive our car sideways but the standard wheel provides real benefit when cornering – lateral friction between the wheels and the

Table 4.1.
Summary of configuration space
characteristics for various robots.
A nonholonomic system is
under-actuated and/or has a
rolling constraint

	dim $\mathcal{C}$	Degrees of freedom	Number of actuators	Actuation	Rolling constraints	Holonomic
Train	1	1	1	full		✓
2-joint robot arm	2	2	2	full		✓
6-joint robot arm	6	6	6	full		✓
10-joint robot arm	10	10	10	over		✓
Hovercraft	3	3	2	under		
Car	3	2	2	under	✓	
Helicopter	6	6	4	under		
Fixed wing aircraft	6	6	4	under		
DEPTHX AUV	6	6	6	full		✓

road provides, for free, the centripetal force which would otherwise require an extra actuator to provide. The hovercraft has many similarities to a car but we can push a hovercraft sideways – we cannot do that with a car. This lateral friction is a distinguishing feature of the car.

The margin note: *The hovercraft, aerial and underwater vehicles are controlled by forces so in this case the constraints are on vehicle acceleration in configuration space not velocity.*

The inability to slip sideways is a constraint, the *rolling* constraint, on the velocity◄ of the vehicle just as under-actuation is. A vehicle with one or more velocity constraints, due to under-actuation or a rolling constraint, is referred to as a nonholonomic system. A key characteristic of these systems is that they cannot move *directly* from one configuration to another – they must perform a maneuver or sequence of motions. A car has a velocity constraint due to its wheels and is also under-actuated.

The margin note: *For example fixing the end of the 10-joint robot arm introduces six holonomic constraints (position and orientation) so the arm would have only 4 degrees of freedom.*

A holonomic constraint restricts the possible configurations that the system can achieve – it can be expressed as an equation written in terms of the configuration variables.◄ A nonholonomic constraint such as Eq. 4.3 and 4.6 is one that restricts the *velocity* (or acceleration) of a system in configuration space – it can only be expressed in terms of the derivatives of the configuration variables.◄ The nonholonomic constraint does not restrict the possible configurations the system can achieve but it does preclude instantaneous velocity or acceleration in certain directions.

The margin note: *The constraint cannot be integrated to a constraint in terms of configuration variables, so such systems are also known as nonintegrable systems.*

In control theoretic terms Brockett's theorem (Brockett 1983) states that nonholonomic systems are controllable but they cannot be stabilized to a desired state using a differentiable, or even continuous, pure state-feedback controller. A time-varying or nonlinear control strategy is required which means that the robot follows some generally nonlinear path. One exception is an under-actuated system moving in 3-dimensional space within a force field, for example a gravity field – gravity acts like an additional actuator and makes the system linearly controllable (but not holonomic), as we showed for the quadrotor example in Sect. 4.2.

Mobility parameters for the various robots that we have discussed here, and earlier in Sect. 2.3.5, are tabulated in Table 4.1. We will discuss under- and over-actuation in the context of arm robots in Chap. 8.

4.4 Wrapping Up

In this chapter we have created and discussed models and controllers for a number of common, but quite different, robot platforms. We first discussed wheeled robots. For car-like vehicles we developed a kinematic model which we used to develop a number of different controllers in order that the platform could perform useful tasks such as driving to a point, driving along a line, following a trajectory or driving to a pose. We then discussed differentially steered vehicles on which many robots are based, and omnidirectional robots based on novel wheel types. Then we we discussed a simple but common flying vehicle, the quadrotor, and developed a dynamic model and a hierarchical control system that allowed the quadrotor to fly a circuit. This hierarchical or nested control approach is described in more detail in Sect. 9.1.7 in the context of robot arms.

We also extended our earlier discussion about configuration space to include the velocity constraints due to under actuation and rolling constraints from wheels.

The next chapters in this part will discuss how to plan paths for robots through complex environments that contain obstacles and then how to determine the location of a robot.

Further Reading

Comprehensive modeling of mobile ground robots is provided in the book by Siegwart et al. (2011). In addition to the models covered here, it presents in-depth discussion of a variety of wheel configurations with different combinations of driven wheels, steered wheels and passive castors. The book by Kelly (2013) also covers vehicle modeling and control. Both books also provide a good introduction to perception, localization and navigation which we will discuss in the coming chapters.

The paper by Martins et al. (2008) discusses kinematics, dynamics and control of differential steer robots. The Handbook of Robotics (Siciliano and Khatib 2016, part E) covers modeling and control of various vehicle types including aerial and underwater. The theory of helicopters with an emphasis on robotics is provided by Mettler (2003) but the definitive reference for helicopter dynamics is the very large book by Prouty (2002). The book by Antonelli (2014) provides comprehensive coverage of modeling and control of underwater robots.

Some of the earliest papers on quadrotor modeling and control are by Pounds, Mahony and colleagues (Hamel et al. 2002; Pounds et al. 2004, 2006). The thesis by Pounds (2007) presents comprehensive aerodynamic modeling of a quadrotor with a particular focus on blade flapping, a phenomenon well known in conventional helicopters but largely ignored for quadrotors. A tutorial introduction to the control of multi-rotor flying robots is given by Mahony, Kumar, and Corke (2012). Quadrotors are now commercially available from many vendors at quite low cost. There are also a number of hardware kits and open-source software projects such as ArduCopter and Mikrokopter.

Mobile ground robots are now a mature technology for transporting parts around manufacturing plants. The research frontier is now for vehicles that operate autonomously in outdoor environments (Siciliano and Khatib 2016, part F). Research into the automation of passenger cars has been ongoing since the 1980s and a number of automative manufacturers are talking about commercial autonomous cars by 2020.

Historical and interesting. The Navlab project at Carnegie-Mellon University started in 1984 and a series of autonomous vehicles, Navlabs, were built and a large body of research has resulted. All vehicles made strong use of computer vision for navigation. In 1995 the supervised autonomous Navlab 5 made a 3 000-mile journey, dubbed "No Hands Across America" (Pomerleau and Jochem 1995, 1996). The vehicle steered itself 98% of the time largely by visual sensing of the white lines at the edge of the road.

In Europe, Ernst Dickmanns and his team at Universität der Bundeswehr München demonstrated autonomous control of vehicles. In 1988 the VaMoRs system, a 5 tonne Mercedes-Benz van, could drive itself at speeds over 90 km h^{-1} (Dickmanns and Graefe 1988b; Dickmanns and Zapp 1987; Dickmanns 2007). The European Prometheus Project ran from 1987–1995 and in 1994 the robot vehicles VaMP and VITA-2 drove more than 1 000 km on a Paris multi-lane highway in standard heavy traffic at speeds up to 130 km h^{-1}. They demonstrated autonomous driving in free lanes, convoy driving, automatic tracking of other vehicles, and lane changes with autonomous passing of other cars. In 1995 an autonomous S-Class Mercedes-Benz made a 1 600 km trip from Munich to Copenhagen and back. On the German Autobahn speeds exceeded 175 km h^{-1} and the vehicle executed traffic maneuvers such as overtaking. The mean time between human interventions was 9 km and it drove up to 158 km without any human intervention. The UK part of the project demonstrated autonomous driving of an XJ6 Jaguar with vision (Matthews et al. 1995) and radar-based sensing for lane keeping and collision avoidance. More recently, in the USA a series of Grand Challenges were run for autonomous cars. The 2005 desert and 2007 urban challenges are comprehensively described in compilations of papers from the various teams in Buehler et al. (2007, 2010). More recent demonstrations of self-driving vehicles are a journey along the fabled silk road described by Bertozzi et al. (2011) and a classic road trip through Germany by Ziegler et al. (2014).

Ackermann's magazine can be found online at http://smithandgosling.wordpress. com/2009/12/02/ackermanns-repository-of-arts and the carriage steering mechanism is published in the March and April issues of 1818. King-Hele (2002) provides a comprehensive discussion about the prior work on steering geometry and Darwin's earlier invention.

Toolbox and MATLAB Notes

In addition to the Simulink `Bicycle` model used in this chapter the Toolbox also provides a MATLAB class which implements these kinematic equations and which we will use in Chap. 6. For example we can create a vehicle model with steer angle and speed limits

```
>> veh = Bicycle('speedmax', 1, 'steermax', 30*pi/180);
```

and evaluate Eq. 4.2 for a particular state and set of control inputs (v, γ)

```
>> veh.deriv([], [0 0 0], [0.3, 0.2])
ans =
    0.3000         0    0.0608
```

The `Unicycle` class is used for a differentially-steered robot and has equivalent methods.

The Robotics System Toolbox™ from The MathWorks has support for differentially-steered mobile robots which can be created using the function `ExampleHelperRobotSimulator`. It also includes a class `robotics.PurePursuit` that implements pure pursuit for a differential steer robot. An example is given in the Toolbox RST folder.

Exercises

1. For a 4-wheel vehicle with $L = 2$ m and width between wheel centers of 1.5 m
   a) What steering wheel angle is needed for a turn rate of 10 deg s^{-1} at a forward speed of 20 km h^{-1}?
   b) compute the difference in wheel steer angle for Ackermann steering around curves of radius 10, 50 and 100 m.
   c) If the vehicle is moving at 80 km h^{-1} compute the difference in back wheel rotation rates for curves of radius 10, 50 and 100 m.
2. Write an expression for turn rate in terms of the angular rotation rate of the two back wheels. Investigate the effect of errors in wheel radius and vehicle width.
3. Consider a car and bus with $L = 4$ and 12 m respectively. To follow a curve with radius of 10, 20 and 50 m determine the respective steered wheel angles.
4. For a number of steered wheel angles in the range -45 to $45°$ and a velocity of 2 m s^{-1} overlay plots of the vehicle's trajectory in the xy-plane.

5. Implement the $\ominus$ operator used in Sect. 4.1.1.1 and check against the code for `angdiff`.
6. Moving to a point (page 101) plot x, y and θ against time.
7. Pure pursuit example (page 104)
   a) Investigate what happens if you vary the look-ahead distance, heading gain or proportional gain in the speed controller.
   b) Investigate what happens when the integral gain in the speed controller is zero.
   c) With integral set to zero, add a constant to the output of the controller. What should the value of the constant be?
   d) Add a velocity feedforward term.
   e) Modify the pure pursuit example so the robot follows a slalom course.
   f) Modify the pure pursuit example to follow a target moving back and forth along a line.
8. Moving to a pose (page 105)
   a) Repeat the example with a different initial orientation.
   b) Implement a parallel parking maneuver. Is the resulting path practical?
   c) Experiment with different control parameters.
9. Use the MATLAB GUI interface to make a simple steering wheel and velocity control, and use this to create a very simple driving simulator. Alternatively interface a gaming steering wheel and pedal to MATLAB.
10. Adapt the various controllers in Sect. 4.1.1 to the differentially steered robot.
11. Derive Eq. 4.4 from the preceding equation.
12. For constant forward velocity, plot v_L and v_R as a function of ICR position along the y-axis. Under what conditions do v_L and v_R have a different sign?
13. Using Simulink implement a controller using Eq. 4.7 that moves the robot in its y-direction. How does the robot's orientation change as it moves?
14. Sketch the design for a robot with three mecanum wheels. Ensure that it cannot roll freely and that it can drive in any direction. Write code to convert from desired vehicle translational and rotational velocity to wheel rotation rates.
15. For the 4-wheel omnidirectional robot of Sect. 4.1.3 write an algorithm that will allow it to move in a circle of radius 0.5 m around a point with its nose always pointed toward the center of the circle.
16. Quadrotor (page 113)
   a) At equilibrium, compute the speed of all the propellers.
   b) Experiment with different control gains. What happens if you reduce the damping gains to zero?
   c) Remove the gravity feedforward and experiment with large altitude gain or a PI controller.
   d) When the vehicle has nonzero roll and pitch angles, the magnitude of the vertical thrust is reduced and the vehicle will slowly descend. Add compensation to the vertical thrust to correct this.
   e) Simulate the quadrotor flying inverted, that is, its z-axis is pointing upwards.
   f) Program a ballistic motion. Have the quadrotor take off at 45 deg to horizontal then remove all thrust.
   g) Program a smooth landing.
   h) Program a barrel roll maneuver. Have the quadrotor fly horizontally in its x-direction and then increase the roll angle from 0 to 2π.
   i) Program a flip maneuver. Have the quadrotor fly horizontally in its x-direction and then increase the pitch angle from 0 to 2π.
   j) Add another four rotors.
   k) Use the function `mstraj` to create a trajectory through ten via points $(X_i, Y_i, Z_i, \theta_y)$ and modify the controller of Fig. 4.21 for smooth pursuit of this trajectory.
   l) Use the MATLAB GUI interface to make a simple joystick control, and use this to create a very simple flying simulator. Alternatively interface a gaming joystick to MATLAB.

5 Navigation

the process of directing a vehicle so as to reach the intended destination
IEEE Standard 172-1983

Robot navigation is the problem of guiding a robot towards a goal. The human approach to navigation is to make maps and erect signposts, and at first glance it seems obvious that robots should operate the same way. However many robotic tasks can be achieved without any map at all, using an approach referred to as *reactive navigation*. For example, navigating by heading towards a light, following a white line on the ground, moving through a maze by following a wall, or vacuuming a room by following a random path. The robot is reacting directly to its environment: the intensity of the light, the relative position of the white line or contact with a wall. Grey Walter's tortoise Elsie from page 93 demonstrated "life-like" behaviors – she *reacted* to her environment and could seek out a light source. Today tens of millions of robotic vacuum cleaners are cleaning floors and most of them do so without using any map of the rooms in which they work. Instead they do the job by making random moves and sensing only that they have made contact with an obstacle as shown in Fig. 5.1.

Human-style *map-based navigation* is used by more sophisticated robots and is also known as motion planning. This approach supports more complex tasks but is itself more complex. It imposes a number of requirements, not the least of which is having a map of the environment. It also requires that the robot's position is always known. In the next chapter we will discuss how robots can determine their position and create maps. The remainder of this chapter discusses the reactive and map-based approaches to robot navigation with a focus on wheeled robots operating in a planar environment.

Fig. 5.1.
Time lapse photograph of a
Roomba robot cleaning a room
(photo by Chris Bartlett)

© Springer Nature Switzerland AG 2022
P. Corke, *Robotics and Control*, Springer Tracts in Advanced Robotics 141,
https://doi.org/10.1007/978-3-030-79179-7_5

Valentino Braitenberg (1926–2011) was an Italian-Austrian neuroscientist and cyberneticist, and former director at the Max Planck Institute for Biological Cybernetics in Tübingen, Germany. His 1986 book *"Vehicles: Experiments in Synthetic Psychology"* (image on right is the cover of this book, published by The MIT Press, ©MIT 1984) describes reactive goal-achieving vehicles, and such systems are now commonly known as Braitenberg Vehicles.

A Braitenberg vehicle is an automaton which combines sensors, actuators and their direct interconnection to produce goal-oriented behaviors. In the book these vehicles are described conceptually as analog circuits, but more recently small robots based on a digital realization of the same principles have been developed. Grey Walter's tortoise predates the use of this term but was nevertheless an example of such a vehicle.

5.1   Reactive Navigation

Surprisingly complex tasks can be performed by a robot even if it has no map and no real *idea* about where it is. As already mentioned robotic vacuum cleaners use only random motion and information from contact sensors to perform a complex task as shown in Fig. 5.1. Insects such as ants and bees gather food and return it to their nest based on input from their senses, they have far too few neurons to create any kind of mental map of the world and plan paths through it. Even single-celled organisms such as flagellate protozoa exhibit goal-seeking behaviors. In this case we need to temporarily modify our earlier definition of a robot to

> *a goal oriented machine that can sense, ~~plan~~ and act.*

Grey Walter's robotic tortoise demonstrated that it could moves toward a light source, a behavior known as phototaxis.▸ This was an important result in the then emerging scientific field of cybernetics.

More generally a *taxis* is the response of an organism to a stimulus gradient.

5.1.1   Braitenberg Vehicles

A very simple class of goal achieving robots are known as Braitenberg vehicles and are characterized by direct connection between sensors and motors. They have no explicit internal representation of the environment in which they operate and nor do they make explicit plans.▸

Consider the problem of a robot moving in two dimensions that is seeking the local maxima of a scalar field – the field could be light intensity or the concentration of some chemical.▸ The Simulink® model

```
>> sl_braitenberg
```

shown in Fig. 5.2 achieves this using a steering signal derived directly from the sensors.▸

This is a fine philosophical point, the plan could be considered to be implicit in the details of the connections between the motors and sensors.

This is similar to the problem of moving to a point discussed in Sect. 4.1.1.1.

This is similar to Braitenberg's Vehicle 4a.

William Grey Walter (1910–1977) was a neurophysiologist and pioneering cyberneticist born in Kansas City, Missouri and studied at King's College, Cambridge. Unable to obtain a research fellowship at Cambridge, he worked on neurophysiological research in hospitals in London and from 1939 at the Burden Neurological Institute in Bristol. He developed electro-encephalographic brain topography which used multiple electrodes on the scalp and a triangulation algorithm to determine the amplitude and location of brain activity.

Walter was influential in the then new field of cybernetics. He built robots to study how complex reflex behavior could arise from neural interconnections. His tortoise Elsie (of the species *Machina Speculatrix*) is shown, without its shell, on page 93. Built in 1948 Elsie was a three-wheeled robot capable of phototaxis that could also find its way to a recharging station. A second generation tortoise (from 1951) is in the collection of the Smithsonian Institution. He published popular articles in "Scientific American" (1950 and 1951) and a book "The Living Brain" (1953). He was badly injured in a car accident in 1970 from which he never fully recovered. (Image courtesy Reuben Hoggett collection)

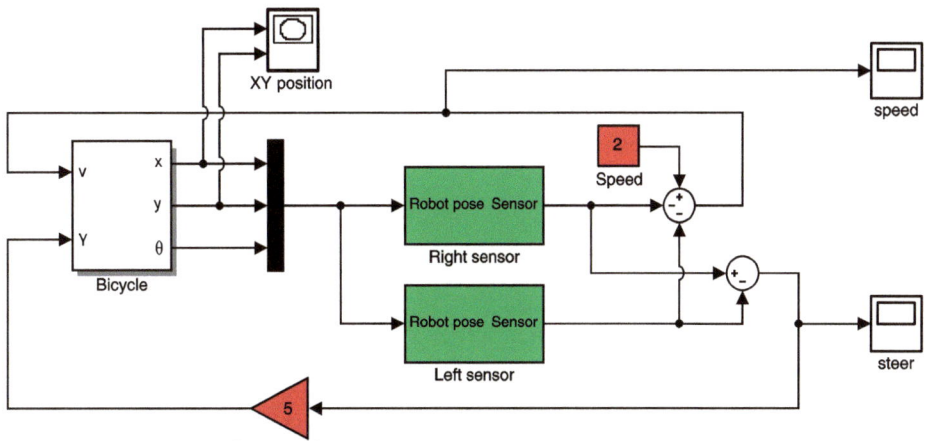

Fig. 5.2.
The Simulink® model
sl_braitenberg drives the
vehicle toward the maxima of
a provided scalar function. The
vehicle plus controller is an example of a Braitenberg vehicle

We can make the measurements simultaneously using two spatially separated
sensors or from one sensor over time as
the robot moves.

To ascend the gradient we need to estimate the gradient direction at the current
location and this requires at least two measurements of the field. ◀ In this example we
use two sensors, bilateral sensing, with one on each side of the robot's body. The sensors are modeled by the green sensor blocks shown in Fig. 5.2 and are parameterized
by the position of the sensor with respect to the robot's body, and the sensing function.
In this example the sensors are at ± 2 units in the vehicle's lateral or y-direction.

The field to be sensed is a simple inverse square field defined by

```
1    function sensor = sensorfield(x, y)
2        xc = 60; yc = 90;
3        sensor = 200./((x-xc).^2 + (y-yc).^2 + 200);
```

which returns the sensor value $s(x, y) \in [0, 1]$ which is a function of the sensor's position in the plane. This particular function has a peak value at the point (60, 90).

The vehicle speed is

$$v = 2 - s_R - s_L$$

where s_R and s_L are the right and left sensor readings respectively. At the goal, where
$s_R = s_L = 1$ the velocity becomes zero.

Steering angle is based on the difference between the sensor readings

$$\gamma = k\left(s_R - s_L\right)$$

Similar strategies are used by moths
whose two antennae are exquisitely
sensitive odor detectors that are used
to steer a male moth toward a phero-
mone emitting female.

so when the field is equal in the left- and right-hand sensors the robot moves straight ahead. ◀

We start the simulation from the Simulink menu or the command line

```
>> sim('sl_braitenberg');
```

and the path of the robot is shown in Fig. 5.3. The starting pose can be changed through
the parameters of the Bicycle block. We see that the robot turns toward the goal and
slows down as it approaches, asymptotically achieving the goal position.

This particular sensor-action control law results in a specific robotic *behavior*. We
could add additional logic to the robot to detect that it had arrived near the goal and
then switch to a stopping behavior. An obstacle would block this robot since its only
behavior is to steer toward the goal, but an additional behavior could be added to handle this case and drive around an obstacle. We could add another behavior to search
randomly for the source if none was visible. Grey Walter's tortoise had four behaviors
and switching was based on light level and a touch sensor.

Multiple behaviors and the ability to switch between them leads to an approach
known as behavior-based robotics. The subsumption architecture was proposed as a

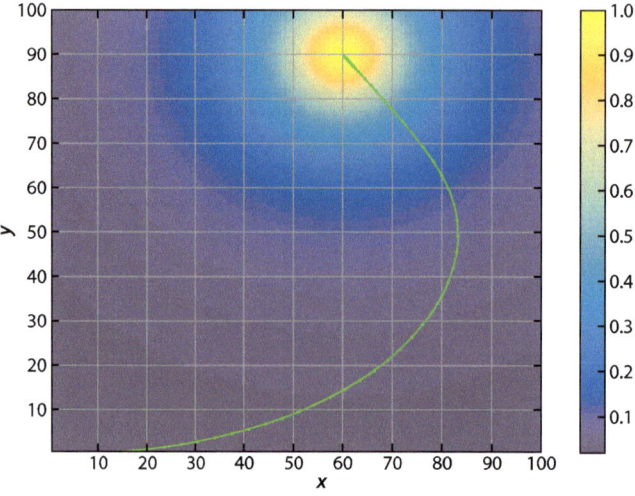

Fig. 5.3.
Path of the Braitenberg vehicle
moving toward the maximum of
a 2D scalar field whose magnitude is shown color coded

means to formalize the interaction between different behaviors. Complex, some might
say *intelligent looking*, behaviors can be manifested by such systems. However as more
behaviors are added the complexity of the system grows rapidly and interactions between behaviors become more complex to express and debug. Ultimately the penalty
of not using a map becomes too great.

5.1.2   Simple Automata

Another class of reactive robots are known as *bugs* – simple automata that perform goal
seeking in the presence of nondriveable areas or obstacles. There are a large number
of *bug* algorithms and they share the ability to sense when they are in proximity to an
obstacle. In this respect they are similar to the Braitenberg class vehicle, but the *bug*
includes a state machine and other logic in between the sensor and the motors. The
automata have memory which our earlier Braitenberg vehicle lacked.▶ In this section
we will investigate a specific *bug* algorithm known as *bug2*.

Braitenberg's book describes a series of
increasingly complex vehicles, some of
which incorporate memory. However the
term *Braitenberg vehicle* has become associated with the simplest vehicles he
described.

We start by loading an obstacle field to challenge the robot

```
>> load house
>> about house
house [double] : 397x596 (1.9 MB)
```

which defines a matrix variable `house` in the workspace. The elements are zero or
one representing free space or obstacle respectively and this is shown in Fig. 5.4. Tools
to generate such maps are discussed on page 129. This matrix is an example of an occupancy grid which will be discussed further in the next section. This command also
loads a list of named places within the house, as elements of a structure

```
>> place
place =
     kitchen: [320 190]
      garage: [500 150]
         br1: [50 220]
           .
           .
```

At this point we state some assumptions. Firstly, the robot operates in a grid world
and occupies one grid cell. Secondly, the robot is capable of omnidirectional motion
and can move to any of its eight neighboring grid cells. Thirdly, it is able to determine its position on the plane which is a nontrivial problem that will be discussed
in detail in Chap. 6. Finally, the robot can only sense its goal and whether adjacent
cells are occupied.

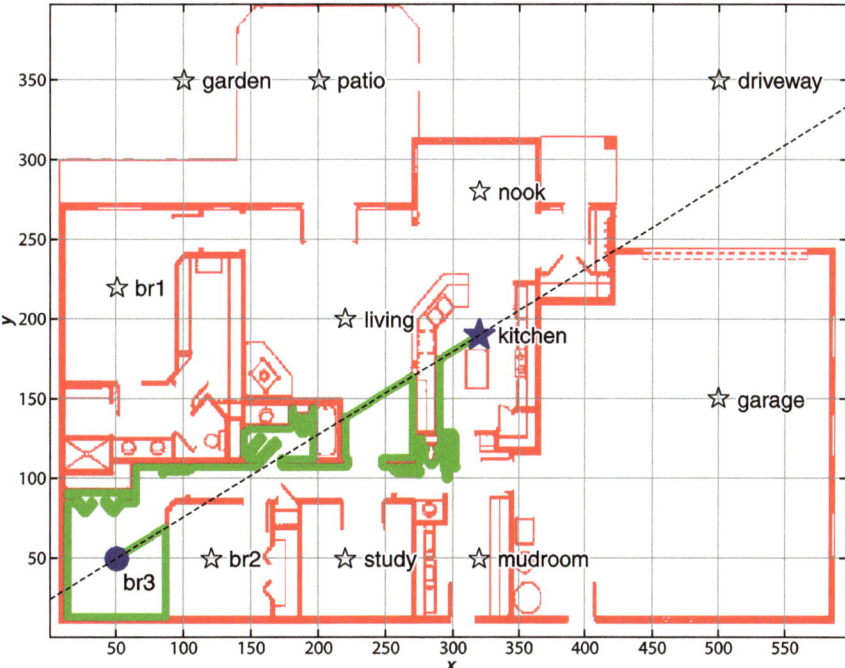

Fig. 5.4.
Obstacles are indicated by *red pixels*. Named places are indicated by *hollow black stars*. Approximate scale is 4.5 cm per cell. The start location is a *solid blue circle* and the goal is a *solid blue star*. The path taken by the bug2 algorithm is marked by a *green line*. The *black dashed line* is the m-line, the direct path from the start to the goal

We create an instance of the bug2 class

```
>> bug = Bug2(house);
```

and pass in the occupancy grid. The *bug2* algorithm does not use the map to plan a path – the map is used by the simulator to provide sensory inputs to the robot. We can display the robot's environment by

```
>> bug.plot();
```

The simulation is run using the query method

```
>> bug.query(place.br3, place.kitchen, 'animate');
```

whose arguments are the start and goal positions of the robot within the house.

The method displays an animation of the robot moving toward the goal and the path is shown as a series of green dots in Fig. 5.4.

The strategy of the *bug2* algorithm is quite simple. It is given a straight line – the m-line – towards its goal. If it encounters an obstacle it turns right and continues until it encounters a point on the m-line that is closer to the goal than when it departed from the m-line. ◄

It could be argued that the m-line represents an explicit plan. Thus *bug* algorithms occupy a position somewhere between Braitenberg vehicles and map-based planning systems in the spectrum of approaches to navigation.

If an output argument is specified

```
>> p = bug.query(place.br3, place.kitchen)
```

it returns the path as a matrix p

```
>> about p
p [double] : 1299x2 (20.8 kB)
```

which has one row per point, and comprises 1 299 points for this example. Invoking the function with an empty matrix

```
>> p = bug.query([], place.kitchen);
```

will prompt for the corresponding point to be selected by clicking on the plot.

In this example the *bug2* algorithm has reached the goal but it has taken a very suboptimal route, traversing the inside of a wardrobe, behind doors and visiting two bathrooms. It would perhaps have been quicker in this case to turn left, rather than right, at the first obstacle but that strategy might give a worse outcome somewhere else. Many variants of the *bug* algorithm have been developed, but while they improve the performance for one type of environment they can degrade performance in others. Fundamentally the robot is limited by not using a map. It cannot see the big picture and therefore takes paths that are locally, rather than globally, optimal.

5.2   Map-Based Planning

The key to achieving the *best* path between points A and B, as we know from everyday life, is to use a map. Typically best means the shortest distance but it may also include some penalty term or cost related to traversability which is how easy the terrain is to drive over – it might be quicker to travel further but faster over better roads. A more sophisticated planner might also consider the size of the robot, the kinematics and dynamics of the vehicle and avoid paths that involve turns that are tighter than the vehicle can execute. Recalling our earlier definition of a robot as a

goal oriented machine that can sense, **plan** and act,

this section concentrates on planning.

There are many ways to represent a map and the position of the vehicle within the map. Graphs, as discussed in Appendix I, can be used to represent places and paths between them. Graphs can be efficiently searched to find a path that minimizes some measure or cost, most commonly the distance traveled. A simpler and very computer-friendly representation is the occupancy grid which is widely used in robotics.

An occupancy grid treats the world as a grid of cells and each cell is marked as occupied or unoccupied. We use zero to indicate an unoccupied cell or free space where the robot can drive. A value of one indicates an occupied or nondriveable cell. The size of the cell depends on the application. The memory required to hold the occupancy grid increases with the spatial area represented and inversely with the cell size. However for modern computers this representation is very feasible. For example a cell size 1×1 m requires▸ just 125 kbyte km^{-2}.

Considering a single bit to represent each cell. The occupancy grid could be compressed or could be kept on a disk with only the local region in memory.

In the remainder of this section we use code examples to illustrate several different planners and all are based on the occupancy grid representation. To create uniformity the planners are all implemented as classes derived from the `Navigation` superclass which is briefly described on page 131. The *bug2* class we used previously was also an instance of this class so the remaining examples follow a familiar pattern.

Once again we state some assumptions. Firstly, the robot operates in a grid world and occupies one grid cell. Secondly, the robot does not have any nonholonomic constraints and can move to any neighboring grid cell. Thirdly, it is able to determine its position on the plane. Fourthly, the robot is able to use the map to compute the path it will take.

In all examples we will use the house map introduced in the last section and find paths from bedroom 3 to the kitchen. These parameters can be varied, and the occupancy grid changed using the tools described above.

5.2.1   Distance Transform

Consider a matrix of zeros with just a single nonzero element representing the goal. The distance transform of this matrix is another matrix, of the same size, but the value of each element is its distance▸ from the original nonzero pixel. For robot path planning we use the default Euclidean distance.

The distance between two points (x_1, y_1) and (x_2, y_2) where $\Delta_x = x_2 - x_1$ and $\Delta_y = y_2 - y_1$ can be Euclidean $\sqrt{\Delta_x^2 + \Delta_y^2}$ or CityBlock (also known as Manhattan) distance $|\Delta_x| + |\Delta_y|$.

Making a map. An occupancy grid is a matrix that corresponds to a region of 2-dimensional space. Elements containing zeros are free space where the robot can move, and those with ones are obstacles where the robot cannot move. We can use many approaches to create a map. For example we could create a matrix filled with zeros (representing all free space)

```
>> map = zeros(100, 100);
```

and use MATLAB operations such as

```
>> map(40:50,20:80) = 1;
```

or the MATLAB builtin matrix editor to create obstacles but this is quite cumbersome. Instead we can use the Toolbox map editor makemap to create more complex maps using an interactive editor

```
>> map = makemap(100)
```

that allows you to add rectangles, circles and polygons to an occupancy grid. In this example the grid is 100 × 100. See online help for details.

The occupancy grid in Fig. 5.4 was derived from a scanned image but online buildings plans and street maps could also be used.

Note that the occupancy grid is a matrix whose coordinates are conventionally expressed as (row, column) and the row is the vertical dimension of a matrix. We use the Cartesian convention of a horizontal x-coordinate first, followed by the y-coordinate therefore the matrix is always indexed as y, x in the code.

To use the distance transform for robot navigation we create a DXform object, which is derived from the Navigation class

```
>> dx = DXform(house);
```

and then create a plan to reach a specific goal

```
>> dx.plan(place.kitchen)
```

which can be visualized

```
>> dx.plot()
```

as shown in Fig. 5.5. We see the obstacle regions in red overlaid on the distance map whose grey level at any point indicates the distance from that point to the goal, in grid cells, taking into account travel *around* obstacles.

The hard work has been done and to find the shortest path from *any* point to the goal we simply consult or query the plan. For example a path from the bedroom to the goal is

For the bug2 algorithm there was no planning step so the query in that case was the simulated robot querying its proximity sensors.

```
>> dx.query(place.br3, 'animate');
```

which displays an animation of the robot moving toward the goal. The path is indicated by a series of green dots as shown in Fig. 5.5.

By convention the plan is based on the goal location and we query for a start location, but we could base the plan on the start position and then query for a goal.

The plan is the distance map. Wherever the robot starts, it moves to the neighboring cell that has the smallest distance to the goal. The process is repeated until the robot reaches a cell with a distance value of zero which is the goal.

If the path method is called with an output argument the path

```
>> p = dx.query(place.br3);
```

is returned as a matrix, one row per point, which we can visualize overlaid on the occupancy grid and distance map

```
>> dx.plot(p)
```

The path comprises

```
>> numrows(p)
ans =
   336
```

points which is considerably shorter than the path found by *bug2*.

This navigation algorithm has exploited its global view of the world and has, through exhaustive computation, found the shortest possible path. In contrast, *bug2* without

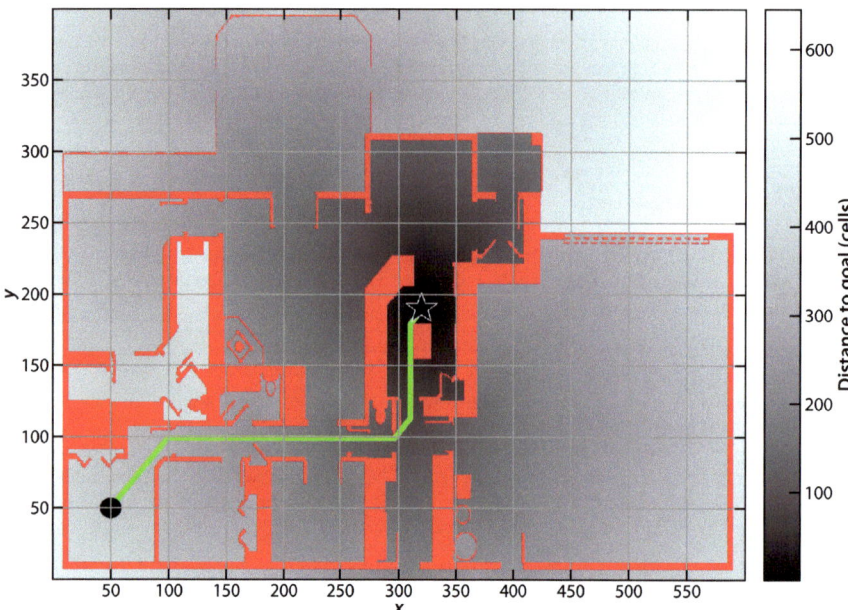

Fig. 5.5.
The distance transform path. Obstacles are indicated by *red cells*. The background grey intensity represents the cell's distance from the goal in units of cell size as indicated by the *scale* on the right-hand side

the global view has just bumped its way through the world. The penalty for achieving the optimal path is computational cost. This particular implementation of the distance transform is iterative. Each iteration has a cost of $O(N^2)$ and the number of iterations is at least $O(N)$, where N is the dimension of the map.

We can visualize the iterations of the distance transform by

```
>> dx.plan(place.kitchen, 'animate');
```

which shows the distance values propagating as a wavefront outward from the goal. The wavefront moves outward, spills through doorways into adjacent rooms and outside the house.▸ Although the plan is expensive to create, once it has been created it can be used to plan a path from *any* initial point to that goal.

We have converted a fairly complex planning problem into one that can now be handled by a Braitenberg-class robot that makes local decisions based on the distance to the goal. Effectively the robot is rolling *downhill* on the distance function which we can plot as a 3D surface

```
>> dx.plot3d(p)
```

shown in Fig. 5.6 with the robot's path and room locations overlaid.

For large occupancy grids this approach to planning will become impractical. The roadmap methods that we discuss later in this chapter provide an effective means to find paths in large maps at greatly reduced computational cost.

The scale associated with this occupancy grid is 4.5 cm per cell and we have assumed the robot occupies a single grid cell – this is a very small robot. The planner could therefore find paths that a larger real robot would be unable to fit through. A common solution to this problem is to *inflate* the occupancy grid – making the obstacles bigger is equivalent to leaving the obstacles unchanged and making the robot bigger. For example, if we inflate the obstacles by 5 cells

```
>> dx = DXform(house, 'inflate', 5);
>> dx.plan(place.kitchen);
>> p = dx.query(place.br3);
>> dx.plot(p)
```

the path shown in Fig. 5.7b now takes the wider corridors to reach its goal. To illustrate how this works we can overlay this new path on the inflated occupancy grid

```
>> dx.plot(p, 'inflated');
```

More efficient algorithms exist such as fast marching methods and Dijkstra's method, but the iterative wavefront method used here is easy to code and to visualize.

and this is shown in Fig. 5.7a. The inflation parameter of 5 has grown the obstacles by 5 grid cells in all directions, a bit like applying a very thick layer of paint.◄ This is equivalent to growing the robot by 5 grid cells in all directions – the robot grows from a single grid cell to a disk with a diameter of 11 cells which is equivalent to around 50 cm.

This is the same as the image processing concept of morphological dilation.

Fig. 5.6.
The distance transform as a 3D function, where height is distance from the goal. Navigation is simply a downhill run. Note the discontinuity in the distance transform where the split wavefronts met

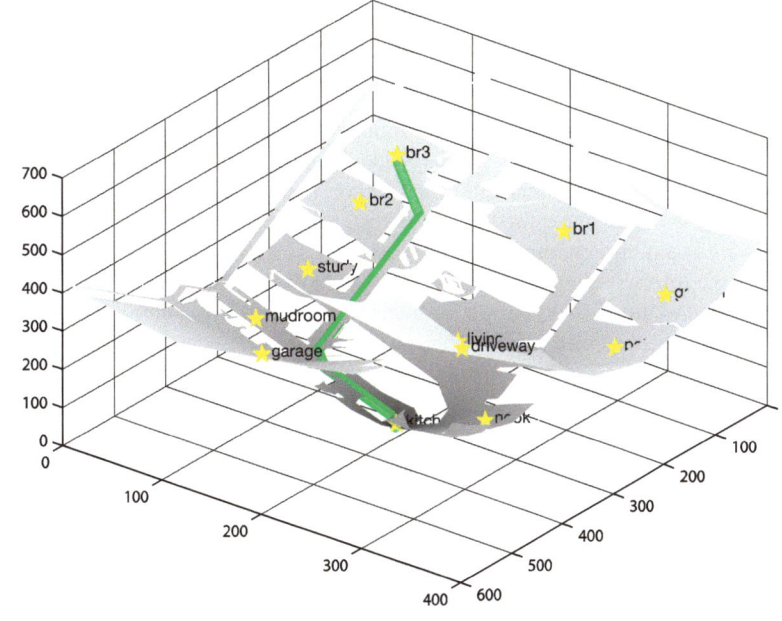

Fig. 5.7. Distance transform path with obstacles inflated by 5 cells. **a** Path shown with inflated obstacles; **b** path computed for inflated obstacles overlaid on original obstacle map, black regions are where no distance was computed
▼

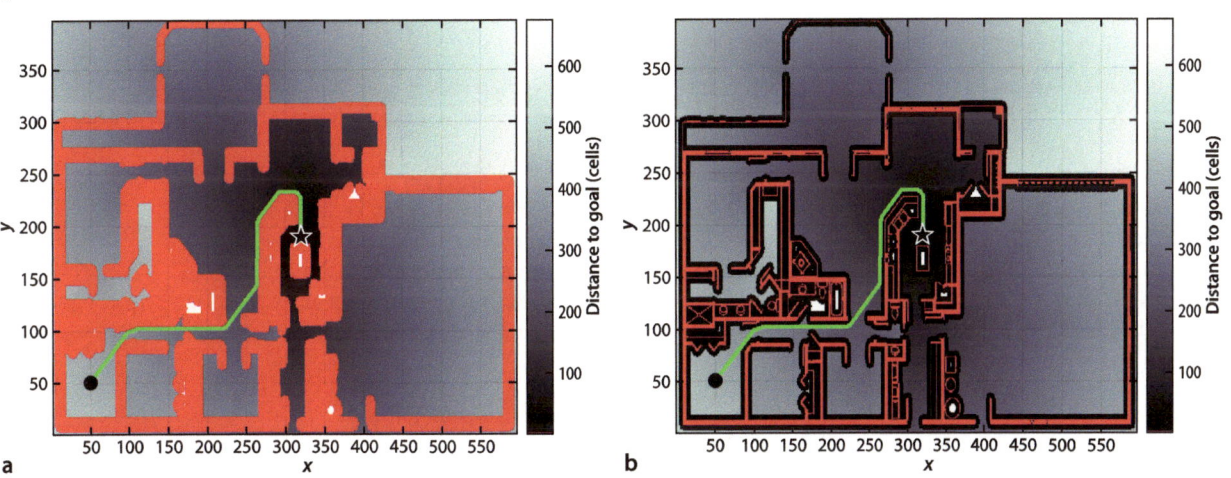

a   b

Navigation superclass. The examples in this chapter are all based on classes derived from the `Navigation` class which is designed for 2D grid-based navigation. Each example consists of essentially the following pattern. Firstly we create an instance of an object derived from the `Navigation` class by calling the class constructor.

```
nav = MyNavClass(map)
```

which is passed the occupancy grid. Then a plan is computed

```
nav.plan()
nav.plan(goal)
```

and depending on the planner the goal may or may not be required. A path from an initial position to the goal is computed by

```
p = nav.query(start, goal)
p = nav.query(start)
```

again depending on whether or not the planner requires a goal. The optional return value `p` is the path, a sequence of points from `start` to `goal`, one row per point, and each row comprises the *x*- and *y*-coordinate. If `start` or `goal` is given as [] the user is prompted to interactively click the point. The 'animate' option causes an animation of the robot's motion to be displayed.

The map and planning information can be visualized by

```
nav.plot()
```

or have a path overlaid

```
nav.plot(p)
```

5.2.2   D*

A popular algorithm for robot path planning is D* which finds the best path▶ through a graph, which it first computes, that corresponds to the input occupancy grid. D* has a number of features that are useful for real-world applications. Firstly, it generalizes the occupancy grid to a cost map which represents the cost $c \in \mathbb{R}$, $c > 0$ of traversing each cell in the horizontal or vertical direction. The cost of traversing the cell diagonally is $c\sqrt{2}$. For cells corresponding to obstacles $c = \infty$ (`Inf` in MATLAB).

Secondly, D* supports incremental replanning. This is important if, while we are moving, we discover that the world is different to our map. If we discover that a route has a higher than expected cost or is completely blocked we can incrementally replan to find a better path. The incremental replanning has a lower computational cost than completely replanning as would be required using the distance transform method just discussed.

D* finds the path which minimizes the total cost of travel. If we are interested in the shortest time to reach the goal then cost is the time to drive across the cell and is inversely related to traversability. If we are interested in minimizing damage to the vehicle or maximizing passenger comfort then cost might be related to the roughness of the terrain within the cell. The costs assigned to cells will also depend on the characteristics of the vehicle: a large 4-wheel drive vehicle may have a finite cost to cross a rough area whereas for a small car that cost might be infinite.

To implement the D* planner using the Toolbox we use a similar pattern and first create a D* navigation object

```
>> ds = Dstar(house);
```

The D* planner converts the passed occupancy grid `map` into a cost map which we can retrieve

```
>> c = ds.costmap();
```

where the elements of `c` will be 1 or ∞ representing free and occupied cells respectively.

A plan for moving to the goal is generated by

```
>> ds.plan(place.kitchen);
```

which creates a very dense directed graph (see Appendix I). Every cell is a graph vertex and has a cost, a distance to the goal, and a link to the neighboring cell that is closest to the goal. Each cell also has a state $t \in \{\text{NEW, OPEN, CLOSED}\}$. Initially every cell is in the NEW state, the cost of the goal cell is zero and its state is OPEN. We can consider the set of all cells in the OPEN state as a wavefront propagating outward from the goal.▶ The cost of

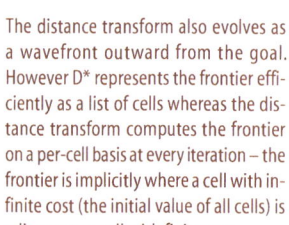

D* is an extension of the A* algorithm for finding minimum cost paths through a graph, see Appendix I.

The distance transform also evolves as a wavefront outward from the goal. However D* represents the frontier efficiently as a list of cells whereas the distance transform computes the frontier on a per-cell basis at every iteration – the frontier is implicitly where a cell with infinite cost (the initial value of all cells) is adjacent to a cell with finite cost.

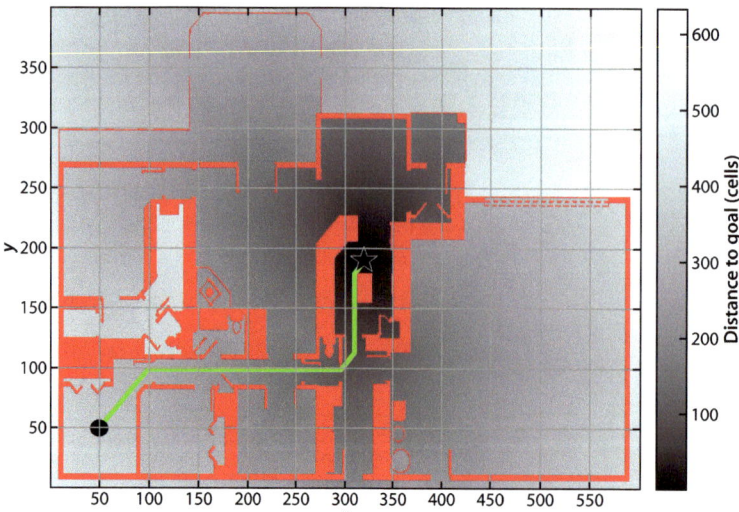

Fig. 5.8.
The D* planner path. Obstacles are indicated by *red cells* and all driveable cells have a cost of 1. The background grey intensity represents the cell's distance from the goal in units of cell size as indicated by the *scale* on the right-hand side

reaching cells that are neighbors of an OPEN cell is computed and these cells in turn are set to OPEN and the original cell is removed from the open list and becomes CLOSED. In MATLAB this initial planning phase is quite slow◄ and takes over a minute and

```
>> ds.niter
ans =
    245184
```

D* is more efficient than the distance transform but it executes more slowly because it is implemented entirely in MATLAB code whereas the distance transform is a MEX-file written in C.

iterations of the planning loop.

The path from an arbitrary starting point to the goal

```
>> ds.query(place.br3);
```

is shown in Fig. 5.8. The robot has again taken a short and efficient path around the obstacles that is almost identical to that generated by the distance transform.

The real power of D* comes from being able to efficiently change the cost map during the mission. This is actually quite a common requirement in robotics since real sensors have a finite range and a robot discovers more of world as it proceeds. We inform D* about changes using the modify_cost method, for example to raise the cost of entering the kitchen via the bottom doorway

```
>> ds.modify_cost( [300,325; 115,125], 5 );
```

we have raised the cost to 5 for a small rectangular region across the doorway. This region is indicated by the yellow dashed rectangle in Fig. 5.9. The other driveable cells have a default cost of 1. The plan is updated by invoking the planning algorithm again

```
>> ds.plan();
```

and this time the number of iterations is only

```
>> ds.niter
ans =
    169580
```

The cost increases with the number of cells modified and the effect those changes have on the distance map. It is possible that incremental replanning takes more time than planning from scratch.

which is 70% of that required to create the original plan.◄ The new path for the robot

```
>> ds.query(place.br3);
```

is shown in Fig. 5.9. The cost change is relatively small but we notice that the increased cost of driving within this region is indicated by a subtle brightening of those cells – in a cost sense these cells are now further from the goal. Compared to Fig. 5.8 the robot has taken a different route to the kitchen and avoided the bottom door. D* allows updates to the map to be made at any time while the robot is moving. After replanning the robot simply moves to the adjacent cell with the lowest cost which ensures continuity of motion even if the plan has changed.

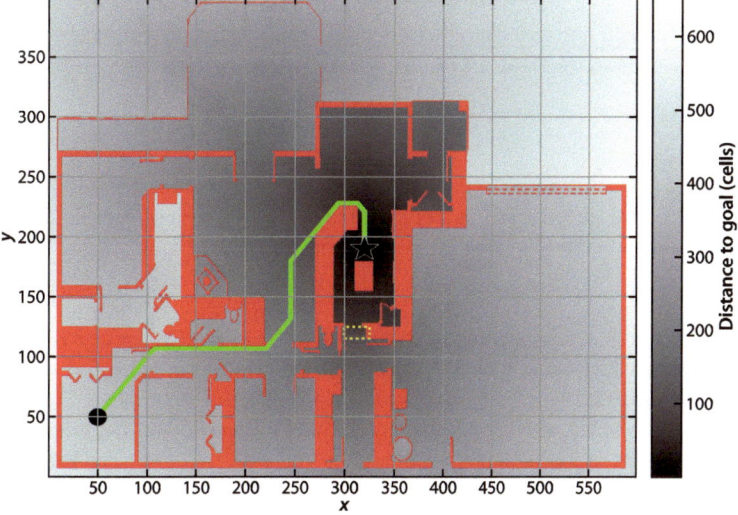

Fig. 5.9.
Path from D* planner with modified map. The higher-cost region is indicated by the *yellow dashed rectangle* and has changed the path compared to Fig. 5.7

A graph is an abstract representation of a set of objects connected by links typically denoted $G(V, E)$ and depicted diagrammatically as shown to the right. The objects shown as circles are called vertices or nodes. The lines that connect some pairs of vertices are called edges or arcs. Edges can be directed (arrows) or undirected as in this case. Edges can have an associated weight or cost associated with moving from one of its vertices to the other. A sequence of edges from one vertex to another is a path. Graphs can be used to represent transport or communications networks and even social relationships, and the branch of mathematics is graph theory. Minimum cost path between two nodes in the graph can be computed using well known algorithms such as Dijstrka's method or A*.

The navigation classes use a simple MATLAB graph class called `PGraph`, see Appendix I.

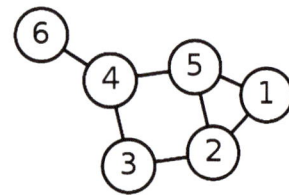

5.2.3 Introduction to Roadmap Methods

In robotic path planning the analysis of the map is referred to as the *planning phase*. The *query phase* uses the result of the planning phase to find a path from A to B. The two previous planning algorithms, distance transform and D*, require a significant amount of computation for the planning phase, but the query phase is very cheap. However the plan depends on the goal. If the goal changes the expensive planning phase must be re-executed. Even though D* allows the path to be recomputed as the costmap changes it does not support a changing goal.

The disparity in planning and query costs has led to the development of roadmap methods where the query can include both the start and goal positions. The planning phase provides analysis that supports changing starting points and changing goals. A good analogy is making a journey by train. We first find a local path to the nearest train station, travel through the train network, get off at the station closest to our goal, and then take a local path to the goal. The train network is invariant and planning a path through the train network is straightforward. Planning paths to and from the entry and exit stations respectively is also straightforward since they are, ideally, short paths. The robot navigation problem then becomes one of building a network of obstacle free paths through the environment which serve the function of the train network. In the literature such a network is referred to as a roadmap. The roadmap need only be computed once and can then be used like the train network to get us from any start location to any goal location.

We will illustrate the principles by creating a roadmap from the occupancy grid's free space using some image processing techniques. The essential steps in creating the roadmap are shown in Fig. 5.10. The first step is to find the free space in the map which is simply the complement of the occupied space

```
>> free = 1 - house
```

and is a matrix with nonzero elements where the robot is free to move. The boundary is also an obstacle so we mark the outermost cells as being not free

```
>> free(1,:) = 0; free(end,:) = 0;
>> free(:,1) = 0; free(:,end) = 0;
```

and this map is shown in Fig. 5.10a where free space is depicted as white.

The topological skeleton of the free space is computed by a morphological image processing algorithm known as thinning▸ applied to the free space of Fig. 5.10a

Also known as skeletonization.

```
>> skeleton = ithin(free);
```

and the result is shown in Fig. 5.10b. We see that the obstacles have grown and the free space, the white cells, have become a thin network of connected white cells which are equidistant from the boundaries of the original obstacles.

Figure 5.10c shows the free space network overlaid on the original map. We have created a network of paths that span the space and which can be used for obstacle-free travel around the map.▸ These paths are the edges of a generalized Voronoi

The junctions in the roadmap are indicated by black dots. The junctions, or triple points, are identified using the morphological image processing function `triplepoint`.

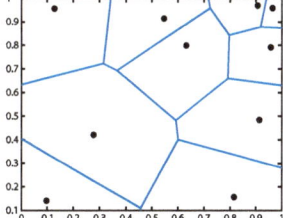

The Voronoi tessellation of a set of planar points, known as sites, is a set of Voronoi cells as shown to the left. Each cell corresponds to a site and consists of all points that are closer to its site than to any other site. The edges of the cells are the points that are equidistant to the two nearest sites. A generalized Voronoi diagram comprises cells defined by measuring distances to objects rather than points. In MATLAB we can generate a Voronoi diagram by

```
>> sites = rand(10,2)
>> voronoi(sites(:,1), sites(:,2))
```

Georgy Voronoi (1868–1908) was a Russian mathematician, born in what is now Ukraine. He studied at Saint Petersburg University and was a student of Andrey Markov. One of his students Boris Delaunay defined the eponymous triangulation which has dual properties with the Voronoi diagram.

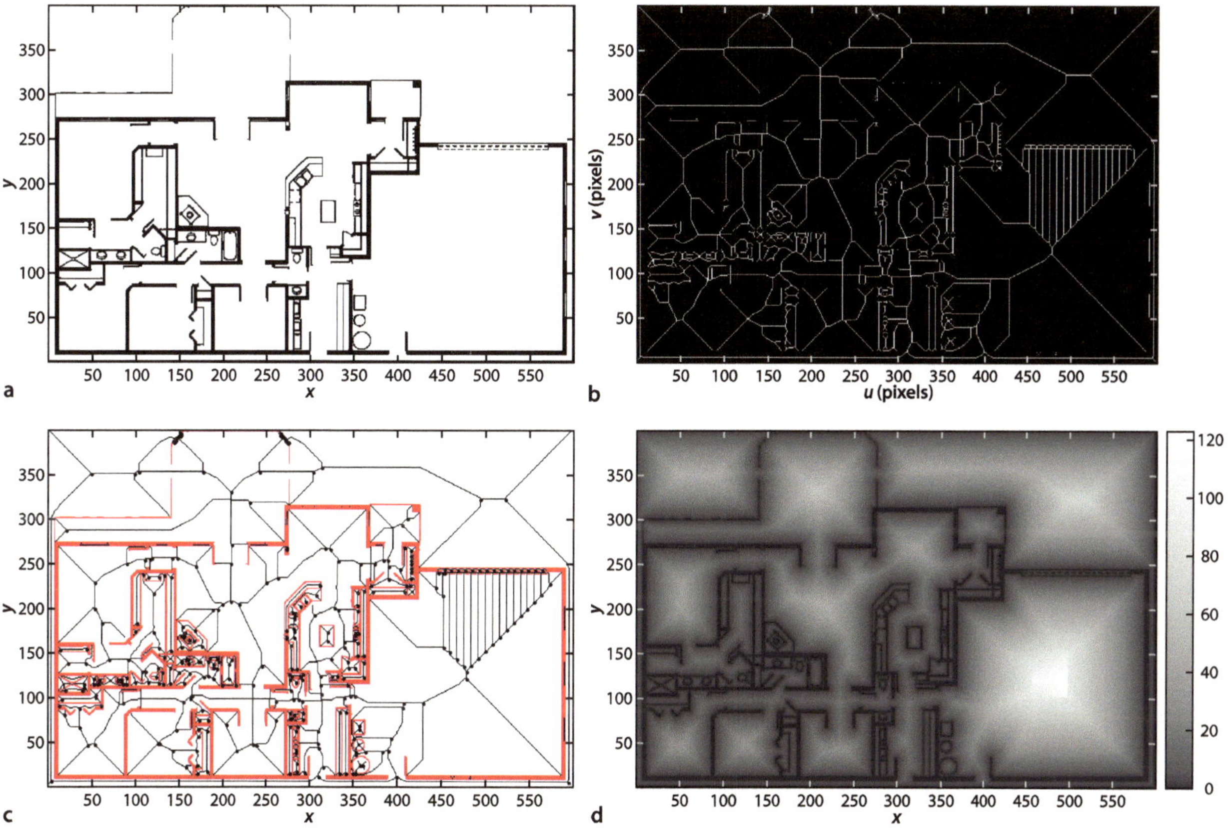

Fig. 5.10. Steps in the creation of a Voronoi roadmap. **a** Free space is indicated by *white cells*; **b** the skeleton of the free space is a network of adjacent cells no more than one cell thick; **c** the skeleton with the obstacles overlaid in red and road-map junction points indicated by *black dots*; **d** the distance transform of the obstacles, pixel values correspond to distance to the nearest obstacle

diagram. We could obtain a similar result by computing the distance transform of the obstacles, Fig. 5.10a, and this is shown in Fig. 5.10d. The value of each pixel is the distance to the nearest obstacle and the ridge lines correspond to the skeleton of Fig. 5.10b. Thinning or skeletonization, like the distance transform, is a computationally expensive iterative algorithm but it illustrates well the principles of finding paths through free space. In the next section we will examine a cheaper alternative.

5.2.4 Probabilistic Roadmap Method (PRM)

The high computational cost of the distance transform and skeletonization methods makes them infeasible for large maps and has led to the development of probabilistic methods. These methods sparsely sample the world map and the most well known of these methods is the probabilistic roadmap or PRM method.

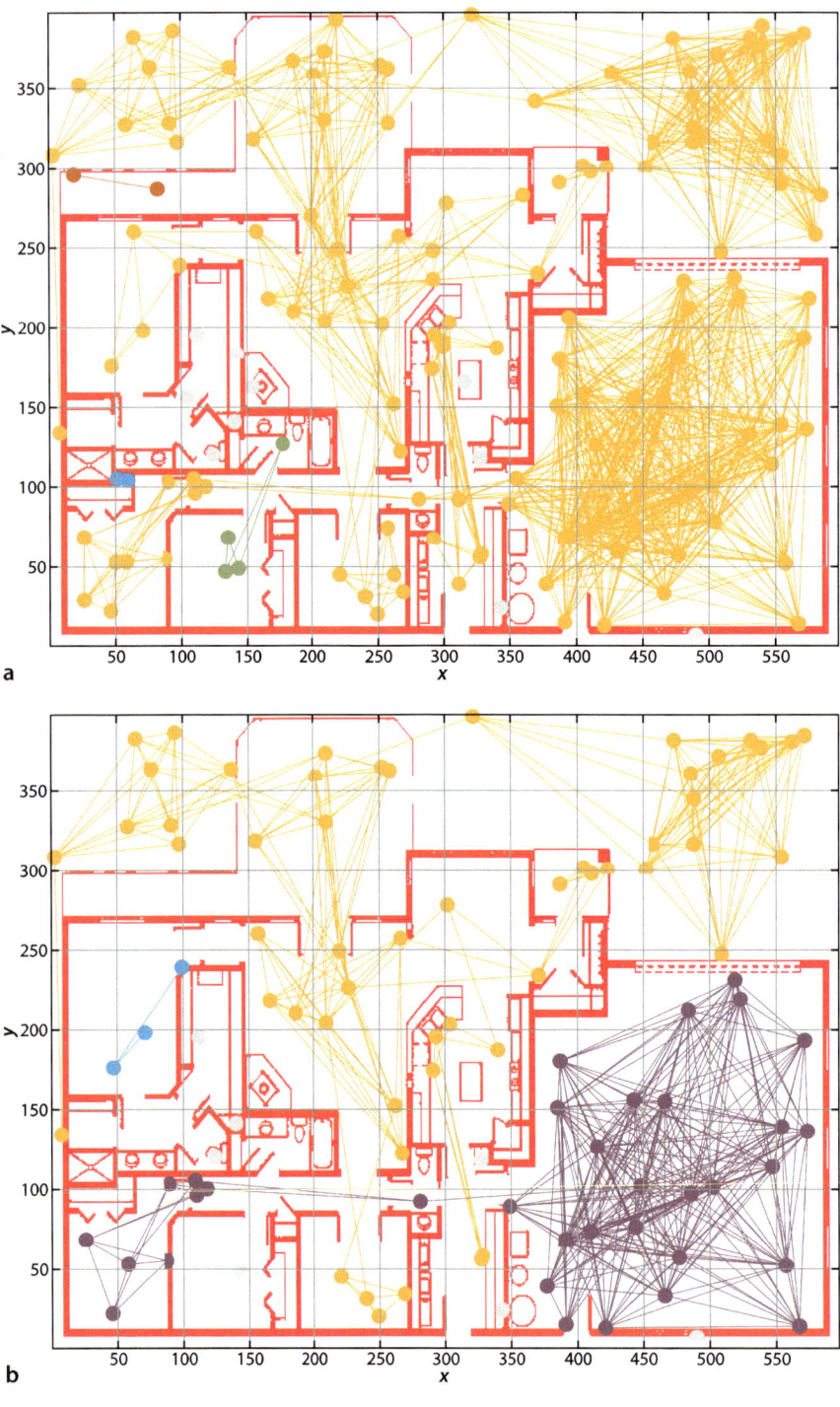

Fig. 5.11.
Probablistic roadmap (PRM)
planner and the random graphs
produced in the planning phase.
a Well connected network with
150 nodes; **b** poorly connected
network with 100 nodes

To use the Toolbox PRM planner for our problem we first create a PRM object

```
>> prm = PRM(house)
```

and then create the plan

```
>> prm.plan('npoints', 150)▶
```

with 150 roadmap nodes. Note that we do not pass the goal as an argument since the
plan is independent of the goal. Creating the path is a two phase process: planning, and

To replicate the following result be sure to
initialize the random number generator
first using randinit. See page 137.

query. The planning phase finds *N* random points, 150 in this case, that lie in free space. Each point is connected to its nearest neighbors by a straight line path that does not cross any obstacles, so as to create a network, or graph, with a minimal number of disjoint components and no cycles. The advantage of PRM is that relatively few points need to be tested to ascertain that the points and the paths between them are obstacle free. The resulting network is stored within the PRM object and a summary can be displayed

```
>> prm
prm =
PRM navigation class:
  occupancy grid: 397x596
  graph size: 150
  dist thresh: 178.8
  2 dimensions
  150 vertices
  1223 edges
  14 components
```

which indicates the number of edges and connected components in the graph. The graph can be visualized

This is derived automatically from the size of the occupancy grid.

```
>> prm.plot()
```

Random numbers. The MATLAB random number generator (used for rand and randn) generates a very long sequence of numbers that are an excellent approximation to a random sequence. The generator maintains an internal state which is effectively the position within the sequence. After startup MATLAB always generates the following random number sequence

```
>> rand
ans =
    0.8147
>> rand
ans =
    0.9058
>> rand
ans =
    0.1270
```

Many algorithms discussed in this book make use of random numbers and this means that the results can never be repeated. Before all such examples in this book is an invisible call to randinit which resets the random number generator to a known state

```
>> randinit
>> rand
ans =
    0.8147
>> rand
ans =
    0.9058
```

and we see that the random sequence has been restarted.

as shown in Fig. 5.11a. The dots represent the randomly selected points and the lines are obstacle-free paths between the points. Only paths less than 178.8 cells long are selected▼ which is the distance threshold parameter of the PRM class. Each edge of the graph has an associated cost which is the distance between its two nodes. The color of the node indicates which component it belongs to and each component is assigned a unique color. In this case there are 14 components but the bulk of nodes belong to a single large component.

The query phase finds a path from the start point to the goal. This is simply a matter of moving to the closest node in the roadmap (the start node), following a minimum cost A* route through the roadmap, getting off at the node closest to the goal and then traveling to the goal. For our standard problem this is

```
>> prm.query(place.br3, place.kitchen)
>> prm.plot()
```

and the path followed is shown in Fig. 5.12. The path that has been found is quite efficient although there are two areas where the path doubles back on itself. Note that we provide the start and the goal position to the query phase. An advantage of this planner is that once the roadmap is created by the planning phase we can change the goal and starting points very cheaply, only the query phase needs to be repeated. The path taken is

```
>> p = prm.query(place.br3, place.kitchen);
>> about p
p [double] : 9x2 (144 bytes)
```

which is a list of the node coordinates that the robot passes through – via points. These could be passed to a trajectory following controller as discussed in Sect. 4.1.1.3.

There are some important tradeoffs in achieving this computational efficiency. Firstly, the underlying random sampling of the free space means that a different roadmap is created every time the planner is run, resulting in different paths and path lengths. Secondly, the planner can fail by creating a network consisting of disjoint components. The roadmap in Fig. 5.11b, with only 100 nodes has several large disconnected components and the nodes in the kitchen and bedrooms belong to different components. If the start and goal nodes are not connected by the roadmap, that is, they are close to different components the query method will report an error. The only solution is to rerun the planner and/or increase the number of nodes. Thirdly, long narrow gaps between obstacles such as corridors are unlikely to be exploited since the probability of randomly choosing points that lie along such spaces is very low.

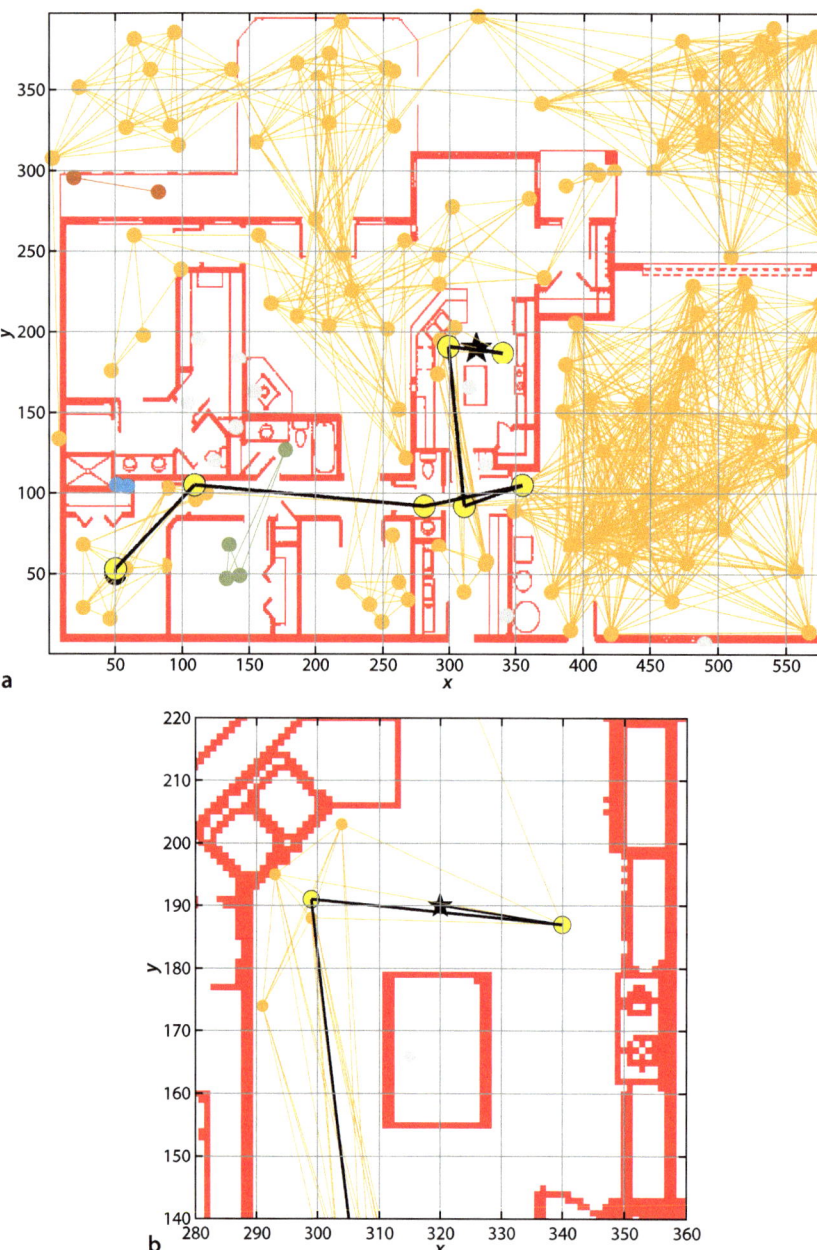

Fig. 5.12.
Probablistic roadmap (PRM)
planner **a** showing the path taken
by the robot via nodes of the
roadmap which are highlighted
in yellow; **b** closeup view of goal
region where the short path from
the final roadmap node to the
goal can be seen

5.2.5   Lattice Planner

The planners discussed so far have generated paths independent of the motion that the vehicle can actually achieve, and we learned in Chap. 4 that wheeled vehicles have significant motion constraints. One common approach is to use the output of the planners we have discussed and move a point along the paths at constant velocity and then follow that point, using techniques such as the trajectory following controller described in Sect. 4.1.1.3.

An alternative is to *design* a path from the outset that we know the vehicle can follow. The next two planners that we introduce take into account the motion model of the vehicle, and relax the assumption we have so far made that the robot is capable of omnidirectional motion.

We consider that the robot is moving between discrete points in its 3-dimensional configuration space. The robot is initially at the origin and can drive forward to the three points shown in black in Fig. 5.13a.◄ Each path is an arc➤ which requires a constant steering wheel setting and the arc radius is chosen so that at the end of each arc the robot's heading direction is some multiple of $\frac{\pi}{2}$ radians.

At the end of each branch we can add the same set of three motions suitably rotated and translated, and this is shown in Fig. 5.13b. The graph now contains 13 nodes and represents 9 paths each 2 segments long. We can create this lattice by using the `Lattice` planner class

```
>> lp = Lattice();
>> lp.plan('iterations', 2)
13 nodes created
>> lp.plot()
```

which will generate a plot like Fig. 5.13b. Each node represents a configuration (x, y, θ), not just a position, and if we rotate the plot we can see in Fig. 5.14 that the paths lie in the 3-dimensional configuration space.

While the paths appear smooth and continuous the curvature is in fact discontinuous – at some nodes the steering wheel angle would have to change instantaneously from hard left to hard right for example.◄

The pitch of the grid is dictated by the turning radius of the vehicle.

Sometimes called Dubins curves.

A real robot would take a finite time to adjust its steering angle and this would introduce some error in the robot path. The steering control system could compensate for this by turning harder later in the segment so as to bring the robot to the end point with the correct orientation.

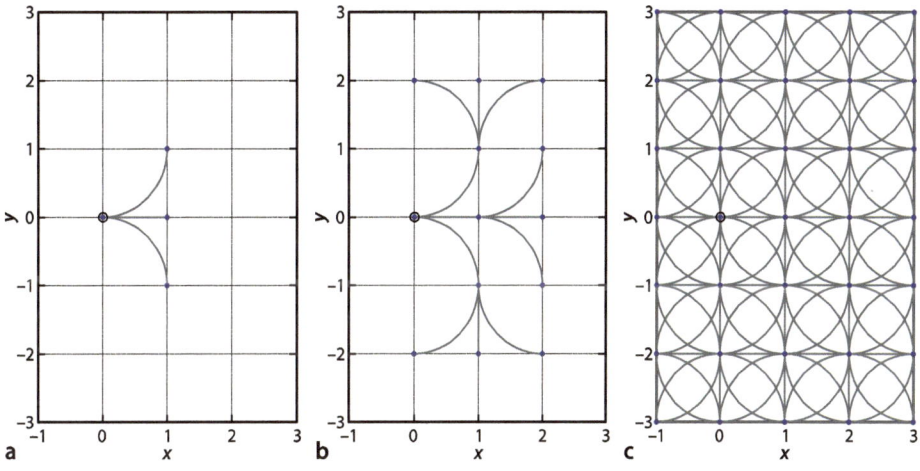

Fig. 5.13.
Lattice plan after 1, 2 and 8 iterations

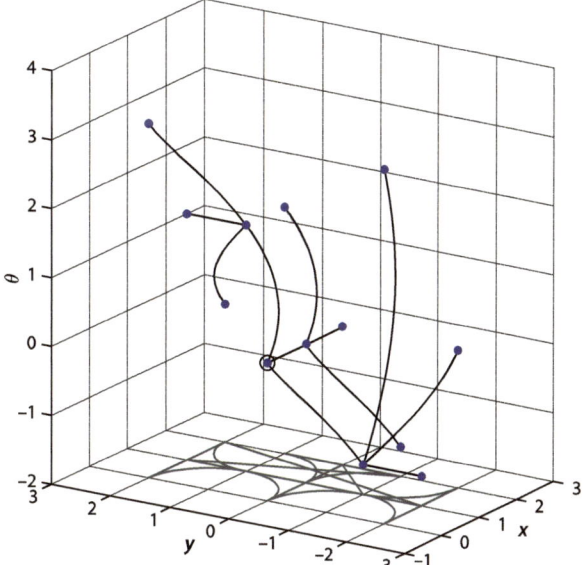

Fig. 5.14.
Lattice plan after 2 iterations shown in 3-dimensional configuration space

By increasing the number of iterations

```
>> lp.plan('iterations', 8)
780 nodes created
>> lp.plot()
```

we can fill in more possible paths as shown in Fig. 5.13c and the paths now extend well beyond the area shown.

Now that we have created the lattice we can compute a path between any two nodes using the query method

```
>> lp.query( [1 2 pi/2], [2 -2 0] );
A* path cost 6
```

where the start and goal are specified as configurations (x, y, θ) and the lowest cost path found by an A* search is reported.▶ We can overlay this on the vertices

Every segment in the lattice has a default cost of 1 so the cost of 6 simply reflects the total number of segments in the path. A* search is introduced in Appendix I.

```
>> lp.plot
```

and is shown in Fig. 5.15a. This is a path that takes into account the fact that the vehicle has an orientation and preferred directions of motion, as do most wheeled robot platforms. We can also access the configuration-space coordinates of the nodes

```
>> p = lp.query( [1 2 pi/2], [2 -2 0] )
A* path cost 6
>> about p
p [double] : 7x3 (168 bytes)
```

where each row represents the configuration-space coordinates (x, y, θ) of a node in the lattice along the path from start to goal configuration.

Implicit in our search for the lowest cost path is the cost of traversing each edge of the graph which by default gives equal cost to the three steering options: straight ahead, turn left and turn right. We can increase the cost associated with turning

```
>> lp.plan('cost', [1 10 10])
>> lp.query([1 2 pi/2], [2 -2 0]);
A* path cost 35
>> lp.plot()
```

and we now have the path shown in Fig. 5.15b which has only 3 turns compared to 5 previously. However the path is longer – having 8 rather than 6 segments.

Consider a more realistic scenario with obstacles in the environment. Specifically we want to find a path to move the robot 2 m in the lateral direction with its final heading angle the same as its initial heading angle

```
>> load road
>> lp = Lattice(road, 'grid', 5, 'root', [50 50 0])
>> lp.plan();
```

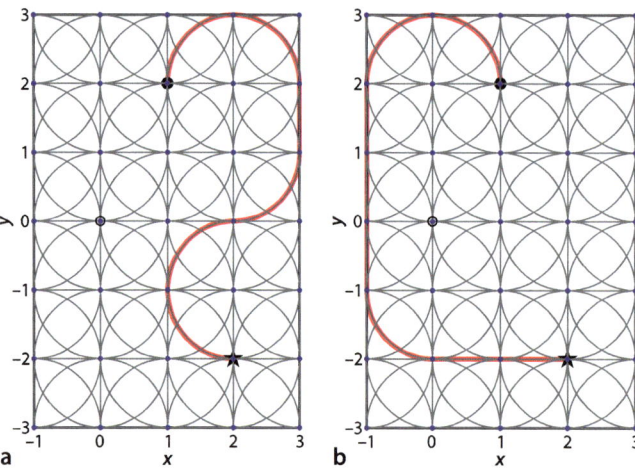

Fig. 5.15.
Paths over the lattice graph.
a With uniform cost; **b** with increased penalty for turns

where we have loaded an obstacle grid that represents a simple parallel-parking scenario and planned a lattice with a grid spacing of 5 units and the root node at a central obstacle-free configuration. In this case the planner continues to iterate until it can add no more nodes to the free space. We query for a path from the road to the parking spot

```
>> lp.query([30 45 0], [50 20 0])
```

and the result is shown in Fig. 5.16.

Paths generated by the lattice planner are inherently driveable by the robot but there are clearly problems driving along a diagonal with this simple lattice. The planner would generate a continual sequence of hard left and right turns which would cause undue wear and tear on a real vehicle and give a very uncomfortable ride. More sophisticated version of lattice planners are able to deal with this by using motion primitives with hundreds of arcs, such as shown in Fig. 5.17, instead of the three shown in these examples.

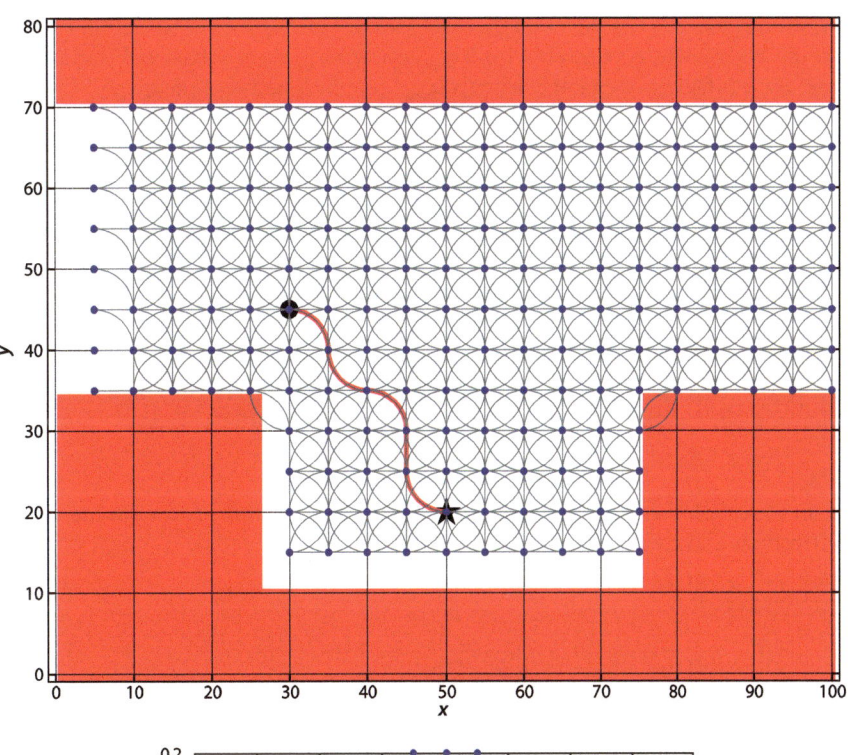

Fig. 5.16.
A simple parallel parking scenario based on the lattice planner with an occupancy grid (cells are 10 cm square)

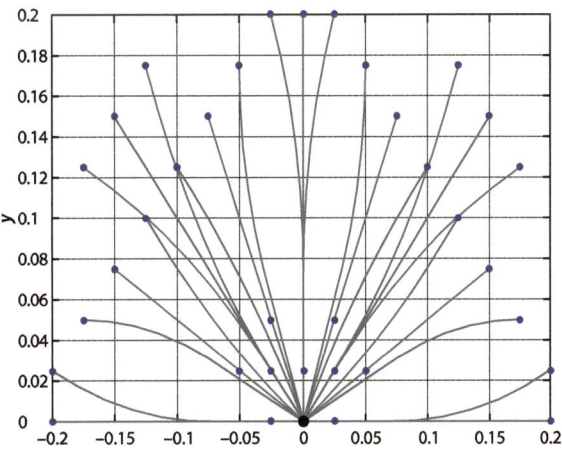

Fig. 5.17.
A more sophisticated lattice generated by the package **sbpl** with 43 paths based on the kinematic model of a unicycle

5.2.6   Rapidly-Exploring Random Tree (RRT)

The final planner that we introduce is also able to take into account the motion model of the vehicle. Unlike the lattice planner which plans over a regular grid, the RRT uses probabilistic methods like the PRM planner.

The underlying insight is similar to that for the lattice planner and Fig. 5.18 shows a family of paths that the bicycle model of Eq. 4.2 would follow in configuration space. The paths are computed over a fixed time interval for discrete values of velocity, forward or backward, and various steering angles. This demonstrates clearly the subset of all possible configurations that a nonholonomic vehicle can reach from a given initial configuration.

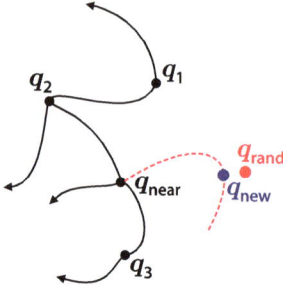

The main steps in creating an RRT are as follows, with the notation shown in the figure to the right. A graph of robot configurations is maintained and each node is a configuration $q \in \mathbb{R}^2 \times \mathbb{S}^1$ which is represented by a 3-vector $q \sim (x, y, \theta)$. The first, or root, node in the graph is the goal configuration of the robot. A random configuration q_{rand} is chosen, and the node with the closest configuration q_{near} is found – this configuration is near in terms of a cost function that includes distance and orientation.▸ A control is computed that moves the robot from q_{near} toward q_{rand} over a fixed path simulation time. The configuration that it reaches is q_{new} and this is added to the graph.

For any desired starting configuration we can find the closest configuration in the graph, and working backward toward the starting configuration we could determine the sequence of steering angles and velocities needed to move from the start to the goal configuration. This has some similarities to the roadmap methods discussed previously, but the limiting factor is the combinatoric explosion in the number of possible poses.

We first of all create a model to describe the vehicle kinematics

```
>> car = Bicycle('steermax', 0.5);
```

and here we have specified a car-like vehicle with a maximum steering angle of 0.5 rad. Following our familiar programming pattern we create an RRT object

```
>> rrt = RRT(car, 'npoints', 1000)
```

for an obstacle free environment which by default extends from –5 to +5 in the x- and y-directions and create a plan

```
>> rrt.plan();
>> rrt.plot();
```

The distance measure must account for a difference in position and orientation and requires appropriate weighting of these quantities. From a consideration of units this is not quite proper since we are adding meters and radians.

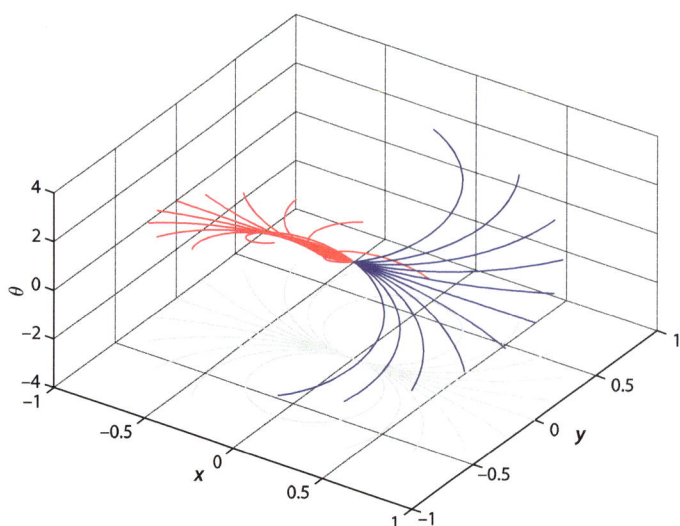

Fig. 5.18.
A set of possible paths that the bicycle model robot could follow from an initial configuration of $(0, 0, 0)$. For $v = \pm 1$, $\alpha \in [-1, 1]$ over a 2 s period. *Red lines* correspond to $v < 0$

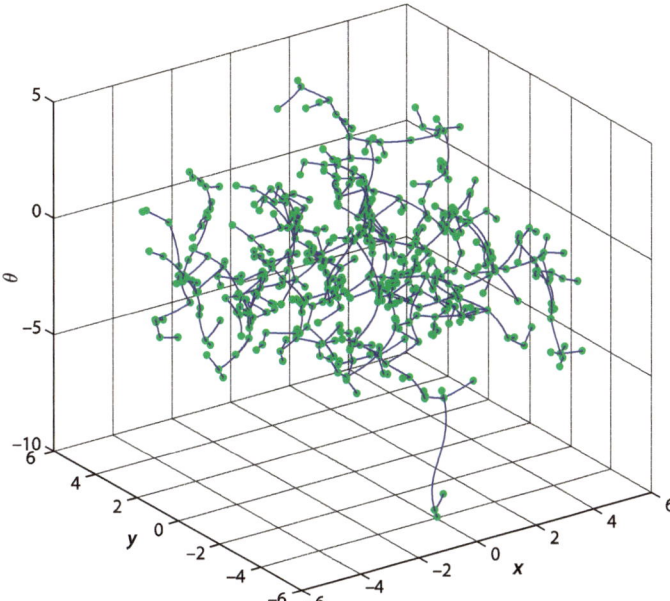

Fig. 5.19.
An RRT computed for the bicycle model with a velocity of
± 1 m s^{-1}, steering angle limits of
± 0.5 rad, integration period of
1 s, and initial configuration of
$(0, 0, 0)$. Each node is indicated by
a *green circle* in the 3-dimension-
al space of vehicle poses (x, y, θ)

The random tree is shown in Fig. 5.19 and we see that the paths have a good coverage of the configuration space, not just in the x- and y-directions but also in orientation, which is why the algorithm is known as *rapidly exploring*.

An important part of the RRT algorithm is computing the control input that moves the robot from an existing configuration in the graph to q_{rand}. From Sect. 4.1 we understand the difficulty of driving a nonholonomic vehicle to a specified configuration. Rather than the complex nonlinear controller of Sect. 4.1.1.4 we will use something simpler that fits with the randomized sampling strategy used in this class of planner. The controller randomly chooses whether to drive forwards or backwards and randomly chooses a steering angle within the limits◄. It then simulates motion of the vehicle model for a fixed period of time, and computes the closest distance to q_{rand}. This is repeated multiple times and the control input with the best performance is chosen. The configuration on its path that was closest to q_{rand} is chosen as q_{near} and becomes a new node in the graph.

Handling obstacles with the RRT is quite straightforward. The configuration q_{rand} is discarded if it lies within an obstacle, and the point q_{near} will not be added to the graph if the path from q_{near} toward q_{rand} intersects an obstacle. The result is a set of paths, a roadmap, that is collision free and driveable by this nonholonomic vehicle.◄

We will repeat the parallel parking example from the last section

```
>> rrt = RRT(car, road, 'npoints', 1000, 'root', [50 22 0], 'simtime', 4)
>> rrt.plan();
```

where we have specified the vehicle kinematic model, an occupancy grid, the number of sample points, the location of the first node, and that each random motion is simulated for 4 seconds. We can query the RRT plan for a path between two configurations

```
>> p = rrt.query([40 45 0], [50 22 0]);
```

and the result is a continuous path

```
>> about p
p [double] : 520x3 (12.5 kB)
```

which will take the vehicle from the street to the parking slot. We can overlay the path on the occupancy grid and RRT

```
>> rrt.plot(p)
```

Uniformly randomly distributed between the steering angle limits.

We have chosen the first node to be the goal configuration, and we search from here toward possible start configurations. However we could also make the first node the start configuration. Alternatively we could choose the start node to be neither the start or goal position, the planner will find a path through the RRT between configurations close to the start and goal.

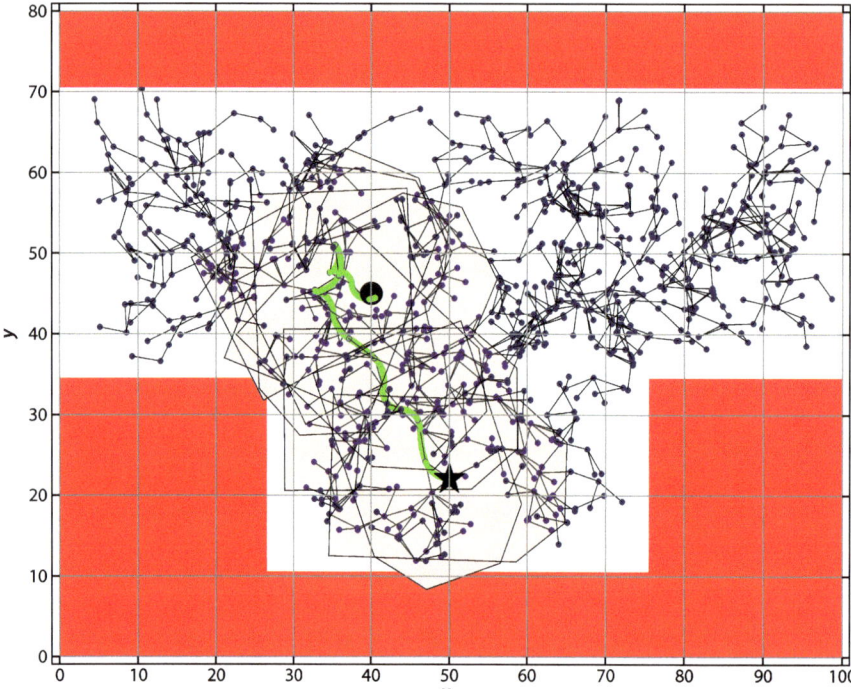

Fig. 5.20.
A simple parallel parking example based on the RRT planner with an occupancy grid (cells are 10 cm square). RRT nodes are shown in *blue*, the initial configuration is a *solid circle* and the goal is a *solid star*. The path through the RRT is shown in *green*, and a few snapshots of the vehicle configuration are overlaid in *pink*

and the result is shown in Fig. 5.20 with some vehicle configurations overlaid. We can also animate the motion along the path

```
>> plot_vehicle(p, 'box', 'size', [20 30], 'fillcolor', 'r', 'alpha', 0.1)
```

where we have specified the vehicle be displayed as a red translucent shape of width 20 and length 30 units.

This example illustrates some important points about the RRT. Firstly, as for the PRM planner, there may be some distance (and orientation) between the start and goal configuration and the nearest node. Minimizing this requires tuning RRT parameters such as the number of nodes and path simulation time. Secondly, the path is feasible but not quite optimal. In this case the vehicle has changed direction twice before driving into the parking slot. This is due to the random choice of nodes – rerunning the planner and/or increasing the number of nodes may help. Finally, we can see that the vehicle body collides with the obstacle, and this is very apparent if you view the animation. This is actually not surprising since the collision check we did when adding a node only tested if the node's position lay in an obstacle – it should properly check if a finite-sized vehicle with that configuration intersects an obstacle. Alternatively, as discussed on page 131 we could inflate the obstacles by the radius of the smallest disk that contains the robot.

5.3   Wrapping Up

Robot navigation is the problem of guiding a robot towards a goal and we have covered a spectrum of approaches. The simplest was the purely reactive Braitenberg-type vehicle. Then we added limited memory to create state machine based automata such as *bug2* which can deal with obstacles, however the paths that it finds are far from optimal.

A number of different map-based planning algorithms were then introduced. The distance transform is a computationally intense approach that finds an optimal path to the goal. D* also finds an optimal path, but supports a more nuanced travel cost – individual cells have a continuous traversability measure rather than being considered as only free space or obstacle. D* also supports computationally cheap incremental replanning for small changes in the map. PRM reduces the computational burden significantly by probabilistic sampling but at the expense of somewhat less optimal paths. In particular it may not discover narrow routes between areas of free space. The lattice planner takes into account the motion constraints of a real vehicle to create paths which are feasible to drive, and can readily account for the orientation of the vehicle as well as its position. RRT is another random sampling method that also generates kinematically feasible paths. All the map-based approaches require a map and knowledge of the robot's location, and these are both topics that we will cover in the next chapter.

Further Reading

Comprehensive coverage of planning for robots is provided by two text books. Choset et al. (2005) covers geometric and probabilistic approaches to planning as well as the application to robots with dynamics and nonholonomic constraints. LaValle (2006) covers motion planning, planning under uncertainty, sensor-based planning, reinforcement learning, nonlinear systems, trajectory planning, nonholonomic planning, and is available online for free at http://planning.cs.uiuc.edu. In particular these books provide a much more sophisticated approach to representing obstacles in configuration space and cover potential-field planning methods which we have not discussed. The powerful planning techniques discussed in these books can be applied beyond robotics to very high order systems such as vehicles with trailers, robotic arms or even the shape of molecules. LaValle (2011a) and LaValle (2011b) provide a concise two-part tutorial introduction. More succinct coverage of planning is given by Kelly (2013), Siegwart et al. (2011), the Robotics Handbook (Siciliano and Khatib 2016, § 7), and also in Spong et al. (2006) and Siciliano et al. (2009).

The *bug1* and *bug2* algorithms were described by Lumelsky and Stepanov (1986). More recently eleven variations of Bug algorithm were implemented and compared for a number of different environments (Ng and Bräunl 2007). The distance transform is well described by Borgefors (1986) and its early application to robotic navigation was explored by Jarvis and Byrne (1988). Efficient approaches to implementing the distance transform include the two-pass method of Hirata (1996), fast marching methods or reframing it as a graph search problem which can be solved using Dijkstra's method; the last two approaches are compared by Alton and Mitchell (2006). The A* algorithm (Nilsson 1971) is an efficient method to find the shortest path through a graph, and we can always compute a graph that corresponds to an occupancy grid map. D* is an extension by Stentz (1994) which allows cheap replanning when the map changes and there have been many further extensions including, but not limited to, Field D* (Ferguson and Stentz 2006) and D* lite (Koenig and Likhachev 2005). D* is used in many real-world robot systems and many implementations exist including open source.

The ideas behind PRM started to emerge in the mid 1990s and it was first described by Kavraki et al. (1996). Geraerts and Overmars (2004) compare the efficacy of a number of subsequent variations that have been proposed to the basic PRM algorithm. Approaches to planning that incorporate the vehicle's dynamics include state-space sampling (Howard et al. 2008), and the RRT which is described in LaValle (1998, 2006) and related resources at http://msl.cs.uiuc.edu. More recently RRT* has been proposed by Karaman et al. (2011). Lattice planners are covered in Pivtoraiko, Knepper, and Kelly (2009).

Historical and interesting. The defining book in cybernetics was written by Wiener in 1948 and updated in 1965 (Wiener 1965). Grey Walter published a number of popular articles (1950, 1951) and a book (1953) based on his theories and experiments with robotic tortoises.

The definitive reference for Braitenberg vehicles is Braitenberg's own book (1986) which is a whimsical, almost poetic, set of thought experiments. Vehicles of increasing complexity (fourteen vehicle families in all) are developed, some including nonlinearities, memory and logic to which he attributes anthropomorphic characteristics such as love, fear, aggression and egotism. The second part of the book outlines the factual basis of these machines in the neural structure of animals.

Early behavior-based robots included the Johns Hopkins Beast, built in the 1960s, and Genghis (Brooks 1989) built in 1989. Behavior-based robotics are covered in the book by Arkin (1999) and the Robotics Handbook (Siciliano and Khatib 2016, § 13). Matarić's Robotics Primer (Matarić 2007) and associated comprehensive web-based resources is also an excellent introduction to reactive control, behavior based control and robot navigation. A rich collection of archival material about early cybernetic machines, including Grey-Walter's tortoise and the Johns Hopkins Beast can be found at the Cybernetic Zoo http://cyberneticzoo.com.

Resources

A very powerful set of motion planners exist in OMPL, the Open MotionPLanning Library (http://ompl.kavrakilab.org) written in C++. It has a Python-based app that provides a convenient means to explore planning problems. Steve LaValle's web site http://msl.cs.illinois.edu/~lavalle/code.html has many code resources (C++ and Python) related to motion planning. Lattice planners are included in the sbpl package from the Search-Based Planning Lab (http://sbpl.net) which has MATLAB tools for generating motion primitives and C++ code for planning over the lattice graphs.

MATLAB Notes

The Robotics System Toolbox™ from The MathWorks Inc. includes functions `Binary-OccupancyGrid` and `PRM` to create occupancy grids and plan paths using probabilistic roadmaps. Other functions support reading and writing ROS navigation and map messages. The Image Processing Toolbox™ function `bwdist` is an efficient implementation of the distance transform.

Exercises

1. Braitenberg vehicles (page 125)
   a) Experiment with different starting configurations and control gains.
   b) Modify the signs on the steering signal to make the vehicle light-phobic.
   c) Modify the `sensorfield` function so that the peak moves with time.
   d) The vehicle approaches the maxima asymptotically. Add a stopping rule so that the vehicle stops when the when either sensor detects a value greater than 0.95.
   e) Create a scalar field with two peaks. Can you create a starting pose where the robot gets confused?
2. Occupancy grids. Create some different occupancy grids and test them on the different planners discussed.
   a) Create an occupancy grid that contains a maze and test it with various planners. See http://rosettacode.org/wiki/Maze_generation.
   b) Create an occupancy grid from a downloaded floor plan.
   c) Create an occupancy grid from a city street map, perhaps apply color segmentation to segment roads from other features. Can you convert this to a cost map for D* where different roads or intersections have different costs?
   d) Experiment with obstacle inflation.
   e) At 1 m cell size how much memory is required to represent the surface of the Earth? How much memory is required to represent just the land area of Earth? What cell size is needed in order for a map of your country to fit in 1 Gbyte of memory?

3. Bug algorithms (page 126)
   a) Using the function `makemap` create a new map to challenge *bug2*. Try different starting points.
   b) Create an obstacle map that contains a maze. Can *bug2* solve the maze?
   c) Experiment with different start and goal locations.
   d) Create a bug trap. Make a hollow box, and start the bug inside a box with the goal outside. What happens?
   e) Modify *bug2* to change the direction it turns when it hits an obstacle.
   f) Implement other bug algorithms such as *bug1* and *tangent bug*. Do they perform better or worse?

4. Distance transform (page 130). Create an obstacle map that contains a maze and solve it using distance transform.

5. D* planner (page 132)
   a) Add a low cost region to the living room. Can you make the robot prefer to take this route to the kitchen?
   b) Block additional doorways to challenge the robot.
   c) Implement D* as a mex-file to speed it up.

6. PRM planner (page 136)
   a) Run the PRM planner 100 times and gather statistics on the resulting path length.
   b) Vary the value of the distance threshold parameter and observe the effect.
   c) Use the output of the PRM planner as input to a pure pursuit planner as discussed in Chap. 4.
   d) Implement a nongrid based version of PRM. The robot is represented by an arbitrary polygon as are the obstacles. You will need functions to determine if a polygon intersects or is contained by another polygon (see the Toolbox `Polygon` class). Test the algorithm on the piano movers problem.

7. Lattice planner (page 138)
   a) How many iterations are required to completely fill the region of interest shown in Fig. 5.13c?
   b) How does the number of nodes and the spatial extent of the lattice increase with the number of iterations?
   c) Given a car with a wheelbase of 4.5 m and maximum steering angles of ± 30 deg what is the smallest possible grid size?
   d) Redo Fig. 5.15b to achieve a path that uses only right hand turns.
   e) Compute curvature as a function of path length for a path through the lattice such as the one shown in Fig. 5.15a.
   f) Design a controller in Simulink that will take a unicycle or bicycle model with a finite steering angle rate (there is a block parameter to specify this) that will drive the vehicle along the three paths shown in Fig. 5.13a.

8. RRT planner (page 142)
   a) Find a path to implement a 3-point turn.
   b) Experiment with RRT parameters such as the number of points, the vehicle steering angle limits, and the path integration time.
   c) Additional information in the node of each graph holds the control input that was computed to reach the node. Plot the steering angle and velocity sequence required to move from start to goal pose.
   d) Add a local planner to move from initial pose to the closest vertex, and from the final vertex to the goal pose.
   e) Determine a path through the graph that minimizes the number of reversals of direction.
   f) The collision test currently only checks that the center point of the robot does not lie in an occupied cell. Modify the collision test so that the robot is represented by a rectangular robot body and check that the entire body is obtacle free.

6 Localization

in order to get somewhere we need to know where we are

In our discussion of map-based navigation we assumed that the robot had a means of knowing its position. In this chapter we discuss some of the common techniques used to estimate the location of a robot in the world – a process known as localization.

Today GPS makes outdoor localization so easy that we often take this capability for granted. Unfortunately GPS is a far from perfect sensor since it relies on very weak radio signals received from distant orbiting satellites. This means that GPS cannot work where there is no *line of sight* radio reception, for instance indoors, underwater, underground, in urban canyons or in deep mining pits. GPS signals are also extremely weak and can be easily jammed and this is not acceptable for some applications.

GPS has only been in use since 1995 yet humankind has been navigating the planet and localizing for many thousands of years. In this chapter we will introduce the *classical* navigation principles such as dead reckoning and the use of landmarks on which modern robotic navigation is founded.

Dead reckoning is the estimation of location based on estimated speed, direction and time of travel with respect to a previous estimate. Figure 6.1 shows how a ship's position is updated on a chart. Given the average compass heading over the previous hour and a distance traveled the position at 3 P.M. can be found using elementary geometry from the position at 2 P.M. However the measurements on which the update is based are subject to both systematic and random error. Modern instruments are quite precise but 500 years ago clocks, compasses and speed measurement were primitive. The recursive nature of the process, each estimate is based on the previous

Fig. 6.1.
Location estimation by dead reckoning. The ship's position at 3 P.M. is based on its position at 2 P.M., the estimated distance traveled since, and the average compass heading

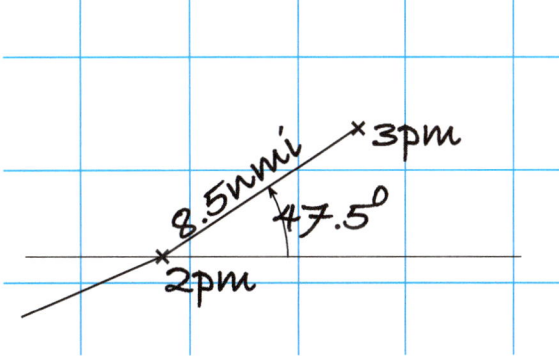

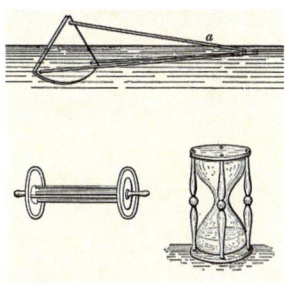

Measuring speed at sea. A ship's log is an instrument that provides an estimate of the distance traveled. The oldest method of determining the speed of a ship at sea was the Dutchman's log – a floating object was thrown into the water at the ship's bow and the time for it to pass the stern was measured using an hourglass. Later came the chip log, a flat quarter-circle of wood with a lead weight on the circular side causing it to float upright and resist towing. It was tossed overboard and a line with knots at 50 foot intervals was payed out. A special hourglass, called a log glass, ran for 30 s, and each knot on the line over that interval corresponds to approximately 1 nmi h^{-1} or 1 knot. A nautical mile (nmi) is now defined as 1.852 km. (Image modified from Text-Book of Seamanship, Commodore S. B. Luce 1891)

© Springer Nature Switzerland AG 2022
P. Corke, *Robotics and Control*, Springer Tracts in Advanced Robotics 141,
https://doi.org/10.1007/978-3-030-79179-7_6

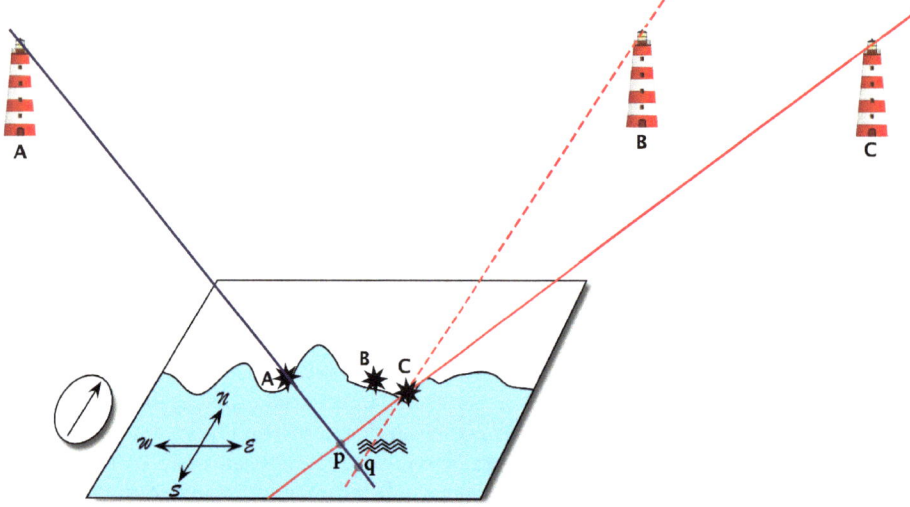

Fig. 6.2.
Location estimation using a
map. Lines of sight from two
light-houses, *A* and *C*, and their
corresponding locations on the
map provide an estimate *p* of
our location. However if we mistake lighthouse *B* for *C* then we
obtain an incorrect estimate *q*

one, means that errors will accumulate over time and for sea voyages of many-years this approach was quite inadequate.

The Phoenicians were navigating at sea more than 4 000 years ago and they did not even have a compass – that was developed 2 000 years later in China. The Phoenicians navigated with crude dead reckoning but wherever possible they used *additional infor-mation* to correct their position estimate – sightings of islands and headlands, primitive maps and observations of the Sun and the Pole Star.

A landmark is a visible feature in the environment whose location is known with respect to some coordinate frame. Figure 6.2 shows schematically a map and a number of lighthouse landmarks. We first of all use a compass to align the north axis of our map with the direction of the north pole. The direction of a single landmark constrains our position to lie along a line on the map. Sighting a second landmark places our position on another constraint line, and our position must be at their intersection – a process known as resectioning.▶ For example lighthouse **A** constrains us to lie along the blue line. Lighthouse **C** constrains us to lie along the red line and the intersection is our true position **p**.

However this process is critically reliant on correctly associating the observed landmark with the feature on the map. If we mistake one lighthouse for another, for example we see **B** but think it is **C** on the map, then the red dashed line leads to a

Resectioning is the estimation of position by measuring the bearing angles to known landmarks. Triangulation is the estimation of position by measuring the bearing angles to the unknown point from each of the landmarks.

Celestial navigation. The position of celestial bodies in the sky is a predictable function of the time and the observer's latitude and longitude. This information can be tabulated and is known as ephemeris (meaning daily) and such data has been published annually in Britain since 1767 as the "*The Nautical Almanac*" by HM Nautical Almanac Office. The elevation of a celestial body with respect to the horizon can be measured using a sextant, a handheld optical instrument.

Time and longitude are coupled, the star field one hour later is the same as the star field 15° to the east. However the northern Pole Star, *Polaris* or the *North Star*, is very close to the celestial pole and its elevation angle is independent of longitude and time, allowing latitude to be determined very conveniently from a single sextant measuremement.

Solving the longitude problem was the greatest scientific challenge to European governments in the eighteenth century since it was a significant impediment to global navigation and maritime supremacy. The British Longitude Act of 1714 created a prize of £20 000 which spurred the development of nautical chronometers, clocks that could maintain high accuracy onboard ships. More than fifty years later a suitable chronometer was developed by John Harrison, a copy of which was used by Captain James Cook on his second voyage of 1772–1775. After a three year journey the error in estimated longitude was just 13 km. With accurate knowledge of time, the elevation angle of stars could be used to estimate latitude and longitude. This technological advance enabled global exploration and trade. (Image courtesy archive.org)

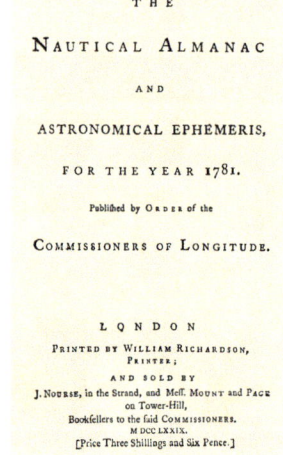

Radio-based localization. One of the earliest systems was LORAN, based on the British World War II GEE system. LORAN transmitters around the world emit synchronized radio pulses and a receiver measures the difference in arrival time between pulses from a pair of radio transmitters. Knowing the identity of two transmitters and the time difference (TD) constrains the receiver to lie along a hyperbolic curve shown on navigation charts as *TD lines*. Using a second pair of transmitters (which may include one of the first pair) gives another hyperbolic constraint curve, and the receiver must lie at the intersection of the two curves.

The Global Positioning System (GPS) was proposed in 1973 but did not become fully operational until 1995. It comprises around 30 active satellites orbiting the Earth in six planes at a distance of 20 200 km. A GPS receiver works by measuring the time of travel of radio signals from four or more satellites whose orbital position is encoded in the GPS signal. With four known points in space and four measured time delays it is possible to compute the (x, y, z) position of the receiver and the time. If the GPS signals are received after reflecting off some surface the dis-

tance traveled is longer and this will introduce an error in the position estimate. This effect is known as multi-pathing and is a common problem in large-scale industrial facilities.

Variations in the propagation speed of radio waves through the atmosphere is the main cause of error in the position estimate. However these errors vary slowly with time and are approximately constant over large areas. This allows the error to be measured at a reference station and transmitted as an augmentation to compatible nearby receivers which can offset the error – this is known as Differential GPS (DGPS). This information can be transmitted via the internet, via coastal radio networks to ships, or by satellite networks such as WAAS, EGNOS or OmniSTAR to aircraft or other users. RTK GPS achieves much higher precision in time measurement by using phase information from the carrier signal. The original GPS system deliberately added error, euphemistically termed selective availability, to reduce its utility to military opponents but this *feature* was disabled in May 2000. Other satellite navigation systems include the Russian GLONASS, the European Galileo, and the Chinese Beidou.

significant error in estimated position – we would believe we were at **q** instead of **p**. This belief would lead us to overestimate our distance from the coastline. If we decided to sail toward the coast we would run aground on rocks and be surprised since they were not where we expected them to be. This is unfortunately a very common error and countless ships have foundered because of this fundamental data association error. This is why lighthouses flash! In the eighteenth century technological advances enabled lighthouses to emit unique flashing patterns so that the identity of the particular lighthouse could be reliably determined and associated with a point on a navigation chart.

Of course for the earliest mariners there were no maps, or lighthouses or even compasses. They had to create maps as they navigated by incrementally adding new nonmanmade features to their maps just beyond the boundaries of what was already known. It is perhaps not surprising that so many early explorers came to grief◄ and that maps were tightly kept state secrets.

Robots operating today in environments without GPS face *exactly* the same problems as ancient navigators and, perhaps surprisingly, borrow heavily from navigational strategies that are centuries old. A robot's estimate of distance traveled will be imperfect and it may have no map, or perhaps an imperfect or incomplete map. Additional information from observed features in the world is critical to minimizing a robot's localization error but the possibility of data association error remains.

We can define the localization problem more formally where x is the true, but unknown, position of the robot and $\hat{x}$ is our best estimate of that position. We also wish to know the *uncertainty* of the estimate which we can consider in statistical terms as the standard deviation associated with the position estimate $\hat{x}$.

It is useful to describe the robot's estimated position in terms of a probability density function (PDF) over all possible positions of the robot.◄ Some example PDFs are shown in Fig. 6.3 where the magnitude of the function at any point is the relative likelihood of the vehicle being at that position. Commonly a Gaussian function is used which can be described succinctly in terms of its mean and standard deviation. The robot is most likely to be at the location of the peak (the mean) and increasingly less likely to be at positions further away from the peak. Figure 6.3a shows a peak with a small standard deviation which indicates that the vehicle's position is very well known. There is an almost zero probability that the vehicle is at the point indicated by the vertical black line. In contrast the peak in Fig. 6.3b has a large standard deviation which means that we are less certain about the location of the vehicle. There is a reasonable probability that the vehicle is at the point indicated by the vertical line.

Magellan's 1519 expedition started with 237 men and 5 ships but most, including Magellan, were lost along the way. Only 18 men and 1 ship returned.

For robot pose (x, y, θ) the PDF is a 4-dimensional surface.

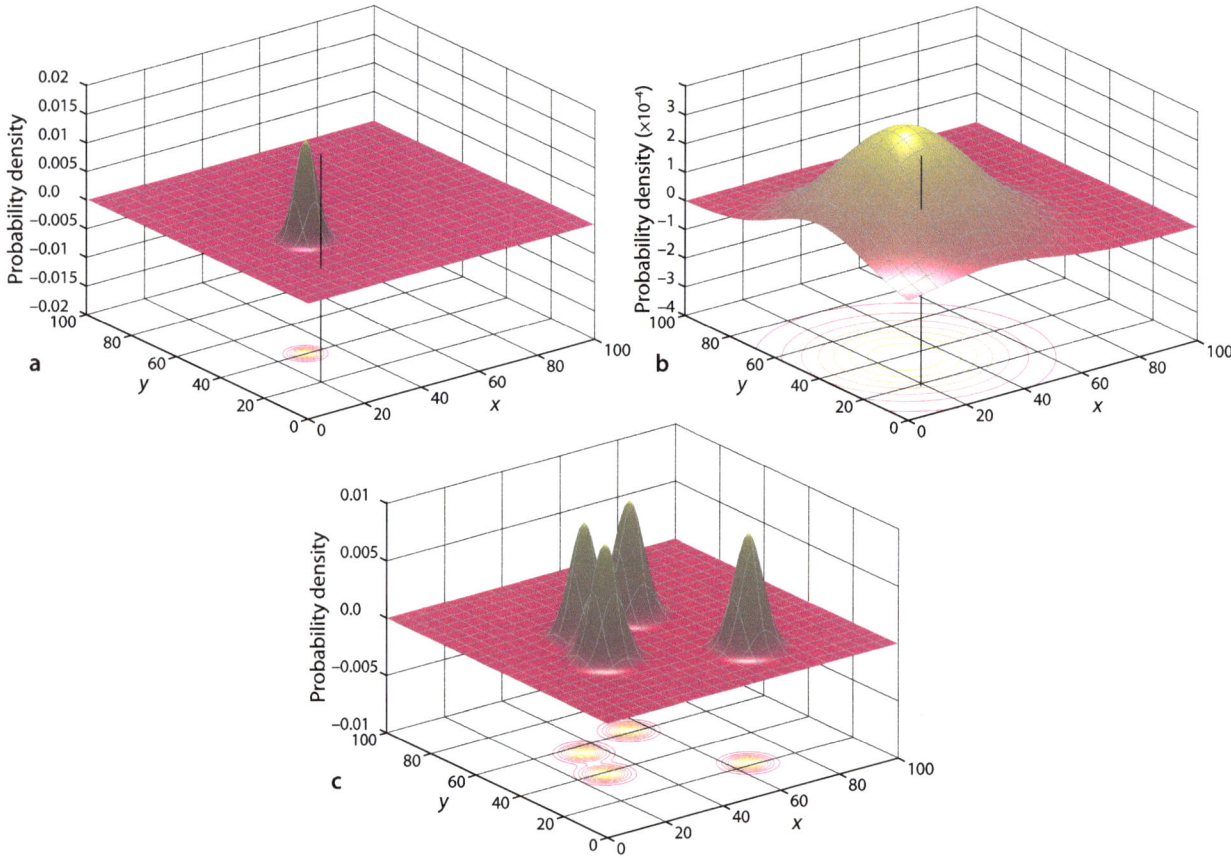

Using a PDF also allows for multiple hypotheses about the robot's position. For example the PDF of Fig. 6.3c describes a robot that is quite certain that it is at one of four places. This is more useful than it seems at face value. Consider an indoor robot that has observed a vending machine and there are four such machines marked on the map. In the absence of any other information the robot must be equally likely to be in the vicinity of *any* of the four vending machines. We will revisit this approach in Sect. 6.7.

Determining the PDF based on knowledge of how the vehicle moves and its observations of the world is a problem in estimation which we can define as:

the process of inferring the value of some quantity of interest, x, by processing data that is in some way dependent on x.

Fig. 6.3. Notions of vehicle position and uncertainty in the xy-plane, where the vertical axis is the relative likelihood of the vehicle being at that position, sometimes referred to as belief or bel(x). Contour lines are displayed on the lower plane. **a** The vehicle has low position uncertainty, $\sigma = 1$; **b** the vehicle has much higher position uncertainty, $\sigma = 20$; **c** the vehicle has multiple hypotheses for its position, each $\sigma = 1$

For example a ship's navigator or a surveyor estimates location by measuring the bearing angles to known landmarks or celestial objects, and a GPS receiver estimates latitude and longitude by observing the time delay from moving satellites whose locations are known.

For our robot localization problem the true and estimated state are vector quantities so uncertainty will be represented as a covariance matrix, see Appendix G. The diagonal elements represent uncertainty of the corresponding states, and the off-diagonal elements represent correlations between states.

> The value of a PDF is *not* the probability of being at that location. Consider a small region of the xy-plane, the volume under that region of the PDF is the probability of being in that region.

6.1   Dead Reckoning

Dead reckoning is the estimation of a robot's pose based on its estimated speed, direction and time of travel with respect to a previous estimate.

An odometer is a sensor that measures distance traveled and sometimes also change in heading direction. For wheeled vehicles this can be determined by measuring the angular rotation of the wheels. The direction of travel can be measured using an electronic compass, or the change in heading can be measured using a gyroscope or differential odometry.◄ These sensors are imperfect due to systematic errors such an incorrect wheel radius or gyroscope bias, and random errors such as slip between wheels and the ground. Robots without wheels, such as aerial and underwater robots, can use visual odometry – a computer vision approach based on observations of the world moving past the robot.

Measuring the difference in angular velocity of a left- and right-hand side wheel.

6.1.1   Modeling the Vehicle

The first step in estimating the robot's pose is to write a function, $f(\cdot)$, that describes how the vehicle's configuration changes from one time step to the next. A vehicle model such as Eq. 4.2 or 4.4 describes the evolution of the robot's configuration as a function of its control inputs, however for real robots we rarely have access to these control inputs. Most robotic platforms have proprietary motion control systems that accept motion commands from the user (speed and direction) and report odometry information.

Instead of using Eq. 4.2 or 4.4 directly we will write a discrete-time model for the evolution of configuration based on odometry where $\delta\langle k \rangle = (\delta_d, \delta_\theta)$ is the distance traveled and change in heading over the preceding interval, and k is the time step. The initial pose is represented in SE(2) as

$$\xi\langle k \rangle \sim \begin{pmatrix} \cos\theta\langle k \rangle & -\sin\theta\langle k \rangle & x\langle k \rangle \\ \sin\theta\langle k \rangle & \cos\theta\langle k \rangle & y\langle k \rangle \\ 0 & 0 & 1 \end{pmatrix}$$

We make a simplifying assumption that motion over the time interval is *small* so the order of applying the displacements is not significant. We choose to move forward in the vehicle x-direction by δ_d, and then rotate by δ_θ giving the new configuration

$$\xi\langle k+1 \rangle \sim \begin{pmatrix} \cos\theta\langle k \rangle & -\sin\theta\langle k \rangle & x\langle k \rangle \\ \sin\theta\langle k \rangle & \cos\theta\langle k \rangle & y\langle k \rangle \\ 0 & 0 & 1 \end{pmatrix} \begin{pmatrix} 1 & 0 & \delta_d \\ 0 & 1 & 0 \\ 0 & 0 & 1 \end{pmatrix} \begin{pmatrix} \cos\delta_\theta & -\sin\delta_\theta & 0 \\ \sin\delta_\theta & \cos\delta_\theta & 0 \\ 0 & 0 & 1 \end{pmatrix}$$

$$\sim \begin{pmatrix} \cos(\theta\langle k \rangle + \delta_\theta) & -\sin(\theta\langle k \rangle + \delta_\theta) & x\langle k \rangle + \delta_d\cos\theta\langle k \rangle \\ \sin(\theta\langle k \rangle + \delta_\theta) & \cos(\theta\langle k \rangle + \delta_\theta) & y\langle k \rangle + \delta_d\sin\theta\langle k \rangle \\ 0 & 0 & 1 \end{pmatrix}$$

which we can represent concisely as a 3-vector $x = (x, y, \theta)$

$$x\langle k+1 \rangle = \begin{pmatrix} x\langle k \rangle + \delta_d\cos\theta\langle k \rangle \\ y\langle k \rangle + \delta_d\sin\theta\langle k \rangle \\ \theta\langle k \rangle + \delta_\theta \end{pmatrix} \tag{6.1}$$

which gives the new configuration in terms of the previous configuration and the odometry.

In practice odometry is not perfect and we model the error by imagining a random number generator that corrupts the output of a perfect odometer. The measured output of the real odometer is the perfect, but unknown, odometry $(\delta_d, \delta_\theta)$ plus the output of the random number generator (v_d, v_θ). Such random errors are often referred to as noise, or more specifically as sensor noise. The random numbers are not known and cannot be measured, but we assume that we know the distribution from which they are drawn.

The robot's configuration at the next time step, including the odometry error, is

$$\boldsymbol{x}\langle k{+}1\rangle = \boldsymbol{f}\big(\boldsymbol{x}\langle k\rangle, \boldsymbol{\delta}\langle k\rangle, \boldsymbol{v}\langle k\rangle\big) = \begin{pmatrix} x\langle k\rangle + (\delta_d + v_d)\cos\theta\langle k\rangle \\ y\langle k\rangle + (\delta_d + v_d)\sin\theta\langle k\rangle \\ \theta\langle k\rangle + \delta_\theta + v_\theta \end{pmatrix} \tag{6.2}$$

where k is the time step, $\boldsymbol{\delta}\langle k\rangle = (\delta_d, \delta_\theta)^T \in \mathbb{R}^{2\times 1}$ is the odometry measurement and $\boldsymbol{v}\langle k\rangle = (v_d, v_\theta)^T \in \mathbb{R}^{2\times 1}$ is the random measurement noise over the preceding interval.▶

In the absence of any information to the contrary we model the odometry noise as $\boldsymbol{v} = (v_d, v_\theta)^T \sim N(0, \boldsymbol{V})$, a zero-mean multivariate Gaussian process▶ with variance

$$\boldsymbol{V} = \begin{pmatrix} \sigma_d^2 & 0 \\ 0 & \sigma_\theta^2 \end{pmatrix}$$

This constant matrix, the covariance matrix, is diagonal which means that the errors in distance and heading are *independent*.▶ Choosing a value for $\boldsymbol{V}$ is not always easy but we can conduct experiments or make some reasonable engineering assumptions. In the examples which follow, we choose $\sigma_d = 2$ cm and $\sigma_\theta = 0.5°$ per sample interval which leads to a covariance matrix of

```
>> V = diag([0.02, 0.5*pi/180].^2);
```

All objects of the Toolbox `Vehicle` superclass provide a method `f()` that implements the appropriate odometry update equation. For the case of a vehicle with bicycle kinematics that has the motion model of Eq. 4.2 and the odometric update Eq. 6.2, we create a `Bicycle` object

```
>> veh= Bicycle('covar', V)
veh =
Bicycle object
  L=1
  Superclass: Vehicle
    max speed=1, max steer input=0.5, dT=0.1, nhist=0
    V=(0.0004, 7.61544e-05)
    configuration: x=0, y=0, theta=0
```

which shows the default parameters such as the vehicle's length, speed, steering limit and the sample interval which defaults to 0.1 s. The object provides a method to simulate motion over one time step

```
>> odo = veh.step(1, 0.3)
odo =
    0.1108    0.0469
```

where we have specified a speed of 1 m s⁻¹ and a steering angle of 0.3 rad. The function updates the robot's true configuration and returns a noise corrupted odometer reading.▶ With a sample interval of 0.1 s the robot reports that is moving approximately 0.1 m each interval and changing its heading by approximately 0.03 rad. The robot's true (but 'unknown') configuration can be seen by

```
>> veh.x'
ans =
    0.1000         0    0.0309
```

Given the reported odometry we can estimate the configuration of the robot after one time step using Eq. 6.2 which is implemented by

```
>> veh.f([0 0 0], odo)
ans =
    0.1106    0.0052    0.0469
```

where the discrepancy with the exact value is due to the use of a noisy odometry measurement.

The odometry noise is *inside* the model of our process (vehicle motion) and is referred to as process noise.

A normal distribution of angles on a circle is actually not possible since $\theta \in \mathbb{S}^1 \notin \mathbb{R}$, that is angles wrap around 2π. However if the covariance for angular states is small this problem is minimal. A normal-like distribution of angles on a circle is the von Mises distribution.

In reality this is unlikely to be the case since odometry distance errors tend to be worse when change of heading is high.

We simulate the odometry noise using MATLAB generated random numbers that have zero-mean and a covariance given by the diagonal of $\mathtt{V}$. The random noise means that repeated calls to this function will return different values.

For the scenarios that we want to investigate we require the simulated robot to drive for a long time period within a defined spatial region. The `RandomPath` class is a *driver* that steers the robot to randomly selected waypoints within a specified region. We create an instance of the driver object and connect it to the robot

```
>> veh.add_driver( RandomPath(10) )
```

where the argument to the `RandomPath` constructor specifies a working region that spans ± 10 m in the *x*- and *y*-directions. We can display an animation of the robot with its driver by

```
>> veh.run()
```

The number of history records is indicated by `nhist=` in the displayed value of the object. The `hist` property is an array of structures that hold the vehicle state at each time step.

which repeatedly calls the `step` method and maintains a history of the true state of the vehicle over the course of the simulation within the `Bicycle` object.◄ The `RandomPath` and `Bicycle` classes have many parameters and methods which are described in the online documentation.

6.1.2 Estimating Pose

The problem we face, just like the ship's navigator, is how to estimate our new pose given the previous pose and noisy odometry. We want the best estimate of where we are and how certain we are about that. The mathematical tool that we will use is the Kalman filter which is described more completely in Appendix H. This filter provides the optimal estimate of the system state, vehicle configuration in this case, assuming that the noise is zero-mean and Gaussian. The filter is a recursive algorithm that updates, at each time step, the optimal estimate of the unknown true configuration and the uncertainty associated with that estimate based on the previous estimate and noisy measurement data. The Kalman filter is formulated for linear systems but our model of the vehicle's motion Eq. 6.2 is nonlinear – the tool of choice is the extended Kalman filter (EKF).

For this problem the state vector is the vehicle's configuration

$$\boldsymbol{x} = \left(x_v, y_v, \theta_v \right)^T$$

The Kalman filter, Fig. 6.6, has two steps: prediction based on the model and update based on sensor data. In this dead-reckoning case we use only the prediction equation.

and the prediction equations◄

$$\hat{\boldsymbol{x}}^+\langle k{+}1\rangle = \boldsymbol{f}\big(\hat{\boldsymbol{x}}\langle k\rangle, \hat{\boldsymbol{u}}\langle k\rangle\big) \tag{6.3}$$

$$\hat{\boldsymbol{P}}^+\langle k{+}1\rangle = \boldsymbol{F_x}\hat{\boldsymbol{P}}\langle k\rangle\boldsymbol{F_x}^T + \boldsymbol{F_v}\hat{\boldsymbol{V}}\boldsymbol{F_v}^T \tag{6.4}$$

describe how the state and covariance evolve with time. The term $\hat{\boldsymbol{x}}^+\langle k{+}1\rangle$ indicates an estimate of $\boldsymbol{x}$ at time $k + 1$ based on information up to, and including, time k. $\hat{\boldsymbol{u}}\langle k\rangle$ is the

Reverend Thomas Bayes (1702–1761) was a nonconformist Presbyterian minister. He studied logic and theology at the University of Edinburgh and lived and worked in Tunbridge-Wells in Kent. There, through his association with the 2nd Earl Stanhope he became interested in mathematics and was elected to the Royal Society in 1742. After his death his friend Richard Price edited and published his work in 1763 as *An Essay towards solving a Problem in the Doctrine of Chances* which contains a statement of a special case of Bayes' theorem. Bayes is buried in Bunhill Fields Cemetery in London.

Bayes' theorem shows the relation between a conditional probability and its inverse: the probability of a hypothesis given observed evidence and the probability of that evidence given the hypothesis. Consider the hypothesis that the robot is at location X and it makes a sensor observation S of a known landmark. The *posterior* probability that the robot is at X given the observation S is

$$P(X|S) = \frac{P(S|X)P(X)}{P(S)}$$

where $P(X)$ is the *prior* probability that the robot is at X (not accounting for any sensory information), $P(S|X)$ is the likelihood of the sensor observation S given that the robot is at X, and $P(S)$ is the prior probability of the observation S. The Kalman filter, and the Monte-Carlo estimator we discuss later in this chapter, are essentially two different approaches to solving this inverse problem.

Rudolf Kálmán (1930–2016) was a mathematical system theorist born in Budapest. He obtained his bachelors and masters degrees in electrical engineering from MIT, and Ph.D. in 1957 from Columbia University. He worked as a Research Mathematician at the Research Institute for Advanced Study, in Baltimore, from 1958–1964 where he developed his ideas on estimation. These were met with some skepticism among his peers and he chose a mechanical (rather than electrical) engineering journal for his paper *A new approach to linear filtering and prediction problems* because "When you fear stepping on hallowed ground with entrenched interests, it is best to go sideways". He has received many awards including the IEEE Medal of Honor, the Kyoto Prize and the Charles Stark Draper Prize.

Stanley F. Schmidt (1926–2015) was a research scientist who worked at NASA Ames Research Center and was an early advocate of the Kalman filter. He developed the first implementation as well as the nonlinear version now known as the extended Kalman filter. This led to its incorporation in the Apollo navigation computer for trajectory estimation. (Extract from Kálmán's famous paper (1960) on the right reprinted with permission of ASME)

input to the process, which in this case is the measured odometry, so $\hat{u}\langle k \rangle = \delta\langle k \rangle$. $\hat{P} \in \mathbb{R}^{3 \times 3}$ is a covariance matrix representing uncertainty in the estimated vehicle configuration. $\hat{V}$ is our estimate of the covariance of the odometry noise which in reality we do not know.

F_x and F_v are Jacobian matrices – the vector version of a derivative. They are obtained by differentiating Eq. 6.2 and evaluating the result at $v = 0$ giving ▶

> Since the noise value cannot actually be measured we use the mean value which is zero.

$$F_x = \left.\frac{\partial f}{\partial x}\right|_{v=0} = \begin{pmatrix} 1 & 0 & -\delta_d \sin\theta_v \\ 0 & 1 & \delta_d \cos\theta_v \\ 0 & 0 & 1 \end{pmatrix} \tag{6.5}$$

$$F_v = \left.\frac{\partial f}{\partial v}\right|_{v=0} = \begin{pmatrix} \cos\theta_v & 0 \\ \sin\theta_v & 0 \\ 0 & 1 \end{pmatrix} \tag{6.6}$$

which are functions of the current state and odometry. ▶ Jacobians are reviewed in Appendix E. All objects of the `Vehicle` superclass provide methods `Fx` and `Fv` to compute these Jacobians, for example

> The time step notation $\langle k \rangle$ is dropped to reduce clutter.

```
>> veh.Fx( [0,0,0], [0.5, 0.1] )
ans =
    1.0000         0   -0.0499
         0    1.0000    0.4975
         0         0    1.0000
```

where the first argument is the state at which the Jacobian is computed and the second is the odometry.

To simulate the vehicle and the EKF using the Toolbox we define the initial covariance to be quite small since, we assume, we have a good idea of where we are to begin with

```
>> P0 = diag([0.005, 0.005, 0.001].^2);
```

and we pass this to the constructor for an `EKF` object

```
>> ekf = EKF(veh, V, P0);
```

Running the filter for 1 000 time steps

```
>> ekf.run(1000);
```

drives the robot as before, along a random path. At each time step the filter updates the state estimate using various methods provided by the `Vehicle` superclass.

We can plot the true path taken by the vehicle, stored within the `Vehicle` superclass object, by

```
>> veh.plot_xy()
```

and the filter's estimate of the path stored within the `EKF` object,

```
>> hold on
>> ekf.plot_xy('r')
```

These are shown in Fig. 6.4 and we see some divergence between the true and estimated robot path.

The covariance at the 700th time step is

```
>> P700 = ekf.history(700).P
P700 =
      1.8929    -0.5575    -0.1851
     -0.5575     3.4184     0.3400
     -0.1851     0.3400     0.0533
```

The matrix is symmetric and the diagonal elements are the estimated variance associated with the states, that is σ_x^2, σ_y^2 and σ_θ^2 respectively. The standard deviation σ_x of the PDF associated with the vehicle's x-coordinate is

```
>> sqrt(P700(1,1))
ans =
    1.3758
```

There is a 95% chance that the robot's x-coordinate is within the $\pm 2\sigma$ bound or ± 2.75 m in this case. We can compute uncertainty for y and θ similarly.

The off-diagonal terms are correlation coefficients and indicate that the uncertainties between the corresponding variables are related. For example the value $P_{1,3} = P_{3,1} = -0.5575$ indicates that the uncertainties in x and θ are related – error in heading angle causes error in x-position and vice versa. Conversely new information about θ can be used to correct θ as well as x. The uncertainty in position is described by the top-left 2×2 covariance submatrix of $\hat{P}$. This can be interpreted as an ellipse defining a confidence bound on position. We can overlay such ellipses on the plot by

```
>> ekf.plot_ellipse('g')
```

as shown in Fig. 6.4. These correspond to the default 95% confidence bound and are plotted by default every 20 time steps. The vehicle started at the origin and as it progresses we see that the ellipses become larger as the estimated uncertainty increases. The ellipses only show x- and y-position but uncertainty in θ also grows.

The total uncertainty,◄ position and heading, is given by $\sqrt{\det(\hat{P})}$ and is plotted as a function of time

```
>> ekf.plot_P();
```

as shown in Fig. 6.5 and we observe that it never decreases. This is because the second term in Eq. 6.4 is positive definite which means that P, the position uncertainty, can never decrease.

The elements of P have different units: m^2 and rad^2. The uncertainty is therefore a mixture of spatial and angular uncertainty with an implicit weighting. If the range of the position variables x, $y \gg \pi$ then positional uncertainty dominates.

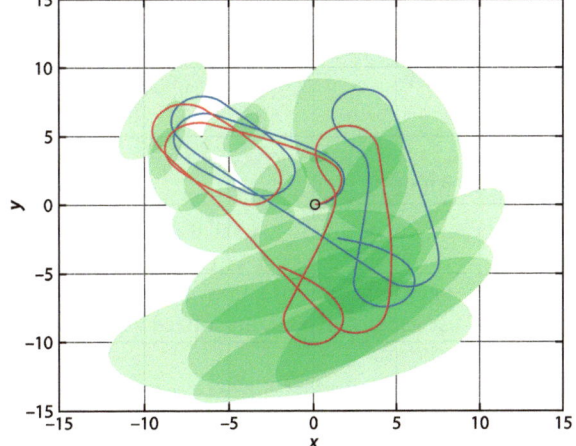

Fig. 6.4.
Deadreckoning using the EKF. The true path of the robot, *blue*, and the path estimated from odometry in *red*. 95% confidence ellipses are indicated in *green*. The robot starts at the origin

Error ellipses. We consider the PDF of the robot's position (ignoring orientation) such as shown in Fig. 6.3 to be a 2-dimensional Gaussian probability density function

$$p(\boldsymbol{x}) = \frac{1}{(2\pi)\det\left(\boldsymbol{P}_{xy}\right)^{1/2}}\exp\left\{-\frac{1}{2}(\boldsymbol{x}-\mu_x)^T\boldsymbol{P}_{xy}^{-1}(\boldsymbol{x}-\mu_x)\right\}$$

where $\boldsymbol{x} = (x, y)^T$ is the position of the robot, $\mu_x = (\hat{x}, \hat{y})^T$ is the estimated mean position and $\boldsymbol{P}_{xy} \in \mathbb{R}^{2\times2}$ is the position covariance matrix, the top left of the covariance matrix P computed by the Kalman filter. A horizontal cross-section is a contour of constant probability which is an ellipse defined by the points $\boldsymbol{x}$ such that

$$\left(\boldsymbol{x}-\mu_x\right)^T\boldsymbol{P}_{xy}^{-1}\left(\boldsymbol{x}-\mu_x\right) = s$$

Such error ellipses are often used to represent positional uncertainty as shown in Fig. 6.4. A large ellipse corresponds to a wider PDF peak and less certainty about position. To obtain a particular confidence contour (eg. 99%) we choose s as the inverse of the χ^2 cumulative distribution function for 2 degrees of freedom, in MATLAB that is `chi2inv(C, 2)` where $C \in [0, 1]$ is the confidence value. Such confidence values can be passed to several `EKF` methods when specifying error ellipses.

A handy scalar measure of total position uncertainty is the area of the ellipse $\pi r_1 r_2$ where the radii $r_i = \sqrt{\lambda_i}$ and λ_i are the eigenvalues of $\boldsymbol{P}_{xy}$. Since $\det(\boldsymbol{P}_{xy}) = \Pi\lambda_i$ the ellipse area – the scalar uncertainty – is proportional to $\sqrt{\det(\boldsymbol{P}_{xy})}$. See also Appendices C.1.4 and G.

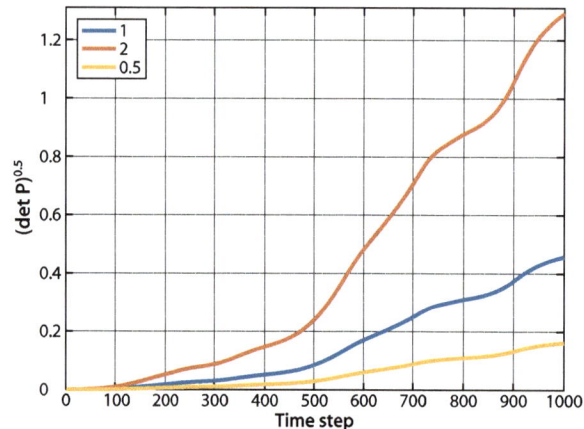

Fig. 6.5.
Overall uncertainty is given by $\sqrt{\det(\hat{\boldsymbol{P}})}$ which shows monotonically increasing uncertainty (*blue*). The effect of changing the magnitude of $\hat{\boldsymbol{V}}$ is to change the rate of uncertainty growth. Curves are shown for $\hat{\boldsymbol{V}} = \alpha V$ where $\alpha = 1/2, 1, 2$

Note that we have used the odometry covariance matrix `V` twice. The first usage, in the `Vehicle` constructor, is the covariance V of the Gaussian noise source that is added to the true odometry to *simulate* odometry error in Eq. 6.2. In a real application this noise is generated by some physical process *hidden inside* the robot and we would not know its parameters. The second usage, in the `EKF` constructor, is $\hat{\boldsymbol{V}}$ which is our best *estimate* of the odometry covariance and is used in the filter's state covariance update equation Eq. 6.4.

The relative values of V and $\hat{\boldsymbol{V}}$ control the rate of uncertainty growth as shown in Fig. 6.5. If $\hat{\boldsymbol{V}} > V$ then P will be larger than it should be and the filter is pessimistic – it overestimates uncertainty and is less certain than it should be. If $\hat{\boldsymbol{V}} < V$ then P will be smaller than it should be and the filter will be *overconfident* of its estimate – the actual uncertainty is greater than the estimated uncertainty. In practice some experimentation is required to determine the appropriate value for the estimated covariance.

6.2   Localizing with a Map

We have seen how uncertainty in position grows without bound using dead-reckoning alone. The solution, as the Phoenicians worked out 4 000 years ago, is to bring in additional information from observations of known features in the world. In the examples that follow we will use a map that contains N fixed but randomly located landmarks whose positions are known.

The Toolbox supports a `LandmarkMap` object

```
>> map = LandmarkMap(20, 10)
```

that in this case contains $N = 20$ landmarks uniformly randomly spread over a region spanning ± 10 m in the x- and y-directions and this can be displayed by

```
>> map.plot()
```

The robot is equipped with a sensor that provides observations of the landmarks *with respect to the robot* as described by

$$z = h(x, p_i) \tag{6.7}$$

where $x = (x_v, y_v, \theta_v)^T$ is the vehicle state, and $p_i = (x_i, y_i)^T$ is the *known* location of the ith landmark in the world frame.

To make this tangible we will consider a common type of sensor that measures the range and bearing angle to a landmark in the environment, for instance a radar or a scanning-laser rangefinder such as shown in Fig. 6.22a. The sensor is mounted on-board the robot so the observation of the ith landmark is

$$z = h(x, p_i) = \begin{pmatrix} \sqrt{(y_i - y_v)^2 + (x_i - x_v)^2} \\ \tan^{-1}(y_i - y_v)/(x_i - x_v) - \theta_v \end{pmatrix} + \begin{pmatrix} w_r \\ w_\beta \end{pmatrix} \tag{6.8}$$

where $z = (r, \beta)^T$ and r is the range, β the bearing angle, and $w = (w_r, w_\beta)^T$ is a zero-mean Gaussian random variable that models errors in the sensor

$$\begin{pmatrix} w_r \\ w_\beta \end{pmatrix} \sim N(0, W), \quad W = \begin{pmatrix} \sigma_r^2 & 0 \\ 0 & \sigma_\beta^2 \end{pmatrix}$$

It also indicates that covariance is independent of range but in reality covariance may increase with range since the strength of the return signal, laser or radar, drops rapidly $(1/r^4)$ with distance (r) to the target.

The constant diagonal covariance matrix indicates that range and bearing errors are independent. ◄

For this example we set the sensor uncertainty to be $\sigma_r = 0.1$ m and $\sigma_\beta = 1°$ giving a sensor covariance matrix

```
>> W = diag([0.1, 1*pi/180].^2);
```

A subclass of Sensor.

We model this type of sensor with a RangeBearingSensor object ◄

```
>> sensor = RangeBearingSensor(veh, map, 'covar', W)
```

The landmark is chosen randomly from the set of visible landmarks, those that are within the field of view and the minimum and maximum range limits. If no landmark is visible i is assigned a value of 0.

which is connected to the vehicle and the map, and the sensor covariance matrix W is specified along with the maximum range and the bearing angle limits. The reading method provides the range and bearing to a randomly selected visible ◄ landmark along with its identity, for example

```
>> [z,i] = sensor.reading()
z =
     9.0905
     1.0334
i =
    17
```

The identity is an integer $i \in [1, 20]$ since the map was created with 20 landmarks. We have avoided the data association problem by assuming that we know the identity of the sensed landmark. The position of landmark 17 can be looked up in the map

```
>> landmark(17)
    -4.4615
    -9.0766
```

Using Eq. 6.8 the robot can estimate the range and bearing angle to the landmark based on its own estimated position and the known position of the landmark from the map. Any difference between the observation $z^{\#}$ and the estimated observation indicates an error in the robot's pose estimate $\hat{x}$ – it isn't where it *thought* it was. However this *difference*

$$\boldsymbol{\nu} = \boldsymbol{z}^{\#}\langle k+1\rangle - \boldsymbol{h}\Big(\hat{\boldsymbol{x}}^+\langle k+1\rangle, \boldsymbol{p}_i\Big) \tag{6.9}$$

has real value and is key to the operation of the Kalman filter. It is called the innovation since it represents *new* information. The Kalman filter uses the innovation to correct the state estimate and update the uncertainty estimate in an optimal way.

The predicted state computed earlier using Eq. 6.3 and Eq. 6.4 is updated by

$$\hat{\boldsymbol{x}}\langle k+1\rangle = \hat{\boldsymbol{x}}^+\langle k+1\rangle + \boldsymbol{K}\boldsymbol{\nu} \tag{6.10}$$

$$\hat{\boldsymbol{P}}\langle k+1\rangle = \hat{\boldsymbol{P}}^+\langle k+1\rangle - \boldsymbol{K}\boldsymbol{H}_x\hat{\boldsymbol{P}}^+\langle k+1\rangle \tag{6.11}$$

which are the Kalman filter update equations. These take the *predicted* values for the next time step denoted with the $^+$ and compute the optimal estimate by applying landmark measurements from time step $k+1$. The innovation is added to the estimated state after multiplying by the Kalman gain matrix $\boldsymbol{K}$ which is defined as

$$\boldsymbol{K} = \boldsymbol{P}^+\langle k+1\rangle \boldsymbol{H}_x^T \boldsymbol{S}^{-1} \tag{6.12}$$

$$\boldsymbol{S} = \boldsymbol{H}_x \boldsymbol{P}^+\langle k+1\rangle \boldsymbol{H}_x^T + \boldsymbol{H}_w \hat{\boldsymbol{W}} \boldsymbol{H}_w^T \tag{6.13}$$

where $\hat{W}$ is the estimated covariance of the sensor noise and $\boldsymbol{H}_x$ and $\boldsymbol{H}_w$ are Jacobians obtained by differentiating Eq. 6.8 yielding

$$\boldsymbol{H}_x = \frac{\partial \boldsymbol{h}}{\partial \boldsymbol{x}}\bigg|_{w=0} = \begin{pmatrix} -\dfrac{x_i - x_v}{r} & -\dfrac{y_i - y_v}{r} & 0 \\ \dfrac{y_i - y_v}{r^2} & -\dfrac{x_i - x_v}{r^2} & -1 \end{pmatrix} \tag{6.14}$$

which is a function of landmark position, vehicle pose and landmark range; and

$$\boldsymbol{H}_w = \frac{\partial \boldsymbol{h}}{\partial \boldsymbol{w}}\bigg|_{w=0} = \begin{pmatrix} 1 & 0 \\ 0 & 1 \end{pmatrix} \tag{6.15}$$

The RangeBearingSensor object above includes methods h to implement Eq. 6.8 and Hx and Hw to compute these Jacobians respectively.

The Kalman gain matrix $\boldsymbol{K}$ in Eq. 6.10 *distributes* the innovation from the landmark observation, a 2-vector, to update every element of the state vector – the position and orientation of the vehicle. Note that the second term in Eq. 6.11 is *subtracted* from the estimated covariance and this provides a means for covariance to decrease which was

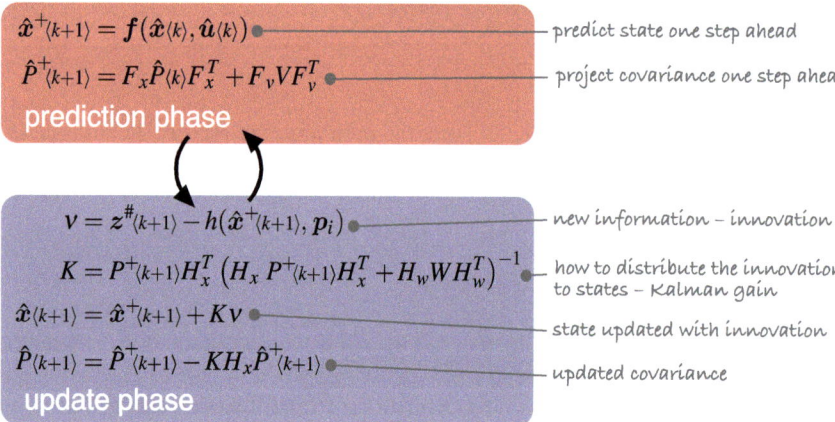

Fig. 6.6.
Summary of extended Kalman filter algorithm showing the prediction and update phases

not possible for the dead-reckoning case of Eq. 6.4. The EKF comprises two phases: prediction and update, and these are summarized in Fig. 6.6.

We now have all the piece to build an estimator that uses odometry and observations of map features. The Toolbox implementation is

```
>> map = LandmarkMap(20);
>> veh = Bicycle('covar', V);
>> veh.add_driver( RandomPath(map.dim) );
>> sensor = RangeBearingSensor(veh, map, 'covar', W, 'angle',↵
[-pi/2 pi/2], 'range', 4, 'animate');
>> ekf = EKF(veh, V, P0, sensor, W, map);
```

The LandmarkMap constructor has a default map dimension of ±10 m which is accessed by its dim property.

Running the simulation for 1 000 time steps

```
>> ekf.run(1000);
```

shows an animation of the robot moving and observations being made to the landmarks. We plot the saved results

```
>> map.plot()
>> veh.plot_xy();
>> ekf.plot_xy('r');
>> ekf.plot_ellipse('k')
```

which are shown in Fig. 6.7a. The error ellipses are now much smaller and many can hardly be seen.

Figure 6.7b shows a zoomed view of the robot's actual and estimated path – the robot is moving from top to bottom. We can see the error ellipses growing as the robot moves and then shrinking, just after a jag in the estimated path. This corresponds to the observation of a landmark. New information, beyond odometry, has been used to correct the state in the Kalman filter update phase.

Figure 6.8a shows that the overall uncertainty is no longer growing monotonically. When the robot sees a landmark it is able to dramatically reduce its estimated covariance. Figure 6.8b shows the error associated with each component of pose and the pink background is the estimated 95% confidence bound (derived from the covariance matrix) and we see that the error is mostly within this envelope. Below this is plotted the landmark observations and we see that the confidence bounds are tight (indicating low uncertainty) while landmarks are being observed but that they start to grow once observations stop. However as soon as an observation is made the uncertainty rapidly decreases.

This EKF framework allows data from many and varied sensors to update the state which is why the estimation problem is also referred to as sensor fusion. For example heading angle from a compass, yaw rate from a gyroscope, target bearing angle from a camera, position

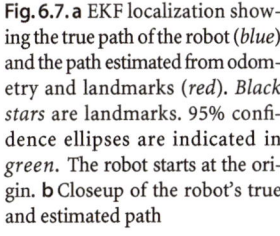

Fig. 6.7. a EKF localization showing the true path of the robot (*blue*) and the path estimated from odometry and landmarks (*red*). *Black stars* are landmarks. 95% confidence ellipses are indicated in *green*. The robot starts at the origin. **b** Closeup of the robot's true and estimated path

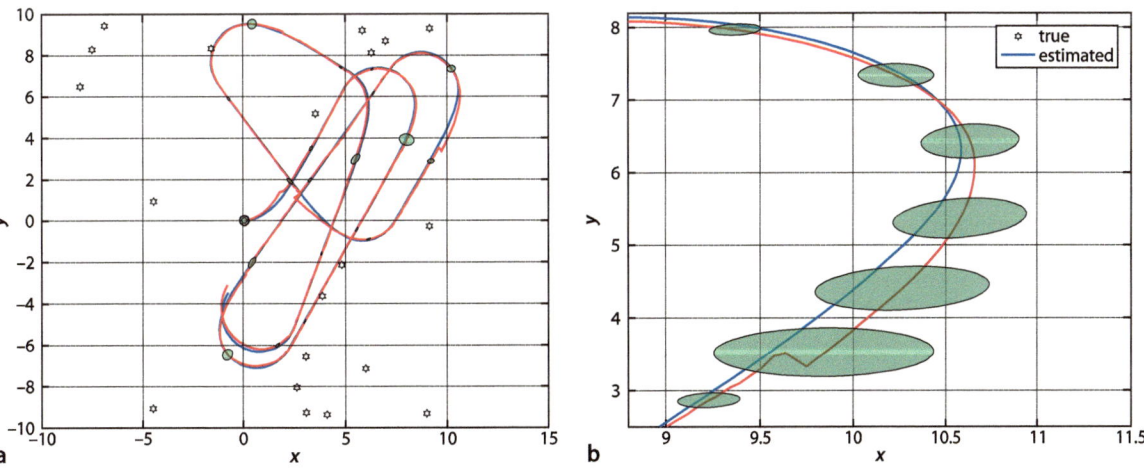

from GPS could all be used to update the state. For each sensor we need only to provide the observation function $h(\cdot)$, the Jacobians H_x and H_w and some estimate of the sensor covariance W. The function $h(\cdot)$ can be nonlinear and even noninvertible – the EKF will do the rest.

> As discussed earlier for V, we also use W twice. The first usage, in the constructor for the `RangeBearingSensor` object, is the covariance W of the Gaussian noise that is added to the computed range and bearing to *simulate* sensor error as in Eq. 6.8. In a real application this noise is generated by some physical process *hidden inside* the sensor and we would not know its parameters. The second usage, $\hat{W}$ is our best *estimate* of the sensor covariance which is used by the Kalman filter Eq. 6.12.

Fig. 6.8. a Covariance magnitude as a function of time. Overall uncertainty is given by $\sqrt{\det(P)}$ and shows that uncertainty does not continually increase with time. **b** Top: pose estimation error with 95% confidence bound shown in *pink*; bottom: observed landmarks the *bar* indicates which landmark is seen at each time step, 0 means no observation

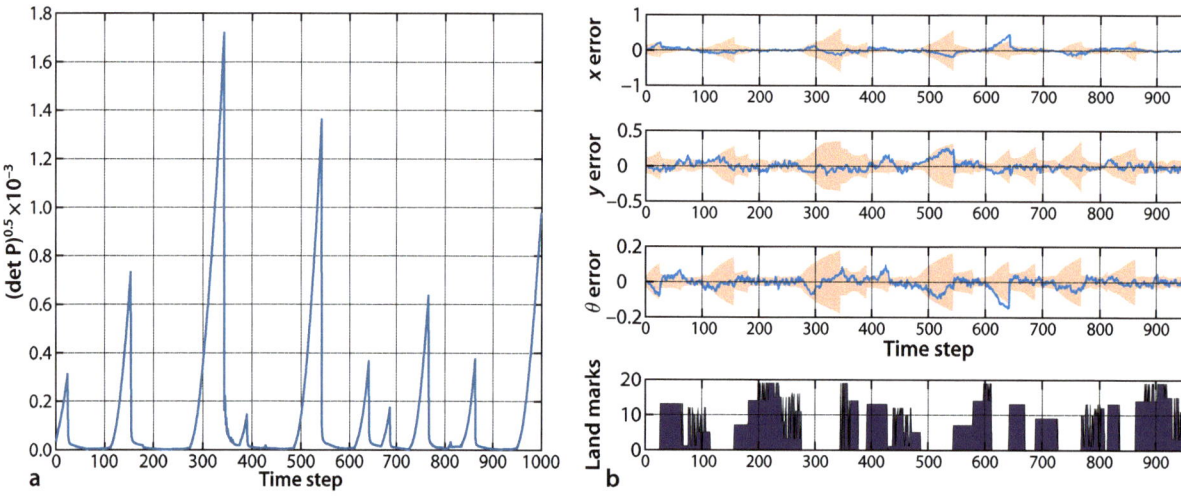

Data association. So far we have assumed that the observed landmark reveals its identity to us, but in reality this is rarely the case. Instead we compare our observation to the predicted position of all currently known landmarks and make a decision as to which landmark it is most likely to be, or whether it is a new landmark. This decision needs to take into account the uncertainty associated with the vehicle's pose, the sensor measurement and the landmarks in the map. This is the data association problem. Errors in this step are potentially catastrophic – incorrect innovation is coupled via the Kalman filter to the state of the vehicle and all the other landmarks which increases the chance of an incorrect data association on the next cycle. In practice, filters only use a landmark when there is a very high confidence in its estimated identity – a process that involves Mahalanobis distance and χ^2 confidence tests. If the situation is ambiguous it is best not to use the landmark – it can do more harm than good.

An alternative is to use a multi-hypothesis estimator, such as the particle filter that we will discuss in Sect. 6.7, which can model the possibility of observing landmark A or landmark B, and future observations will reinforce one hypothesis and weaken the others. The extended Kalman filter uses a Gaussian probability model, with just one peak, which limits it to holding only one hypothesis about the robot's pose. (Picture: the wreck of the Tararua, 1881)

Simple landmarks. For educational purposes it might be appropriate to use artificial landmarks that can be cheaply sensed by a camera. These need to be not only visually distinctive in the environment but also encode an identity. 2-dimensional bar codes such as QR codes or ARTags are well suited for this purpose. The Toolbox supports a variant called AprilTags, shown to the right, and

```
>> tags = apriltags(im);
```

returns a vector of `AprilTag` objects whose elements correspond to tags found in the image `im`. The centroid of the tag (`centre` property) can be used to determine relative bearing (see page 159), and the length of the edges (from the `corners` property) is a

A landmark might be some easily identifiable pattern such as this April tag (36h11) which can be detected in an image. Its position and size in the image encodes the bearing angle and range. The pattern itself encodes a number between 0 and 586 which could be used to uniquely identify the landmark in a map.

function of distance. The tag object also includes an homography (`H` property) which encodes information about the orientation of the plane of the April tag. More details about April tags can be found at http://april.eecs.umich.edu.

6.3   Creating a Map

So far we have taken the existence of the map for granted, an understandable mindset given that maps today are common and available for free via the internet. Nevertheless somebody, or something, has to create the maps we will use. Our next example considers the problem of a robot moving in an environment with landmarks and creating a map of their locations.

As before we have a range and bearing sensor mounted on the robot which measures, imperfectly, the position of landmarks with respect to the robot. There are a total of N landmarks in the environment and as for the previous example we assume that the sensor can determine the identity of each observed landmark. However for this case we assume that the robot knows its own location perfectly – it has ideal localization. This is unrealistic but this scenario is an important stepping stone to the next section. ◀

Since the vehicle pose is known perfectly we do not need to estimate it, but we do need to estimate the coordinates of the landmarks. For this problem the state vector comprises the estimated coordinates of the M landmarks that have been observed so far

$$\hat{x} = \left(x_1, y_1, x_2, y_2, \cdots x_M, y_M \right)^T \in \mathbb{R}^{2M \times 1}$$

The corresponding estimated covariance $\hat{P}$ will be a $2M \times 2M$ matrix. The state vector has a variable length since we do not know in advance how many landmarks exist in the environment. Initially $M = 0$ and is incremented every time a previously unseen landmark is observed.

The prediction equation is straightforward in this case since the landmarks are assumed to be stationary

$$\hat{x}^+ \langle k{+}1 \rangle = \hat{x} \langle k \rangle \tag{6.16}$$

$$\hat{P}^+ \langle k{+}1 \rangle = \hat{P} \langle k \rangle \tag{6.17}$$

We introduce the function $g(\cdot)$ which is the inverse of $h(\cdot)$ and gives the coordinates of the observed landmark based on the known vehicle pose and the sensor observation

$$g(x, z) = \begin{pmatrix} x_v + r \cos(\theta_v + \beta) \\ y_v + r \sin(\theta_v + \beta) \end{pmatrix}$$

Since $\hat{x}$ has a variable length we need to extend the state vector and the covariance matrix whenever we encounter a landmark we have not previously seen. The state vector is extended by the function $y(\cdot)$

$$x \langle k \rangle' = y\big(x \langle k \rangle, z \langle k \rangle, x_v \langle k \rangle \big) \tag{6.18}$$

$$= \begin{pmatrix} x \langle k \rangle \\ g\big(x_v \langle k \rangle, z \langle k \rangle \big) \end{pmatrix} \tag{6.19}$$

which appends the sensor-based estimate of the new landmark's coordinates to those already in the map. The order of feature coordinates within $\hat{x}$ therefore depends on the order in which they are observed.

The covariance matrix also needs to be extended when a new landmark is observed and this is achieved by

$$\hat{P} \langle k \rangle' = Y_z \begin{pmatrix} \hat{P} \langle k \rangle & 0 \\ 0 & \hat{W} \end{pmatrix} Y_z^T$$

where Y_z is the insertion Jacobian

$$Y_z = \frac{\partial y}{\partial z} = \begin{pmatrix} I_{n \times n} & & 0_{n \times 2} \\ G_x & 0_{2 \times n-3} & G_z \end{pmatrix} \tag{6.20}$$

A close and realistic approximation would be a high-end RTK GPS+INS system operating in an environment with no buildings or hills to obscure satellites.

that relates the rate of change of the extended state vector to the new observation. n is the dimension of $\hat{P}$ prior to it being extended and

$$G_x = \frac{\partial g}{\partial x} = \begin{pmatrix} 0 & 0 & 0 \\ 0 & 0 & 0 \end{pmatrix} \tag{6.21}$$

$$G_z = \frac{\partial g}{\partial z} = \begin{pmatrix} \cos(\theta_v + \beta) & -r\sin(\theta_v + \beta) \\ \sin(\theta_v + \beta) & r\cos(\theta_v + \beta) \end{pmatrix} \tag{6.22}$$

G_x is zero since $g(\cdot)$ is independent of the map in x. An additional Jacobian for $h(\cdot)$ is

$$H_{p_i} = \frac{\partial h}{\partial p_i} = \begin{pmatrix} \frac{x_i - x_v}{r} & \frac{y_i - y_v}{r} \\ -\frac{y_i - y_v}{r^2} & \frac{x_i - x_v}{r^2} \end{pmatrix} \tag{6.23}$$

which describes how the landmark observation changes with respect to landmark position for a particular robot pose, and is implemented by the method `Hp`.

For the mapping case the Jacobian H_x used in Eq. 6.11 describes how the landmark observation changes with respect to the full state vector. However the observation depends only on the position of that landmark so this Jacobian is mostly zeros

$$H_x = \frac{\partial h}{\partial x}\bigg|_{w=0} = \left(0 \cdots H_{p_i} \cdots 0\right) \in \mathbb{R}^{2 \times 2M} \tag{6.24}$$

where H_{p_i} is at the location in the vector corresponding to the state p_i. This structure represents the fact that observing a particular landmark provides information to estimate the position of that landmark, but no others.

The Toolbox implementation is

```
>> map = LandmarkMap(20);
>> veh = Bicycle();  % error free vehicle
>> veh.add_driver( RandomPath(map.dim) );
>> W = diag([0.1, 1*pi/180].^2);
>> sensor = RangeBearingSensor(veh, map, 'covar', W);
>> ekf = EKF(veh, [], [], sensor, W, []);
```

the empty matrices passed to `EKF` indicate respectively that there is no estimated odometry covariance for the vehicle (the estimate is perfect), no initial vehicle state covariance, and the map is unknown. We run the simulation for 1 000 time steps

```
>> ekf.run(1000);
```

Fig. 6.9. EKF mapping results. **a** The estimated landmarks are indicated by *black dots* with 95% confidence ellipses (*green*), the true location (*black ✿-marker*) and the robot's path (*blue*). The landmark estimates have not fully converged on their true values and the estimated covariance ellipses can only be seen by zooming; **b** the nonzero elements of the final covariance matrix

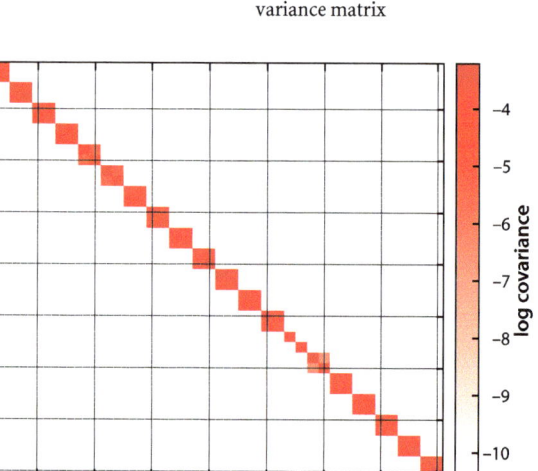

a   b

and see an animation of the robot moving and the covariance ellipses associated with the map features evolving over time. The estimated landmark positions

```
>> map.plot();
>> ekf.plot_map('g');
>> veh.plot_xy('b');
```

are shown in Fig. 6.9a as 95% confidence ellipses along with the true landmark positions and the path taken by the robot. The covariance matrix has a block diagonal structure which is shown graphically in Fig. 6.9b. The off-diagonal elements are zero, which implies that the landmark estimates are uncorrelated or independent. This is to be expected since observing one landmark provides no new information about any other landmark.

Internally the EKF object maintains a table to relate the landmark's identity, returned by the RangeBearingSensor, to the position of that landmark's coordinates in the state vector. For example the landmark with identity 6

```
>> ekf.landmarks(:,6)
ans =
    19
    71
```

was seen a total of 71 times during the simulation and comprises elements 19 and 20 of $\hat{x}$

```
>> ekf.x_est(19:20)'
ans =
    -6.4803    9.6233
```

which is its estimated location. Its estimated covariance is a submatrix within $\hat{P}$

```
>> ekf.P_est(19:20,19:20)
ans =
   1.0e-03 *
    0.2913    0.1814
    0.1814    0.3960
```

Fig. 6.10. Map of the New Holland coast (now eastern Australia) by Captain James Cook in 1770. The path of the ship and the map of the coast were determined at the same time. *Numbers* indicate depth in fathoms (1.83 m) (National Library of Australia, MAP NK 5557 A)

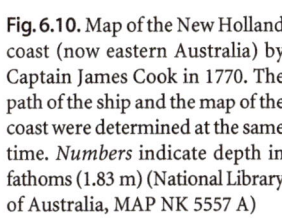

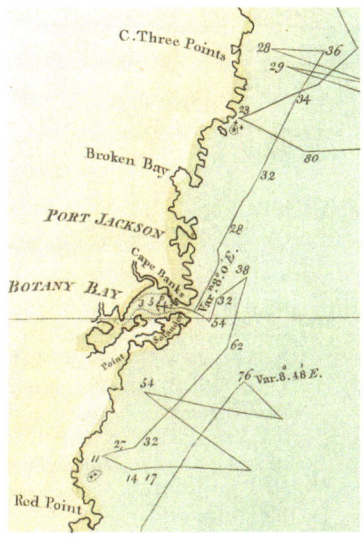

6.4   Localization and Mapping

Finally we tackle the problem of determining our position and creating a map at the same time. This is an old problem in marine navigation and cartography – incrementally extending maps while also using the map for navigation. Figure 6.10 shows what can be done without GPS from a moving ship with poor odometry and infrequent celestial position "fixes". In robotics this problem is known as simultaneous localization and mapping (SLAM) or concurrent mapping and localization (CML). This is often considered to be a "chicken and egg" problem – we need a map to localize and we need to localize to make the map. However based on what we have learned in the previous sections this problem is now quite straightforward to solve.

The state vector comprises the vehicle configuration *and* the coordinates of the M landmarks that have been observed so far

$$\hat{x} = \left(x_v, y_v, \theta_v, x_1, y_1, x_2, y_2, \cdots x_M, y_M\right)^T \in \mathbb{R}^{2M+3 \times 1}$$

The estimated covariance is a $(2M + 3) \times (2M + 3)$ matrix and has the structure

$$\hat{P} = \begin{pmatrix} \hat{P}_{vv} & \hat{P}_{vm} \\ \hat{P}_{vm}^T & \hat{P}_{mm} \end{pmatrix}$$

where $\hat{P}_{vv}$ is the covariance of the vehicle pose, $\hat{P}_{mm}$ the covariance of the map landmark positions, and $\hat{P}_{vm}$ is the correlation between vehicle and landmark states.

The predicted vehicle state and covariance are given by Eq. 6.3 and Eq. 6.4 and the sensor-based update is given by Eq. 6.10 to 6.15. When a new feature is observed the state vector is updated using the insertion Jacobian Y_z given by Eq. 6.20 but in this case G_x is nonzero

$$G_x = \frac{\partial g}{\partial x} = \begin{pmatrix} 1 & 0 & -r\sin(\theta_v + \beta) \\ 0 & 1 & r\cos(\theta_v + \beta) \end{pmatrix} \tag{6.25}$$

since the estimate of the new landmark depends on the state vector which now contains the vehicle's pose.

For the SLAM case the Jacobian H_x used in Eq. 6.11 describes how the landmark observation changes with respect to the state vector. The observation will depend on the position of the vehicle and on the position of the observed landmark and is

$$H_x = \frac{\partial h}{\partial x}\bigg|_{w=0} = \left(H_{x_v} \cdots 0 \cdots H_{p_i} \cdots 0 \right) \in \mathbb{R}^{2 \times (2M+3)} \tag{6.26}$$

where H_{p_i} is at the location corresponding to the landmark p_i. This is similar to Eq. 6.24 but with an extra nonzero block H_{x_v} given by Eq. 6.14.

The Kalman gain matrix K *distributes* innovation from the landmark observation, a 2-vector, to update *every* element of the state vector – the pose of the vehicle *and* the position of *every* landmark in the map.

The Toolbox implementation is by now quite familiar

```
>> P0 = diag([.01, .01, 0.005].^2);
>> map = LandmarkMap(20);
>> veh = Bicycle('covar', V);
>> veh.add_driver( RandomPath(map.dim) );
>> sensor = RangeBearingSensor(veh, map, 'covar', W);
>> ekf = EKF(veh, V, P0, sensor, W, []);
```

and the empty matrix passed to `EKF` indicates that the map is unknown. `P0` is the initial 3×3 covariance for the vehicle state.

We run the simulation for 1 000 time steps

```
>> ekf.run(1000);
```

and as usual an animation is shown of the vehicle moving. We also see the covariance ellipses associated with the map features evolving over time. We can plot the results

```
>> map.plot();
>> ekf.plot_map('g');
>> ekf.plot_xy('r');
>> veh.plot_xy('b');
```

which are shown in Fig. 6.11.

Figure 6.12a shows that uncertainty is decreasing over time. The final covariance matrix is shown graphically in Fig. 6.12b and we see a complex structure. Unlike the mapping case af Fig. 6.9 $\hat{P}_{mm}$ is not block diagonal, and the finite off-diagonal terms

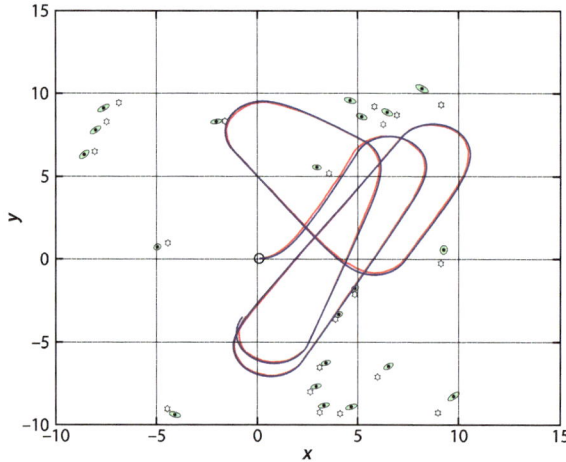

Fig. 6.11.
Simultaneous localization and mapping showing the true (*blue*) and estimated (*red*) robot path superimposed on the true map (*black ✿-marker*). The estimated map features are indicated by *black dots* with 95% confidence ellipses (*green*)

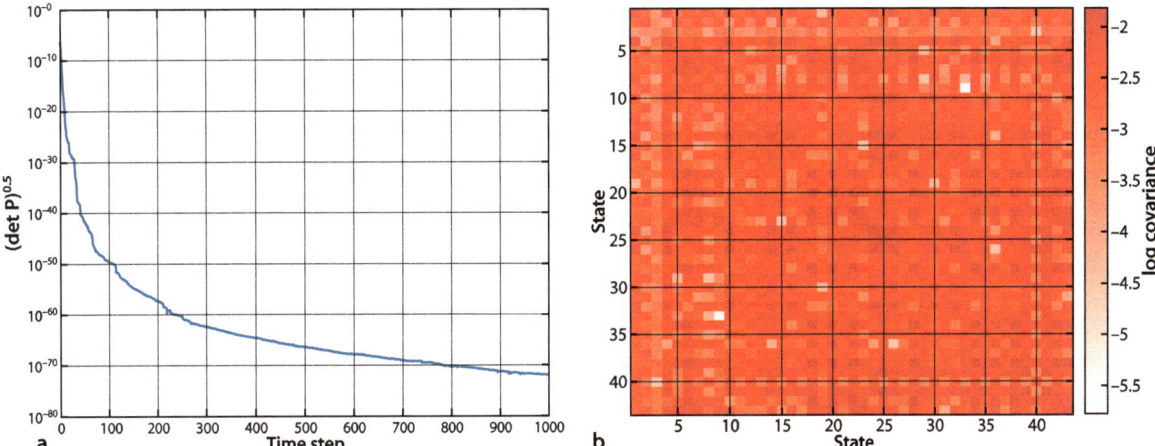

Such as the particle filter that we will discuss in Sect. 6.7.

Fig. 6.12. Simultaneous localization and mapping. **a** Covariance versus time; **b** the final covariance matrix

represent correlation *between* the landmarks in the map. The landmark uncertainties never increase, the position prediction model is that they do not move, but they also never drop below the initial uncertainty of the vehicle which was set in P_0. The block $\hat{P}_{vm}$ is the correlation between errors in the vehicle pose and the landmark locations. A landmark's location estimate is a function of the vehicle's location and errors in the vehicle location appear as errors in the landmark location – and vice versa.

The correlations are used by the Kalman filter to connect the observation of any landmark to an improvement in the estimate of every other landmark in the map as well as the vehicle pose. Conceptually it is as if all the states were connected by springs and the movement of any one affects all the others.

The extended Kalman filter introduced here has a number of drawbacks. Firstly the size of the matrices involved increase with the number of landmarks and can lead to memory and computational bottlenecks as well as numerical problems. The underlying assumption of the Kalman filter is that all errors are Gaussian and this is far from true for sensors like laser rangefinders which we will discuss later in this chapter. We also need good estimates of covariance of the noise sources which in practice is challenging.

6.5 Rao-Blackwellized SLAM

We will briefly and informally introduce the underlying principle of Rao-Blackwellized SLAM of which FastSLAM is a popular and well known instance. The approach is motivated by the fact that the size of the covariance matrix for EKF SLAM is quadratic in the number of landmarks, and for large-scale environments becomes computationally intractable.

If we compare the covariance matrices shown in Fig. 6.9b and 6.12b we notice a stark difference. In both cases we were creating a map of unknown landmarks but Fig. 6.9b is mostly zero with a finite block diagonal structure whereas Fig. 6.12b has no zero values at all. The difference is that for Fig. 6.9b we assumed the robot trajectory was known exactly and that makes the landmark estimates *independent* – observing one landmark provides information about *only* that landmark. The landmarks are *uncorrelated*, hence all the zeros in the covariance matrix. If the robot trajectory is not known, the case for Fig. 6.12b, then the landmark estimates are correlated – error in one landmark position is related to errors in robot pose and other landmark positions. The Kalman filter uses the correlation information so that a measurement of any one landmark provides information to improve the estimate of all the other landmarks and the robot's pose.

In practice we don't know the true pose of the robot but imagine a multi-hypothesis estimator◄ where every hypothesis represents a robot trajectory that we *assume* is correct. This means that the covariance matrix will be block diagonal like Fig. 6.9b – rather than a filter with a $2N \times 2N$ covariance matrix we can have N simple filters which are each *independently* estimating the position of a single landmark and have a 2×2 covariance matrix. Independent estimation leads to a considerable saving in both memory and computation. Importantly though, we are only able to do this because we *assumed* that the robot's estimated trajectory is correct.

Each hypothesis also holds an estimate of the robot's trajectory to date. Those hypotheses that best explain the landmark measurements are retained and propagated while those that don't are removed and recycled. If there are M hypotheses the overall computational burden falls from $O(N^2)$ for the EKF SLAM case to $O(M \log N)$ and in practice works well for M in the order of tens to hundreds but can work for a value as low as $M = 1$.

6.6   Pose Graph SLAM

An alternative approach to the SLAM problem is to separate it into two components: a front end and a back end, connected by a pose graph as shown in Fig. 6.13. The robot's path is considered to be a sequence of distinct poses and the task is to estimate those poses. Constraints between the unknown poses are based on measurements from a variety of sensors including odometry, laser scanners and cameras. The problem is formulated as a directed graph as shown in Fig. 6.14. A node corresponds to a robot pose or a landmark position. An edge between two nodes represents a spatial constraint between the nodes derived from some sensor data.

As the robot progresses it compounds an increasing number of uncertain relative poses so that the cumulative error in the pose of the nodes will increase – the problem with dead reckoning we discussed earlier. This is shown in exaggerated fashion in Fig. 6.14 where the robot is traveling around a square. By the time the robot reaches node 4 the error is significant. However when it makes a measurement of node 1 a constraint is added – the dashed edge – indicating that the nodes are closer than the estimated relative pose based on the chain of relative poses from odometry: ${}^1\xi_2^\# \oplus {}^2\xi_3^\# \oplus {}^3\xi_4^\#$. The back-end algorithm will then *pull* all the nodes closer to their correct pose.

The front end adds new nodes as the robot travels▶ as well as edges that define constraints between poses. For example, when moving from one place to another wheel odometry gives an estimate of distance and change in orientation which is a constraint. In addition the robot's exteroceptive sensors may observe the relative position of a landmark and this also adds a constraint. Every measurement adds a constraint – an edge in the graph. There is no limit to the number of edges entering or leaving a node.

Typically a new place is declared every meter or so of travel, or after a sharp turn.

The back end adjusts the poses of the nodes▶ so that the constraints are satisfied as well as possible, that is, that the sensor observations are best explained.

Also the positions of landmarks as we discuss later in this section.

Figure 6.15 shows the notation associated with two poses in the graph. Coordinate frames $\{i\}$ and $\{j\}$ are associated with robot poses i and j respectively and we seek to estimate ${}^0\xi_i$ and ${}^0\xi_j$ in the world coordinate frame. The robot makes a measurement of the relative pose ${}^i\xi_j^\#$ which will, in general, be different to the relative pose ${}^i\xi_j$ inferred from the poses ${}^0\xi_i$ and ${}^0\xi_j$. This difference, or innovation, is caused by error in the sensor measurement ${}^i\xi_j^\#$ and/or the node poses ${}^0\xi_i$ and ${}^0\xi_j$ and we use it to adjust the poses of the nodes. However there is insufficient information to determine where the error lies so naively adjusting ${}^0\xi_i$ and ${}^0\xi_j$ to better explain the measurement might increase

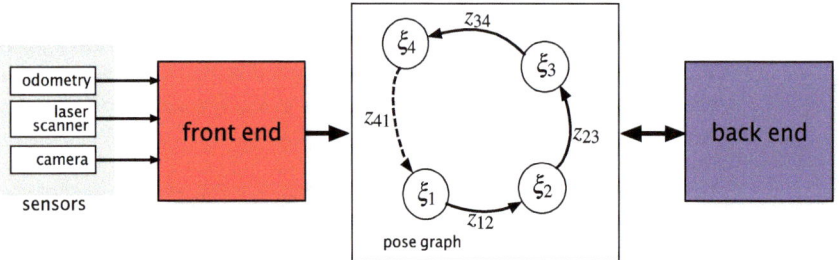

Fig. 6.13.
Pose-graph SLAM system. The front end creates nodes as the robot travels, and creates edges based on sensor data. The back end adjusts the node positions to minimize total error

the error in another part of the graph – we need to minimize the error consistently over the whole graph.

The first step is to express the error associated with the graph edge in terms of the sensor measurement and our best estimates of the node poses with respect to the world frame◀

We have used our pose notation here but in the literature measurements are typically denoted by z, error by e and pose or position by x.

$$\xi_\varepsilon = \ominus\, {}^i\xi_j^{\#} \ominus {}^0\hat{\xi}_i \oplus {}^0\hat{\xi}_j \in \mathbf{SE}(2) \tag{6.27}$$

which is ideally zero.

We can formulate this as a minimization problem and attempt to find the poses of all the nodes $x = \{\xi_1, \xi_2 \cdots \xi_N\}$ that minimizes the error across all the edges

$$x^* = \arg\min_x \sum_k F_k(x) \tag{6.28}$$

where x is the state of the pose graph and contains the pose of every node, and $F_k(x)$ is a nonnegative scalar cost associated with the edge k connecting node i to node j.

We convert the edge pose error in Eq. 6.27 to a vector representation $\xi_\varepsilon \sim (x, y, \theta)$ which is a function $f_k(x) \in \mathbb{R}^3$ of the state. The scalar cost can be obtained from a quadratic expression

In practice this matrix is diagonal reflecting confidence in the x-, y- and θ-directions. The "bigger" (in a matrix sense) Ω is, the more the edge *matters* in the optimization procedure. Different sensors have different accuracy and this must be taken into account. Information from a high-quality sensor should be given more weight than information from a low-quality sensor.

$$F_k(x) = f_k^T(x)\,\Omega_k\, f_k(x) \tag{6.29}$$

where Ω_k is a positive-definite information matrix used as a weighting term.◀ Although Eq. 6.29 is written as a function of all poses x, it in fact depends only on the pose of its two vertices ξ_i and ξ_j and the measurement ${}^i\xi_j^{\#}$. Solving Eq. 6.28 is a complex optimization problem which does not have a closed-form solution, but this kind of nonlinear least squares problem can be solved numerically if we have a good initial estimate of x. Specifically this is a sparse nonlinear least squares problem which is discussed in Sect. F.2.4.

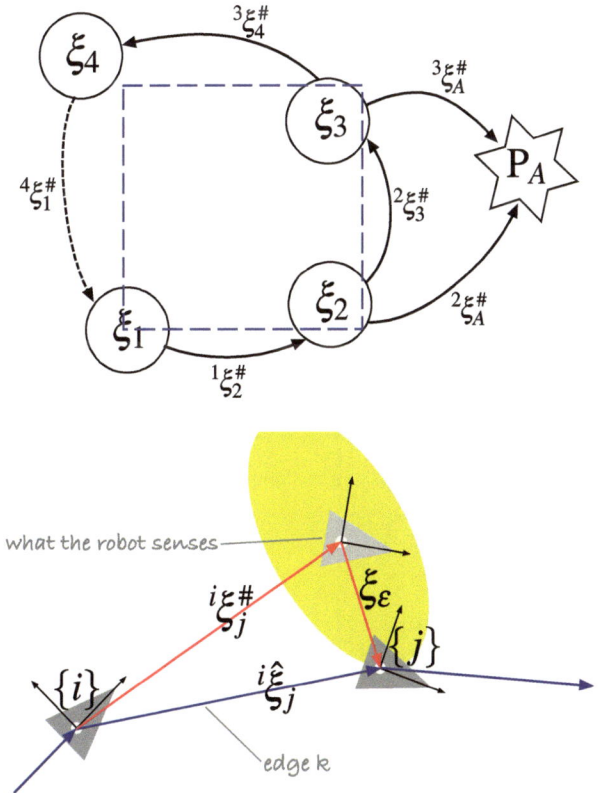

Fig. 6.14.
Pose-graph SLAM example. Places are shown as *circular nodes* and have an associated pose. Landmarks are shown as *star-shaped nodes* and have an associated position. Edges represent a measurement of a relative pose or position with respect to the node at the tail of the arrow

Fig. 6.15.
Pose graph notation. The *light grey robot* is the estimated pose of {j} based on the sensor measurement ${}^i\xi_j^{\#}$. The *yellow ellipse* indicates uncertainty associated with that measurement

The edge error $f_k(x)$ can be linearized about the current state x_0 of the pose graph

$$f_k'(\Delta) \approx f_{0,k} + J_k\Delta$$

where $f_{0,k} = f_k(x_0)$ and

$$J_k = \frac{\partial f_k(x)}{\partial x} \in \mathbb{R}^{3 \times 3N}$$

is a Jacobian matrix which depends only on the pose of its two vertices ξ_i and ξ_j so it is mostly zeros

$$J_k = \left(0 \cdots A_i \cdots B_j \cdots 0\right), \quad \text{where } A_i = \frac{\partial f_k(x)}{\partial \xi_i} \in \mathbb{R}^{3 \times 3}, B_j = \frac{\partial f_k(x)}{\partial \xi_j} \in \mathbb{R}^{3 \times 3}$$

and more details are provided in Appendix E.

There are many ways to compute the Jacobians but here will demonstrate use of the MATLAB Symbolic Math Toolbox™

```
>> syms xi yi ti xj yj tj xm ym tm assume real
>> xi_e = inv( SE2(xm, ym, tm) ) * inv( SE2(xi, yi, ti) ) * SE2(xj, yj, tj);
>> fk = simplify(xi_e.xyt);
```

and the Jacobian which describes how the function f_k varies with respect to ξ_i is

```
>> jacobian( fk, [xi yi ti] );
>> Ai = simplify(ans)
Ai =
[ -cos(ti+tm), -sin(ti+tm), yj*cos(ti+tm)-yi*cos(ti+tm)+xi*sin(ti+tm)-xj*sin(ti+tm)]
[  sin(ti+tm), -cos(ti+tm), xi*cos(ti+tm)-xj*cos(ti+tm)+yi*sin(ti+tm)-yj*sin(ti+tm)]
[      0,           0,                                                           -1]
```

and we follow a similar procedure for B_j.

It is quite straightforward to solve this type of pose-graph problem with the Toolbox. We load a simple pose graph, similar to Fig. 6.14, from a data file▶

The file format is one used by the popular posegraph optimization package g²o which you can find at http://openslam.org.

```
>> pg = PoseGraph('pg1.g2o')
loaded g2o format file: 4 nodes, 4 edges in 0.00 sec
```

which returns a Toolbox `PoseGraph` object that describes the pose graph. We can visualize this by▶

The nodes have an orientation which is in the z-direction, rotate the graph to see this.

```
>> pg.plot()
```

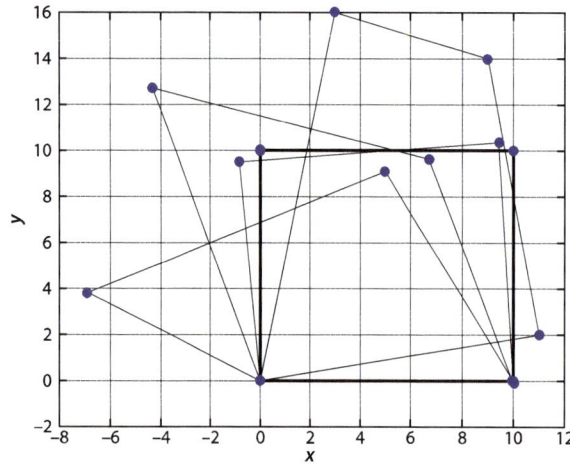

Fig. 6.16.
Pose graph optimization showing the result over consecutive iterations, the final configuration is the *square* shown in *bold*

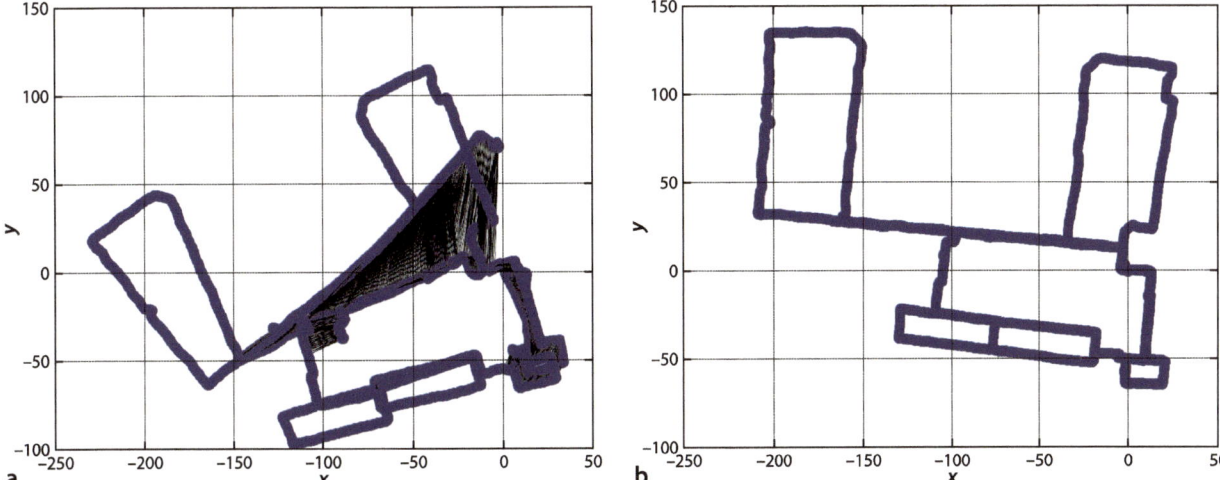

Fig. 6.17. Pose graph with 1 941 nodes and 3 995 edges from the MIT Killian Court dataset. **a** Initial configuration; **b** final configuration after optimization

The optimization reduces the error in the network while animating the changing pose of the nodes

```
>> pg.optimize('animate')
solving....done in 0.075 sec.   Total cost 316.88
solving....done in 0.0033 sec.   Total cost 47.2186
  .
  .
solving....done in 0.0023 sec.   Total cost 3.14139e-11
```

The displayed text indicates that the total cost is decreasing while the graphics show the nodes moving into a configuration that minimizes the overall error in the network. The pose graph configurations are overlaid and shown in Fig. 6.16.

Now let's look a much larger example based on real robot data

```
>> pg = PoseGraph('killian-small.toro');
loaded TORO/LAGO format file: 1941 nodes, 3995 edges in 0.68 sec
```

There are a lot of nodes and this takes a few seconds.

which we can plot◄

```
>> pg.plot()
```

and this is shown in Fig. 6.17a. Note the mass of edges in the center of the graph, and if you zoom in you can see these in detail. We optimize the pose graph by

```
>> pg.optimize()
solving....done in 0.91 sec.   Total cost 1.78135e+06
  .
  .
solving....done in 1.1 sec.   Total cost 5.44567
```

and the final configuration is shown in Fig. 6.17b. The original pose graph had severe pose errors from accumulated odometry error which meant that two trips along the corridor were initially very poorly aligned.

The pose graph can also include landmarks as shown in Fig. 6.18. Landmarks have a position $P_j \in \mathbb{R}^2$ not a pose, and therefore differ from the nodes discussed so far. To accomodate this we redefine the state vector to be $x = \{\xi_1, \xi_2 \cdots \xi_N \mid P_1, P_2 \cdots P_M\}$ which includes N robot poses and M landmark positions. The robot at pose i observes landmark j at range and bearing $z^\# = (r^\#, \beta^\#)$ which is converted to Cartesian form in frame $\{i\}$

$$^i P_j^\# = \left(r^\# \cos\beta^\#, r^\# \sin\beta^\#\right) \in \mathbb{R}^2$$

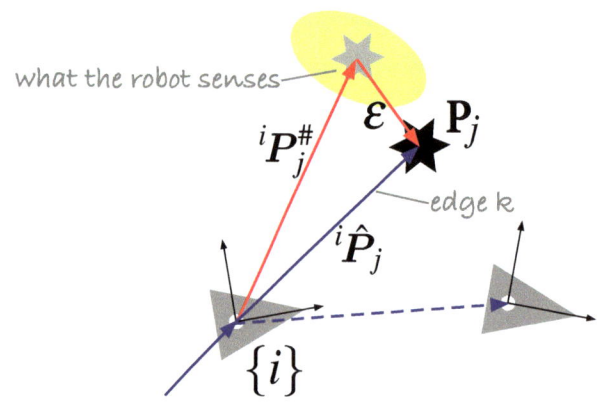

Fig. 6.18.
Notation for a pose graph with a landmark indicated by the *star-shaped symbol*. The measured position of landmark j with respect to robot pose i is ${}^iP_j^{\#}$. The *yellow ellipse* indicates uncertainty associated with that measurement

The estimated position of the landmark in frame $\{i\}$ is

$$ {}^i\hat{P}_j = \left(\ominus {}^0\xi_i\right) \bullet \hat{P}_j \in \mathbb{R}^2 $$

and the error vector is

$$ f_k(x) = \varepsilon = {}^i\hat{P}_j - {}^iP_j^{\#} \in \mathbb{R}^2 $$

We follow a similar approach as earlier but the Jacobian matrix is now

$$ J_k = \frac{\partial f_k(x)}{\partial x} \in \mathbb{R}^{2\times(3N+2M)} $$

which again is mostly zero but the two nonzero blocks now have different widths

$$ A_i = \frac{\partial f_k(x)}{\partial \xi_i} \in \mathbb{R}^{2\times 3}, \ B_j = \frac{\partial f_k(x)}{\partial P_j} \in \mathbb{R}^{2\times 2} $$

and the solution can be achieved as before, see Sect. F.2.3 for more details.

Pose graph optimization results in a graph that has optimal relative poses and positions between the nodes but the absolute poses and positions are not necessarily correct. To remedy this we can fix or *anchor* one or more nodes (poses or landmarks) and not update them during the optimization, and this is discussed in Sect. F.2.4.

In practice the front and back ends can operate asynchronously. The graph is continually extended by the front end while the back end runs periodically to opti-

> **Monte Carlo methods** are a class of computational algorithms that rely on repeated random sampling to compute their results. An early example of this idea is Buffon's needle problem posed in the eighteenth century by Georges-Louis Leclerc (1707–1788), Comte de Buffon: *Suppose we have a floor made of parallel strips of wood of equal width t, and a needle of length l is dropped onto the floor. What is the probability that the needle will lie across a line between the strips?* If n needles are dropped and h cross the lines, the probability can be shown to be $h/n = 2l/\pi t$ and in 1901 an Italian mathematician Mario Lazzarini performed the experiment, tossing a needle 3 408 times, and obtained the estimate $\pi \approx 355/113$ (3.14159292).
>
> Monte Carlo methods are often used when simulating systems with a large number of coupled degrees of freedom with significant uncertainty in inputs. Monte Carlo methods tend to be used when it is infeasible or impossible to compute an exact result with a deterministic algorithm. Their reliance on repeated computation and random or pseudo-random numbers make them well suited to calculation by a computer. The method was developed at Los Alamos as part of the Manhattan project during WW II by the mathematicians John von Neumann, Stanislaw Ulam and Nicholas Metropolis. The name Monte Carlo alludes to games of chance and was the code name for the secret project.

mize the pose graph. Since the graph is only ever extended in a local region it is possible to optimize just a local subset of the pose graph and less frequently optimize the entire graph. If nodes are found to be equivalent after optimization they can be merged. The parallel tracking and mapping system (PTAM) is a vision-based SLAM system that has two parallel computational threads. One is the map builder which performs the front- and back-end tasks, adding landmarks to the pose graph based on estimated camera (vehicle) pose and performing graph optimization. The other thread is the localizer which matches observed landmarks to the estimated map to estimate the camera pose.

6.7   Sequential Monte-Carlo Localization

The estimation examples so far have assumed that the error in sensors such as odometry and landmark range and bearing have a Gaussian probability density function. In practice we might find that a sensor has a one sided distribution (like a Poisson distribution) or a multi-modal distribution with several peaks. The functions we used in the Kalman filter such as Eq. 6.2 and Eq. 6.7 are strongly nonlinear which means that sensor noise with a Gaussian distribution will not result in a Gaussian error distribution on the value of the function – this is discussed further in Appendix H. The probability density function associated with a robot's configuration may have multiple peaks to reflect several hypotheses that equally well explain the data from the sensors as shown for example in Fig. 6.3c.

The Monte-Carlo estimator that we discuss in this section makes no assumptions about the distribution of errors. It can also handle multiple hypotheses for the state of the system. The basic idea is disarmingly simple. We maintain many *different* values of the vehicle's configuration or state vector. When a new measurement is available we score how well each particular value of the state explains what the sensor just observed. We keep the best fitting states and randomly sample from the prediction distribution to form a new generation of states. Collectively these many possible states and their scores form a discrete approximation of the probability density function of the state we are trying to estimate. There is never any assumption about Gaussian distributions nor any need to linearize the system. While computationally expensive it is quite feasible to use this technique with today's standard computers. If we plot these state vectors as points in the state space we have a cloud of particles hence this type of estimator is often referred to as a particle filter.

We will apply Monte-Carlo estimation to the problem of localization using odometry and a map. Estimating only three states $x = (x, y, \theta)$ is computationally tractable to solve with straightforward MATLAB code. The estimator is initialized by creating N particles x_i, $i \in [1, N]$ distributed randomly over the configuration space of the vehicle. All particles have the same initial weight or likelihood $w_i = 1 / N$. The steps in the main iteration of the algorithm are:

1. Apply the state update to each particle

$$x_i^+ \langle k+1 \rangle = f\big(x_i\langle k \rangle, u\langle k \rangle + q\langle k \rangle\big)$$

where $\hat{u}\langle k \rangle$ is the input to the system or the measured odometry $\hat{u}\langle k \rangle = \delta\langle k \rangle$. We also add a random vector $q\langle k \rangle$ which represents uncertainty in the model or the odometry. Often q is drawn from a Gaussian random variable with covariance Q but any physically meaningful distribution can be used. The state update is often simplified to

$$x_i^+ \langle k+1 \rangle = f\big(x_i\langle k \rangle, u\langle k \rangle\big) + q\langle k \rangle$$

where $q\langle k \rangle$ represents uncertainty in the pose of the vehicle.

2. We make an observation $z^{\#}$ of landmark j which has, according to the map, coordinate p_j. For each particle we compute the innovation

$$\boldsymbol{\nu}_i = \boldsymbol{h}\left(\boldsymbol{x}_i^{+}, \boldsymbol{p}_j\right) - \boldsymbol{z}^{\#}$$

which is the error between the predicted and actual landmark observation. A likelihood function provides a scalar measure of how well the particular particle explains this observation. In this example we choose a likelihood function

$$w_i = \mathrm{e}^{-\boldsymbol{\nu}_i^T L^{-1} \boldsymbol{\nu}_i} + w_0$$

where w is referred to as the *importance* or *weight* of the particle, L is a covariance-like matrix, and $w_0 > 0$ ensures that there is a finite probability of a particle being retained despite sensor error. We use a quadratic exponential function only for convenience, the function does not need to be smooth or invertible but only to adequately describe the likelihood of an observation.▶

> In this bootstrap type filter the weight is computed at each step, with no dependence on previous values.

3. Select the particles that best explain the observation, a process known as resampling▶ or importance sampling. A common scheme is to randomly select particles according to their weight. Given N particles $\boldsymbol{x}_i$ with corresponding weights w_i we first normalize the weights $w_i' = w_i / \Sigma_{i=1}^N w_i$ and construct a cumulative histogram $c_j = \Sigma_{i=1}^j w_i'$. We then draw a uniform random number $r \in [0, 1]$ and find

$$\arg\min_i r < c_i$$

where particle i is selected for the next generation. The process is repeated N times.

> There are many resampling strategies for particle filters, both the resampling algorithm and the resampling frequency. Here we use the simplest strategy known variously as multinomial resampling, simple random resampling or select with replacement, at every time step. This is sometimes referred to as bootstrap particle filtering or condensation.

Particles with a large weight will correspond to a larger fraction of the vertical span of the cumulative histogram and therefore be more likely to be chosen. The result will have the same number of particles, some will have been copied▶ multiple times, others not at all. Resampling is a critical component of the particle filter without which the filter would quickly produce a degenerate set of particles where a few have high weights and the bulk have almost zero weight.

> Step 1 of the next iteration will *spread out* these copies through the addition of $\boldsymbol{q}_{\langle k \rangle}$.

These steps are summarized in Fig. 6.19. The Toolbox implementation is broadly similar to the previous examples. We create a map

```
>> map = LandmarkMap(20);
```

and a robot with noisy odometry and an initial condition

```
>> W = diag([0.1, 1*pi/180].^2);
>> veh = Bicycle('covar', V);
>> veh.add_driver( RandomPath(10) );
```

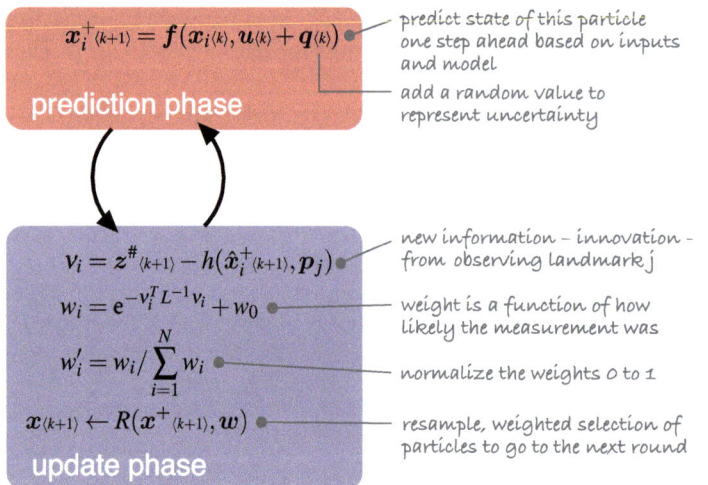

predict state of this particle one step ahead based on inputs and model

add a random value to represent uncertainty

new information – innovation – from observing landmark j

weight is a function of how likely the measurement was

normalize the weights 0 to 1

resample, weighted selection of particles to go to the next round

Fig. 6.19.
The particle filter estimator showing the prediction and update phases

and then a sensor with noisy readings

```
>> V = diag([0.005, 0.5*pi/180].^2);
>> sensor = RangeBearingSensor(veh, map, 'covar', W);
```

For the particle filter we need to define two covariance matrices. The first is the covariance of the random noise added to the particle states at each iteration to represent uncertainty in configuration. We choose the covariance values to be comparable with those of W

```
>> Q = diag([0.1, 0.1, 1*pi/180]).^2;
```

and the covariance of the likelihood function applied to innovation

```
>> L = diag([0.1 0.1]);
```

Finally we construct a `ParticleFilter` estimator

```
>> pf = ParticleFilter(veh, sensor, Q, L, 1000);
```

which is configured with 1 000 particles. The particles are initially uniformly distributed over the 3-dimensional configuration space.

We run the simulation for 1 000 time steps

```
>> pf.run(1000);
```

and watch the animation, two snapshots of which are shown in Fig. 6.20. We see the particles move about as their states are updated by odometry and random perturbation. The initially randomly distributed particles begin to aggregate around those regions of the configuration space that best *explain* the sensor observations that are made. In Darwinian fashion these particles become more highly weighted and survive the resampling step while the lower weight particles are extinguished.

The particles approximate the probability density function of the robot's configuration. The most likely configuration is the expected value or mean of all the particles. A measure of uncertainty of the estimate is the spread of the particle cloud or its standard deviation. The `ParticleFilter` object keeps the history of the mean and standard deviation of the particle state at each time step, taking into account the particle weighting▸. As usual we plot the results of the simulation

```
>> map.plot();
>> veh.plot_xy('b');
```

and overlay the mean of the particle cloud

```
>> pf.plot_xy('r');
```

Here we take statistics over all particles. Other strategies are to estimate the kernel density at every particle – the sum of the weights of all neighbors within a fixed radius – and take the particle with the largest value.

Fig. 6.20. Particle filter results showing the evolution of the particle cloud (*green dots*) over time. The vehicle is shown as a *blue triangle*. The *red diamond* is a waypoint, or temporary goal. When the simulation is running this is actually a 3D plot with orientation plotted in the *z*-direction, rotate the plot to see this dimension

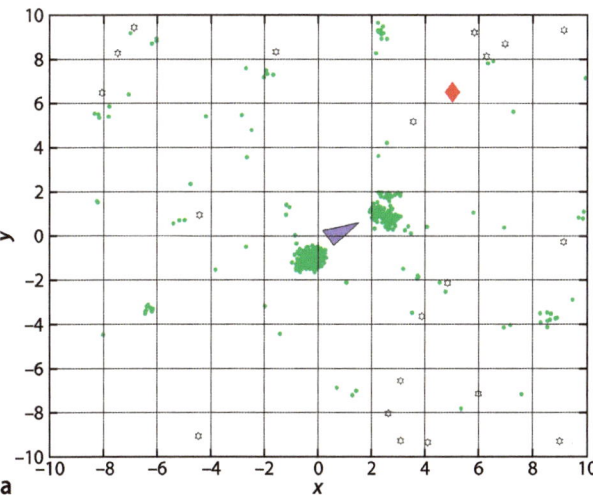
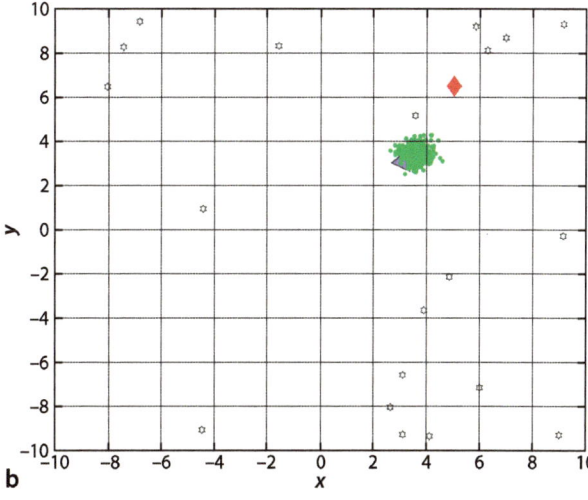

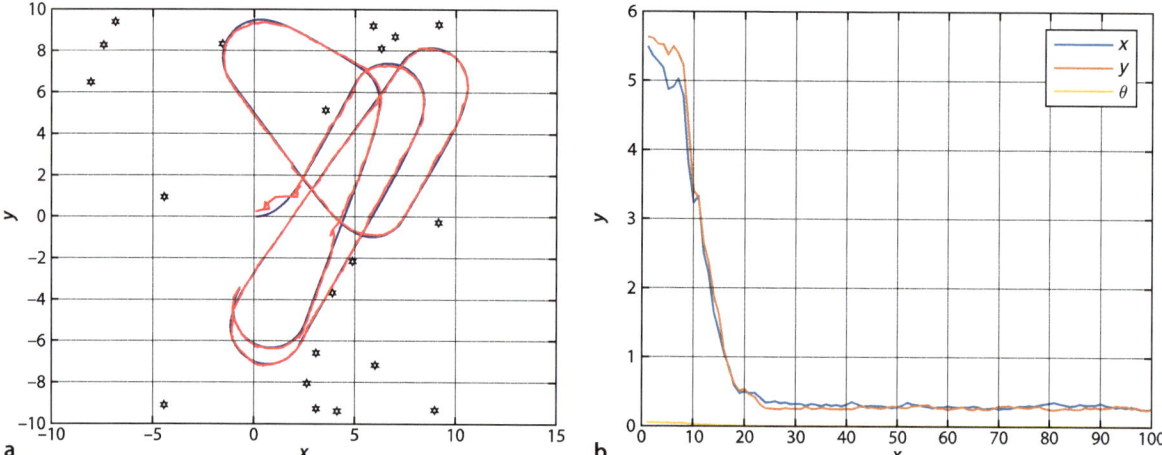

Fig. 6.21. Particle filter results. **a** True (*blue*) and estimated (*red*) robot path; **b** standard deviation of the particles versus time

which is shown in Fig. 6.21. The initial part of the estimated path has quite high standard deviation since the particles have not converged on the true configuration. We can plot the standard deviation against time

```
>> plot(pf.std(1:100,:))
```

and this is shown in Fig. 6.21b. We can see the sudden drop between timesteps 10–20 as the particles that are distant from the true solution are eliminated. As mentioned at the outset the particles are a sampled approximation to the PDF and we can display this as

```
>> pf.plot_pdf()
```

The problem we have just solved is known in robotics as the kidnapped robot problem where a robot is placed in the world with no idea of its initial location. To represent this large uncertainty we uniformly distribute the particles over the 3-dimensional configuration space and their sparsity can cause the particle filter to take a long time to converge unless a very large number of particles is used. It is debatable whether this is a realistic problem. Typically we have some approximate initial pose of the robot and the particles would be initialized to that part of the configuration space. For example, if we know the robot is in a corridor then the particles would be placed in those areas of the map that are corridors, or if we know the robot is pointing north then set all particles to have that orientation.

Setting the parameters of the particle filter requires a little experience and the best way to learn is to experiment. For the kidnapped robot problem we set Q and the number of particles high so that the particles explore the configuration space but once the filter has converged lower values could be used. There are many variations on the particle filter in the shape of the likelihood function and the resampling strategy.

6.8   Application: Scanning Laser Rangefinder

As we have seen, robot localization is informed by measurements of range and bearing to landmarks. Sensors that measure range can be based on many principles such as laser rangefinding (Fig. 6.22a, 6.22b), ultrasonic ranging (Fig. 6.22c), computer vision or radar.

A laser rangefinder emits short pulses of infra-red laser light and measures how long it takes for the reflected pulse to return. Operating range can be up to 50 m with an accuracy of the order of centimeters.

A *scanning* laser rangefinder, as shown in Fig. 6.22a, contains a rotating laser rangefinder and typically emits a pulse every quarter, half or one degree over an angular range of 180 or 270 degrees and returns a planar cross-section of the world in polar coordinate form $\{(r_i, \theta_i), i \in 1 \cdots N\}$. Some scanning laser rangefinders also measure

Fig. 6.22.
Robot rangefinders. **a** A scanning laser rangefinder with a maximum range of 30 m, an angular range of 270 deg in 0.25 deg intervals at 40 scans per second (courtesy of Hokuyo Automatic Co. Ltd.); **b** a low-cost time-of-flight rangefinder with maximum range of 20 cm at 10 measurements per second (VL6180 courtesy of SparkFun Electronics); **c** a low-cost ultrasonic rangefinder with maximum range of 6.5 m at 20 measurements per second (LV-MaxSonar-EZ1 courtesy of SparkFun Electronics)

the return signal strength, remission, which is a function of the infra-red reflectivity of the surface. The rangefinder is typically configured to scan in a plane parallel to, and slightly above, the ground.

Laser rangefinders have advantages and disadvantages compared to cameras and computer vision. On the positive side laser scanners provide metric data, that is, the actual range to points in the world in units of meters, and they can work in the dark. However laser rangefinders work less well than cameras outdoors since the returning laser pulse is overwhelmed by infra-red light from the sun. Other disadvantages include providing only a linear cross section of the world, rather than an area as a camera does; inability to discern fine texture or color; having moving parts; as well as being bulky, power hungry and expensive compared to cameras.

Laser Odometry

A common application of scanning laser rangefinders is laser odometry, estimating the change in robot pose using laser scan data rather than wheel encoder data. We will illustrate this with laser scan data from a real robot

```
>> pg = PoseGraph('killian.g2o', 'laser');
loaded g2o format file: 3873 nodes, 4987 edges in 1.78 sec
  3873 laser scans: 180 beams, fov -90 to 90 deg, max range 50
```

and each scan is associated with a vertex of this already optimized pose graph. The range and bearing data for the scan at node 2 580 is

```
>> [r, theta] = pg.scan(2580);
>> about r theta
r [double] : 1x180 (1.4 kB)
theta [double] : 1x180 (1.4 kB)
```

represented by two vectors each of 180 elements. We can plot these in polar form

```
>> polar(theta, r)
```

or convert them to Cartesian coordinates and plot them

```
>> [x,y] = pol2cart(theta, r);
>> plot(x, y, '.')
```

The method `scanxy` is a simpler way to perform these operations. We load scans from two closely spaced nodes

```
>> p2580 = pg.scanxy(2580);
>> p2581 = pg.scanxy(2581);
>> about p2580
p2580 [double] : 2x180 (2.9 kB)
```

Note that the points close to the laser, at coordinate (0,0) in this sensor reference frame are much more tightly clustered and this is a characteristic of laser scanners where the points are equally spaced in angle not over an area.

which creates two matrices whose columns are Cartesian point coordinates and these are overlaid in Fig. 6.23a.◀

To determine the change in pose of the robot between the two scans we need to align these two sets of points and this can be achieved with iterated closest-point-matching

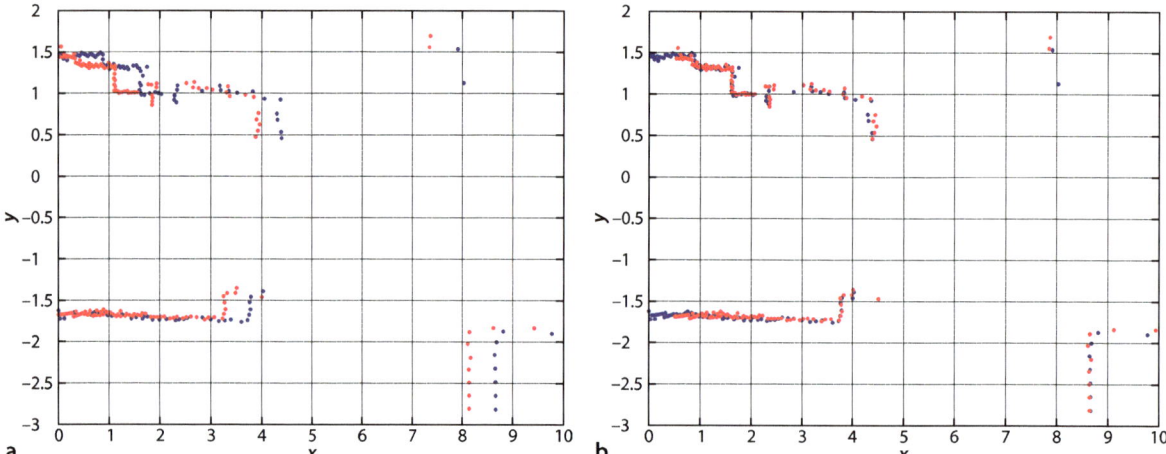

or ICP. This is implemented by the Toolbox function `icp`◀ and we pass in the second and first set of points, each organized as a $2 \times N$ matrix

```
>> T = icp( p2581, p2580, 'verbose' , 'T0', transl2(0.5, 0), 'distthresh', 3)
[1]: n=132/180, d=   0.466, t = (   0.499   -0.006), th = (  -0.0) deg
[2]: n=130/180, d=   0.429, t = (   0.500   -0.009), th = (   0.0) deg

   .

   .

[6]: n=130/180, d=   0.425, t = (   0.503   -0.011), th = (   0.0) deg

T =
    1.0000   -0.0002    0.5032
    0.0002    1.0000   -0.0113
         0         0    1.0000
```

and the algorithm converges after a few iterations with an estimate of $T \sim {}^{2580}\xi_{2581}$ $\in$ **SE**(2).▶ This transform maps points from the second scan so that they are as close as possible to the points in the first scan. Figure 6.23b shows the first set of points transformed and overlaid on the second set and we see good alignment. The translational part of this transform is an estimate of the robot's motion between scans – around 0.50 m in the x-direction. The nodes of the graph also hold time stamp information and these two scans were captured

```
>> pg.time(2581)-pg.time(2580)
ans =
    1.7600
```

seconds apart which indicates that the robot is moving quite slowly – a bit under 0.3 m s^{-1}.

At each iteration ICP assigns each point in the second set to the closest point in the first set and then computes a transform that minimizes the sum of distances between all corresponding points. Some points may not actually be corresponding but as long as enough are, the algorithm will converge. The `'verbose'` option causes data about each iteration to be displayed and `d` is the total distance between corresponding points which is decreasing but does not reach zero. This is due to many factors. The beams from the laser at the two different poses will not strike the walls at the same location so ICP's assumption about point correspondence is not actually valid.▶

In practice there are additional challenges. Some laser pulses will not return to the sensor if they fall on a surface with low reflectivity or on an oblique polished surface that specularly reflects the pulse away from the sensor – in these cases the sensor typically reports its maximum value. People moving through the environment change the shape of the world and temporarily cause a shorter range to be reported. In very large spaces all the walls may be beyond the maximum range of the sensor. Outdoors the beams can be reflected from rain drops, absorbed by fog or smoke and the return pulse can be overwhelmed by ambient sunlight. Finally the laser rangefinder, like all sensors, has measurement noise.

Fig. 6.23. Laser scan matching. **a** Laser scans from location 2 580 (*blue*) and 2 581 (*red*); **b** location 2 580 points (*blue*) and transformed points from location 2 581 (*red*)

We demonstrate the principle using ICP but in practice more robust algorithms are used. Here we provide an initial estimate of the translation between frames, based on odometry, so as to avoid getting stuck in a local minimum. ICP works poorly in plain corridors where the points lie along lines – this example was deliberately chosen because it has wall segments in orthogonal directions.

To remove invalid correspondences we pass the `'distthresh'` option to `icp()`. This causes any correspondences that involve a distance more than three times the median distance between all corresponding points to be dropped. In the `icp()` output the notation `132/180` means that 132 out of 180 possible correspondences met this test, 48 were rejected.

Laser-Based Map Building

If the robot pose is sufficiently well known, through some localization process, then we can transform all the laser scans to a global coordinate frame and build a map. Various map representations are possible but here we will outline how to build an occupancy grid as discussed in Chap. 5.

For a robot at a given pose, each beam in the scan is a ray and tells us several things. From the range measurement we can determine the coordinates of a cell that contains an obstacle but we can tell nothing about cells further along the ray. It is also implicit that all the cells between the sensor and the obtacle must be obstacle free. A maximum distance value, 50 m in this case, is the sensor's way of indicating that there was no returning laser pulse so we ignore all such measurements. We create the occupancy grid as a matrix and use the Bresenham algorithm to find all the cells along the ray based on the robot's pose and the laser range and bearing measurement, then a simple voting scheme to determine whether cells are free or occupied

```
>> pg.scanmap()
>> pg.plot_occgrid()
```

and the result is shown in Fig. 6.24. More sophisticated approaches treat the beam as a wedge of finite angular width and employ a probabilistic model of sensor return versus range. The principle can be extended to creating 3-dimensional point clouds from a scanning laser rangefinder on a moving vehicle as shown in Fig. 6.25.

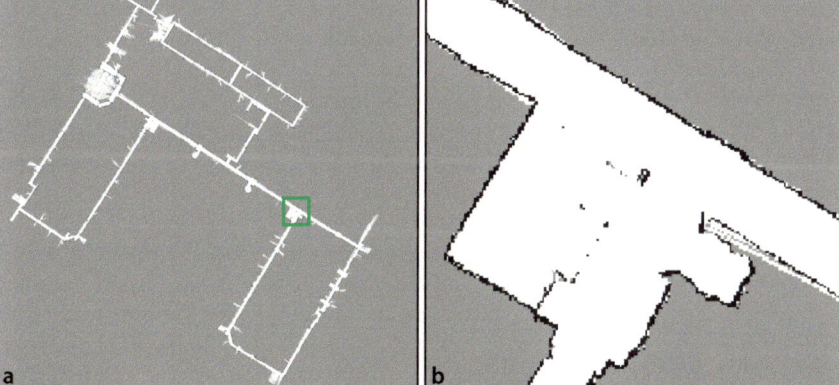

Fig. 6.24.
a Laser scans rendered into an occupancy grid, the area enclosed in the *green square* is dispayed in **b**. *White cells* are free space, *black cells* are occupied and *grey cells* are unknown. Grid cell size is 10 cm

Fig. 6.25.
3D point cloud created by integrating multiple scans from a vehicle-mounted scanning laser rangefinder, where the scans are in a vertical plane normal to the vehicle's forward axis. This is sometimes called a "2.5D" representation since only the front surfaces of objects are described – note the range shadows on the walls behind cars. Note also that the density of laser points is not constant across the map, for example the point density on the road surface is much greater than it is high on the walls of buildings (image courtesy Alex Stewart; Stewart 2014)

Laser-Based Localization

We have mentioned landmarks a number of times in this chapter but avoided concrete examples of what they are. They could be distinctive visual features or artificial markers as discussed on page 162. If we consider a laser scan such as shown in Fig. 6.23a or 6.24b we see a fairly distinctive arrangement of points – a geometric signature – which we can use as a landmark. In many cases the signature will be ambiguous and of little value, for example a long corridor where all the points are collinear, but some signatures will be highly unique and can serve as a useful landmark. Naively we could match the current laser scan against all others and if the fit is good (the ICP error is low) we could add another constraint to the pose graph. However this strategy would be expensive with a large number of scans so typically only scans in the vicinity of the robot's estimated position are checked, and this once again raises the data association problem.

6.9   Wrapping Up

In this chapter we learned about two very different ways of estimating a robot's position: by dead reckoning, and by observing landmarks whose true position is known from a map. Dead reckoning is based on the integration of odometry information, the distance traveled and the change in heading angle. Over time errors accumulate leading to increased uncertainty about the pose of the robot.

We modeled the error in odometry by adding noise to the sensor outputs. The noise values are drawn from some distribution that describes the errors of that particular sensor. For our simulations we used zero-mean Gaussian noise with a specified covariance, but only because we had no other information about the specific sensor. The most realistic noise model available should be used. We then introduced the Kalman filter which provides an optimal estimate, in the least-squares sense, of the true configuration of the robot based on noisy measurements. The Kalman filter is however only optimal for the case of zero–mean Gaussian noise and a linear model. The model that describes how the robot's configuration evolves with time can be nonlinear in which case we approximate it with a linear model which included some partial derivatives expressed as Jacobian matrices – an approach known as extended Kalman filtering.

The Kalman filter also estimates uncertainty associated with the pose estimate and we see that the magnitude can never decrease and typically grows without bound. Only additional sources of information can reduce this growth and we looked at how observations of landmarks, with known locations, relative to the robot can be used. Once again we use the Kalman filter but in this case we use both the prediction and the update phases of the filter. We see that in this case the uncertainty can be decreased by a landmark observation, and that over the longer term the uncertainty does not grow. We then applied the Kalman filter to the problem of estimating the positions of the landmarks given that we knew the precise position of the vehicle. In this case, the state vector of the filter was the coordinates of the landmarks themselves.

Next we brought all this together and estimated the vehicle's position, the position of the landmarks and their uncertainties – simultaneous localization and mapping. The state vector in this case contained the configuration of the robot and the coordinates of the landmarks.

An important problem when using landmarks is data association, being able to determine which landmark has been known or observed by the sensor so that its position can be looked up in a map or in a table of known or estimated landmark positions. If the wrong landmark is looked up then an error will be introduced in the robot's position.

The Kalman filter scales poorly with an increasing number of landmarks and we introduced two alternative approaches: Rao-Blackwellized SLAM and pose-graph SLAM. The latter involves solving a large but sparse nonlinear least squares problem, turning the problem from one of (Kalman) filtering to one of optimization.

We finished our discussion of localization methods with Monte-Carlo estimation and introduced the particle filter. This technique is computationally intensive but makes no assumptions about the distribution of errors from the sensor or the linearity of the vehicle model, and supports multiple hypotheses. Particles filters can be considered as providing an approximate solution to the true system model, whereas a Kalman filter provides an exact solution to an approximate system model.

Finally we introduced laser rangefinders and showed how they can be applied to robot navigation, odometry and creating detailed floor plan maps.

Further Reading

Localization and SLAM. The tutorials by Bailey and Durrant-Whyte (2006) and Durrant-Whyte and Bailey (2006) are a good introduction to this topic, while the textbook *Probabilistic Robotics* (Thrun et al. 2005) is a readable and comprehensive coverage of all the material touched on in this chapter.

The book by Siegwart et al. (2011) also has a good treatment of robot localization. FastSLAM (Montemerlo et al. 2003; Montemerlo and Thrun 2007) is a state-of-the-art algorithm for Rao-Blackwellized SLAM.

Particle filters are described by Thrun et al. (2005), Stachniss and Burgard (2014) and the tutorial introduction by Rekleitis (2004). There are many variations such as fixed or adaptive number of particles and when and how to resample – and Li et al. (2015) provide a comprehensive review of resampling strategies. Determining the most likely pose was demonstrated by taking the weighted mean of the particles but many more approaches have been used. The kernel density approach takes the particle with the highest weight of neighboring particles within a fixed-size surrounding hypersphere.

Pose graph optimization, also known as GraphSLAM, has a long history starting with Lu and Milios (1997). There has been significant recent interest with many publications and open-source tools including g^{2}o (Kümmerle et al. 2011), $\sqrt{\text{SAM}}$ (Dellaert and Kaess 2006), iSAM (Kaess et al. 2007) and factor graphs. Agarwal et al. (2014) provides a gentle introduction to pose-graph SLAM and discusses the connection to land-based geodetic survey which is centuries old. Parallel Tracking and Mapping (PTAM) was described in Klein and Murray (2007), the code is available on github and there is also a blog.

There are many online resources related to SLAM. A collection of open-source SLAM implementations such as gmapping and iSam is available from OpenSLAM at http://www.openslam.org. An implementation of smoothing and mapping using factor graphs is available at https://bitbucket.org/gtborg/gtsam and has C++ and MATLAB bindings. MATLAB implementations include a 6DOF SLAM system at http://www.iri.upc.edu/people/jsola/JoanSola/eng/toolbox.html and the now dated CAS Robot Navigation Toolbox for planar SLAM at http://www.cas.kth.se/toolbox. Tim Bailey's website http://www-personal.acfr.usyd.edu.au/tbailey has MATLAB implementations of various SLAM and scan matching algorithms.

Many of the SLAM summer schools have websites that host excellent online resources such as lecture notes and practicals. Great teaching resources available online include Giorgio Grisetti's site http://www.dis.uniroma1.it/~grisetti and Paul Newman's *C4B Mobile Robots and Estimation Resources* ebook at https://www.free-ebooks.net/ebook/C4B-Mobile-Robotics.

Scan matching and map making. Many versions and variants of the ICP algorithm exist. Improved convergence and accuracy can be obtained using the normal distribution transform (NDT), originally proposed for 2D by Biber and Straßer (2003), extended to 3D by Magnusson et al. (2007) and implementations are available at pointclouds.org. A comparison of ICP and NDT for a field robotic application is described by Magnusson et al. (2009). A fast and popular approach to laser scan matching is that of Censi (2008).

When attempting to match a local geometric signature in a large point cloud (2D or 3D) to determine loop closure we often wish to limit our search to a local spatial region. An efficient way to achieve this is to organize the data using a kd-tree which is provided in MATLAB's Statistics and Machine Learning Toolbox™ and various contributions on File Exchange. FLANN (Muja and Lowe 2009) is a fast approximation which is available on github and has a MATLAB binding, and is also included in the VLFeat package.

For creating a map from robotic laser scan data in Sect. 6.8 we used a naive approach – a more sophisticated technique is the beam model or likelihood field as described in Thrun et al. (2005).

Kalman filtering. There are many published and online resources for Kalman filtering. Kálmán's original paper, Kálmán (1960), over 50 years old, is quite readable. The book by Zarchan and Musoff (2005) is a very clear and readable introduction to Kalman filtering. I have always found the classic book, recently republished, Jazwinski (2007) to be very readable. Bar-Shalom et al. (2001) provide comprehensive coverage of estimation theory and also the use of GPS. Groves (2013) also covers Kalman filtering. Welch and Bishop's online resources at http://www.cs.unc.edu/~welch/kalman have pointers to papers, courses, software and links to other relevant web sites.

A significant limitation of the EKF is its first-order linearization, particularly for processes with strong nonlinearity. Alternatives include the iterated EKF described by Jazwinski (2007) or the Unscented Kalman Filter (UKF) (Julier and Uhlmann 2004) which uses discrete sample points (sigma points) to approximate the PDF. Some of these topics are covered in the Handbook (Siciliano and Khatib 2016, §5 and §35). The information filter is an equivalent filter that maintains an inverse covariance matrix which has some useful properties, and is discussed in Thrun et al. (2005) as the sparse extended information filter.

Data association. SLAM techniques are critically dependent on accurate data association between observations and mapped landmarks, and a review of data association techniques is given by Neira and Tardós (2001). FastSLAM (Montemerlo and Thrun 2007) is capable of estimating data association as well as landmark position. The April tag which can be used as an artificial landmark is described in Olson (2011) and is supported by the Toolbox function `apriltags`. Mobile robots can uniquely identify places based on their visual appearance using tools such as OpenFABMAP (Glover et al. 2012).

Data association for Kalman filtering is covered in the Robotics Handbook (Siciliano and Khatib 2016). Data association in the tracking context is covered in considerable detail in, the now very old, book by Bar-Shalom and Fortmann (1988).

Sensors. The book by Kelly (2013) has a good coverage of sensors particularly laser range finders. For flying and underwater vehicles, odometry cannot be determined from wheel motion and an alternative, also suitable for wheeled vehicles, is visual odometry (VO). This is introduced in the tutorials by Fraundorfer and Scaramuzza (2012) and Scaramuzza and Fraundorfer (2011). The Robotics Handbook (Siciliano and Khatib 2016) has good coverage of a wide range of robotic sensors. The principles of GPS and other radio-based localization systems are covered in some detail in the book by Groves (2013), and a number of links to GPS technical data are provided from this book's web site. The SLAM problem can be formulated in terms of bearing-only or range-only measurements. A camera is effectively a bearing-only sensor, giving the direction to a feature in the world. A VSLAM system is one that performs SLAM using bearing-only visual information, just a camera, and an introduction to the topic is given by Neira et al. (2008) and the associated special issue. Interestingly the robotic VSLAM problem is the same as the bundle adjustment problem known to the computer vision community.

The book by Borenstein et al. (1996) although dated has an excellent discussion of robotic sensors in general and odometry in particular. It is out of print but can be found

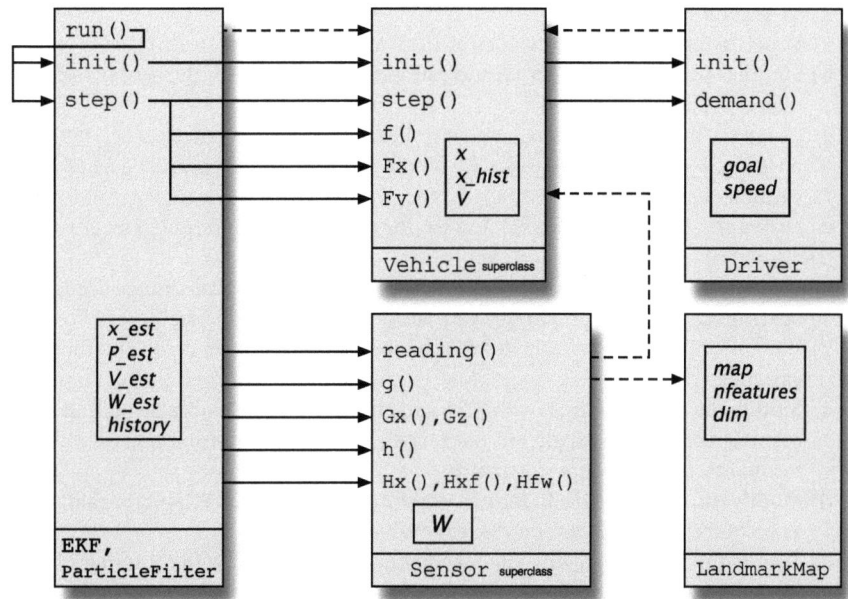

Fig. 6.26.
Toolbox class relationship for localization and mapping. Each class is shown as a *rectangle*, method calls are shown as *arrows* from caller to callee, properties are *boxed*, and *dashed lines* represent object references

online. The book by Everett (1995) covers odometry, range and bearing sensors, as well as radio, ultrasonic and optical localization systems. Unfortunately the discussion of range and bearing sensors is now quite dated since this technology has evolved rapidly over the last decade.

General interest. Bray (2014) gives a very readable account of the history of techniques to determine our location on the planet. If you ever wondered how to navigate by the stars or use a sextant Blewitt (2011) is a slim book that provides an easy to understand introduction.

The book *Longitude* (Sobel 1996) is a very readable account of the longitude problem and John Harrison's quest to build a marine chronometer.

Toolbox and MATLAB Notes

This chapter has introduced a number of Toolbox classes to solve mapping and localization problems. The principle was to decompose the problem into clear functional subsystems and implement these as a set of cooperating classes, and this allows quite complex problems to be expressed in very few lines of code.

The relationships between the objects and their methods and properties are shown in Fig. 6.26. As always more documentation is available through the online help system or comments in the code. `Vehicle` is a superclass and concrete subclasses include `Unicycle` and `Bicycle`.

The MATLAB Computer Vision System Toolbox™ includes a fast version of ICP called `pcregrigid`. The Robotics System Toolbox™ contains a generic particle filter class `ParticleFilter` and a particle filter based localizer class `MonteCarloLocalization`.

Exercises

1. What is the value of the Longitude Prize in today's currency?
2. Implement a driver object (page 155) that drives the robot around inside a circle with specified center and radius.
3. Derive an equation for heading change in terms of the rotational rate of the left and right wheels for the car-like and differential-steer vehicle models.

4. Dead-reckoning (page 154)
   a) Experiment with different values of P_0, V and $\hat{V}$.
   b) Figure 6.4 compares the actual and estimated position. Plot the actual and estimated heading angle.
   c) Compare the variance associated with heading to the variance associated with position. How do these change with increasing levels of range and bearing angle variance in the sensor?
   d) Derive the Jacobians in Eq. 6.5 and 6.6 for the case of a differential steer robot.

5. Using a map (page 161)
   a) Vary the characteristics of the sensor (covariance, sample rate, range limits and bearing angle limits) and investigate the effect on performance
   b) Vary W and $\hat{W}$ and investigate what happens to estimation error and final covariance.
   c) Modify the `RangeBearingSensor` to create a bearing-only sensor, that is, as a sensor that returns angle but not range. The implementation includes all the Jacobians. Investigate performance.
   d) Modify the sensor model to return occasional errors (specify the error rate) such as incorrect range or beacon identity. What happens?
   e) Modify the EKF to perform data association instead of using the landmark identity returned by the sensor.
   f) Figure 6.7 compares the actual and estimated position. Plot the actual and estimated heading angle.
   g) Compare the variance associated with heading to the variance associated with position. How do these change with increasing levels of range and bearing angle variance in the sensor?

6. Making a map (page 164)
   a) Vary the characteristics of the sensor (covariance, sample rate, range limits and bearing angle limits) and investigate the effect on performance.
   b) Use the bearing-only sensor from above and investigate performance relative to using a range and bearing sensor.
   c) Modify the EKF to perform data association instead of using identity returned by the sensor.

7. Simultaneous localization and mapping (page 166)
   a) Vary the characteristics of the sensor (covariance, sample rate, range limits and bearing angle limits) and investigate the effect on performance.
   b) Use the bearing-only sensor from above and investigate performance relative to using a range and bearing sensor.
   c) Modify the EKF to perform data association instead of using the landmark identity returned by the sensor.
   d) Figure 6.11 compares the actual and estimated position. Plot the actual and estimated heading angle.
   e) Compare the variance associated with heading to the variance associated with position. How do these change with increasing levels of range and bearing angle variance in the sensor?

8. Modify the pose-graph optimizer and test using the simple graph `pg1.g2o`
   a) anchor one node at a particular pose.
   b) add one or more landmarks. You will need to derive the relevant Jacobians first and add the landmark positions, constraints and information matrix to the data file.

9. Create a simulator for Buffon's needle problem, and estimate π for 10, 100, 1 000 and 10 000 needle throws. How does convergence change with needle length?

10. Particle filter (page 174)
    a) Run the filter numerous times. Does it always converge?
    b) Vary the parameters Q, L, w_0 and N and understand their effect on convergence speed and final standard deviation.

c) Investigate variations to the kidnapped robot problem. Place the initial particles around the initial pose. Place the particles uniformly over the xy-plane but set their orientation to its actual value.

d) Use a different type of likelihood function, perhaps inverse distance, and compare performance.

11. Experiment with April tags. Print some tags and extract them from images using the `apriltags` function.

12. Implement a laser odometer and test it over the entire path saved in `killian.g2o`. Compare your odometer with the relative pose changes in the file.

13. In order to measure distance using laser rangefinding what timing accuracy is required to achieve 1cm depth resolution?

14. Reformulate the localization, mapping and SLAM problems using a bearing-only landmark sensor.

15. Implement a localization or SLAM system using an external simulator such as V-REP or Gazebo. Obtain range measurements from the simulated robot, do laser odometry and landmark recognition, and send motion commands to the robot. You can communicate with these simulators from MATLAB using the ROS protocol if you have the Robotics System Toolbox. Alternatively you can communicate with V-REP using the Toolbox `VREP` class, see the documentation.

Part III | Arm-Type Robots

Chapter 7   **Robot Arm Kinematics**

Chapter 8   **Velocity Relationships**

Chapter 9   **Dynamics and Control**

III Arm-Type Robots

Arm-type robots or robot manipulators are a very common and familiar type of robot. We are used to seeing pictures or video of them at work in factories doing jobs such as assembly, welding and handling tasks, or even in operating rooms doing surgery. The first robot manipulators started work nearly 60 years ago and have been enormously successful in practice – many millions of robot manipulators are working in the world today. Many products we buy have been assembled, packed or handled by a robot.

Unlike the mobile robots we discussed in the previous part, robot manipulators do not move through the world. They have a static base and therefore operate within a limited workspace. Many different types of robot manipulator have been created and Fig. III.1 shows some of the diversity. The most common is the 6DOF arm-type of robot comprising a series of rigid-links and actuated joints. The SCARA (Selective Compliance Assembly Robot Arm) is rigid in the vertical direction and compliant in the horizontal plane which is an advantage for planar tasks such as electronic circuit board assembly. A gantry robot has one or two degrees of freedom of motion along overhead rails which gives it

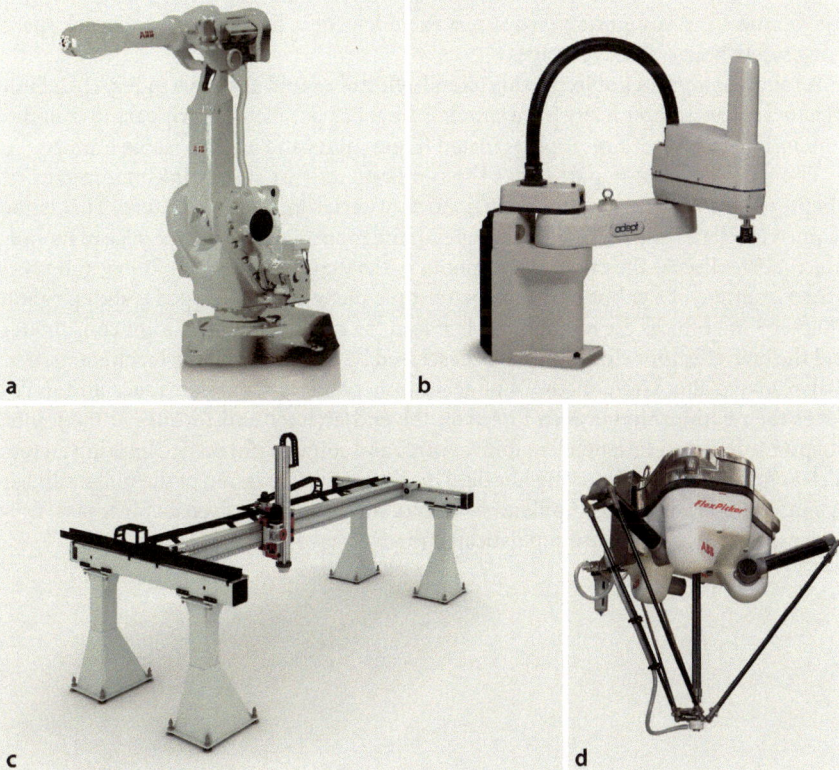

Fig. III.1.
a A 6DOF serial-link manipulator. General purpose industrial manipulator (source: ABB).
b SCARA robot which has 4DOF, typically used for electronic assembly (photo of Adept Cobra s600 SCARA robot courtesy of Adept Technology, Inc.).
c A gantry robot; the arm moves along an overhead rail (image courtesy of Güdel AG Switzerland | Mario Rothenbühler | www.gudel.com). **d** A parallel-link manipulator, the end-effector is driven by 6 parallel links (source: ABB)

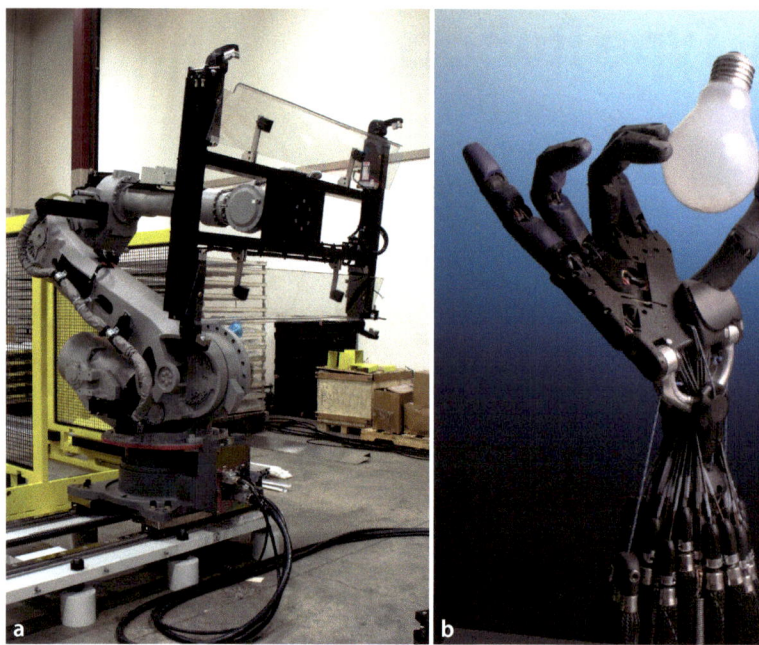

Fig. III.2.
Robot end-effectors. **a** A vacuum gripper holds a sheet of glass. **b** A human-like robotic hand (© Shadow Robot Company 2008)

a very large working volume. A parallel-link manipulator has its links connected in parallel to the tool which brings a number of advantages such as having all the motors on the base and providing a very stiff structure. The focus of this part is serial-link arm-type robot manipulators.

These nonmobile robots allow some significant simplifications to problems such as perception and safety. The work environment for a factory robot can be made very orderly so the robot can be fast and precise and *assume* the location of objects that it is working with. The safety problem is simplified since the robot has a limited working volume – it is straightforward to just exclude people from the robot's work space using safety barriers or even cages.

A robot manipulates objects using its end-effector or tool as shown in Fig. III.2. End-effectors range in complexity from simple 2-finger or parallel-jaw grippers to complex human-like hands with multiple actuated finger joints and an opposable thumb.

The chapters in this part cover the fundamentals of serial-link manipulators. Chapter 7 is concerned with the kinematics of serial-link manipulators. This is the geometric relationship between the angles of the robot's joints and the pose of its end-effector. We discuss the creation of smooth paths that the robot can follow and present an example of a robot drawing a letter on a plane and a 4-legged walking robot. Chapter 8 introduces the relationship between the rate of change of joint coordinates and the end-effector velocity which is described by the manipulator Jacobian matrix. It also covers alternative methods of generating paths in Cartesian space and introduces the relationship between forces on the end-effector and torques at the joints. Chapter 9 discusses independent joint control and some performance limiting factors such as gravity load and varying inertia. This leads to a discussion of the full nonlinear dynamics of serial-link manipulators – effects such as inertia, gyroscopic forces, friction and gravity – and more sophisticated model-based control approaches.

7

Robot Arm Kinematics

Take to kinematics. It will repay you.
It is more fecund than geometry; it adds a fourth dimension to space.
Chebyshev to Sylvester 1873

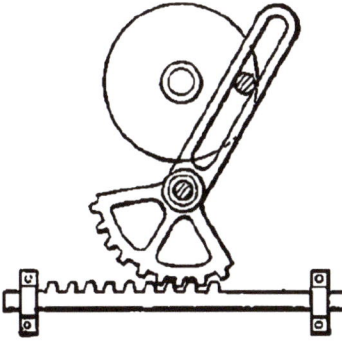

From the Greek word for motion.

Kinematics is the branch of mechanics that studies the motion of a body, or a system of bodies, without considering its mass or the forces acting on it.

A robot arm, more formally a serial-link manipulator, comprises a chain of rigid links and joints. Each joint has one degree of freedom, either translational (a sliding or prismatic joint) or rotational (a revolute joint). Motion of the joint changes the relative pose of the links that it connects. One end of the chain, the base, is generally fixed and the other end is free to move in space and holds the tool or end-effector that does the useful work.

Figure 7.1 shows two modern arm-type robots that have six and seven joints respectively. Clearly the pose of the end-effector will be a complex function of the state of each joint and Sect. 7.1 describes how to compute the pose of the end-effector. Section 7.2 discusses the inverse problem, how to compute the position of each joint given the end-effector pose. Section 7.3 describes methods for generating smooth paths for the end-effector. The remainder of the chapter covers advanced topics and two complex applications: writing on a plane surface and a four-legged walking robot whose legs are simple robotic arms.

7.1 Forward Kinematics

Forward kinematics is the mapping from joint coordinates, or robot configuration, to end-effector pose. We start in Sect. 7.1.1 with conceptually simple robot arms that move in 2-dimensions in order to illustrate the principles, and in Sect. 7.1.2 extend this to more useful robot arms that move in 3-dimensions.

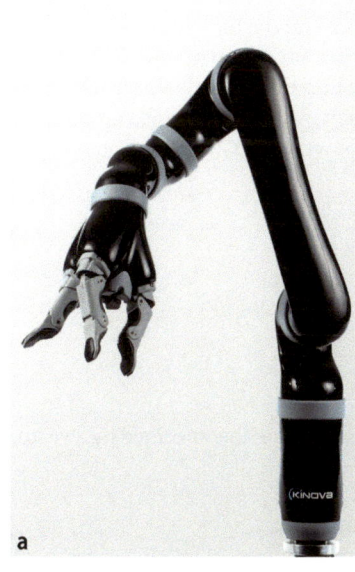

Fig. 7.1.
a Mico 6-joint robot with 3-fingered hand (courtesy of Kinova Robotics). **b** Baxter 2-armed robotic coworker, each arm has 7 joints (courtesy of Rethink Robotics)

© Springer Nature Switzerland AG 2022
P. Corke, *Robotics and Control*, Springer Tracts in Advanced Robotics 141,
https://doi.org/10.1007/978-3-030-79179-7_7

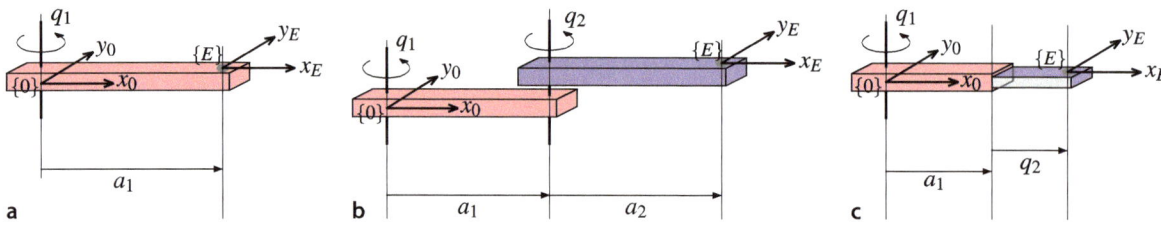

7.1.1   2-Dimensional (Planar) Robotic Arms

Consider the simple robot arm shown in Fig. 7.2a which has a single rotational joint. We can describe the pose of its end-effector – frame {E} – by a sequence of relative poses: a rotation about the joint axis and then a translation by a_1 along the rotated x-axis◄

$$\xi_E(q) = \mathscr{R}_z(q_1) \oplus \mathscr{T}_x(a_1)$$

The Toolbox allows us to express this, for the case $a_1 = 1$, by

```
>> import ETS2.*
>> a1 = 1;
>> E = Rz('q1') * Tx(a1)
```

which is a sequence of `ETS2` class objects. The argument to `Rz` is a string which indicates that its parameter is a joint variable whereas the argument to `Tx` is a constant numeric robot dimension.

The forward kinematics for a *particular* value of $q_1 = 30$ deg

```
>> E.fkine( 30, 'deg')
ans =
    0.8660   -0.5000    0.866
    0.5000    0.8660    0.5
         0         0      1
```

is an **SE**(2) homogeneous transformation matrix representing the pose of the end-effector – coordinate frame {E}.

An easy and intuitive way to understand how this simple robot behaves is interactively

```
>> E.teach
```

which generates a graphical representation of the robot arm as shown in Fig. 7.3. The rotational joint is indicated by a grey vertical cylinder and the link by a red horizontal pipe. You can adjust the joint angle q_1 using the slider and the arm pose and the displayed end-effector position and orientation will be updated. Clearly this is not a very useful robot arm since its end-effector can only reach points that lie on a circle.

Consider now a robot arm with two joints as shown in Fig. 7.2b. The pose of the end-effector is

$$\xi_E(q) = \underbrace{\mathscr{R}_z(q_1)}_{\text{joint 1}} \oplus \underbrace{\mathscr{T}_x(a_1)}_{\text{link 1}} \oplus \underbrace{\mathscr{R}_z(q_2)}_{\text{joint 2}} \oplus \underbrace{\mathscr{T}_x(a_2)}_{\text{link 2}} \tag{7.1}$$

We can represent this using the Toolbox as

```
>> a1 = 1; a2 = 1;
>> E = Rz('q1') * Tx(a1) * Rz('q2') * Tx(a2)
```

When computing the forward kinematics the joint angles are now specified by a vector

```
>> E.fkine( [30, 40], 'deg')
ans =
    0.3420   -0.9397    1.208
    0.9397    0.3420    1.44
         0         0      1
```

We use the symbols $\mathscr{R}, \mathscr{T}_x, \mathscr{T}_y$ to denote relative poses in SE(2) that are respectively pure rotation and pure translation in the x- and y-directions.

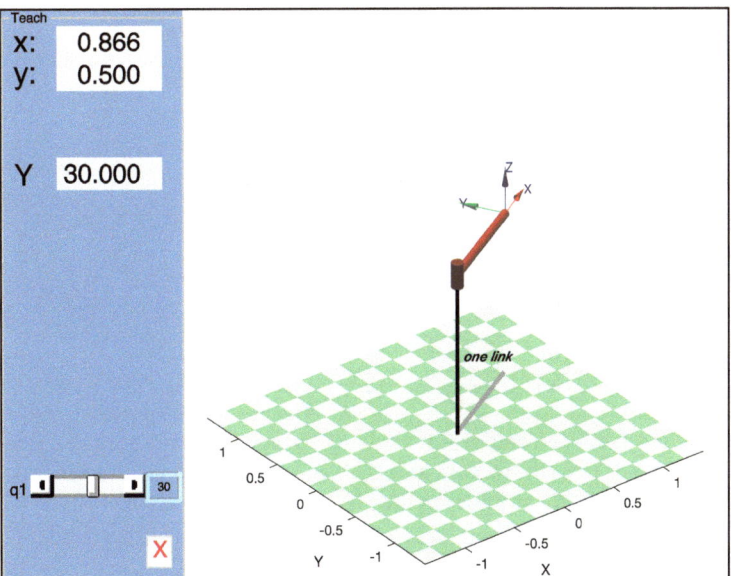

Fig. 7.3.
Toolbox depiction of 1-joint planar robot using the `teach` method. The *blue panel* contains the joint angle slider and displays the position and orientation (yaw angle) of the end-effector (in degrees)

and the result is the end-effector pose when $q_1 = 30$ and $q_2 = 40$ deg. We could display the robot interactively as in the previous example, or noninteractively by

```
>> E.plot( [30, 40], 'deg')
```

The joint structure of a robot is often referred to by a shorthand comprising the letters R (for revolute) or P (for prismatic) to indicate the number and types of its joints. For this robot

```
>> E.structure
ans =
RR
```

indicates a revolute-revolute sequence of joints. The notation underneath the terms in Eq. 7.1 describes them in the context of a physical robot manipulator which comprises a series of joints and links.

You may have noticed a few characteristics of this simple planar robot arm. Firstly, most end-effector positions can be reached with two *different* joint angle vectors. Secondly, the robot can position the end-effector at any point within its reach but we cannot specify an arbitrary orientation. This robot has 2 degrees of freedom and its configuration space is $\mathcal{C} = \mathbb{S}^1 \times \mathbb{S}^1$. This is sufficient to achieve positions in the task space $\mathcal{T} \subset \mathbb{R}^2$ since dim $\mathcal{C} = $ dim $\mathcal{T}$. However if our task space includes orientation $\mathcal{T} \subset \mathbf{SE}(2)$ then it is under-actuated since dim $\mathcal{C} < $ dim $\mathcal{T}$ and the robot can access only a subset of the task space.

So far we have only considered revolute joints but we could use a prismatic joint instead as shown in Fig. 7.2c. The end-effector pose is

$$\xi_E(q) = \underbrace{\mathcal{R}_z(q_1)}_{\text{joint 1}} \oplus \underbrace{\mathcal{T}_x(a_1)}_{\text{link 1}} \oplus \underbrace{\mathcal{T}_x(q_2)}_{\text{joint 2}}$$

Prismatic joints. Robot joints are commonly revolute (rotational) but can also be prismatic (linear, sliding, telescopic, etc.). The SCARA robot of Fig. III.1b has a prismatic third joint while the gantry robot of Fig. III.1c has three prismatic joints for motion in the x-, y- and z-directions.

The Stanford arm shown here has a prismatic third joint. It was developed at the Stanford AI Lab in 1972 by robotics pioneer Victor Scheinman who went on to design the PUMA robot arms. This type of arm supported a lot of important early research work in robotics and one can be seen in the Smithsonian Museum of American History, Washington DC. (Photo courtesy Oussama Khatib)

and the Toolbox representation follows a familiar pattern

```
>> a1 = 1;
>> E = Rz('q1') * Tx(a1) * Tx('q2')
```

and the arm structure is now

```
>> E.structure
ans =
RP
```

which is commonly called a polar-coordinate robot arm.
   We can easily add a third joint

$$\xi_e(\boldsymbol{q}) = \underbrace{\mathscr{R}_z(q_1)}_{\text{joint 1}} \oplus \underbrace{\mathscr{T}_x(a_1)}_{\text{link 1}} \oplus \underbrace{\mathscr{R}_z(q_2)}_{\text{joint 2}} \oplus \underbrace{\mathscr{T}_x(a_2)}_{\text{link 2}} \oplus \underbrace{\mathscr{R}_z(q_3)}_{\text{joint 3}} \oplus \underbrace{\mathscr{T}_x(a_3)}_{\text{link 3}}$$

and use the now familiar Toolbox functionality to represent and work with this arm. This robot has 3 degrees of freedom and is able to access all points in the task space $\mathcal{T} \subset \mathbf{SE}(2)$, that is, achieve any pose in the plane (limited by reach).

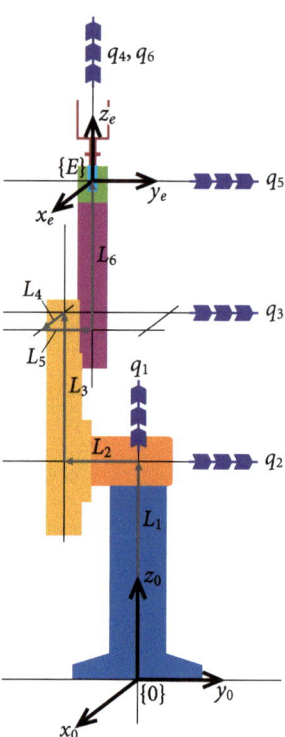

Fig. 7.4. Puma robot in the zero-joint-angle configuration showing dimensions and joint axes (indicated by *blue triple arrows*) (after Corke 2007)

We use the symbols $\mathscr{R}_i, \mathscr{T}_i, i \in \{x, y, z\}$ to denote relative poses in $\mathbf{SE}(3)$ that are respectively pure rotation about, or pure translation along, the i-axis.

7.1.2   3-Dimensional Robotic Arms

Truly useful robots have a task space $\mathcal{T} \subset \mathbf{SE}(3)$ enabling arbitrary position and orientation of the end-effector. This requires a robot with a configuration space dim $\mathcal{C} \geq$ dim $\mathcal{T}$ which can be achieved by a robot with six or more joints. In this section we will use the Puma 560 as an exemplar of the class of all-revolute six-axis robot manipulators with $\mathcal{C} \subset (\mathbb{S}^1)^6$.
   We can extend the technique from the previous section for a robot like the Puma 560 whose dimensions are shown in Fig. 7.4. Starting with the world frame {0} we move up, rotate about the waist axis (q_1), rotate about the shoulder axis (q_2), move to the left, move up and so on. As we go, we write down the transform expression◄

$$\xi_E = \mathscr{T}_z(L_1) \oplus \mathscr{R}_z(q_1) \oplus \mathscr{R}_y(q_2) \oplus \mathscr{T}_y(L_2) \oplus \mathscr{T}_z(L_3) \oplus \mathscr{R}_y(q_3)$$
$$\oplus \mathscr{T}_x(L_4) \oplus \mathscr{T}_y(L_5) \oplus \mathscr{T}_z(L_6) \oplus \underbrace{\mathscr{R}_z(q_4) \oplus \mathscr{R}_y(q_5) \oplus \mathscr{R}_z(q_6)}_{\text{wrist}}$$

The marked term represents the kinematics of the robot's wrist and should be familiar to us as a *ZYZ* Euler angle sequence from Sect. 2.2.1.2 – it provides an arbitrary orientation but is subject to a singularity when the middle angle $q_5 = 0$.
   We can represent this using the 3-dimensional version of the Toolbox class we used previously

```
>> import ETS3.*
>> L1 = 0; L2 = -0.2337; L3 = 0.4318; L4 = 0.0203; L5 = 0.0837; L6 = 0.4318;
>> E3 = Tz(L1) * Rz('q1') * Ry('q2') * Ty(L2) * Tz(L3) * Ry('q3') ↵
    * Tx(L4) * Ty(L5) * Tz(L6) * Rz('q4') * Ry('q5') * Rz('q6');
```

We can use the interactive teach facility or compute the forward kinematics

```
>> E3.fkine([0 0 0 0 0 0])
ans =
       1         0         0     0.0203
       0         1         0     -0.15
       0         0         1     0.8636
       0         0         0         1
```

While this notation is intuitive it does becomes cumbersome as the number of robot joints increases. A number of approaches have been developed to more concisely describe a serial-link robotic arm: Denavit-Hartenberg notation and product of exponentials.

Table 7.1.
Denavit-Hartenberg parameters: their physical meaning, symbol and formal definition

Joint angle	θ_j	the angle between the x_{j-1} and x_j axes about the z_{j-1} axis	revolute joint variable
Link offset	d_j	the distance from the origin of frame $j-1$ to the x_j axis along the z_{j-1} axis	prismatic joint variable
Link length	a_j	the distance between the z_{j-1} and z_j axes along the x_j axis; for intersecting axes is parallel to $\hat{z}_{j-1} \times \hat{z}_j$	constant
Link twist	α_j	the angle from the z_{j-1} axis to the z_j axis about the x_j axis	constant
Joint type	σ_j	$\sigma = R$ for a revolute joint, $\sigma = P$ for a prismatic joint	constant

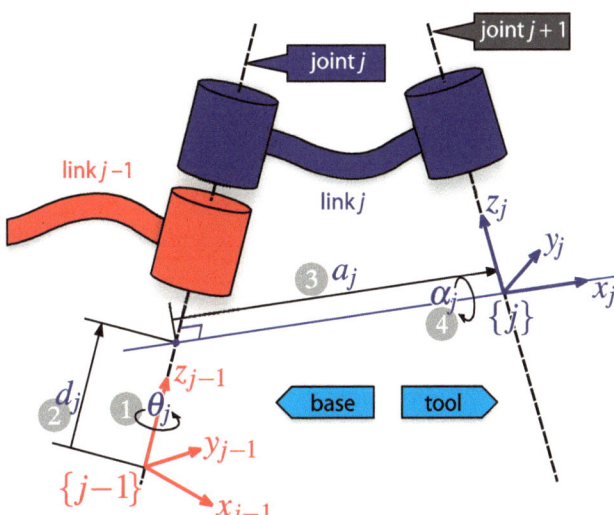

Fig. 7.5.
Definition of standard Denavit and Hartenberg link parameters. The colors red and blue denote all things associated with links $j-1$ and j respectively. The numbers in circles represent the order in which the elementary transforms are applied. x_j is parallel to $z_{j-1} \times z_j$ and if those two axes are parallel then d_j can be arbitrarily chosen

7.1.2.1 Denavit-Hartenberg Parameters

One systematic way of describing the geometry of a serial chain of links and joints is Denavit-Hartenberg notation.

> For a manipulator with N joints numbered from 1 to N, there are $N+1$ links, numbered from 0 to N. Joint j connects link $j-1$ to link j and moves them relative to each other. It follows that link ℓ connects joint ℓ to joint $\ell + 1$. Link 0 is the base of the robot, typically fixed and link N, the last link of the robot, carries the end-effector or tool.

In Denavit-Hartenberg, notation a link defines the spatial relationship between two neighboring joint axes as shown in Fig. 7.5. A link is specified by four parameters. The relationship between two link coordinate frames would ordinarily entail six parameters, three each for translation and rotation. For Denavit-Hartenberg notation there are only four parameters but there are also two constraints: axis x_j intersects z_{j-1} and axis x_j is perpendicular to z_{j-1}. One consequence of these constraints is that sometimes the link coordinate frames are not actually located on the physical links of the robot. Another consequence is that the robot must be placed into a particular configuration – the zero-angle configuration – which is discussed further in Sect. 7.4.1. The Denavit-Hartenberg parameters are summarized in Table 7.1.

The coordinate frame $\{j\}$ is attached to the far (distal) end of link j. The z-axis of frame $\{j\}$ is aligned with the axis of joint $j+1$.

The transformation from link coordinate frame $\{j-1\}$ to frame $\{j\}$ is defined in terms of elementary rotations and translations as

$$^{j-1}\xi_j\left(\theta_j, d_j, a_j, \alpha_j\right) = \mathscr{R}_z\left(\theta_j\right) \oplus \mathscr{T}_z\left(d_j\right) \oplus \mathscr{T}_x\left(a_j\right) \oplus \mathscr{R}_x\left(\alpha_j\right) \tag{7.2}$$

which can be expanded in homogeneous matrix form as

$$^{j-1}\mathbf{A}_j = \begin{pmatrix} \cos\theta_j & -\sin\theta_j\cos\alpha_j & \sin\theta_j\sin\alpha_j & a_j\cos\theta_j \\ \sin\theta_j & \cos\theta_j\cos\alpha_j & -\cos\theta_j\sin\alpha_j & a_j\sin\theta_j \\ 0 & \sin\alpha_j & \cos\alpha_j & d_j \\ 0 & 0 & 0 & 1 \end{pmatrix} \tag{7.3}$$

The parameters α_j and a_j are always constant. For a revolute joint, θ_j is the joint variable and d_j is constant, while for a prismatic joint, d_j is variable, θ_j is constant and $\alpha_j = 0$. In many of the formulations that follow, we use generalized joint coordinates q_j

$$\text{if } \sigma_j = \begin{cases} R: & \theta_j \leftarrow q_j \\ P: & d_j \leftarrow q_j \end{cases}$$

For an N-axis robot, the generalized joint coordinates $q \in \mathcal{C}$ where $\mathcal{C} \subset \mathbb{R}^N$ is called the joint space or configuration space.► For the common case of an all-revolute robot $\mathcal{C} \subset (\mathbb{S}^1)^N$ the joint coordinates are referred to as joint angles. The joint coordinates are also referred to as the *pose of the manipulator* which is different to the *pose of the end-effector* which is a Cartesian pose $\xi \in \mathbf{SE}(3)$. The term *configuration* is shorthand for *kinematic configuration* which will be discussed in Sect. 7.2.2.1.

This is the same concept as was introduced for mobile robots in Sect. 2.3.5.

Within the Toolbox a robot revolute joint and link can be created by

```
>> L = Revolute('a', 1)
L =
Revolute(std): theta=q, d=0, a=1, alpha=0, offset=0
```

which is a revolute-joint object of type `Revolute` which is a subclass of the generic `Link` object. The displayed value of the object shows the kinematic parameters (most of which have defaulted to zero), the joint type and that standard Denavit-Hartenberg convention is used (the tag `std`).►

A slightly different notation, *modifed Denavit-Hartenberg* notation is discussed in Sect. 7.4.3.

Jacques Denavit and **Richard Hartenberg** introduced many of the key concepts of kinematics for serial-link manipulators in a 1955 paper (Denavit and Hartenberg 1955) and their later classic text *Kinematic Synthesis of Linkages* (Hartenberg and Denavit 1964).

Jacques Denavit (1930–2012) was born in Paris where he studied for his Bachelor degree before pursuing his masters and doctoral degrees in mechanical engineering at Northwestern University, Illinois. In 1958 he joined the Department of Mechanical Engineering and Astronautical Science at Northwestern where the collaboration with Hartenberg was formed. In addition to his interest in dynamics and kinematics Denavit was also interested in plasma physics and kinetics. After the publication of the book he moved to Lawrence Livermore National Lab, Livermore, California, where he undertook research on computer analysis of plasma physics problems. He retired in 1982.

Richard Hartenberg (1907–1997) was born in Chicago and studied for his degrees at the University of Wisconsin, Madison. He served in the merchant marine and studied aeronautics for two years at the University of Göttingen with space-flight pioneer Theodore von Kármán. He was Professor of mechanical engineering at Northwestern University where he taught for 56 years. His research in kinematics led to a revival of interest in this field in the 1960s, and his efforts helped put kinematics on a scientific basis for use in computer applications in the analysis and design of complex mechanisms. He also wrote extensively on the history of mechanical engineering.

A `Link` object has many parameters and methods which are described in the online documentation, but the most common ones are illustrated by the following examples. The link transform Eq. 7.3 for $q = 0.5$ rad is an `SE3` object

```
>> L.A(0.5)
ans =
    0.8776   -0.4794        0    0.8776
    0.4794    0.8776       -0    0.4794
         0         0        1         0
         0         0        0         1
```

representing the homogeneous transformation due to this robot link with the particular value of θ. Various link parameters can be read or altered, for example

```
>> L.type
ans =
    R
```

indicates that the link is revolute and

```
>> L.a
ans =
    1.0000
```

returns the kinematic parameter a. Finally a link can contain an offset

```
>> L.offset = 0.5;
>> L.A(0)
ans =
    0.8776   -0.4794        0    0.8776
    0.4794    0.8776       -0    0.4794
         0         0        1         0
         0         0        0         1
```

which is added to the joint variable before computing the link transform and will be discussed in more detail in Sect. 7.4.1.

The forward kinematics is a function of the joint coordinates and is simply the composition of the relative pose due to each link

$$
{}^{0}\xi_N = \mathcal{K}(q; \theta, d, a, \alpha, \sigma) = {}^{0}\xi_1 \oplus {}^{1}\xi_2 \cdots {}^{N-1}\xi_N \tag{7.4}
$$

In this notation link 0 is the base of the robot and commonly for the first link $d_1 = 0$ but we could set $d_1 > 0$ to represent the height of the first joint above the world coordinate frame. The final link, link N, carries the tool – the parameters d_N, a_N and α_N provide a limited means to describe the tool-tip pose with respect to the {N} frame. By convention the robot's tool points in the z-direction as shown in Fig. 2.16.

We have used *W* to denote the world frame in this case since 0 designates link 0, the base link.

More generally we add two extra transforms to the chain◀

$$
{}^{W}\xi_E = \underbrace{{}^{W}\xi_0}_{\xi_B} \oplus {}^{0}\xi_1 \oplus {}^{1}\xi_2 \cdots {}^{N-1}\xi_N \oplus \underbrace{{}^{N}\xi_E}_{\xi_T}
$$

The base transform ξ_B puts the base of the robot arm at an arbitrary pose within the world coordinate frame. In a manufacturing system the base is usually fixed to the environment but it could be mounted on a mobile ground, aerial or underwater robot, a truck, or even a space shuttle.

The frame {N} is often defined as the center of the spherical wrist mechanism, and the tool transform ξ_T describes the pose of the tool tip with respect to that. In practice ξ_T might consist of several components. Firstly, a transform to a tool-mounting flange on the physical end of the robot. Secondly, a transform from the flange to the end of the tool that is bolted to it, where the tool might be a gripper, screwdriver or welding torch.

In the Toolbox we connect `Link` class objects in series using the `SerialLink` class

```
>> robot = SerialLink( [ Revolute('a', 1) Revolute('a', 1) ],↵
 'name', 'my robot')
robot =
my robot:: 2 axis, RR, stdDH
+---+-----------+-----------+-----------+-----------+-----------+
| j |    theta  |        d  |        a  |    alpha  |    offset |
+---+-----------+-----------+-----------+-----------+-----------+
| 1 |       q1|0 |        |1 |        |0 |        |0 |          |
| 2 |       q2|0 |        |1 |        |0 |        |0 |          |
+---+-----------+-----------+-----------+-----------+-----------+
```

We have just recreated the 2-robot robot we looked at earlier, but now it is embedded in **SE**(3). The forward kinematics are

```
>> robot.fkine([30 40], 'deg')
ans =
    0.3420   -0.9397         0    1.208
    0.9397    0.3420         0    1.44
         0         0         1         0
         0         0         0         1
```

The Toolbox contains a large number of robot arm models defined in this way and they can be listed by

```
>> models
ABB, IRB140, 6DOF, standard_DH (mdl_irb140)
Aldebaran, NAO, humanoid, 4DOF, standard_DH (mdl_nao)
Baxter, Rethink Robotics, 7DOF, standard_DH (mdl_baxter)
   ...
```

where the name of the Toolbox script to load the model is given in parentheses at the end of each line, for example

```
>> mdl_irb140
```

The `models` function also supports searching by keywords and robot arm type. You can adjust the parameters of any model using the editing method, for example

```
>> robot.edit
```

Determining the Denavit-Hartenberg parameters for a particular robot is described in more detail in Sect. 7.4.2.

7.1.2.2   Product of Exponentials

In Chap. 2 we introduced twists. A twist is defined by a screw axis direction and pitch, and a point that the screw axis passes through. In matrix form the twist $S \in \mathbb{R}^6$

$$T' = e^{[S]\theta}T$$

rotates the coordinate frame described by the pose T about the screw axis by an angle θ.► This is exactly the case of the single-joint robot of Fig. 7.2a, where the screw axis *is* the joint axis and T is the pose of the end-effector when $q_1 = 0$. We can therefore write the forward kinematics as

For a prismatic twist, the motion is a displacement of θ along the screw axis. Here we are working in the plane so $T \in$ **SE**(2) and $S \in \mathbb{R}^3$.

$$T_E = e^{[S_1]q_1}T_E(0)$$

where $T_E(0)$ is the end-effector pose in the zero-angle joint configuration: $q_1 = 0$.
For the 2-joint robot of Fig. 7.2b we would write

$$T_E = e^{[S_1]q_1} \underbrace{e^{[S_2]q_2}T_E(0)}$$

where S_1 and S_2 are the screws defined by the joint axes and $T_E(0)$ is the end-effector pose in the zero-angle joint configuration: $q_1 = q_2 = 0$. The indicated term is similar to the single-joint robot above, and the first twist rotates that joint and link about S_1. In MATLAB we define the link lengths and compute $T_E(0)$

```
>> a1 = 1; a2 = 1;
>> TE0 = SE2(a1+a2, 0, 0);
```

define the two twists, in **SE**(2), for this example

```
>> S1 = Twist( 'R', [0 0] );
>> S2 = Twist( 'R', [a1 0] );
```

and apply them to $T_E(0)$

```
>> TE = S1.T(30, 'deg') * S2.T(40, 'deg') * TE0
TE =
    0.3420   -0.9397    1.208
    0.9397    0.3420    1.44
         0         0       1
```

For a general robot that moves in 3-dimensions we can write the forward kinematics in product of exponential (PoE) form as

$$\xi_E \sim {}^0T_E = e^{[S_1]q_1} \cdots e^{[S_N]q_N} \, {}^0T_E(0)$$

where ${}^0T_E(0)$ is the end-effector pose when the joint coordinates are all zero and S_j is the twist for joint j expressed in the world frame.◄ This can also be written as

$$\xi_E \sim {}^0T_E = {}^0T_E(0) e^{\left[{}^E S_1\right]q_1} \cdots e^{\left[{}^E S_N\right]q_N}$$

The tool and base transform are effectively included in ${}^0T_E(0)$, but an explicit base transform could be added if the screw axes are defined with respect to the robot's base rather than the world coordinate frame, or use the adjoint matrix to transform the screw axes from base to world coordinates.

and ${}^E S_j$ is the twist for joint j expressed in the end-effector frame which is related to the twists above by ${}^E S_j = \text{Ad}({}^E\xi_0)S_j$.

A serial-link manipulator can be succinctly described by a table listing the 6 screw parameters for each joint as well as the zero-joint-coordinate end-effector pose.

7.1.2.3   6-Axis Industrial Robot

Truly useful robots have a task space $\mathcal{T} \subset \textbf{SE}(3)$ enabling arbitrary position and attitude of the end-effector – the task space has six spatial degrees of freedom: three translational and three rotational. This requires a robot with a configuration space $\mathcal{C} \subset \mathbb{R}^6$ which can be achieved by a robot with six joints. In this section we will use the Puma 560 as an example of the class of all-revolute six-axis robot manipulators. We define an instance of a Puma 560 robot using the script

```
>> mdl_puma560
```

which creates a `SerialLink` object, `p560`, in the workspace. Displaying the variable shows the table of its Denavit-Hartenberg parameters

```
>> p560
Puma 560 [Unimation]:: 6 axis, RRRRRR, stdDH, slowRNE
 - viscous friction; params of 8/95;
+---+----------+----------+----------+----------+----------+
| j |   theta  |       d  |       a  |   alpha  |  offset  |
+---+----------+----------+----------+----------+----------+
| 1 |      q1  |       0  |       0  |   1.571  |       0  |
| 2 |      q2  |       0  |  0.4318  |       0  |       0  |
| 3 |      q3  |    0.15  |  0.0203  |  -1.571  |       0  |
| 4 |      q4  |  0.4318  |       0  |   1.571  |       0  |
| 5 |      q5  |       0  |       0  |  -1.571  |       0  |
| 6 |      q6  |       0  |       0  |       0  |       0  |
+---+----------+----------+----------+----------+----------+
```

The **Puma 560 robot** (Programmable Universal Manipulator for Assembly) released in 1978 was the first modern industrial robot and became enormously popular. It featured an anthropomorphic design, electric motors and a spherical wrist – the archetype of all that followed. It can be seen in the Smithsonian Museum of American History, Washington DC.

The Puma 560 catalyzed robotics research in the 1980s and it was a very common laboratory robot. Today it is obsolete and rare but in homage to its important role in robotics research we use it here. For our purposes the advantages of this robot are that it has been well studied and its parameters are very well known – it has been described as the "white rat" of robotics research.

Most modern 6-axis industrial robots are very similar in structure and can be accomodated simply by changing the Denavit-Hartenberg parameters. The Toolbox has kinematic models for a number of common industrial robots from manufacturers such as Rethink, Kinova, Motoman, Fanuc and ABB. (Puma photo courtesy Oussama Khatib)

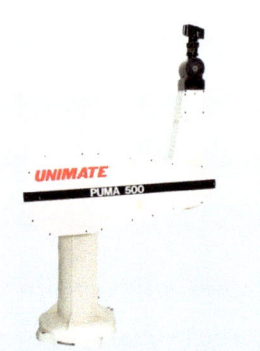

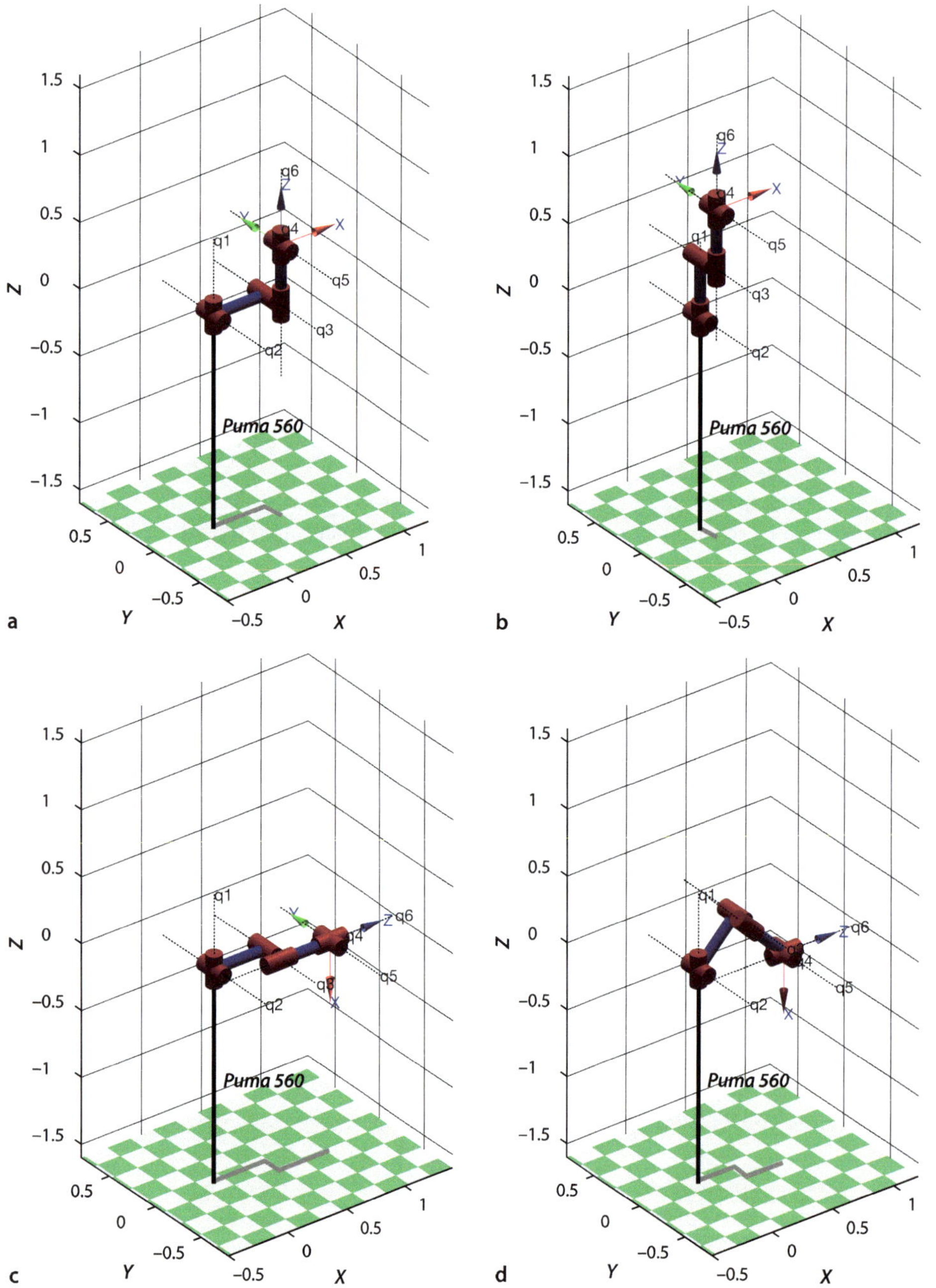

Fig. 7.6.
The Puma robot in 4 different poses. **a** *Zero angle*; **b** *ready pose*; **c** *stretch*; **d** *nominal*

> **Anthropomorphic** means having human-like characteristics. The Puma 560 robot was designed to have approximately the dimensions and reach of a human worker. It also had a spherical joint at the wrist just as humans have.
>
> Roboticists also tend to use anthropomorphic terms when describing robots. We use words like waist, shoulder, elbow and wrist when describing serial link manipulators. For the Puma these terms correspond respectively to joint 1, 2, 3 and 4–6.

Note that a_j and d_j are in SI units which means that the translational part of the forward kinematics will also have SI units.

The script `mdl_puma560` also creates a number of joint coordinate vectors in the workspace which represent the robot in some canonic configurations:

`qz`	$(0, 0, 0, 0, 0, 0)$	*zero angle*
`qr`	$(0, \frac{\pi}{2}, -\frac{\pi}{2}, 0, 0, 0)$	*ready*, the arm is straight and vertical
`qs`	$(0, 0, -\frac{\pi}{2}, 0, 0, 0)$	*stretch*, the arm is straight and horizontal
`qn`	$(0, \frac{\pi}{4}, -\pi, 0, \frac{\pi}{4}, 0)$	*nominal*, the arm is in a dextrous working pose◀

Well away from singularities, which will be discussed in Sect. 7.3.4.

and these are shown graphically in Fig. 7.6. These plots are generated using the `plot` method, for example

```
>> p560.plot(qz)
```

which shows a skeleton of the robot with pipes that connect the link coordinate frames as defined by the Denavit-Hartenberg parameters. The `plot` method has many options for showing the joint axes, wrist coordinate frame, shadows and so on. More realistic-looking plots such as shown in Fig. 7.7 can be created by the `plot3d` method for a limited set of Toolbox robot models.

Forward kinematics can be computed as before

```
>> TE = p560.fkine(qz)
TE =
     1.0000         0         0    0.4521
          0    1.0000         0   -0.1500
          0         0    1.0000    0.4318
          0         0         0    1.0000
```

By the Denavit-Hartenberg parameters of the model in the `mdl_puma560` script.

where the joint coordinates are given as a row vector. This returns the homogeneous transformation corresponding to the end-effector pose. The origin of this frame, the link-6 coordinate frame {6}, is defined◀ as the point of intersection of the axes of the last 3 joints – physically this point is inside the robot's wrist mechanism. We can define a tool transform, from the T_6 frame to the actual tool tip by

```
>> p560.tool = SE3(0, 0, 0.2);
```

Alternatively we could change the kinematic parameter d_6. The tool transform approach is more general since the final link kinematic parameters only allow setting of d_6, a_6 and α_6 which provide z-axis translation, x-axis translation and x-axis rotation respectively.

in this case a 200 mm extension in the T_6 z-direction.◀ The pose of the tool tip, often referred to as the tool center point or TCP, is now

```
>> p560.fkine(qz)
ans =
     1.0000         0         0    0.4521
          0    1.0000         0   -0.1500
          0         0    1.0000    0.6318
          0         0         0    1.0000
```

The kinematic definition we have used considers that the base of the robot is the intersection point of the waist and shoulder axes which is a point inside the structure of the robot. The Puma 560 robot includes a "30-inch" tall pedestal. We can shift the origin of the robot from the point inside the robot to the base of the pedestal using a base transform

```
>> p560.base = SE3(0, 0, 30*0.0254);
```

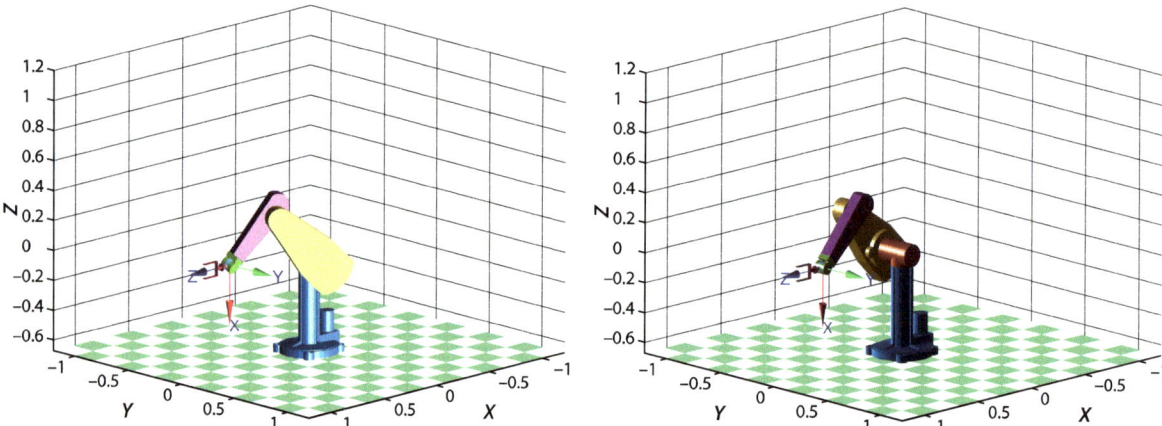

where for consistency we have converted the pedestal height to SI units. Now, with both base and tool transform, the forward kinematics are

```
>> p560.fkine(qz)
ans =
    1.0000         0         0    0.4521
         0    1.0000         0   -0.1500
         0         0    1.0000    1.3938
         0         0         0    1.0000
```

Fig. 7.7. These two different robot configurations result in the same end-effector pose. They are called the left- and right-handed configurations, respectively. These graphics, produced using the `plot3d` method, are available for a limited subset of robot models

and we can see that the z-coordinate of the tool is now greater than before.

We can also do more interesting things, for example

```
>> p560.base = SE3(0,0,3) * SE3.Rx(pi);
>> p560.fkine(qz)
ans =
    1.0000         0         0    0.4521
         0   -1.0000   -0.0000    0.1500
         0    0.0000   -1.0000    2.3682
         0         0         0    1.0000
```

which positions the robot's origin 3 m above the world origin with its coordinate frame rotated by 180° about the x-axis. This robot is now hanging from the ceiling!

The Toolbox supports joint angle time series, or trajectories, such as

```
>> q
q =
         0         0         0         0         0         0
         0    0.0365   -0.0365         0         0         0
         0    0.2273   -0.2273         0         0         0
         0    0.5779   -0.5779         0         0         0
         0    0.9929   -0.9929         0         0         0
         0    1.3435   -1.3435         0         0         0
         0    1.5343   -1.5343         0         0         0
         0    1.5708   -1.5708         0         0         0
```

where each row represents the joint coordinates at a different timestep and the columns represent the joints.▸ In this case the method `fkine`

Generated by the `jtraj` function, which is discussed in Sect. 7.3.1.

```
>> T = p560.fkine(q);
```

returns an array of SE3 objects

```
>> about T
T [SE3] : 1x8 (1.0 kB)
```

one per timestep. The homogeneous transform corresponding to the joint coordinates in the fourth row of `q` is

```
>> T(4)
ans =
    1.0000         0         0     0.382
         0        -1         0      0.15
         0         0   -1.0000     2.132
         0         0         0         1
```

Creating trajectories will be covered in Sect. 7.3.

7.2   Inverse Kinematics

We have shown how to determine the pose of the end-effector given the joint coordinates and optional tool and base transforms. A problem of real practical interest is the inverse problem: given the desired pose of the end-effector ξ_E what are the required joint coordinates? For example, if we know the Cartesian pose of an object, what joint coordinates does the robot need in order to reach it? This is the inverse kinematics problem which is written in functional form as

$$q = \mathcal{K}^{-1}(\xi) \tag{7.5}$$

and in general this function is not unique, that is, several joint coordinate vectors q will result in the same end-effector pose.

Two approaches can be used to determine the inverse kinematics. Firstly, a closed-form or analytic solution can be determined using geometric or algebraic approaches. However this becomes increasingly challenging as the number or robot joints increases and for some serial-link manipulators no closed-form solution exists. Secondly, an iterative numerical solution can be used. In Sect. 7.2.1 we again use the simple 2-dimensional case to illustrate the principles and then in Sect. 7.2.2 extend these to robot arms that move in 3-dimensions.

7.2.1   2-Dimensional (Planar) Robotic Arms

We will illustrate inverse kinematics for the 2-joint robot of Fig. 7.2b in two ways: algebraic closed-form and numerical.

7.2.1.1   Closed-Form Solution

We start by computing the forward kinematics algebraically as a function of joint angles. We can do this easily, and in a familiar way

```
>> import ETS2.*
>> a1 = 1; a2 = 1;
>> E = Rz('q1') * Tx(a1) * Rz('q2') * Tx(a2)
```

but now using the MATLAB Symbolic Math Toolbox™ we define some real-valued symbolic variables to represent the joint angles

```
>> syms q1 q2 real
```

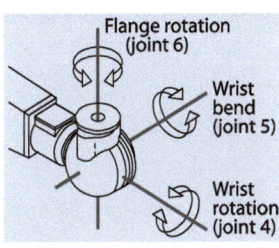

Flange rotation (joint 6)

Wrist bend (joint 5)

Wrist rotation (joint 4)

Spherical wrists are a key component of almost all modern arm-type robots. They have three axes of rotation that are orthogonal and intersect at a common point. This is a gimbal-like mechanism, and as discussed in Sect. 2.2.1.3 and will have a singularity.

The robot end-effector pose, position and an orientation, is defined at the center of the wrist. Since the wrist axes intersect at a common point they cause zero translation, therefore the position of the end-effector is a function only of the first three joints. This is a critical simplification that makes it possible to find closed-form inverse kinematic solutions for 6-axis industrial robots. An arbitrary end-effector orientation is achieved independently by means of the three wrist joints.

and then compute the forward kinematics

```
>> TE = E.fkine( [q1, q2] )
TE =
[ cos(q1 + q2), -sin(q1 + q2), cos(q1 + q2) + cos(q1)]
[ sin(q1 + q2),  cos(q1 + q2), sin(q1 + q2) + sin(q1)]
[            0,             0,                      1]
```

which is an algebraic representation of the robot's forward kinematics – the end-effector pose as a function of the joint variables.

We can define two more symbolic variables to represent the desired end-effector position (x, y)

```
>> syms x y real
```

and equate them with the results of the forward kinematics►

```
>> e1 =  x == TE.t(1)
e1 =
x ==  cos(q1 + q2) + cos(q1)
>> e2 =  y == TE.t(2)
e2 =
y == sin(q1 + q2) + sin(q1)
```

With the MATLAB Symbolic Math Toolbox™ the == operator denotes equality, as opposed to = which denotes assignment.

which gives two scalar equations that we can solve simultaneously

```
>> [s1,s2] = solve( [e1 e2], [q1 q2] )
```

where the arguments are respectively the set of equations and the set of unknowns to solve for. The outputs are the solutions for q_1 and q_2 respectively. We observed in Sect. 7.1.1 that two different sets of joint angles give the same end-effector position, and this means that the inverse kinematics does not have a unique solution. Here MATLAB has returned

```
>> length(s2)
ans =
     2
```

indicating two solutions. One solution for q_2 is

```
>> s2(1)
ans =
-2*atan((-(x^2 + y^2)*(x^2 + y^2 - 4))^(1/2)/(x^2 + y^2))
```

and would be used in conjunction with the corresponding element of the solution vector for q_1 which is s1(1).

As mentioned earlier the complexity of algebraic solution increases dramatically with the number of joints and more sophisticated symbolic solution approaches need to be used. The `SerialLink` class has a method `ikine_sym` that generates symbolic inverse kinematics solutions for a limited class of robot manipulators.

7.2.1.2   Numerical Solution

We can think of the inverse kinematics problem as one of adjusting the joint coordinates until the forward kinematics matches the desired pose. More formally this is an optimization problem – to minimize the error between the forward kinematic solution and the desired pose ξ^*

$$q^* = \underset{q}{\arg\min}\left\|\mathcal{K}(q) \ominus \xi^*\right\|$$

For our simple 2-link example the error function comprises only the error in the end-effector position, not its orientation

$$E(q) = \left\|\left[\mathcal{K}(q)\right]_t - \left(x^* \; y^*\right)^T\right\|$$

We can solve this using the builtin MATLAB multi-variable minimization function `fminsearch`

```
>> pstar = [0.6; 0.7];
>> q = fminsearch( @(q) norm( E.fkine(q).t - pstar ), [0 0] )
q =
   -0.2295    2.1833
```

where the first argument is the error function, expressed here as a MATLAB anonymous function, that incorporates the desired end-effector position; and the second argument is the initial guess at the joint coordinates. The computed joint angles indeed give the desired end-effector position

```
>> E.fkine(q).print
t = (0.6, 0.7), theta = 111.9 deg
```

As already discussed there are two solutions for *q* but the solution that is found using this approach depends on the initial choice of *q*.

7.2.2 3-Dimensional Robotic Arms

7.2.2.1 Closed-Form Solution

Closed-form solutions have been developed for most common types of 6-axis industrial robots and many are included in the Toolbox. A necessary condition for a closed-form solution of a 6-axis robot is a spherical wrist mechanism. We will illustrate closed-form inverse kinematics using the Denavit-Hartenberg model for the Puma robot

```
>> mdl_puma560
```

At the *nominal* joint coordinates shown in Fig. 7.6d

```
>> qn
qn =
        0    0.7854    3.1416         0    0.7854         0
```

the end-effector pose is

```
>> T = p560.fkine(qn)
T =
   -0.0000    0.0000    1.0000    0.5963
   -0.0000    1.0000   -0.0000   -0.1501
   -1.0000   -0.0000   -0.0000   -0.0144
         0         0         0    1.0000
```

The method `ikine6s` checks the Denavit-Hartenberg parameters to determine if the robot meets these criteria.

Since the Puma 560 is a 6-axis robot arm with a spherical wrist we use the method `ikine6s` to compute the inverse kinematics using a closed-form solution.◄ The required joint coordinates to achieve the pose `T` are

```
>> qi = p560.ikine6s(T)
qi =
    2.6486   -3.9270    0.0940    2.5326    0.9743    0.3734
```

Surprisingly, these are quite different to the joint coordinates we started with. However if we investigate a little further

```
>> p560.fkine(qi)
ans =
   -0.0000    0.0000    1.0000    0.5963
    0.0000    1.0000   -0.0000   -0.1500
   -1.0000    0.0000   -0.0000   -0.0144
         0         0         0    1.0000
```

we see that these two different sets of joint coordinates result in the *same* end-effector pose and these are shown in Fig. 7.7. The shoulder of the Puma robot is horizontally offset from the waist, so in one solution the arm is to the left of the waist, in the other

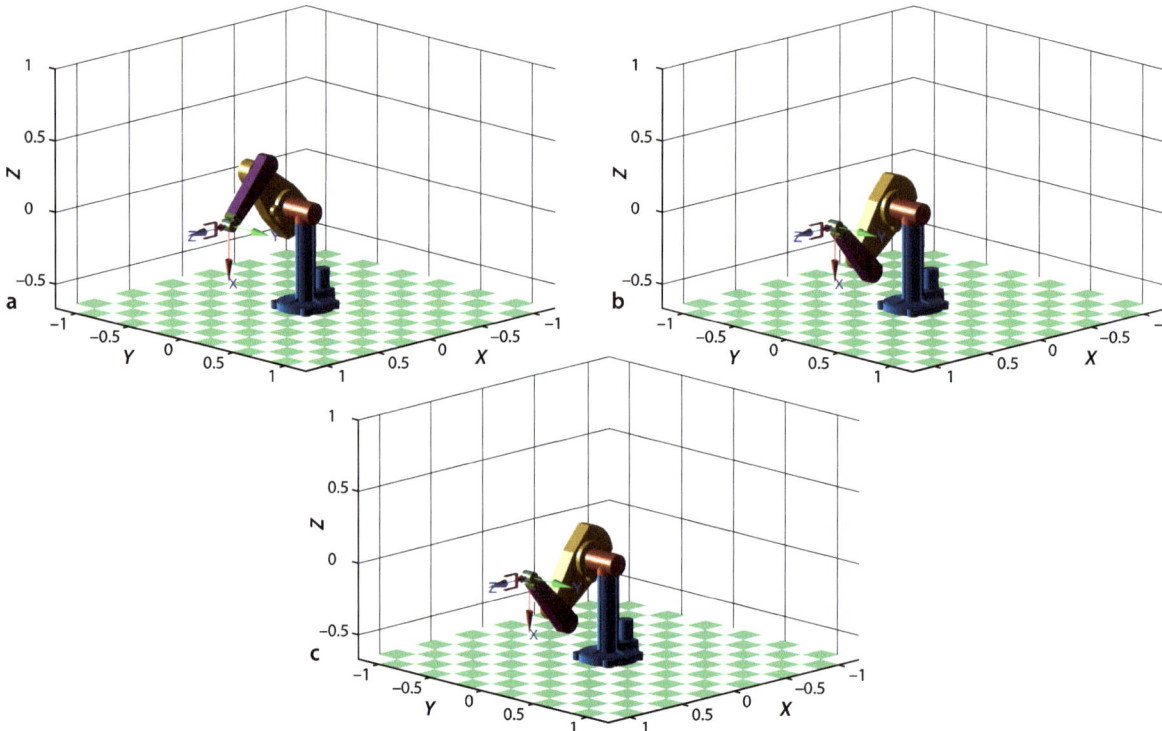

Fig. 7.8. Different configurations of the Puma 560 robot. **a** Right-up-noflip; **b** right-down-noflip; **c** right-down-flip

it is to the right. These are referred to as the left- and right-handed kinematic configurations respectively. In general there are eight sets of joint coordinates that give the same end-effector pose – as mentioned earlier the inverse solution is not unique.

We can *force* the right-handed solution

```
>> qi = p560.ikine6s(T, 'ru')
qi =
   -0.0000    0.7854    3.1416    0.0000    0.7854   -0.0000
```

which gives the original set of joint angles by specifying a *right handed* configuration with the elbow *up*.

In addition to the left- and right-handed solutions, there are solutions with the elbow either up or down,► and with the wrist flipped or not flipped. For the Puma 560 robot the wrist joint, θ_4, has a large rotational range and can adopt one of two angles that differ by π radians.

More precisely the elbow is above or below the shoulder.

Some different various kinematic configurations are shown in Fig. 7.8. The kinematic configuration returned by `ikine6s` is controlled by one or more of the options:

left or right handed	`'l'`, `'r'`
elbow up or down	`'u'`, `'d'`
wrist flipped or not flipped	`'f'`, `'n'`

Due to mechanical limits on joint angles and possible collisions between links not all eight solutions are physically achievable. It is also possible that no solution can be achieved. For example

```
>> p560.ikine6s( SE3(3, 0, 0) )
Warning: point not reachable
ans =
    NaN    NaN    NaN    NaN    NaN    NaN
```

has failed because the arm is simply not long enough to reach this pose.

A pose may also be unachievable due to singularity where the alignment of axes reduces the effective degrees of freedom (the gimbal lock problem again). The Puma 560 has a wrist singularity when q_5 is equal to zero and the axes of joints 4 and 6 become aligned. In this case the best that `ikine6s` can do is to constrain $q_4 + q_6$ but their individual values are arbitrary. For example consider the configuration

```
>> q = [0 pi/4 pi 0.1 0 0.2];
```

for which $q_4 + q_6 = 0.3$. The inverse kinematic solution is

```
>> p560.ikine6s(p560.fkine(q), 'ru')
ans =
   -0.0000    0.7854    3.1416   -3.0409    0.0000   -2.9423
```

which has quite different values for q_4 and q_6 but their sum

```
>> q(4)+q(6)
ans =
    0.3000
```

remains the same.

7.2.2.2   Numerical Solution

For the case of robots which do not have six joints and a spherical wrist we need to use an iterative numerical solution. Continuing with the example of the previous section we use the method `ikine` to compute the general inverse kinematic solution

```
>> T = p560.fkine(qn)
ans =
   -0.0000    0.0000    1.0000    0.5963
   -0.0000    1.0000   -0.0000   -0.1501
   -1.0000   -0.0000   -0.0000   -0.0144
         0         0         0    1.0000
>> qi = p560.ikine(T)
qi =
    0.0000   -0.8335    0.0940   -0.0000   -0.8312    0.0000
```

which is different to the original value

```
>> qn
qn =
         0    0.7854    3.1416         0    0.7854         0
```

but does result in the correct tool pose

```
>> p560.fkine(qi)
ans =
   -0.0000    0.0000    1.0000    0.5963
   -0.0000    1.0000   -0.0000   -0.1501
   -1.0000   -0.0000   -0.0000   -0.0144
         0         0         0    1.0000
```

Plotting the pose

```
>> p560.plot(qi)
```

shows clearly that `ikine` has found the elbow-down configuration.

A limitation of this general numeric approach is that it does not provide explicit control over the arm's kinematic configuration as did the analytic approach – the only control is implicit via the initial value of joint coordinates (which defaults to zero). If we specify the initial joint coordinates

```
>> qi = p560.ikine(T, 'q0', [0 0 3 0 0 0])
qi =
    0.0000    0.7854    3.1416    0.0000    0.7854   -0.0000
```

When solving for a trajectory as on p. 202 the inverse kinematic solution for one point is used to initialize the solution for the next point on the path.

we have forced the solution to converge on the elbow-up configuration. ◄

As would be expected the general numerical approach of `ikine` is considerably slower than the analytic approach of `ikine6s`. However it has the great advantage of being able to work with manipulators at singularities and manipulators with less than six or more than six joints. Details of the principle behind `ikine` is provided in Sect. 8.6.

7.2.2.3   Under-Actuated Manipulator

An under-actuated manipulator is one that has fewer than six joints, and is therefore limited in the end-effector poses that it can attain. SCARA robots such as shown on page 189 are a common example. They typically have an x-y-z-θ task space, $\mathcal{T} \subset \mathbb{R}^3 \times \mathbb{S}^1$ and a configuration space $\mathcal{C} \subset (\mathbb{S}^1)^3 \times \mathbb{R}$.

We will load a model of SCARA robot

```
>> mdl_cobra600
>> c600
c600 =
Cobra600 [Adept]:: 4 axis, RRPR, stdDH
+---+-----------+-----------+-----------+-----------+-----------+
| j |   theta   |     d     |     a     |   alpha   |  offset   |
+---+-----------+-----------+-----------+-----------+-----------+
| 1|        q1|      0.387|      0.325|          0|          0|
| 2|        q2|          0|      0.275|      3.142|          0|
| 3|         0|         q3|          0|          0|          0|
| 4|        q4|          0|          0|          0|          0|
+---+-----------+-----------+-----------+-----------+-----------+
```

and then define a desired end-effector pose

```
>> T = SE3(0.4, -0.3, 0.2) * SE3.rpy(30, 40, 160, 'deg')
```

where the end-effector approach vector is pointing downward but is not vertically aligned. This *pose* is over-constrained for the 4-joint SCARA robot – the tool physically cannot meet the orientation requirement for an approach vector that is not vertically aligned. Therefore we require the `ikine` method to not consider rotation about the x- and y-axes when computing the end-effector pose error. We achieve this by specifying a mask vector as the fourth argument

```
>> q = c600.ikine(T, 'mask', [1 1 1 0 0 1])
q =
   -0.1110   -1.1760    0.1870   -0.8916
```

The elements of the mask vector correspond respectively to the three translations and three orientations: $t_x, t_y, t_z, r_x, r_y, r_z$ in the end-effector coordinate frame. In this example we specified that rotation about the x- and y-axes are to be ignored (the zero elements). The resulting joint angles correspond to an achievable end-effector pose

```
>> Ta = c600.fkine(q);
>> Ta.print('xyz')
t = (0.4, -0.3, 0.2), RPY/xyz = (22.7, 0, 180) deg
```

which has the desired translation and yaw angle, but the roll and pitch angles are incorrect, as *we allowed* them to be. They are what the robot mechanism actually permits. We can also compare the desired and achievable poses graphically

```
>> trplot(T, 'color', 'b')
>> hold on
>> trplot(Ta, 'color', 'r')
```

7.2.2.4   Redundant Manipulator

A redundant manipulator is a robot with more than six joints. As mentioned previously, six joints is theoretically sufficient to achieve any desired pose in a Cartesian taskspace $\mathcal{T} \subset \mathbf{SE}(3)$. However practical issues such as joint limits and singularities mean that not all poses within the robot's reachable space can be achieved. Adding additional joints is one way to overcome this problem but results in an infinite number of joint-coordinate solutions. To find a single solution we need to introduce constraints – a common one is the minimum-norm constraint which returns a solution where the joint-coordinate vector elements have the smallest magnitude.

We will illustrate this with the Baxter robot shown in Fig. 7.1b. This is a two armed robot, and each arm has 7 joints. We load the Toolbox model

```
>> mdl_baxter
```

which defines two SerialLink objects in the workspace, one for each arm. We will work with the left arm

```
>> left
left =
Baxter LEFT [Rethink Robotics]:: 7 axis, RRRRRRR, stdDH
+---+-----------+-----------+-----------+-----------+-----------+
| j |   theta   |     d     |     a     |   alpha   |  offset   |
+---+-----------+-----------+-----------+-----------+-----------+
| 1|        q1|      0.27|     0.069|    -1.571|         0|
| 2|        q2|         0|         0|     1.571|     1.571|
| 3|        q3|     0.364|     0.069|    -1.571|         0|
| 4|        q4|         0|         0|     1.571|         0|
| 5|        q5|     0.374|      0.01|    -1.571|         0|
| 6|        q6|         0|         0|     1.571|         0|
| 7|        q7|      0.28|         0|         0|         0|
+---+-----------+-----------+-----------+-----------+-----------+
base:    t = (0.064614,0.25858,0.119), RPY/xyz = (0, 0, 45) deg
```

which we can see has a base offset that reflects where the arm is attached to Baxter's torso. We want the robot to move to this pose

```
>> TE = SE3(0.8, 0.2, -0.2) * SE3.Ry(pi);
```

which has its approach vector downward. The required joint angles are obtained using the numerical inverse kinematic solution and

```
>> q = left.ikine(TE)
q =
    0.0895   -0.0464   -0.4259    0.6980   -0.4248    1.0179    0.2998
```

is the joint-angle vector with the smallest norm that results in the desired end-effector pose. We can verify this by computing the forward kinematics or plotting

```
>> left.fkine(q).print('xyz')
t = (0.8, 0.2, -0.2), RPY/xyz = (180, 180, 180) deg
>> left.plot(q)
```

7.3 Trajectories

One of the most common requirements in robotics is to move the end-effector smoothly from pose A to pose B. Building on what we learned in Sect. 3.3 we will discuss two approaches to generating such trajectories: straight lines in configuration space and straight lines in task space. These are known respectively as joint-space and Cartesian motion.

7.3.1 Joint-Space Motion

In this robot configuration, similar to Fig. 7.6d, we specify the pose to include a rotation so that the end-effector z-axis is not pointing straight up in the world z-direction. For the Puma 560 robot this would be physically impossible to achieve in the elbow-up configuration.

Consider the end-effector moving between two Cartesian poses◄

```
>> T1 = SE3(0.4,  0.2, 0) * SE3.Rx(pi);
>> T2 = SE3(0.4, -0.2, 0) * SE3.Rx(pi/2);
```

which describe points in the xy-plane with different end-effector orientations. The joint coordinate vectors associated with these poses are

```
>> q1 = p560.ikine6s(T1);
>> q2 = p560.ikine6s(T2);
```

and we require the motion to occur over a time period of 2 seconds in 50 ms time steps

```
>> t = [0:0.05:2]';
```

A joint-space trajectory is formed by smoothly interpolating between the joint configurations `q1` and `q2`. The scalar interpolation functions `tpoly` or `lspb` from Sect. 3.3.1 can be used in conjunction with the multi-axis *driver* function `mtraj`

```
>> q = mtraj(@tpoly, q1, q2, t);
```

or

```
>> q = mtraj(@lspb, q1, q2, t);
```

which each result in a 50 × 6 matrix `q` with one row per time step and one column per joint. From here on we will use the equivalent `jtraj` convenience function

```
>> q = jtraj(q1, q2, t); ▶
```

▶ This is equivalent to `mtraj` with `tpoly` interpolation but optimized for the multi-axis case and also allowing initial and final velocity to be set using additional arguments.

For `mtraj` and `jtraj` the final argument can be a time vector, as here, or an integer specifying the number of time steps.

We can obtain the joint velocity and acceleration vectors, as a function of time, through optional output arguments

```
>> [q,qd,qdd] = jtraj(q1, q2, t);
```

An even more concise way to achieve the above steps is provided by the `jtraj` method of the `SerialLink` class

```
>> q = p560.jtraj(T1, T2, t)
```

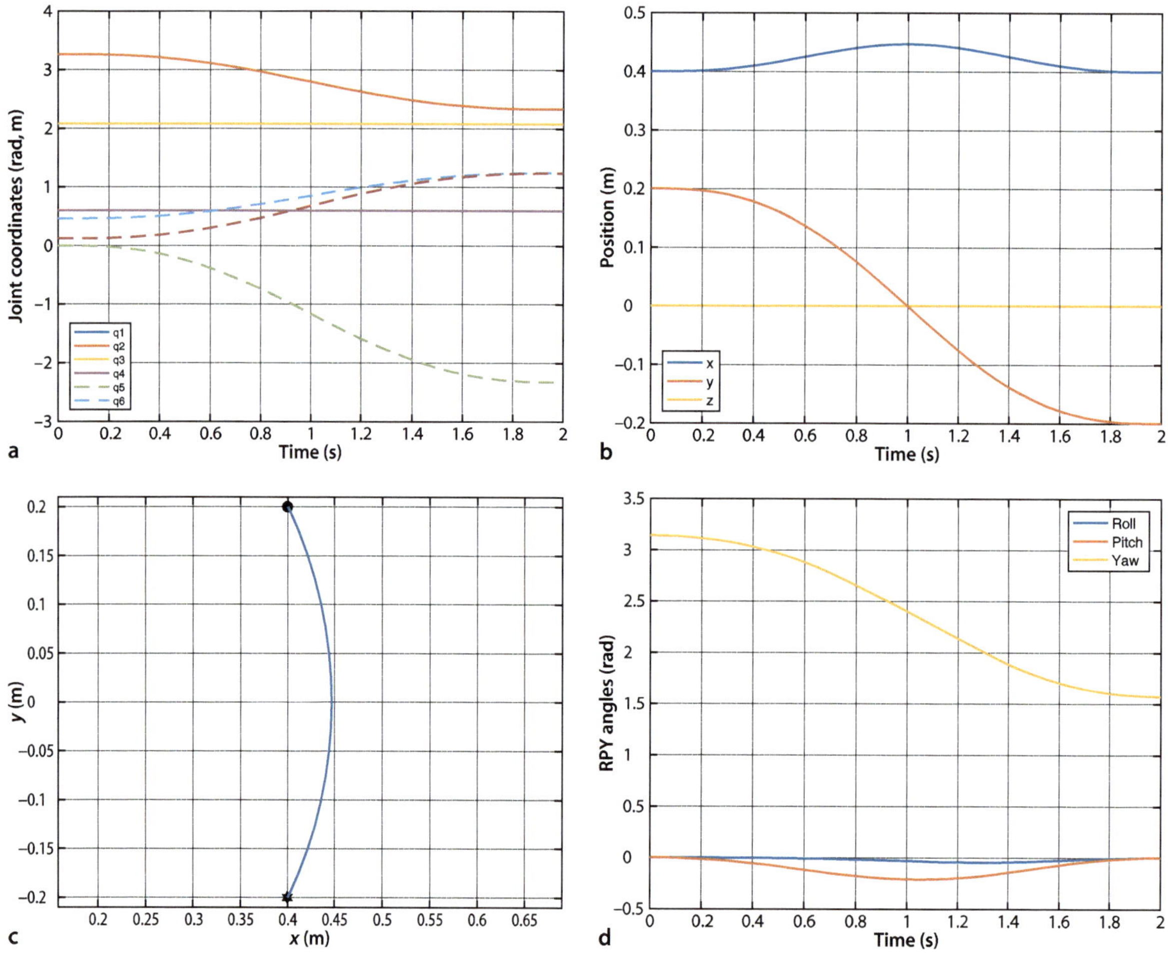

Fig. 7.9. Joint-space motion. **a** Joint coordinates versus time; **b** Cartesian position versus time; **c** Cartesian position locus in the *xy*-plane **d** roll-pitch-yaw angles versus time

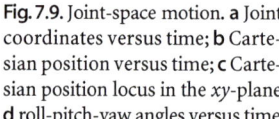

The trajectory is best viewed as an animation

```
>> p560.plot(q)
```

but we can also plot the joint angle, for instance q_2, versus time

```
>> plot(t, q(:,2))
```

or all the angles versus time

```
>> qplot(t, q);
```

as shown in Fig. 7.9a. The joint coordinate trajectory is smooth but we do not know how the robot's end-effector will move in Cartesian space. However we can easily determine this by applying forward kinematics to the joint coordinate trajectory

```
>> T = p560.fkine(q);
```

which results in an array of SE3 objects. The translational part of this trajectory is

```
>> p = T.transl;
```

which is in matrix form

Fig. 7.10. Cartesian motion. **a** Joint coordinates versus time; **b** Cartesian position versus time; **c** Cartesian position locus in the *xy*-plane; **d** roll-pitch-yaw angles versus time

```
>> about(p)
p [double] : 41x3 (984 bytes)
```

and has one column per time step, and each column is the end-effector position vector. This is plotted against time in Fig. 7.9b. The path of the end-effector in the *xy*-plane

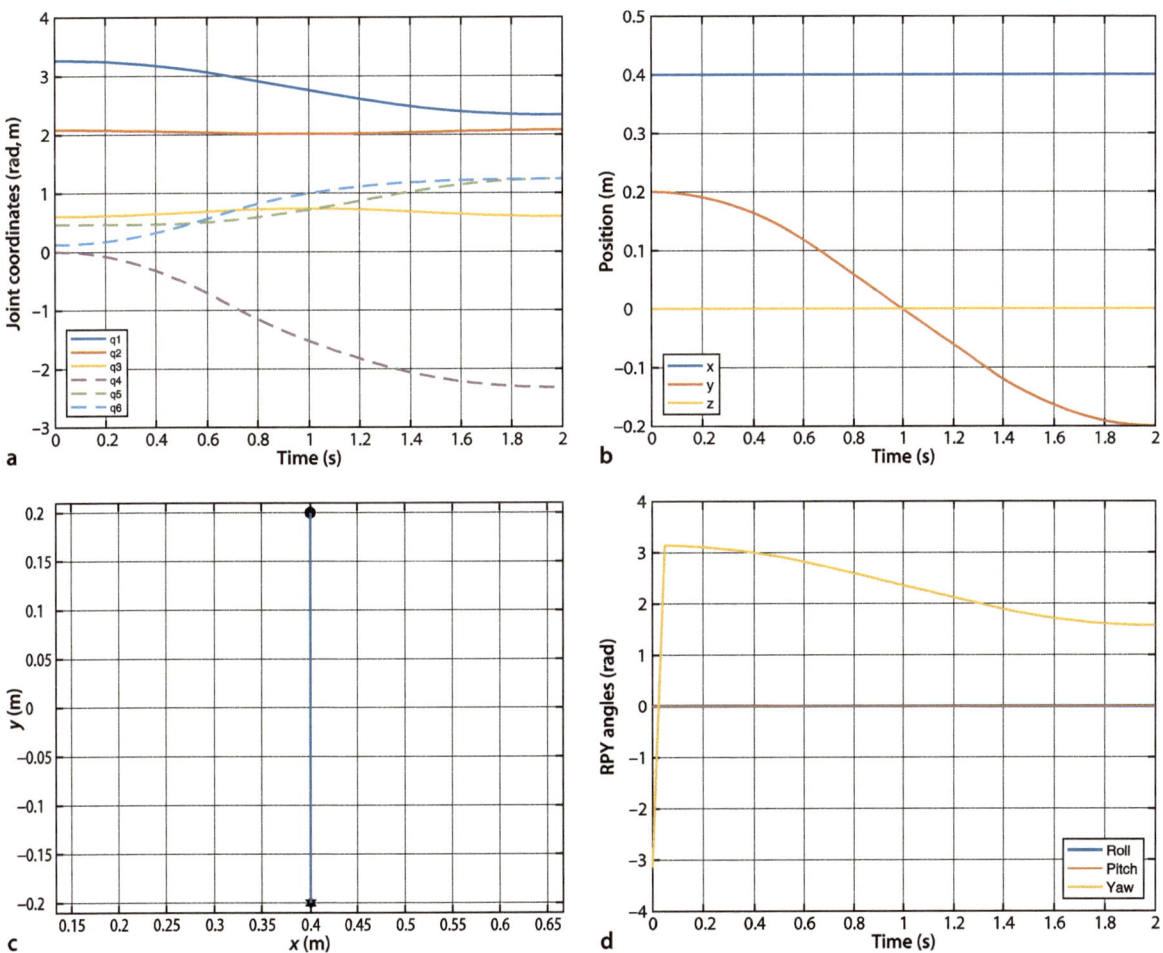

```
>> plot(p(1,:), p(2,:))
```

is shown in Fig. 7.9c and it is clear that the path is not a straight line. This is to be expected since we only specified the Cartesian coordinates of the end-points. As the robot rotates about its waist joint during the motion the end-effector will naturally follow a circular arc. In practice this could lead to collisions between the robot and nearby objects even if they do not lie on the path between poses A and B. The orientation of the end-effector, in XYZ roll-pitch-yaw angle form, can also be plotted against time

```
>> plot(t, T.torpy('xyz'))
```

as shown in Fig. 7.9d. Note that the yaw angle▶ varies from 0 to $\frac{\pi}{2}$ radians as we specified. However while the roll and pitch angles have met their boundary conditions they have varied along the path.

▶ Rotation about *x*-axis for a robot end-effector from Sect. 2.2.1.2.

7.3.2 Cartesian Motion

For many applications we require straight-line motion in Cartesian space which is known as Cartesian motion. This is implemented using the Toolbox function `ctraj` which was introduced in Sect. 3.3.5. Its usage is very similar to `jtraj`

```
>> Ts = ctraj(T1, T2, length(t));
```

where the arguments are the initial and final pose and the *number of* time steps and it returns the trajectory as an array of `SE3` objects.

As for the previous joint-space example we will extract and plot the translation

```
>> plot(t, Ts.transl);
```

and orientation components

```
>> plot(t, Ts.torpy('xyz'));
```

of this motion which is shown in Fig. 7.10 along with the path of the end-effector in the *xy*-plane. Compared to Fig. 7.9 we note some important differences. Firstly the end-effector follows a straight line in the *xy*-plane as shown in Fig. 7.10c. Secondly the roll and pitch angles shown in Fig. 7.10d are constant at zero along the path.

The corresponding joint-space trajectory is obtained by applying the inverse kinematics

```
>> qc = p560.ikine6s(Ts);
```

and is shown in Fig. 7.10a. While broadly similar to Fig. 7.9a the minor differences are what result in the straight line Cartesian motion.

7.3.3 Kinematics in Simulink

We can also implement this example in Simulink®

```
>> sl_jspace
```

and the block diagram model is shown in Fig. 7.11. The parameters of the `jtraj` block are the initial and final values for the joint coordinates and the time duration of the motion segment. The smoothly varying joint angles are wired to a `plot` block which will animate a robot in a separate window, and to an `fkine` block to compute the forward kinematics. Both the `plot` and `fkine` blocks have a parameter which is a `SerialLink` object, in this case p560. The Cartesian position of the end-effector pose is extracted using the `T2xyz` block which is analogous to the Toolbox function `transl`. The `XY Graph` block plots `y` against `x`.

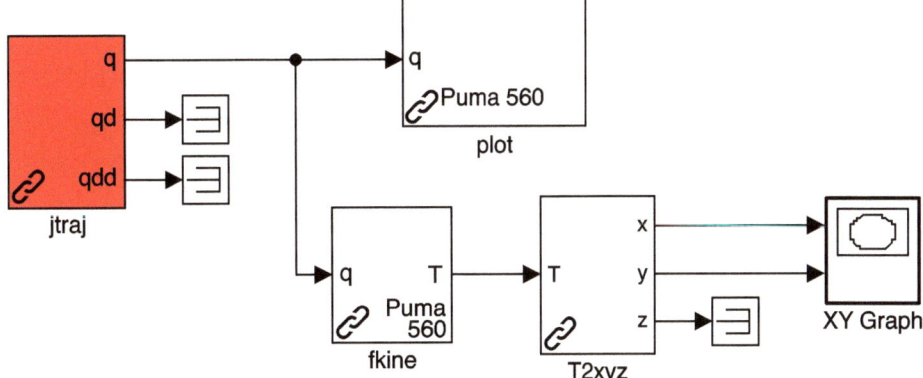

Fig. 7.11.
Simulink model sl_jspace
for joint-space motion

7.3.4 Motion through a Singularity

We have already briefly touched on the topic of singularities (page 207) and we will revisit it again in the next chapter. In the next example we deliberately choose a trajectory that moves through a robot wrist singularity. We change the Cartesian endpoints of the previous example to

```
>> T1 = SE3(0.5,  0.3, 0.44) * SE3.Ry(pi/2);
>> T2 = SE3(0.5, -0.3, 0.44) * SE3.Ry(pi/2);
```

which results in motion in the y-direction with the end-effector z-axis pointing in the world x-direction. The Cartesian path is

```
>> Ts = ctraj(T1, T2, length(t));
```

which we convert to joint coordinates

```
>> qc = p560.ikine6s(Ts)
```

q_6 has increased rapidly, while q_4 has decreased rapidly and wrapped around from $-\pi$ to π. This counter-rotational motion of the two joints means that the gripper does not rotate but the two motors are working hard.

and is shown in Fig. 7.12a. At time $t \approx 0.7$ s we observe that the rate of change of the wrist joint angles q_4 and q_6 has become very high.◀ The cause is that q_5 has become almost zero which means that the q_4 and q_6 rotational axes are almost aligned – another gimbal lock situation or singularity.

The joint axis alignment means that the robot has lost one degree of freedom and is now effectively a 5-axis robot. Kinematically we can only solve for the sum $q_4 + q_6$ and there are an infinite number of solutions for q_4 and q_6 that would have the same sum. From Fig. 7.12b we observe that the generalized inverse kinematics method `ikine` handles the singularity with far less unnecessary joint motion. This is a consequence of the minimum-norm solution which has returned the smallest magnitude q_4 and q_6 which have the correct sum. The joint-space motion between the two poses, Fig. 7.12c, is immune to this problem since it is does not involve inverse kinematics. However it will not maintain the orientation of the tool in the x-direction for the whole path – only at the two end points.

The dexterity of a manipulator, its ability to move easily in any direction, is referred to as its manipulability. It is a scalar measure, high is good, and can be computed for each point along the trajectory

```
>> m = p560.maniplty(qc);
```

and is plotted in Fig. 7.12d. This shows that manipulability was almost zero around the time of the rapid wrist joint motion. Manipulability and the generalized inverse kinematics function are based on the manipulator's Jacobian matrix which is the topic of the next chapter.

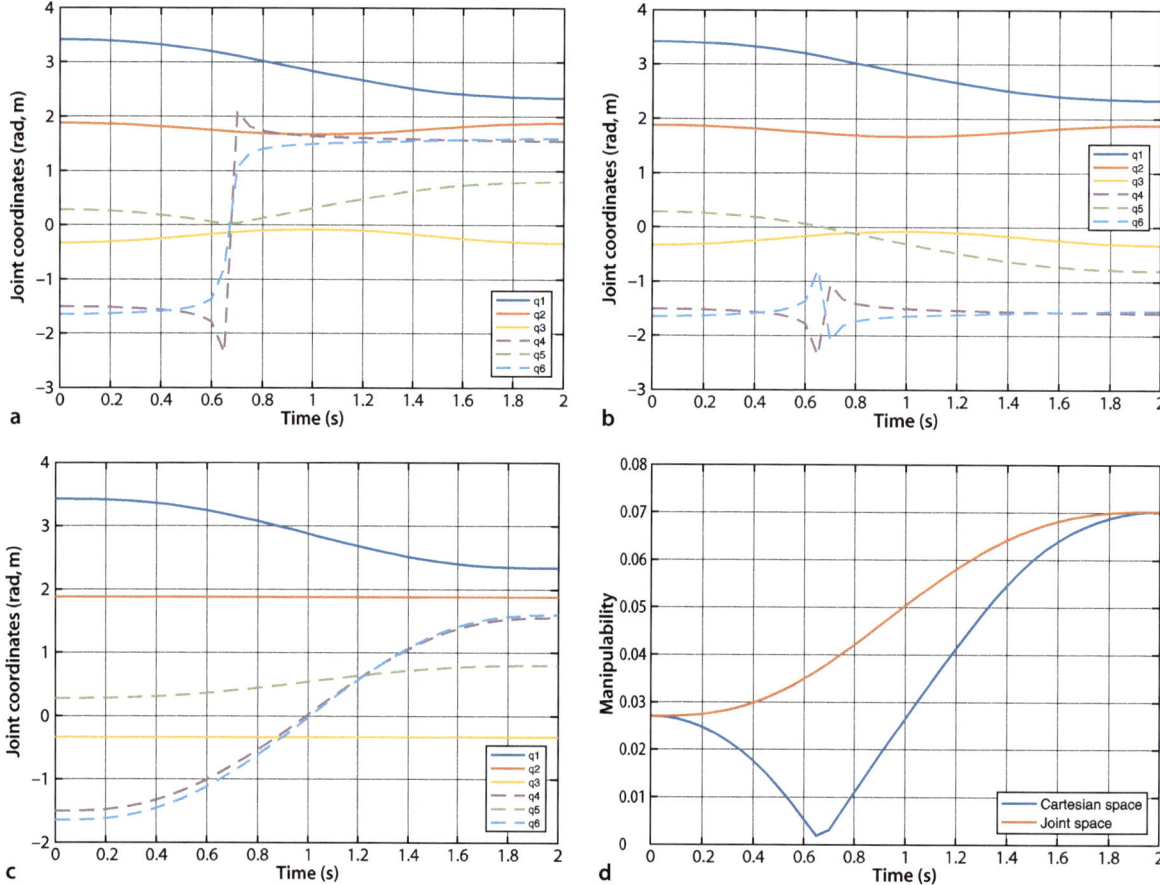

Fig. 7.12. Cartesian motion through a wrist singularity. **a** Joint coordinates computed by inverse kinematics (`ikine6s`); **b** joint coordinates computed by numerical inverse kinematics (`ikine`); **c** joint coordinates for joint-space motion; **d** manipulability

7.3.5   Configuration Change

Earlier (page 206) we discussed the kinematic configuration of the manipulator arm and how it can work in a left- or right-handed manner and with the elbow up or down. Consider the problem of a robot that is working for a while left-handed at one work station, then working right-handed at another. Movement from one configuration to another ultimately results in no change in the end-effector pose since both configuration have the same forward kinematic solution – therefore we *cannot* create a trajectory in Cartesian space. Instead we must use joint-space motion.

For example to move the robot arm from the right- to left-handed configuration we first define some end-effector pose

```
>> T = SE3(0.4, 0.2, 0) * SE3.Rx(pi);
```

and then determine the joint coordinates for the right- and left-handed elbow-up configurations

```
>> qr = p560.ikine6s(T, 'ru');
>> ql = p560.ikine6s(T, 'lu');
```

and then create a joint-space trajectory between these two joint coordinate vectors

```
>> q = jtraj(qr, ql, t);
```

Although the initial and final end-effector pose is the same, the robot makes some quite significant joint space motion as shown in Fig. 7.13 – in the real world you need to be careful the robot doesn't hit something. Once again, the best way to visualize this is in animation

```
>> p560.plot(q)
```

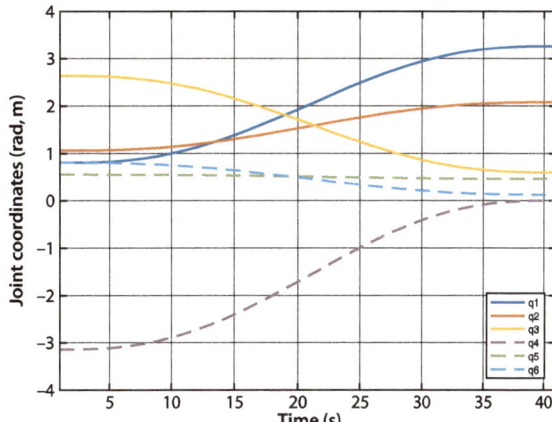

Fig. 7.13.
Joint space motions for configuration change from right-handed
to left-handed

7.4   Advanced Topics

7.4.1   Joint Angle Offsets

The pose of the robot with zero joint angles is an arbitrary decision of the robot
designer and might even be a mechanically unachievable pose. For the Puma robot
the zero-angle pose is a nonobvious *L-shape* with the upper arm horizontal and the
lower arm vertically upward as shown in Fig. 7.6a. This is a consequence of constraints imposed by the Denavit-Hartenberg formalism.

The joint coordinate offset provides a mechanism to set an arbitrary configuration for the zero joint coordinate case. The offset vector, q_0, is added to the user
specified joint angles before any kinematic or dynamic function is invoked,◄ for
example

$$\xi_E = \mathcal{K}(q + q_0) \tag{7.6}$$

Similarly it is subtracted after an operation such as inverse kinematics

$$q = \mathcal{K}^{-1}(\xi_E) - q_0 \tag{7.7}$$

The offset is set by assigning the `offset` property of the `Link` object, or giving the
`'offset'` option to the `SerialLink` constructor.

It is actually implemented within the `Link` object.

7.4.2   Determining Denavit-Hartenberg Parameters

The classical method of determining Denavit-Hartenberg parameters is to system-
atically assign a coordinate frame to each link. The link frames for the Puma robot
using the standard Denavit-Hartenberg formalism are shown in Fig. 7.14. However
there are strong constraints on placing each frame since joint rotation must be
about the z-axis and the link displacement must be in the x-direction. This in turn
imposes constraints on the placement of the coordinate frames for the base and the
end-effector, and ultimately dictates the zero-angle pose just discussed. Determining
the Denavit-Hartenberg parameters and link coordinate frames for a completely
new mechanism is therefore more difficult than it should be – even for an experienced roboticist.

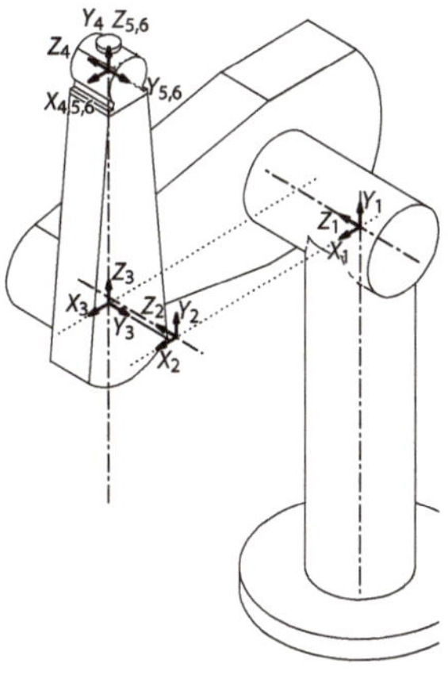

An alternative approach, supported by the Toolbox, is to simply describe the manipulator as a series of elementary translations and rotations from the base to the
tip of the end-effector as we discussed in Sect. 7.1.2. Some of the elementary operations are constants such as translations that represent link lengths or offsets, and

Fig. 7.14. Puma 560 robot coordinate frames. Standard Denavit-
Hartenberg link coordinate frames
for Puma in the zeroangle pose
(Corke 1996b)

some are functions of the generalized joint coordinates q_j. Unlike the conventional approach we impose no constraints on the axes about which these rotations or translations can occur.

For the Puma robot shown in Fig. 7.4 we first define a convenient coordinate frame at the base and then write down the sequence of translations and rotations, from "toe to tip", in a string▸

```
>> s = 'Tz(L1) Rz(q1) Ry(q2) Ty(L2) Tz(L3) Ry(q3) Tx(L4) Ty(L5)
          Tz(L6) Rz(q4) Ry(q5) Rz(q6)'
```

All lengths must start with *L* and negative signs cannot be used in the string, but the values themselves can be negative. You can generate this string from an ETS3 sequence (page 194) using its `string` method.

Note that we have described the second joint as `Ry(q2)`, a rotation about the *y*-axis, which is not permissible using the Denavit-Hartenberg formalism.

This string is input to a symbolic algebra function▸

```
>> dh = DHFactor(s);
```

Written in Java, the MATLAB® Symbolic Math Toolbox™ is not required.

which returns a `DHFactor` object▸ that holds the kinematic structure of the robot that has been factorized into Denavit-Hartenberg parameters. We can display this in a human-readable form

Actually a Java object.

```
>> dh
dh =
DH(q1, L1, 0, -90).DH(q2+90, 0, -L3, 0).DH(q3-90, L2+L5, L4, 90).
DH(q4, L6, 0, -90).DH(q5, 0, 0, 90).DH(q6, 0, 0, 0)
```

which shows the Denavit-Hartenberg parameters for each joint in the order θ, d, a and α. Joint angle offsets (the constants added to or subtracted from joint angle variables such as `q2` and `q3`) are generated automatically, as are base and tool transformations. The object can generate a string that is a complete Toolbox command to create the robot named "puma"

```
>> cmd = dh.command('puma')
cmd =
SerialLink([0, L1, 0, -pi/2, 0; 0, 0, -L3, 0, 0; 0, L2+L5, L4, ↵
pi/2, 0; 0, L6, 0, -pi/2, 0; 0, 0, 0, pi/2, 0; 0, 0, 0, 0, 0; ], ...
  'name', 'puma', ...
  'base', eye(4,4), 'tool', eye(4,4), ...
  'offset', [0 pi/2 -pi/2 0 0 0 ])
```

which can be executed

```
>> robot = eval(cmd)
```

to create a workspace variable called `robot` that is a `SerialLink` object.▸

The length parameters `L1` to `L6` must be defined in the workspace first.

7.4.3   Modified Denavit-Hartenberg Parameters

The Denavit-Hartenberg notation introduced in this chapter is commonly used and described in many robotics textbooks. Craig (2005) first introduced the modified Denavit-Hartenberg parameters where the link coordinate frames shown in Fig. 7.15 are attached to the near (proximal), rather than the far (distal) end of each link. This modified notation is in some ways clearer and tidier and is also now commonly used. However its introduction has increased the scope for confusion, particularly for those who are new to robot kinematics. The root of the problem is that the algorithms for kinematics, Jacobians and dynamics depend on the kinematic conventions used. According to Craig's convention the link transform matrix is

$$^{j-1}\xi_j\left(\alpha_{j-1}, a_{j-1}, d_j, \theta_j\right) = \mathscr{R}_x\left(\alpha_{j-1}\right) \oplus \mathscr{T}_x\left(a_{j-1}\right) \oplus \mathscr{T}_z\left(d_j\right) \oplus \mathscr{R}_z\left(\theta_j\right) \quad (7.8)$$

denoted in that book as $^{j-1}_j A$. This has the same terms as Eq. 7.2 but in a different order – remember rotations are not commutative – and this is the nub of the problem. a_j is

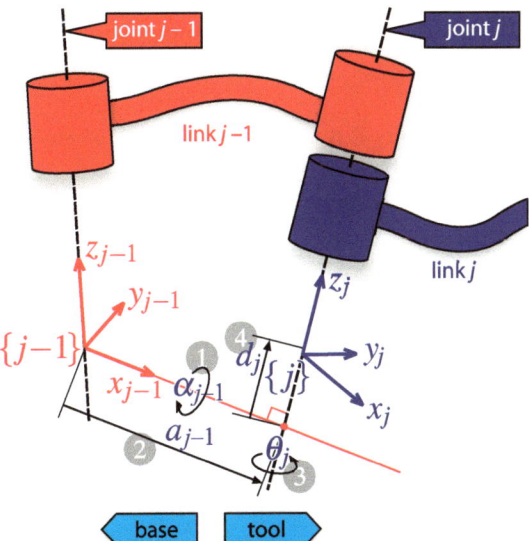

Fig. 7.15.
Definition of modified Denavit
and Hartenberg link parameters.
The colors *red* and *blue* denote
all things associated with links
$j - 1$ and j respectively. The
numbers in circles represent the
order in which the elementary
transforms are applied

always the length of link j, but it is the displacement between the origins of frame $\{j\}$ and frame $\{j + 1\}$ in one convention, and frame $\{j - 1\}$ and frame $\{j\}$ in the other.

> If you intend to build a Toolbox robot model from a table of kinematic parameters provided in a research paper it is really important to know which convention the author of the table used. Too often this important fact is not mentioned. An important clue lies in the column headings. If they all have the same subscript, i.e. θ_j, d_j, a_j and α_j then this is standard Denavit-Hartenberg notation. If half the subscripts are different, i.e. θ_j, d_j, a_{j-1} and α_{j-1} then you are dealing with modified Denavit-Hartenberg notation. In short, you must know which kinematic convention your Denavit-Hartenberg parameters conform to.
>
> You can also help the field when publishing by stating clearly which kinematic convention is used for your parameters.

The Toolbox can handle either form, it only needs to be specified, and this is achieved using variant classes when creating a link object

```
>> L1 = RevoluteMDH('d', 1)
L1 =
Revolute(mod): theta=q, d=1, a=0, alpha=0, offset=0
```

Everything else from here on, creating the robot object, kinematic and dynamic functions works as previously described.

The two forms can be interchanged by considering the link transform as a string of elementary rotations and translations as in Eq. 7.2 or Eq. 7.8. Consider the transformation chain for standard Denavit-Hartenberg notation

$$\underbrace{\mathscr{R}_z(\theta_1) \oplus \mathscr{T}_z(d_1) \oplus \mathscr{T}_x(a_1) \oplus \mathscr{R}_x(\alpha_1)}_{\text{DH}_1} \oplus \underbrace{\mathscr{R}_z(\theta_2) \oplus \mathscr{T}_z(d_2) \oplus \mathscr{T}_x(a_2) \oplus \mathscr{R}_x(\alpha_2)}_{\text{DH2}} \cdots$$

which we can regroup as

$$\underbrace{\mathscr{T}_x(0) \oplus \mathscr{R}_x(0) \oplus \mathscr{R}_z(\theta_1) \oplus \mathscr{T}_z(d_1)}_{\text{MDH}_1} \oplus \underbrace{\mathscr{T}_x(a_1) \oplus \mathscr{R}_x(\alpha_1) \oplus \mathscr{R}_z(\theta_2) \oplus \mathscr{T}_z(d_2)}_{\text{MDH}_2} \oplus \cdots$$

where the terms marked as MDH$_j$ have the form of Eq. 7.8 taking into account that translation along, and rotation about the same axis *is* commutative, that is, $\mathscr{R}_i(\theta) \oplus \mathscr{T}_i(d) = \mathscr{T}_i(d) \oplus \mathscr{R}_i(\theta)$ for $i \in \{x, y, z\}$.

7.5   Applications

7.5.1   Writing on a Surface [examples/drawing.m]

Our goal is to create a trajectory that will allow a robot to draw a letter. The Toolbox comes with a preprocessed version of the Hershey font▸

```
>> load hershey
```

as a cell array of character descriptors. For an upper-case 'B'

```
>> B = hershey{'B'}
B =
    stroke: [2x23 double]
     width: 0.8400
       top: 0.8400
    bottom: 0
```

the structure describes the dimensions of the character, vertically from 0 to 0.84 and horizontally from 0 to 0.84▸. The path to be drawn is

```
>> B.stroke
ans =
  Columns 1 through 11
    0.1600    0.1600       NaN    0.1600    0.5200    0.6400    ...
    0.8400         0       NaN    0.8400    0.8400    0.8000    ...
```

where the rows are the x- and y-coordinates respectively, and a column of NaNs indicates the end of a segment – the pen is lifted and placed down again at the beginning of the next segment. We perform some processing

```
>> path = [ 0.25*B.stroke; zeros(1,numcols(B.stroke))];
>> k = find(isnan(path(1,:)));
>> path(:,k) = path(:,k-1); path(3,k) = 0.2;
```

to scale the path by 0.25 so that the character is around 20 cm tall, append a row of zeros (add z-coordinates to this 2-dimensional path), find the columns that contain NaNs and replace them with the preceding column but with the z-coordinate set to 0.2 in order to lift the pen off the surface.

Next we convert this to a continuous trajectory

```
>> traj = mstraj(path(:,2:end)', [0.5 0.5 0.5], [], path(:,1)',↵
   0.02, 0.2);
```

where the second argument is the maximum speed in the x-, y- and z-directions, the fourth argument is the initial coordinate followed by the sample interval and the acceleration time. The number of steps in the interpolated path is

```
>> about(traj)
traj [double] : 487x3 (11.7 kB)
```

and will take

```
>> numrows(traj) * 0.02
ans =
    9.7400
```

seconds to execute at the 20 ms sample interval. The trajectory can be plotted

```
>> plot3(traj(:,1), traj(:,2), traj(:,3))
```

as shown in Fig. 7.16.

We now have a sequence of 3-dimensional points but the robot end-effector has a pose, not just a position, so we need to attach a coordinate frame to every point. We assume that the robot is writing on a horizontal surface so these frames must have their approach vector pointing downward, that is, $a = [0, 0, -1]$, with the gripper ar-

Developed by Dr. Allen V. Hershey at the Naval Weapons Laboratory in 1967, data from http://paulbourke.net/data-formats/hershey.

This is a variable-width font, and all characters fit within a unit-height rectangle.

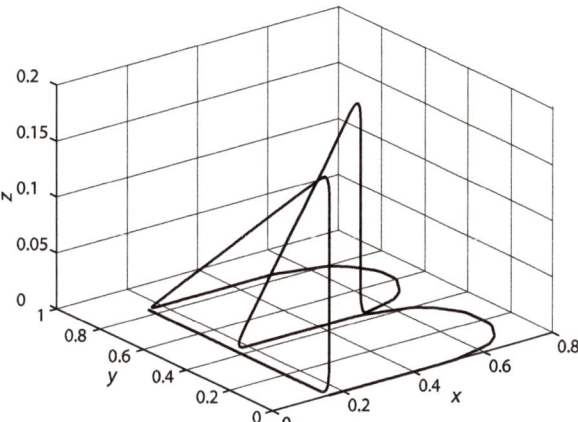

Fig. 7.16.
The end-effector path drawing
the letter 'B'

bitrarily oriented in the *y*-direction with $o = [0, 1, 0]$. The character is also placed at $(0.6, 0, 0)$ in the workspace, and all this is achieved by

```
>> Tp = SE3(0.6, 0, 0) * SE3(traj) * SE3.oa( [0 1 0], [0 0 -1]);
```

Now we can apply inverse kinematics

```
>> q = p560.ikine6s(Tp);
```

to determine the joint coordinates and then animate it

> We have not considered the force that the robot-held pen exerts on the paper, we cover force control in Chap. 9. In a real implementation of this example it would be prudent to use a spring to push the pen against the paper with sufficient force to allow it to write.

```
>> p560.plot(q)
```

The Puma is drawing the letter 'B', and lifting its pen in between strokes! The approach is quite general and we could easily change the size of the letter, write whole words and sentences, write on an arbitrary plane or use a robot with quite different kinematics. ◄

7.5.2 A Simple Walking Robot [examples/walking.m]

Four legs good, two legs bad!
Snowball the pig, Animal Farm by George Orwell

Our goal is to create a four-legged walking robot. We start by creating a 3-axis robot *arm* that we use as a leg, plan a trajectory for the leg that is suitable for walking, and then instantiate four instances of the leg to create the walking robot.

Kinematics

Kinematically a robot leg is much like a robot arm. For this application a three joint serial-link manipulator is sufficient since the foot has point contact with the ground and orientation is not important. Determining the Denavit-Hartenberg parameters, even for a simple robot like this, is an involved procedure and the zero-angle offsets need to be determined in a separate step. Therefore we will use the procedure introduced in Sect. 7.4.2.

As always we start by defining our coordinate frame. This is shown in Fig. 7.17 along with the robot leg in its zero-angle pose. We have chosen the aerospace coordinate convention which has the *x*-axis forward and the *z*-axis downward, constraining the *y*-axis to point to the right-hand side. The first joint will be hip motion, forward and backward, which is rotation about the *z*-axis or $R_z(q1)$. The second joint is hip motion up and down, which is rotation about the *x*-axis, $R_x(q_2)$. These form a

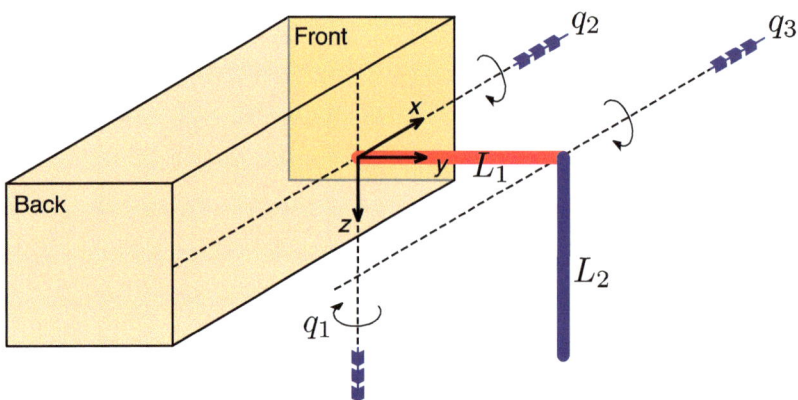

Fig. 7.17.
The coordinate frame and axis rotations for the simple leg. The leg is shown in its zero angle pose

spherical hip joint since the axes of rotation intersect. The knee is translated by L_1 in the y-direction or $T_y(L_1)$. The third joint is knee motion, toward and away from the body, which is $R_x(q_3)$. The foot is translated by L_2 in the z-direction or $T_z(L_2)$. The transform sequence of this robot, from hip to toe, is therefore $R_z(q1)R_x(q_2)T_y(L_1)R_x(q_3)T_z(L_2)$.

Using the technique of Sect. 7.4.2 we write this sequence as the string

```
>> s = 'Rz(q1) Rx(q2) Ty(L1) Rx(q3) Tz(L2)';
```

The string can be automatically manipulated into Denavit-Hartenberg factors

```
>> dh = DHFactor(s)
DH(q1+90, 0, 0, 90).DH(q2, 0, L1, 0).DH(q3-90, 0, -L2, 0)↵
.Rz(+90).Rx(-90).Rz(-90)
```

The last three terms in this factorized sequence is a tool transform

```
   >> dh.tool
ans =
trotz(pi/2)*trotx(-pi/2)*trotz(-pi/2)
```

that changes the orientation of the frame at the foot. However for this problem the foot is simply a point that contacts the ground so we are not concerned about its orientation. The method dh.command generates a string that is the Toolbox command to create a SerialLink object

```
>> dh.command('leg')
ans =
SerialLink([0, 0, 0, pi/2, 0; 0, 0, L1, 0, 0; 0, 0, -L2, 0, 0; ],↵
'name', 'leg', 'base', eye(4,4),↵
'tool', trotz(pi/2)*trotx(-pi/2)*trotz(-pi/2),↵
'offset', [pi/2 0 -pi/2 ])
```

which is input to the MATLAB eval command

```
>> L1 = 0.1; L2 = 0.1;
>> leg = eval( dh.command('leg') )
>> leg
leg =
leg:: 3 axis, RRR, stdDH, slowRNE
+---+-----------+-----------+-----------+-----------+-----------+
| j |     theta |         d |         a |     alpha |    offset |
+---+-----------+-----------+-----------+-----------+-----------+
|  1|        q1|         0|         0|    1.5708|    1.5708|
|  2|        q2|         0|       0.1|         0|         0|
|  3|        q3|         0|      -0.1|         0|   -1.5708|
+---+-----------+-----------+-----------+-----------+-----------+
tool:    t = (0, 0, 0), RPY/zyx = (0, -90, 0) deg
```

after first setting the length of each leg segment to 100 mm in the MATLAB workspace.

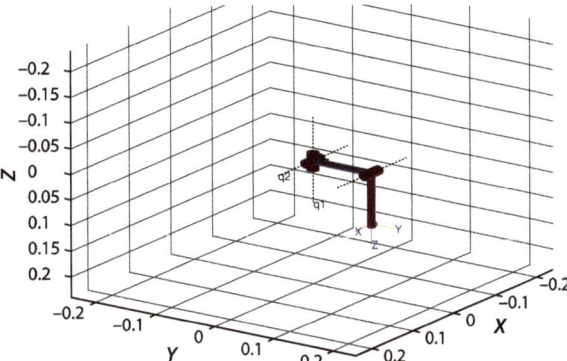

Fig. 7.18.
Robot leg in its zero angle pose.
Note that the z-axis points
downward

We perform a quick sanity check of our robot. For zero joint angles the foot is at

```
>> transl( leg.fkine([0,0,0]) )
ans =
         0    0.1000    0.1000
```

as we designed it. We can visualize the zero-angle pose

```
>> leg.plot([0,0,0], 'nobase', 'noshadow', 'notiles')
>> set(gca, 'Zdir', 'reverse'); view(137,48);
```

which is shown in Fig. 7.18. Now we should test that the other joints result in the expected motion. Increasing q_1

```
>> transl( leg.fkine([0.2,0,0]) )
ans =
   -0.0199    0.0980    0.1000
```

results in motion in the xy-plane, and increasing q_2

```
>> transl( leg.fkine([0,0.2,0]) )
ans =
   -0.0000    0.0781    0.1179
```

results in motion in the yz-plane, as does increasing q_3

```
>> transl( leg.fkine([0,0,0.2]) )
ans =
   -0.0000    0.0801    0.0980
```

We have now created and verified a simple robot leg.

Motion of One Leg

The next step is to define the path that the end-effector of the leg, its foot, will follow. The first consideration is that the end-effector of all feet move backwards at the same speed in the ground plane – propelling the robot's body forward without its feet slipping. Each leg has a limited range of movement so it cannot move backward for very long. At some point we must reset the leg – lift the foot, move it forward and place it on the ground again. The second consideration comes from static stability – the robot must have at least three feet on the ground at all times so each leg must take its turn to reset. This requires that any leg is in contact with the ground for ¾ of the cycle and is resetting for ¼ of the cycle. A consequence of this is that the leg has to move much faster during reset since it has a longer path and less time to do it in.

The required trajectory is defined by the via points

```
>> xf = 50; xb = -xf;  y = 50; zu = 20; zd = 50;
>> path = [xf y zd; xb y zd; xb y zu; xf y zu; xf y zd] * 1e-3;
```

where xf and xb are the forward and backward limits of leg motion in the x-direction (in units of mm), y is the distance of the foot from the body in the y-direction, and zu and zd

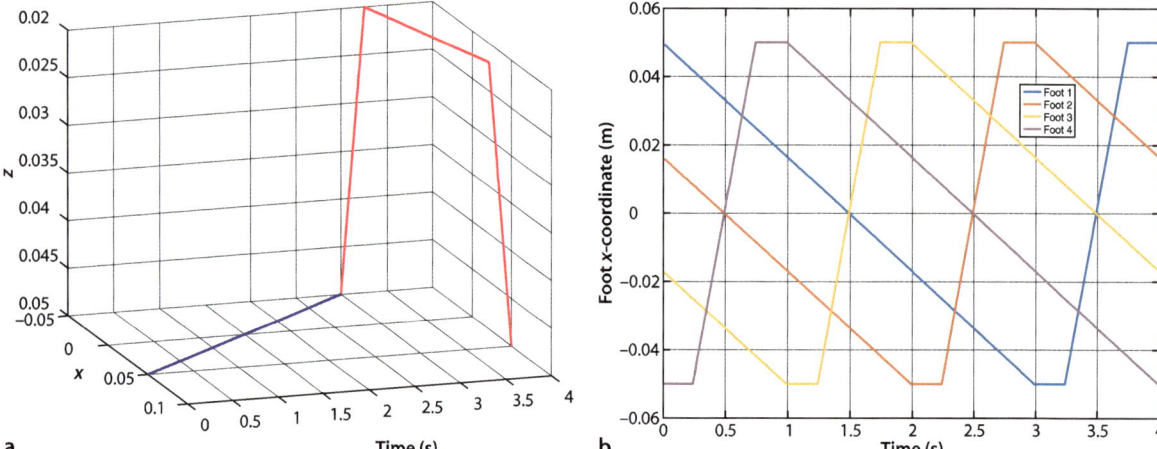

are respectively the height of the foot motion in the z-direction for foot up and foot down. In this case the foot moves from 50 mm forward of the hip to 50 mm behind. When the foot is down, it is 50 mm below the hip and it is raised to 20 mm below the hip during reset. The points in `path` define a complete cycle: the start of the stance phase, the end of stance, top of the leg lift, top of the leg return and the start of stance. This is shown in Fig. 7.19a.

Next we sample the multi-segment path at 100 Hz

```
>> p = mstraj(path, [], [0, 3, 0.25, 0.5, 0.25]', path(1,:), 0.01, 0);
```

In this case we have specified a vector of desired segment times rather than maximum joint velocities.► The final three arguments are the initial leg configuration, the sample interval and the acceleration time. This trajectory has a total time of 4 s and therefore comprises 400 points.

We apply inverse kinematics to determine the joint angle trajectories required for the foot to follow the computed Cartesian trajectory. This robot is under-actuated so we use the generalized inverse kinematics `ikine` and set the mask so as to solve only for end-effector translation

```
>> qcycle = leg.ikine( SE3(p), 'mask', [1 1 1 0 0 0] );
```

We can view the motion of the leg in animation

```
>> leg.plot(qcycle, 'loop')
```

to verify that it does what we expect: slow motion along the ground, then a rapid lift, forward motion and foot placement. The `'loop'` option displays the trajectory in an endless loop and you need to type control-C to stop it.

Motion of Four Legs

Our robot has width and length

```
>> W = 0.1; L = 0.2;
```

We create multiple instances of the leg by cloning the `leg` object we created earlier, and providing different base transforms so as to attach the legs to different points on the body

```
>> legs(1) = SerialLink(leg, 'name', 'leg1');
>> legs(2) = SerialLink(leg, 'name', 'leg2', 'base', SE3(-L, 0, 0));
>> legs(3) = SerialLink(leg, 'name', 'leg3', 'base', SE3(-L, -W, 0) ↵
   *SE3.Rz(pi));
>> legs(4) = SerialLink(leg, 'name', 'leg4', 'base', SE3(0, -W, 0) ↵
   *SE3.Rz(pi));
```

Fig. 7.19. a Trajectory taken by a single foot. Recall from Fig. 7.17 that the z-axis is downward. The red segments are the leg reset. **b** The x-direction motion of each leg (offset vertically) to show the gait. The leg reset is the period of high x-direction velocity

This way we can ensure that the reset takes exactly one quarter of the cycle.

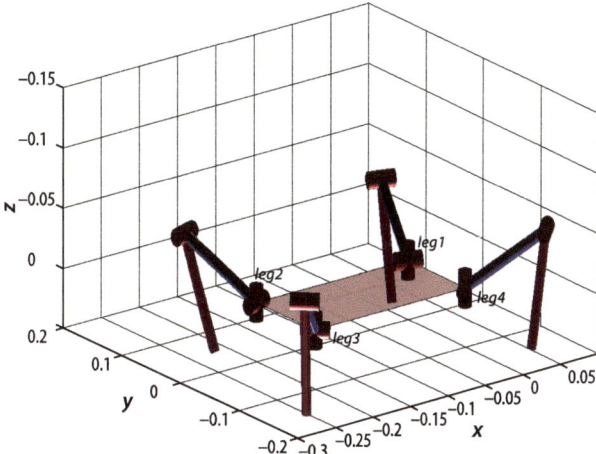

Fig. 7.20.
The walking robot

The result is a vector of `SerialLink` objects. Note that legs 3 and 4, on the left-hand side of the body have been rotated about the z-axis so that they point away from the body.

As mentioned earlier each leg must take its turn to reset. Since the trajectory is a cycle, we achieve this by having each leg run the trajectory with a phase shift equal to one quarter of the total cycle time. Since the total cycle has 400 points, each leg's trajectory is offset by 100, and we use modulo arithmetic to index into the cyclic gait for each leg. The result is the gait pattern shown in Fig. 7.19b.

The core of the walking program is

```
clf; k = 1;
while 1
    legs(1).plot( gait(qcycle, k, 0,    false) );
    if k == 1, hold on; end
    legs(2).plot( gait(qcycle, k, 100, false) );
    legs(3).plot( gait(qcycle, k, 200, true) );
    legs(4).plot( gait(qcycle, k, 300, true) );
    drawnow
    k = k+1;
end
```

where the function

```
gait(q, k, ph, flip)
```

returns the k+ph[th] element of q with modulo arithmetic that considers q as a cycle. The argument `flip` reverses the sign of the joint 1 motion for the legs on the left-hand side of the robot. A snapshot from the simulation is shown in Fig. 7.20. The entire implementation, with some additional refinement, is in the file `examples/walking.m` and detailed explanation is provided by the comments.

7.6 Wrapping Up

In this chapter we have learned how to determine the forward and inverse kinematics of a serial-link manipulator arm. Forward kinematics involves compounding the relative poses due to each joint and link, giving the pose of the robot's end-effector relative to its base. Commonly the joint and link structure is expressed in terms of Denavit-Hartenberg parameters for each link. Inverse kinematics is the problem of determining the joint coordinates given the end-effector pose. For simple robots, or those with six joints and a spherical wrist we can compute the inverse kinematics using an analytic solution. This inverse is not unique and the robot may have several joint configurations that result in the same end-effector pose.

For robots which do not have six joints and a spherical wrist we can use an iterative numerical approach to solving the inverse kinematics. We showed how this could be applied to an under-actuated 4-joint SCARA robot and a redundant 7-link robot. We also touched briefly on the topic of singularities which are due to the alignment of joint axes.

We also learned about creating paths to move the end-effector smoothly between poses. Joint-space paths are simple to compute but in general do not result in straight line paths in Cartesian space which may be problematic for some applications. Straight line paths in Cartesian space can be generated but singularities in the workspace may lead to very high joint rates.

Further Reading

Serial-link manipulator kinematics are covered in all the standard robotics textbooks such as the Robotics Handbook (Siciliano and Khatib 2016), Siciliano et al. (2009), Spong et al. (2006) and Paul (1981). Craig's text (2005) is also an excellent introduction to robot kinematics and uses the modified Denavit-Hartenberg notation, and the examples in the third edition are based on an older version of the Robotics Toolbox. Lynch and Park (2017) and Murray et al. (1994) cover the product of exponential approach. An emerging alternative to Denavit-Hartenberg notation is URDF (unified robot description format) which is described at http://wiki.ros.org/urdf.

Siciliano et al. (2009) provide a very clear description of the process of assigning Denavit-Hartenberg parameters to an arbitrary robot. The alternative approach described here based on symbolic factorization was described in detail by Corke (2007). The definitive values for the parameters of the Puma 560 robot are described in the paper by Corke and Armstrong-Hélouvry (1995).

Robotic walking is a huge field in its own right and the example given here is very simplistic. Machines have been demonstrated with complex gaits such as running and galloping that rely on dynamic rather than static balance. A good introduction to legged robots is given in the Robotics Handbook (Siciliano and Khatib 2016, § 17). Robotic hands, grasping and manipulation is another large topic which we have not covered – there is a good introduction in the Robotics Handbook (Siciliano and Khatib 2016, §37, 38).

Parallel-link manipulators were mentioned only briefly on page 190 and have advantages such as increased actuation force and stiffness (since the actuators form a truss-like structure). For this class of mechanism the inverse kinematics is usually closed-form and it is the forward kinematics that requires numerical solution. Useful starting points for this class of robots are the handbook (Siciliano and Khatib 2016, §18), a brief section in Siciliano et al. (2009) and Merlet (2006).

Closed-form inverse kinematic solutions can be derived symbolically by writing down a number of kinematic relationships and solving for the joint angles, as described in Paul (1981). Software packages to automatically generate the forward and inverse kinematics for a given robot have been developed and these include Robotica (Nethery and Spong 1994) which is now obsolete, and SYMORO (Khalil and Creusot 1997) which is now available as open-source.

Historical. The original work by Denavit and Hartenberg was their 1955 paper (Denavit and Hartenberg 1955) and their textbook (Hartenberg and Denavit 1964). The book has an introduction to the field of kinematics and its history but is currently out of print, although a version can be found online. The first full description of the kinematics of a six-link arm with a spherical wrist was by Paul and Zhang (1986).

MATLAB and Toolbox Notes

The workhorse of the Toolbox is the `SerialLink` class which has considerable functionality and very many methods – we will use it extensively for the remainder of Part III. The classes `ETS2` and `ETS3` used in the early parts of this chapter were designed to illustrate principles as concisely as possible and have limited functionality, but the names of their methods are the same as their equivalents in the `SerialLink` class.

The `plot` method draws a *stick figure* robot and needs only Denavit-Hartenberg parameters. However the joints depicted are associated with the link frames and don't necessarily correspond to physical joints on the robot, but that is a limitation of the Denavit-Hartenberg parameters. A small number of robots have more realistic 3-dimensional models defined by STL files and these can be displayed using the `plot3d`. The models shipped with the Toolbox were created by Arturo Gil and are also shipped with his ARTE Toolbox.

The numerical inverse kinematics method `ikine` can handle over- and underactuated robot arms, but does not handle joint coordinate limits which can be set in the `SerialLink` object. The alternative inverse kinematic method `ikcon` respects joint limits but requires the MATLAB Optimization Toolbox™.

The MATLAB Robotics System Toolbox™ provides a `RigidBodyTree` class to represent a serial-link manipulator, and this also supports branched mechanisms such as a humanoid robot. It also provides a general `InverseKinematics` class to solve inverse kinematic problems and can handle joint limits.

Exercises

1. Forward kinematics for planar robot from Sect. 7.1.1.
   a) For the 2-joint robot use the `teach` method to determine the two sets of joint angles that will position the end-effector at (0.5, 0.5).
   b) Experiment with the three different models in Fig. 7.2 using the `fkine` and `teach` methods.
   c) Vary the models: adjust the link lengths, create links with a translation in the *y*-direction, or create links with a translation in the *x*- and *y*-direction.
2. Experiment with the `teach` method for the Puma 560 robot.
3. Inverse kinematics for the 2-link robot on page 204.
   a) Compute forward and inverse kinematics with a_1 and a_2 as symbolic rather than numeric values.
   b) What happens to the solution when a point is out of reach?
   c) Most end-effector positions can be reached by two different sets of joint angles. What points can be reached by only one set?
4. Compare the solutions generated by `ikine6s` and `ikine` for the Puma 560 robot at different poses. Is there any difference in accuracy? How much slower is `ikine`?
5. For the Puma 560 at configuration `qn` demonstrate a configuration change from elbow up to elbow down.
6. For a Puma 560 robot investigate the errors in end-effector pose due to manufacturing errors.
   a) Make link 2 longer by 0.5 mm. For 100 random joint configurations what is the mean and maximum error in the components of end-effector pose?
   b) Introduce an error of 0.1 degrees in the joint 2 angle and repeat the analysis above.
7. Investigate the redundant robot models `mdl_hyper2d` and `mdl_hyper3d`. Manually control them using the `teach` method, compute forward kinematics and numerical inverse kinematics.

8. If you have the MATLAB Optimization Toolbox™ experiment with the `ikcon` method which solves inverse kinematics for the case where the joint coordinates have limits (modeling mechanical end stops). Joint limits are set with the `qlim` property of the `Link` class.

9. Drawing a 'B' (page 218)
   a) Change the size of the letter.
   b) Write a word or sentence.
   c) Write on a vertical plane.
   d) Write on an inclined plane.
   e) Change the robot from a Puma 560 to the Fanuc 10L.
   f) Write on a sphere. Hint: Write on a tangent plane, then project points onto the sphere's surface.
   g) This writing task does not require 6DOF since the rotation of the pen about its axis is not important. Remove the final link from the Puma 560 robot model and repeat the exercise.

10. Walking robot (page 219)
   a) Shorten the reset trajectory by reducing the leg lift during reset.
   b) Increase the stride of the legs.
   c) Figure out how to steer the robot by changing the stride length on one side of the body.
   d) Change the gait so the robot moves sideways like a crab.
   e) Add another pair of legs. Change the gait to reset two legs or three legs at a time.
   f) Currently in the simulation the legs move but the body does not move forward. Modify the simulation so the body moves.
   g) A robot hand comprises a number of fingers, each of which is a small serial-link manipulator. Create a model of a hand with 2, 3 or 5 fingers and animate the finger motion.

11. Create a simulation with two robot arms next to each other, whose end-effectors are holding a basketball at diametrically opposite points in the horizontal plane. Write code to move the robots so as to rotate the ball about the vertical axis.

12. Create STL files to represent your own robot and integrate them into the Toolbox. Check out the code in `SerialLink.plot3d`.

8

Manipulator Velocity

A robot's end-effector moves in Cartesian space with a translational and rotational velocity – a spatial velocity. However that velocity is a consequence of the velocities of the individual robot joints. In this chapter we introduce the relationship between the velocity of the joints and the spatial velocity of the end-effector.

The 3-dimensional end-effector pose $\xi_E \in \mathbf{SE}(3)$ has a rate of change which is represented by a 6-vector known as a spatial velocity that was introduced in Sect. 3.1. The joint rate of change and the end-effector velocity are related by the manipulator Jacobian matrix which is a function of manipulator configuration.

Section 8.1 uses a simple planar robot to introduce the manipulator Jacobian and then extends this concept to more general robots. Section 8.2 discusses the numerical properties of the Jacobian matrix which are shown to provide insight into the dexterity of the manipulator – the directions in which it can move easily and those in which it cannot. In Sect. 8.3 we use the inverse Jacobian to generate Cartesian paths without requiring inverse kinematics, and this can be applied to over- and under-actuated robots which are discussed in Sect. 8.4. Section 8.5 demonstrates how the Jacobian transpose is used to transform forces from the end-effector to the joints and between coordinate frames. Finally, in Sect. 8.6 the numeric inverse kinematic solution, used in the previous chapter, is introduced and its dependence on the Jacobian matrix is fully described.

8.1 Manipulator Jacobian

In the last chapter we discussed the relationship between joint coordinates and end-effector pose – the manipulator kinematics. Now we investigate the relationship between the rate of change of these quantities – between joint *velocity* and *velocity* of the end-effector. This is called the velocity or differential kinematics of the manipulator.

8.1.1 Jacobian in the World Coordinate Frame

We illustrate the basics with our now familiar 2-dimensional example, see Fig. 8.1, this time defined using Denavit-Hartenberg notation

```
>> mdl_planar2_sym
>> p2
two link:: 2 axis, RR, stdDH
+---+-----------+-----------+-----------+-----------+-----------+
| j |   theta   |      d    |      a    |   alpha   |   offset  |
+---+-----------+-----------+-----------+-----------+-----------+
| 1 |        q1 |        0  |       a1  |        0  |        0  |
| 2 |        q2 |        0  |       a2  |        0  |        0  |
+---+-----------+-----------+-----------+-----------+-----------+
```

© Springer Nature Switzerland AG 2022
P. Corke, *Robotics and Control*, Springer Tracts in Advanced Robotics 141,
https://doi.org/10.1007/978-3-030-79179-7_8

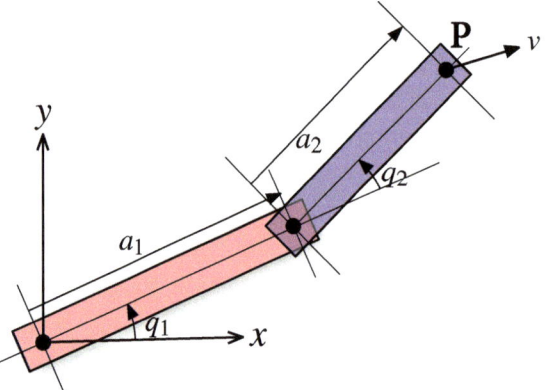

Fig. 8.1.
Two-link robot showing the
end-effector position $p = (x, y)$
and the Cartesian velocity vector
$\nu = \mathrm{d}p\,/\,\mathrm{d}t$

and define two real-valued symbolic variables to represent the joint angles

```
>> syms q1 q2 real
```

and then compute the forward kinematics using these

```
>> TE = p2.fkine( [q1 q2] );
```

The position of the end-effector $p = (x, y) \in \mathbb{R}^2$ is ▶

The Toolbox considers robot pose in 3-dimensions using SE(3). This robot operates in a plane, a subset of SE(3), so we select $p = (x, y)$.

```
>> p = TE.t;   p = p(1:2)
p =
  a2*cos(q1 + q2) + a1*cos(q1)
  a2*sin(q1 + q2) + a1*sin(q1)
```

and we compute the derivative of p with respect to the joints variables q. Since p and q are both vectors the derivative

$$\frac{\mathrm{d}p}{\mathrm{d}q} = J(q) \tag{8.1}$$

will be a matrix – a Jacobian matrix

```
>> J = jacobian(p, [q1 q2])
J =
[ - a2*sin(q1 + q2) - a1*sin(q1), -a2*sin(q1 + q2)]
[   a2*cos(q1 + q2) + a1*cos(q1),  a2*cos(q1 + q2)]
```

which is typically denoted by the symbol J and in this case is 2×2.

To determine the relationship between joint *velocity* and end-effector *velocity* we rearrange Eq. 8.1 as

$$\mathrm{d}p = J(q)\mathrm{d}q$$

and divide through by $\mathrm{d}t$ to obtain

$$\frac{\mathrm{d}p}{\mathrm{d}t} = J(q)\frac{\mathrm{d}q}{\mathrm{d}t}$$
$$\dot{p} = J(q)\dot{q}$$

The Jacobian matrix maps velocity from the joint coordinate or configuration space to the end-effector's Cartesian coordinate space and is itself a function of the joint coordinates.

More generally we write the forward kinematics in functional form, Eq. 7.4, as

$$^0\xi = \mathcal{K}(q)$$

A Jacobian is the matrix equivalent of the derivative – the derivative of a vector-valued function of a vector with respect to a vector. If $y = f(x)$ and $x \in \mathbb{R}^n$ and $y \in \mathbb{R}^m$ then the Jacobian is the $m \times n$ matrix

$$J = \frac{\partial f}{\partial x} = \begin{pmatrix} \frac{\partial y_1}{\partial x_1} & \cdots & \frac{\partial y_1}{\partial x_n} \\ \vdots & \ddots & \vdots \\ \frac{\partial y_m}{\partial x_1} & \cdots & \frac{\partial y_m}{\partial x_n} \end{pmatrix}$$

The Jacobian is named after Carl Jacobi, and more details are given in Appendix E.

and taking the derivative we write

$$^{0}v = {}^{0}J(q)\dot{q} \tag{8.2}$$

where $^{0}\nu = (v_x, v_y, v_z, \omega_x, \omega_y, \omega_z) \in \mathbb{R}^6$ is the spatial velocity, as discussed in Sect. 3.1.1, of the end-effector in the world frame and comprises translational and rotational velocity components. The matrix $^{0}J(q) \in \mathbb{R}^{6 \times N}$ is the manipulator Jacobian or the geometric Jacobian. This relationship is sometimes referred to as the instantaneous forward kinematics.

For a realistic 3-dimensional robot this Jacobian matrix can be numerically computed by the `jacob0` method of the `SerialLink` object, based on its Denavit-Hartenberg parameters. For the Puma robot in the pose shown in Fig. 8.2 the Jacobian is

```
>> J = p560.jacob0(qn)
J =
    0.1501    0.0144    0.3197         0         0         0
    0.5963    0.0000    0.0000         0         0         0
         0    0.5963    0.2910         0         0         0
         0   -0.0000   -0.0000    0.7071   -0.0000   -0.0000
         0   -1.0000   -1.0000   -0.0000   -1.0000   -0.0000
    1.0000    0.0000    0.0000   -0.7071    0.0000   -1.0000
```

τ is the task space of the robot, typically $\tau \subset$ **SE**(3), and $\mathcal{C} \subset \mathbb{R}^N$ is the configuration or joint space of the robot where N is the number of robot joints.

and is a matrix with dimensions dim $\mathcal{T} \times$ dim $\mathcal{C}$ – in this case 6×6◀. Each row corresponds to a Cartesian degree of freedom. Each column corresponds to a joint – it is the end-effector spatial velocity created by unit velocity of the corresponding joint. In this configuration, motion of joint 1, the first column, causes motion in the world x- and y-directions and rotation about the z-axis. Motion of joints 2 and 3 cause motion in the world x- and z-directions and negative rotation about the y-axis.

Physical insight comes from Fig. 8.2 which shows the joint axes in space. Alternatively you could use the `teach` method

```
>> p560.teach(qn)
```

and jog the various joint angles and observe the change in end-effector pose.

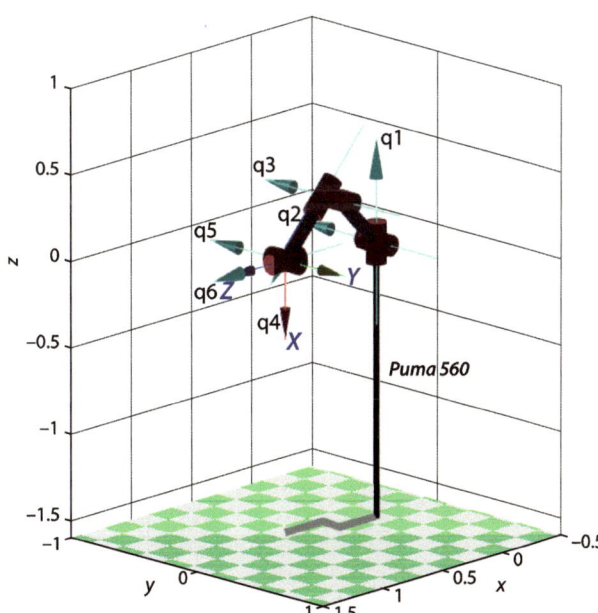

Fig. 8.2.
Puma robot in its nominal pose qn. The end-effector z-axis points in the world x-direction, and the x-axis points downward

Carl Gustav Jacob Jacobi (1804–1851) was a Prussian mathematician. He obtained a Doctor of Philosophy degree from Berlin University in 1825. In 1827 he was appointed professor of mathematics at Königsberg University and held this position until 1842 when he suffered a breakdown from overwork.

Jacobi wrote a classic treatise on elliptic functions in 1829 and also described the derivative of m functions of n variables which bears his name. See Appendix E. He was elected a foreign member of the Royal Swedish Academy of Sciences in 1836. He is buried in Cementary I of the Trinity Church (Dreifaltigkeitskirche) in Berlin.

The 3×3 block of zeros in the top right indicates that motion of the wrist joints have no effect on the end-effector translational motion – this is a consequence of the spherical wrist and the default zero-length tool.

8.1.2   Jacobian in the End-Effector Coordinate Frame

The Jacobian computed by the method `jacob0` maps joint velocity to the end-effector spatial velocity expressed in the *world coordinate* frame. To obtain the spatial velocity in the end-effector coordinate frame we introduce the velocity transformation Eq. 3.4 from the world frame to the end-effector frame which is a function of the end-effector pose

$$
{}^E v = {}^E J_0 \left({}^E \xi_0 \right) {}^0 J(q) \dot{q} = \begin{pmatrix} {}^E R_0 & 0_{3\times3} \\ 0_{3\times3} & {}^E R_0 \end{pmatrix} {}^0 J(q)\, \dot{q} = {}^E J(q) \dot{q}
$$

which results in a new Jacobian for end-effector velocity.▶

In the Toolbox this Jacobian is computed by the method `jacobe` and for the Puma robot at the pose used above is

> For historical reasons the Toolbox implementation computes the end-effector Jacobian directly and applies a velocity transform for the world frame Jacobian.

```
>> p560.jacobe(qn)
ans =
   -0.0000   -0.5963   -0.2910         0         0         0
    0.5963    0.0000    0.0000         0         0         0
    0.1500    0.0144    0.3197         0         0         0
   -1.0000         0         0    0.7071         0         0
   -0.0000   -1.0000   -1.0000   -0.0000   -1.0000         0
   -0.0000    0.0000    0.0000    0.7071    0.0000    1.0000
```

8.1.3   Analytical Jacobian

In Eq. 8.2 the spatial velocity was expressed in terms of translational and angular velocity vectors, however angular velocity is not a very intuitive concept. For some applications it can be more intuitive to consider the rotational velocity in terms of rates of change of roll-pitch-yaw angles or Euler angles. Analytical Jacobians are those where the rotational velocity is expressed in a representation other than angular velocity, commonly in terms of triple-angle rates.

Consider the case of XYZ roll-pitch-yaw angles $\varGamma = (\theta_r, \theta_p, \theta_y)$ for which the rotation matrix is

$$
\begin{aligned}
R &= R_x\left(\theta_y\right) R_y\left(\theta_p\right) R_z\left(\theta_r\right) \\[6pt]
&= \begin{pmatrix}
c\theta_p c\theta_r & -c\theta_p s\theta_r & s\theta_p \\
c\theta_y s\theta_r + c\theta_r s\theta_p s\theta_y & -s\theta_p s\theta_y s\theta_r + c\theta_y c\theta_r & -c\theta_p s\theta_y \\
s\theta_y s\theta_r - c\theta_y c\theta_r s\theta_p & c\theta_y s\theta_p s\theta_r + c\theta_r s\theta_y & c\theta_p c\theta_y
\end{pmatrix}
\end{aligned}
$$

where we use the shorthand $c\theta$ and $s\theta$ to mean $\cos\theta$ and $\sin\theta$ respectively. With some effort we can write the derivative $\dot{R}$ and recalling Eq. 3.1

$$\dot{R} = [\boldsymbol{\omega}]_\times R$$

we can solve for ω in terms of roll-pitch-yaw angles and their rates to obtain

$$\begin{pmatrix} \omega_x \\ \omega_y \\ \omega_z \end{pmatrix} = \begin{pmatrix} s\theta_p\dot{\theta}_r + \dot{\theta}_y \\ -c\theta_ps\theta_y\dot{\theta}_r + c\theta_y\dot{\theta}_p \\ c\theta_pc\theta_y\dot{\theta}_r + s\theta_y\dot{\theta}_p \end{pmatrix}$$

which can be factored as

$$\omega = \begin{pmatrix} s\theta_p & 0 & 1 \\ -c\theta_ps\theta_y & c\theta_y & 0 \\ c\theta_pc\theta_y & s\theta_y & 0 \end{pmatrix} \begin{pmatrix} \dot{\theta}_r \\ \dot{\theta}_p \\ \dot{\theta}_y \end{pmatrix}$$

and written concisely as

$$\omega = A(\boldsymbol{\Gamma})\dot{\boldsymbol{\Gamma}}$$

This matrix A is itself a Jacobian that maps XYZ roll-pitch-yaw angle rates to angular velocity. It can be computed by the Toolbox function

```
>> rpy2jac(0.1, 0.2, 0.3)
ans =
      0.1987           0    1.0000
     -0.2896      0.9553           0
      0.9363      0.2955           0
```

where the arguments are the roll, pitch and yaw angles. The analytical Jacobian is

$$J_a(\boldsymbol{q}) = \begin{pmatrix} \boldsymbol{I}_{3\times3} & \boldsymbol{0}_{3\times3} \\ \boldsymbol{0}_{3\times3} & A^{-1}(\boldsymbol{\Gamma}) \end{pmatrix} J(\boldsymbol{q})$$

provided that A is not singular. A is singular when $\cos\phi = 0$ or pitch angle $\phi = \pm\frac{\pi}{2}$ and is referred to as a representational singularity. A similar approach can be taken for Euler angles using the corresponding function `eul2jac`.

The analytical Jacobian can be computed by passing an extra argument to the Jacobian function `jacob0`, for example

```
>> p560.jacob0(qn, 'eul');
```

to specify the Euler angle analytical form.

Another useful analytical Jacobian expresses angular rates as the rate of change of exponential coordinates $\boldsymbol{s} = \hat{\boldsymbol{v}}\theta \in \mathbf{so}(3)$

$$\omega = A(\boldsymbol{s})\dot{\boldsymbol{s}}$$

where

$$A(\boldsymbol{s}) = \boldsymbol{I} - \frac{1-\cos\theta}{\theta}[\hat{\boldsymbol{v}}]_\times + \frac{\theta-\sin\theta}{\theta}[\hat{\boldsymbol{v}}]_\times^2$$

Implemented by the Toolbox functions `trlog` and `tr2rotvec` or the SE3 method `torotvec`.

and $\hat{\boldsymbol{v}}$ and θ can be determined from the end-effector rotation matrix via the matrix logarithm. ◀

8.2   Jacobian Condition and Manipulability

We have discussed how the Jacobian matrix maps joint rates to end-effector Cartesian velocity but the inverse problem has strong practical use – what joint velocities are needed to achieve a required end-effector Cartesian velocity? We can invert Eq. 8.2 and write

$$\dot{q} = J(q)^{-1}\nu \tag{8.3}$$

provided that J is square and nonsingular. The Jacobian is a dim $\mathcal{T}$ × dim $\mathcal{C}$ matrix so in order to achieve a square Jacobian matrix a robot operating in the task space $\mathcal{T} \subset \mathbf{SE}(3)$, which has 6 spatial degrees-of-freedom, requires a robot with 6 joints.

8.2.1   Jacobian Singularities

A robot configuration q at which $\det(J(q)) = 0$ is described as singular or degenerate. Singularities occur when the robot is at maximum reach or when one or more axes become aligned resulting in the loss of degrees of freedom – the gimbal lock problem again.

For example at the Puma's *ready* pose the Jacobian

```
>> J = p560.jacob0(qr)
J =
    0.1500   -0.8636   -0.4318        0        0        0
    0.0203    0.0000    0.0000        0        0        0
         0    0.0203    0.0203        0        0        0
         0         0         0        0        0        0
         0   -1.0000   -1.0000        0  -1.0000        0
    1.0000    0.0000    0.0000   1.0000   0.0000   1.0000
```

is singular

```
>> det(J)
ans =
     0
```

Digging a little deeper we see that the Jacobian rank is only

```
>> rank(J)
ans =
     5
```

compared to a maximum of six for a 6 × 6 matrix. The rank deficiency of one means that one column is equal to a linear combination of other columns. Looking at the Jacobian it is clear that columns 4 and 6 are identical meaning that two of the wrist joints (joints 4 and 6) are aligned. This leads to the loss of one degree of freedom since motion of these joints results in the same Cartesian velocity, leaving motion in one Cartesian direction unaccounted for.► The function `jsingu` performs this analysis automatically, for example

```
>> jsingu(J)
1 linearly dependent joints:
  q6 depends on: q4
```

indicating velocity of q_6 can be expressed completely in terms of the velocity of q_4.

However if the robot is close to, but not actually at, a singularity we encounter problems where some Cartesian end-effector velocities require very high joint rates► – at the singularity those rates will go to infinity. We can illustrate this by choosing a configuration slightly away from `qr` which we just showed was singular. We set q_5 to a small but nonzero value of 5 deg

```
>> qns = qr; qns(5) = 5 * pi/180
qns =
         0    1.5708   -1.5708        0    0.0873        0
```

For the Puma 560 robot arm joints 4 and 6 are the only ones that can become aligned and lead to singularity. The offset distances, d_j and a_j, between links prevents other axes becoming aligned.

We observed this effect in Fig. 7.12.

and the Jacobian is now

```
>> J=p560.jacob0(qns);
```

To achieve relatively slow end-effector motion of 0.1 m s^{-1} in the z-direction requires

```
>> qd = inv(J)*[0 0 0.1 0 0 0]' ;
>> qd'
ans =   -0.0000   -4.9261    9.8522    0.0000   -4.9261         0
```

very high-speed motion of the shoulder and elbow – the elbow would have to move at 9.85 rad s^{-1} or nearly 600 deg s^{-1}. The reason is that although the robot is no longer at a singularity, the determinant of the Jacobian is still very small

```
>> det(J)
ans =
  -1.5509e-05
```

Alternatively we can say that its condition number is very high

```
>> cond(J)
ans =
  235.2498
```

and the Jacobian is *poorly conditioned*.

However for some motions, such as rotation in this case, the poor condition of the Jacobian is not problematic. If we wished to rotate the tool about the y-axis then

```
>> qd = inv(J)*[0 0 0 0 0.2 0]';
>> qd'
ans =    0.0000   -0.0000         0    0.0000   -0.2000         0
```

the required joint rates are very modest.

This particular joint configuration is therefore good for certain motions but poor for others.

8.2.2   Manipulability

Consider the set of generalized joint velocities with a unit norm

$$\dot{q}^T \dot{q} = 1$$

which lie on the surface of a hypersphere in the N-dimensional joint velocity space. Substituting Eq. 8.3 we can write

$$\nu^T \left(J(q)J(q)^T \right)^{-1} \nu = 1 \tag{8.4}$$

which is the equation of points on the surface of an ellipsoid within the dim $\mathcal{T}$-dimensional end-effector velocity space. If this ellipsoid is close to spherical, that is, its radii are of the same order of magnitude then all is well – the end-effector can achieve arbitrary Cartesian velocity. However if one or more radii are very small this indicates that the end-effector cannot achieve velocity in the directions corresponding to those small radii.

We will load the numerical, rather than symbolic model, for the planar robot arm of Fig. 8.1

```
>> mdl_planar2
```

which allows us to plot the velocity ellipse for an arbitrary pose

```
>> p2.vellipse([30 40], 'deg')
```

We can also interactively explore how its shape changes with configuration by

```
>> p2.teach('callback', @(r,q) r.vellipse(q), 'view', 'top')
```

which is shown in Fig. 8.3.

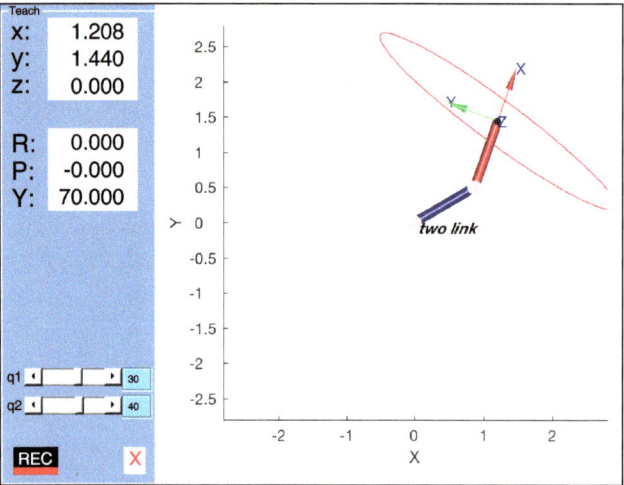

Fig. 8.3.
Two-link robot with overlaid velocity ellipse

For a robot with a task space $\mathcal{T} \subset \mathbf{SE}(3)$ Eq. 8.4 describes a 6-dimensional ellipsoid which is impossible to visualize. However we can extract that part of the Jacobian relating to *translational* velocity▶ in the world frame

> Since we can only plot three dimensions.

```
>> J = p560.jacob0(qns);
>> J = J(1:3, :);
```

and plot the corresponding velocity ellipsoid

```
>> plot_ellipse(J*J')
```

which is shown in Fig. 8.4a. The Toolbox provides a shorthand for this

```
>> p560.vellipse(qns, 'trans');
```

We see that the end-effector can achieve higher velocity in the y- and z-directions than in the x-direction. Ellipses and ellipsoids are discussed in more detail in Sect. C.1.4.

The *rotational* velocity ellipsoid for the near singular case

```
>> p560.vellipse(qns, 'rot')
```

is shown in Fig. 8.4b and is an elliptical plate with almost zero thickness.▶ This indicates an inability to rotate about the direction corresponding to the small radius, which in this case is rotation about the x-axis. This is the degree of freedom that was lost – both joints 4 and 6 provide rotation about the world z-axis, joint 5 provides provides rotation about the world y-axis, but none allow rotation about the world x-axis.

> This is much easier to see if you change the viewpoint interactively.

The shape of the ellipsoid describes how *well-conditioned* the manipulator is for making certain motions. Manipulability is a succinct scalar measure that describes how spherical the ellipsoid is, for instance the ratio of the smallest to the largest radius.▶ The Toolbox method `maniplty` computes Yoshikawa's manipulability measure

> The radii are the square roots of the eigenvalues of the $J(q)J(q)^T$ as discussed in Sect. C.1.4.

$$m = \sqrt{\det\left(\boldsymbol{JJ}^T\right)}$$

which is proportional to the volume of the ellipsoid. For example

```
>> m = p560.maniplty(qr)
m =
    0
```

indicates a singularity. If the method is called with no output arguments it displays the volume of the translational and rotational velocity ellipsoids

```
>> p560.maniplty(qr)
Manipulability: translation 0.00017794, rotation 0
```

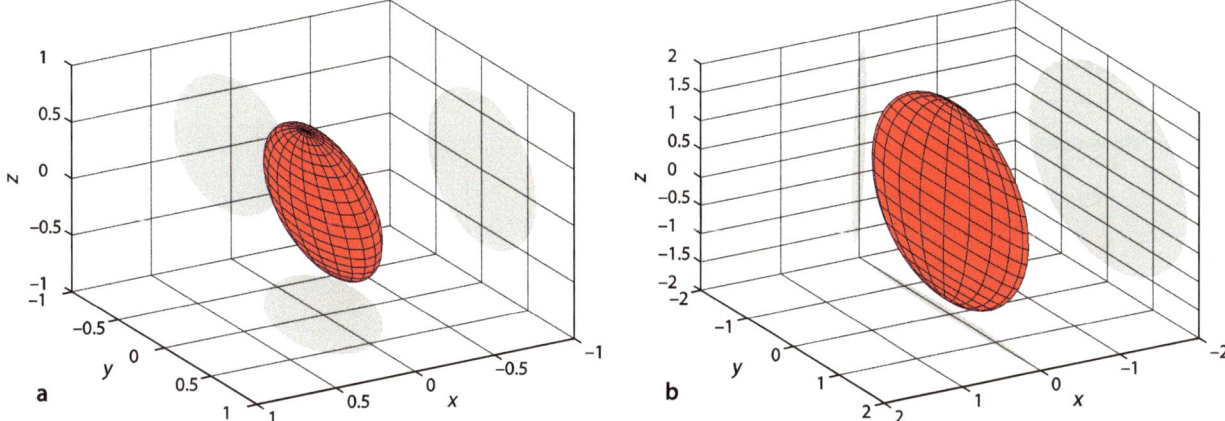

Fig. 8.4. End-effector velocity ellipsoids. **a** Translational velocity ellipsoid for the nominal pose (m s^{-1}); **b** rotational velocity ellipsoid for a near singular pose (rad s^{-1}), the ellipsoid is an elliptical plate

The manipulability measure combines translational and rotational velocity information which have different units. The options `'trans'` and `'rot'` can be used to compute manipulability on just the translational or rotational velocity respectively.

which indicates very poor manipulability for translation and zero for rotation. At the nominal pose the manipulability is higher◄

```
>> p560.maniplty(qn)
Manipulability: translation 0.111181, rotation 2.44949
```

In practice we find that the seemingly large workspace of a robot is greatly reduced by joint limits, self collision, singularities and regions of reduced manipulability. The manipulability measure discussed here is based only on the kinematics of the mechanism. The fact that it is easier to move a small wrist joint than the larger waist joint suggests that mass and inertia should be taken into account and such manipulability measures are discussed in Sect. 9.2.7.

8.3   Resolved-Rate Motion Control

Resolved-rate motion control is a simple and elegant algorithm to generate straight line motion by exploiting Eq. 8.3

$$\dot{q} = J(q)^{-1}\nu$$

to map or *resolve* desired Cartesian velocity to joint velocity without explicitly requiring inverse kinematics as we used earlier. For now we will assume that the Jacobian is square (6×6) and nonsingular but we will relax these constraints later.

The motion control scheme is typically implemented in discrete-time form as

$$\dot{q}^*\langle k\rangle = J(q\langle k\rangle)^{-1}\nu^* \tag{8.5}$$

$$q^*\langle k+1\rangle \leftarrow q\langle k\rangle + \delta_t \dot{q}^*\langle k\rangle$$

where δ_t is the sample interval. The first equation computes the required joint velocity as a function of the current joint configuration and the desired end-effector velocity ν^*. The second performs forward rectangular integration to give the desired joint angles for the next time step, $q^*\langle k+1\rangle$.

An example of the algorithm is implemented by the Simulink® model

```
>> sl_rrmc
```

In this model we assume that the robot is perfect, that is, the actual joint angles are equal to the desired joint angles q^*. The issue of tracking error is discussed in Sect. 9.1.7.

shown in Fig. 8.5. The Cartesian velocity is a constant 0.05 m s^{-1} in the y-direction. The `Jacobian` block has as its input the current manipulator joint angles and outputs a 6×6 Jacobian matrix. This is inverted and multiplied by the desired velocity to form the desired joint rates. The robot is modeled by a discrete-time integrator – an ideal velocity controller.◄

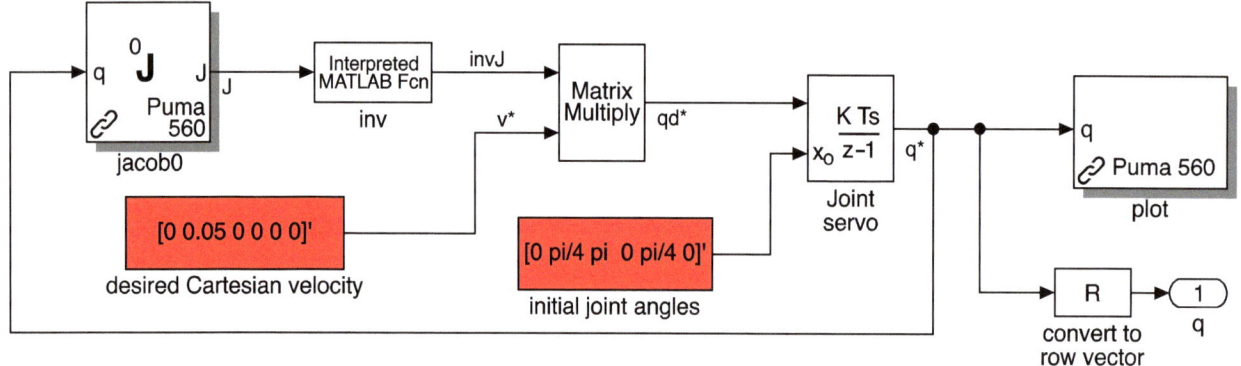

Running the simulation

```
>> r = sim('sl_rrmc');
```

Fig. 8.5. The Simulink® model
sl_rrmc for resolved-rate motion control for constant end-effector velocity

we see an animation of the manipulator end-effector moving at constant velocity in Cartesian space. Simulation results are returned in the simulation object `r` from which we extract time and joint coordinates

```
>> t = r.find('tout');
>> q = r.find('yout');
```

We apply forward kinematics to determine the end-effector position

```
>> T = p560.fkine(q);
>> xyz = transl(T);
```

which we then plot▶ as a function of time

```
>> mplot(t, xyz(:,1:3))
```

The function mplot is a Toolbox utility that plots columns of a matrix in separate subgraphs.

which is shown in Fig. 8.6a. The Cartesian motion is 0.05 m s^{-1} in the y-direction as demanded but we observe some small and unwanted motion in the x- and z-directions.

The motion of the first three joints

```
>> mplot(t, q(:,1:3))
```

is shown in Fig. 8.6b and are not linear with time – reflecting the changing kinematic configuration of the arm.

The approach just described, based purely on integration, suffers from an accumulation of error which we observed as the unwanted x- and z-direction motion in Fig. 8.6a. We can eliminate this by changing the algorithm to a *closed-loop* form based on the difference between the desired and actual pose

$$\dot{\boldsymbol{q}}^{*}\langle k\rangle \leftarrow K_{p}\boldsymbol{J}\big(\boldsymbol{q}\langle k\rangle\big)^{-1}\Delta\big(\mathcal{K}(\boldsymbol{q}\langle k\rangle), \xi^{*}\langle k\rangle\big) \tag{8.6}$$

where K_p is a proportional gain, $\Delta(\cdot) \in \mathbb{R}^6$ is a spatial displacement▶ and the desired pose $\xi^*\langle k\rangle$ is a function of time.

See Sect. 3.1.4 for definition.

A Simulink example to demonstrate this for a circular path is

```
>> sl_rrmc2
```

shown in Fig. 8.7 and the tool of a Puma 560 robot traces out a circle of radius 50 mm. The x-, y- and z-coordinates as a function of time are computed and converted to a homogeneous transformation by the blocks in the grey area. The *difference* between the desired pose and the current pose from forward kinematics using the $\Delta(\cdot)$ operator is computed by the tr2delta block. The result is a spatial displacement, a translation and a rotation described by a 6-vector which is used as

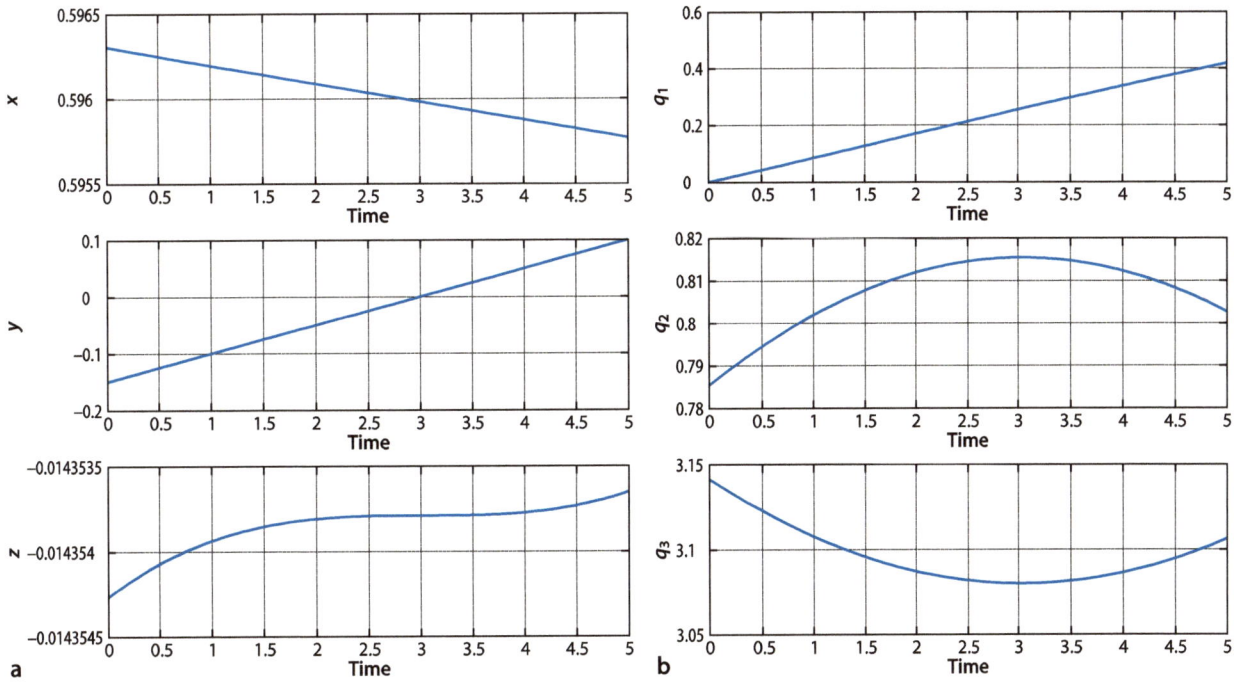

Fig. 8.6. Resolved-rate motion control, Cartesian and joint coordinates versus time. **a** Cartesian end-effector motion. Note the small, but unwanted motion in the x- and z-directions; **b** joint motion

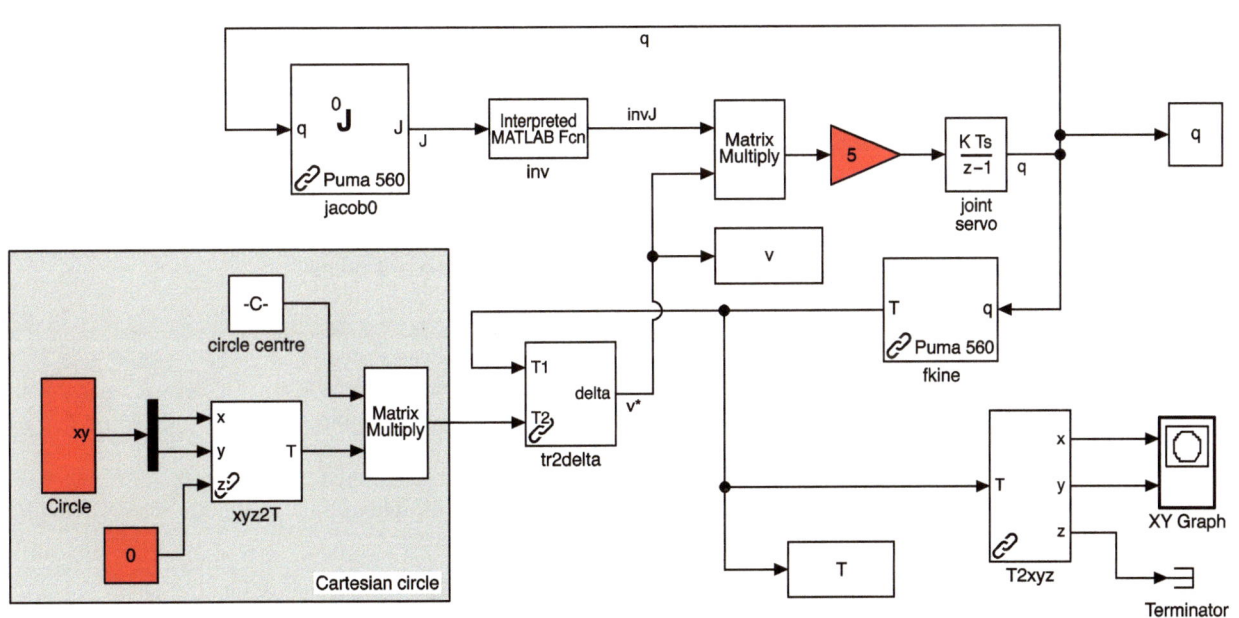

Fig. 8.7. The Simulink® model `sl_rrmc2` for closed-loop resolved-rate motion control with circular end-effector motion

the desired spatial velocity to drive the end-effector toward the desired pose. The Jacobian matrix is computed from the current manipulator joint angles and is inverted so as to transform the desired spatial velocity to joint angle rates. These are scaled by a proportional gain, to become the desired joint-space velocity that will correct any Cartesian error.

8.3.1   Jacobian Singularity

For the case of a square Jacobian where $\det(\boldsymbol{J}(\boldsymbol{q})) = 0$ we cannot solve Eq. 8.3 direct-ly. One strategy to deal with singularity is to replace the inverse with the damped inverse

$$\dot{\boldsymbol{q}} = \left(\boldsymbol{J}(\boldsymbol{q}) + \lambda \boldsymbol{I}\right)^{-1}\boldsymbol{\nu}$$

where λ is a small constant added to the diagonal which places a *floor* under the de-terminant. This will introduces some error in $\dot{\boldsymbol{q}}$, which integrated over time could lead to a significant discrepancy in tool position but the closed-loop resolved-rate motion scheme of Eq. 8.6 would minimize this.

An alternative is to use the pseudo-inverse of the Jacobian $\boldsymbol{J}^{+}$ which has the property

$$\boldsymbol{J}^{+}\boldsymbol{J} = \boldsymbol{I}$$

just as the inverse does. It is defined as

$$\boldsymbol{J}^{+} = \left(\boldsymbol{J}^{T}\boldsymbol{J}\right)^{-1}\boldsymbol{J}^{T}$$

and is readily computed using the MATLAB® builtin function `pinv`.► The solution

$$\dot{\boldsymbol{q}} = \boldsymbol{J}(\boldsymbol{q})^{+}\boldsymbol{\nu}$$

provides a least-squares solution for which $\|\boldsymbol{J}\dot{\boldsymbol{q}} - \boldsymbol{\nu}\|$ is smallest.►

Yet another approach is to delete from the Jacobian all those columns that are lin-early dependent on other columns. This is effectively locking the joints correspond-ing to the deleted columns and we now have an under-actuated system which we treat as per the next section.

> This is the left generalized- or pseudoin-verse, see Sect. F.1.1 for more details.

> A matrix expression like $\boldsymbol{v} = \boldsymbol{J}\dot{\boldsymbol{q}}$ is a sys-tem of scalar equations which we can solve for $\dot{\boldsymbol{q}}$. At singularity some of the equations are the same, leading to more unknowns than equations, and therefore an infinite number of solutions. The pseudo-inverse computes a solution that satisfies the equation and has the minumum norm.

8.4   Under- and Over-Actuated Manipulators

So far we have assumed that the Jacobian is square. For the nonsquare cases it is help-ful to consider the velocity relationship

$$\boldsymbol{\nu} = \boldsymbol{J}(\boldsymbol{q})\dot{\boldsymbol{q}}$$

in the diagrammatic form shown in Fig. 8.8. The Jacobian is a $6 \times N$ matrix, the joint velocity is an N-vector, and $\boldsymbol{\nu}$ is a 6-vector.

The case of $N < 6$ is referred to as an under-actuated robot, and $N > 6$ is over-ac-tuated or redundant. The under-actuated case cannot be solved because the system of equations is under-constrained but the system can be *squared up* by deleting some rows of $\boldsymbol{\nu}$ and $\boldsymbol{J}$ – accepting that some Cartesian degrees of freedom are not controllable given the low number of joints. For the over-actuated case the system of equations is under-constrained and the best we can do is find a least-squares solution as described in the previous section. Alternatively we can *square up* the Jacobian to make it invert-ible by deleting some columns – effectively *locking* the corresponding joints.

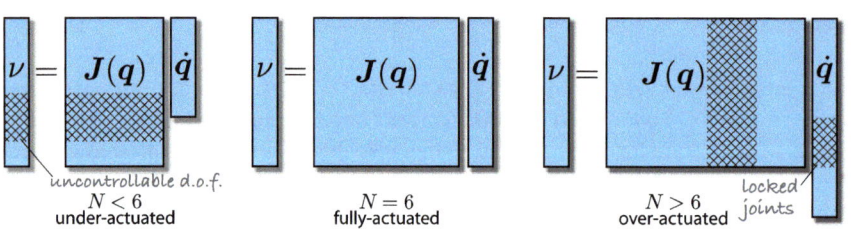

Fig. 8.8.
Schematic of Jacobian, $\boldsymbol{\nu}$ and $\dot{\boldsymbol{q}}$ for different cases of N. The *hatched* areas represent matrix regions that could be deleted in order to create a square sub-system capable of solution

8.4.1   Jacobian for Under-Actuated Robot

An under-actuated robot has $N < 6$, and a Jacobian that is taller than it is wide. For example a 2-joint manipulator at a nominal pose

```
>> mdl_planar2
>> qn = [1 1];
```

has the Jacobian

```
>> J = p2.jacob0(qn)
J =
   -1.7508   -0.9093
    0.1242   -0.4161
         0         0
         0         0
         0         0
    1.0000    1.0000
```

We cannot solve the inverse problem Eq. 8.3 using the pseudo-inverse since it will attempt to satisfy motion constraints that the manipulator cannot meet. For example the desired motion of $0.1\ \mathrm{m\ s^{-1}}$ in the x-direction gives the required joint velocity

```
>> qd = pinv(J) * [0.1 0 0 0 0 0]'
qd =
   -0.0698
    0.0431
```

which results in end-effector velocity

```
>> xd = J*qd;
>> xd'
ans =
    0.0829   -0.0266         0         0         0   -0.0266
```

This has the desired motion in the x-direction but undesired motion in y-axis translation and z-axis rotation. The end-effector rotation cannot be independently controlled (since it is a function of q_1 and q_2) yet this solution has taken it into account in the least-squares solution.

We have to confront the reality that we have *only* two degrees of freedom which we will use to control just v_x and v_y. We rewrite Eq. 8.2 in partitioned form as

$$\begin{pmatrix} v_x \\ v_y \\ \hline v_z \\ \omega_x \\ \omega_y \\ \omega_z \end{pmatrix} = \left(\frac{\boldsymbol{J}_{xy}}{\boldsymbol{J}_0} \right) \begin{pmatrix} \dot{q}_1 \\ \dot{q}_2 \end{pmatrix}$$

and taking the top partition, the first two rows, we write

$$\begin{pmatrix} v_x \\ v_y \end{pmatrix} = \boldsymbol{J}_{xy} \begin{pmatrix} \dot{q}_1 \\ \dot{q}_2 \end{pmatrix}$$

where $\boldsymbol{J}_{xy}$ is a 2×2 matrix. We invert this

$$\begin{pmatrix} \dot{q}_1 \\ \dot{q}_2 \end{pmatrix} = \boldsymbol{J}_{xy}^{-1} \begin{pmatrix} v_x \\ v_y \end{pmatrix}$$

which we can solve if $\det(\boldsymbol{J}_{xy}) \neq 0$.

```
>> Jxy = J(1:2,:);
>> qd = inv(Jxy)* [0.1 0]'
qd =
   -0.0495
   -0.0148
```

which results in end-effector velocity

```
>> xd = J*qd;
>> xd'
ans =
     0.1000    0.0000        0         0         0   -0.0642
```

We have achieved the desired x-direction motion with no unwanted motion apart from the z-axis rotation which is unavoidable – we have used the two degrees of freedom to control x- and y-translation, not z-rotation.

8.4.2   Jacobian for Over-Actuated Robot

An over-actuated or redundant robot has $N > 6$, and a Jacobian that is wider than it is tall. In this case we rewrite Eq. 8.3 to use the left pseudo-inverse

$$\dot{q} = J(q)^+ \nu \tag{8.7}$$

which, of the infinite number of solutions possible, will yield the one for which $\|\dot{q}\|$ is smallest – the minimum-norm solution.

We will demonstrate this for the left arm of the Baxter robot from Sect. 7.2.2.4 at a nominal pose

```
>> mdl_baxter
>> TE = SE3(0.8, 0.2, -0.2) * SE3.Ry(pi);
>> q = left.ikine(TE)
```

and its Jacobian

```
>> J = jacob0(left, q);
>> about J
J [double] : 6x7 (336 bytes)
```

is a 6×7 matrix. Now consider that we want the end-effector to move at 0.2 m s^{-1} in the x-, y- and z-directions. Using Eq. 8.7 we compute the required joint rates

```
>> xd = [0.2 0.2 0.2 0 0 0]';
>> qd = pinv(J) * xd;
>> qd'
ans =
     0.0895   -0.0464   -0.4259    0.6980   -0.4248    1.0179    0.2998
```

We see that all joints have nonzero velocity and contribute to the desired end-effector motion.▶

This Jacobian has seven columns and a rank of six

```
>> rank(J)
ans =
     6
```

and therefore a null space▶ whose basis has just one vector

```
>> N = null(J)
N =
    -0.2244
    -0.1306
     0.6018
     0.0371
    -0.7243
     0.0653
     0.2005
```

In the case of a Jacobian matrix any joint velocity that is a linear combination of its null-space vectors will result in *no* end-effector motion. For this robot there is only one vector and we can show that this *null-space joint motion* causes no end-effector motion

```
>> norm( J * N(:,1))
ans =
     2.6004e-16
```

If the robot end-effector follows a repetitive path using RRMC the joint angles may *drift* over time and *not* follow a repetitive path, potentially moving toward joint limits. We can use null-space control to provide additional constraints to prevent this.

See Appendix B.

This is remarkably useful because it allows Eq. 8.7 to be written as

$$\dot{q} = \underbrace{J(q)^{+}\nu}_{\text{end-effector motion}} + \underbrace{NN^{+}\dot{q}_{\text{null}}}_{\text{null-space motion}} \tag{8.8}$$

where the matrix $NN^{+} \in \mathbb{R}^{N \times N}$ *projects* the desired joint motion into the null space so that it will not affect the end-effector Cartesian motion, allowing the two motions to be superimposed.

Null-space motion can be used for highly-redundant robots to avoid collisions between the links and obstacles (including other links), or to keep joint coordinates away from their mechanical limit stops. Consider that in addition to the desired Cartesian velocity xd we wish to simultaneously increase joint 5 in order to move the arm away from some obstacle. We set a desired joint velocity

```
>> qd_null = [0 0 0 0 1 0 0]';
```

and project it into the null space

```
>> qp = N * pinv(N) * qd_null;
>> qp'
    0.1625     0.0946    -0.4359    -0.0269     0.5246    -0.0473    -0.1452
```

A scaling has been introduced but this joint velocity, or a scaled-up version of, this will increase the joint 5 angle without changing the end-effector pose. Other joints move as well – they provide the required compensating motion in order that the end-effector pose is not disturbed as shown by

```
>> norm( J * qp)
ans =
    1.9541e-16
```

A highly redundant snake robot like that shown in Fig. 8.9 would have a null space with 14 dimensions (20-6). This can be used to control the shape of the arm which is critical when moving within confined spaces.

Fig. 8.9.
20-DOF snake-robot arm: 2.5 m reach, 90 mm diameter and payload capacity of 25 kg (image courtesy of OC Robotics)

8.5   Force Relationships

In Sect. 3.2.2 we introduced wrenches $W = (f_x, f_y, f_z, m_x, m_y, m_z) \in \mathbb{R}^6$ which are a vector of forces and moments.

8.5.1   Transforming Wrenches to Joint Space

The manipulator Jacobian transforms joint velocity to an end-effector spatial velocity according to Eq. 8.2 and the Jacobian transpose transforms a wrench applied at the end-effector to torques and forces experienced at the joints ▶

Derived through the principle of virtual work, see for instance Spong et al. (2006, sect. 4.10).

$$Q = {}^0J(q)^T \, {}^0W \qquad (8.9)$$

where W is a wrench in the world coordinate frame and Q is the generalized joint force vector. The elements of Q are joint torque or force for revolute or prismatic joints respectively.

The mapping for velocity, from end-effector to joints, involves the inverse Jacobian which can potentially be singular. The mapping of forces and torques, from end-effector to joints, is different – it involves the transpose of the Jacobian which can never be singular. We exploit this property in the next section to solve the inverse-kinematic problem numerically.

If the wrench is defined in the end-effector coordinate frame then we use instead

$$Q = {}^EJ(q)^T \, {}^EW \qquad (8.10)$$

For the Puma 560 robot in its nominal pose, see Fig. 8.2, a force of 20 N in the world y-direction results in joint torques of

```
>> tau = p560.jacob0(qn)' * [0 20 0 0 0 0]';
>> tau'
ans =
   11.9261    0.0000    0.0000       0         0         0
```

The force pushes the arm *sideways* and only the waist joint will rotate in response – experiencing a torque of 11.93 N m due to a lever arm effect. A force of 20 N applied in the world x-direction results in joint torques of

```
>> tau = p560.jacob0(qn)' * [20 0 0 0 0 0]';
>> tau'
ans =
    3.0010    0.2871    6.3937       0         0         0
```

which is pulling the end-effector away from the base which results in torques being applied to the first three joints.

8.5.2   Force Ellipsoids

In Sect. 8.2.2 we introduced the velocity ellipse and ellipsoid which describe the directions in which the end-effector is best able to move. We can perform a similar analysis for the forces and torques at the end-effector – the end-effector wrench. We start with a set of generalized joint forces with a unit norm

$$Q^T Q = 1$$

and substituting Eq. 8.9 we can write

$$W^T \left(J(q)J(q)^T \right) W = 1$$

which is the equation of points on the surface of a 6-dimensional ellipsoid in the end-effector wrench space. For the planar robot arm of Fig. 8.1 we can plot this ellipse

```
>> p2.fellipse([30 40], 'deg')
```

or we can interactively explore how its shape changes with configuration by

```
>> p2.teach(qn, 'callback', @(r,q) r.fellipse(q), 'view', 'top')
```

If this ellipsoid is close to spherical, that is, its radii are of the same order of magnitude then the end-effector can achieve an arbitrary wrench. However if one or more radii are very small this indicates that the end-effector cannot exert a force along, or a moment about, the axes corresponding to those small radii.

The force and velocity ellipsoids provide complementary information about how well suited the configuration of the arm is to a particular task. We know from personal experience that to throw an object quickly we have our arm outstretched and orthogonal to the throwing direction, whereas to lift something heavy we hold our arms close in to our body.

8.6 Inverse Kinematics: a General Numerical Approach

In Sect. 7.2.2.1 we solved the inverse kinematic problem using an explicit solution that required the robot to have 6 joints and a spherical wrist. For the case of robots which do not meet this specification, for example those with more or less than 6 joints, we need to consider a numerical solution. Here we will develop an approach based on the forward kinematics and the Jacobian transpose which we can compute for any manipulator configuration since these functions have no singularities.

8.6.1 Numerical Inverse Kinematics

The principle is shown in Fig. 8.10 where the robot in its current configuration is drawn solidly and the desired configuration is faint. From the overlaid pose graph the error between actual ξ_E and desired pose ξ_E^* is ξ_Δ which can be described by a spatial displacement as discussed in Sect. 3.1.4

$$^E\boldsymbol{\Delta} = \Delta\left(\xi_E, \xi_E^*\right) = (\boldsymbol{t}, \hat{\boldsymbol{v}}\theta) \in \mathbb{R}^6$$

where the current pose is computed using forward kinematics $\xi_E = \mathcal{K}(\boldsymbol{q})$.

Imagine a *special* spring between the end-effector of the two poses which is pulling (and twisting) the robot's end-effector toward the desired pose with a wrench proportional to the spatial displacement

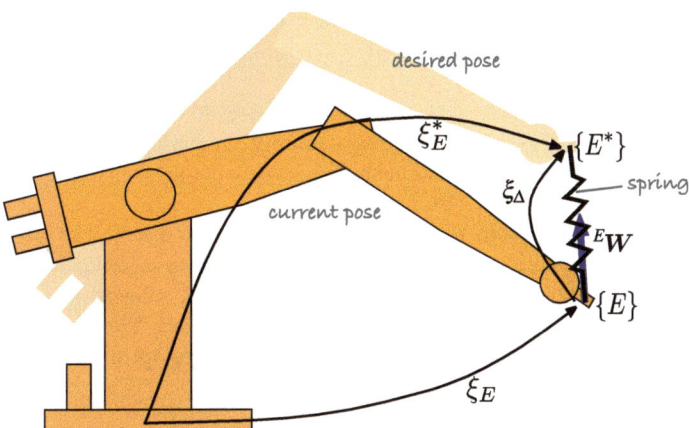

Fig. 8.10.
Schematic of the numerical inverse kinematic approach, showing the current ξ_E and the desired ξ_E^* manipulator pose

$$^{E}\boldsymbol{W} = \gamma \, ^{E}\boldsymbol{\Delta} \tag{8.11}$$

which is *resolved* to generalized joint forces

$$\boldsymbol{Q} = \, ^{E}\boldsymbol{J}(\boldsymbol{q})^{T} \, ^{E}\boldsymbol{W}$$

using the Jacobian transpose Eq. 8.10. We assume that this virtual robot has no joint motors only viscous dampers so the joint velocity will be proportional to the applied forces

$$\dot{\boldsymbol{q}} = \boldsymbol{Q}/B$$

where B is the joint damping coefficient (assuming all dampers are the same). Putting all this together we can write

$$\dot{\boldsymbol{q}} = \frac{1}{B}\boldsymbol{J}(\boldsymbol{q})^{T} \, \Delta\big(\mathcal{K}(\boldsymbol{q}), \xi_{E}^{*}\big)$$

which gives the joint velocities that will drive the forward kinematic solution toward the desired end-effector pose. This can be solved iteratively by

$$\delta_{q}\langle k\rangle = \alpha \, \boldsymbol{J}\big(\boldsymbol{q}\langle k\rangle\big)^{T} \Delta\big(\mathcal{K}\big(\boldsymbol{q}\langle k\rangle\big), \xi_{E}^{*}\big) \tag{8.12}$$

$$\boldsymbol{q}\langle k+1\rangle \leftarrow \boldsymbol{q}\langle k\rangle + \delta_{q}\langle k\rangle$$

until the norm of the update $\|\delta_{q}\langle k\rangle\|$ is sufficiently small and where $\alpha > 0$ is a well-chosen constant. Since the solution is based on the Jacobian transpose rather than inverse the algorithm works when the Jacobian is nonsquare or singular. In practice however this algorithm is slow to converge and very sensitive to the choice of α.

More practically we can formulate this as a least-squares problem in the world coordinate frame and minimize the scalar cost

$$E = \boldsymbol{\Delta}^{T}\boldsymbol{M}\boldsymbol{\Delta}$$

where $\boldsymbol{M} = \mathrm{diag}\,(\boldsymbol{m}) \in \mathbb{R}^{6\times 6}$ and $\boldsymbol{m}$ is the mask vector introduced in Sect. 7.2.2.3. The update becomes

$$\delta_{q}\langle k\rangle = \Big(\boldsymbol{J}\big(\boldsymbol{q}\langle k\rangle\big)^{T}\boldsymbol{M}\boldsymbol{J}\big(\boldsymbol{q}\langle k\rangle\big)\Big)^{-1}\boldsymbol{J}\big(\boldsymbol{q}\langle k\rangle\big)^{T}\boldsymbol{M}\,\Delta\big(\mathcal{K}\big(\boldsymbol{q}\langle k\rangle\big), \xi_{E}^{*}\big)$$

which is much faster to converge but can behave poorly near singularities. We remedy this by introducing a damping constant λ

$$\delta_{q}\langle k\rangle = \Big(\boldsymbol{J}\big(\boldsymbol{q}\langle k\rangle\big)^{T}\boldsymbol{M}\boldsymbol{J}\big(\boldsymbol{q}\langle k\rangle\big) + \lambda\boldsymbol{I}_{N\times N}\Big)^{-1}\boldsymbol{J}\big(\boldsymbol{q}\langle k\rangle\big)^{T}\boldsymbol{M}\,\Delta\big(\mathcal{K}\big(\boldsymbol{q}\langle k\rangle\big), \xi_{E}^{*}\big)$$

which ensures that the term being inverted can never be singular.

An effective way to choose λ is to test whether or not an iteration reduces the error, that is if $\|\delta_{q}\langle k\rangle\| < \|\delta_{q}\langle k-1\rangle\|$. If the error is reduced we can decrease λ in order to speed convergence. If the error has increased we revert to our previous estimate of $\boldsymbol{q}\langle k\rangle$ and increase λ. This adaptive damping factor scheme is the basis of the well-known Levenberg-Marquardt optimization algorithm.

This algorithm is implemented by the `ikine` method and works well in practice. As with all optimization algorithms it requires a reasonable initial estimate of $\boldsymbol{q}$ and this can be explicitly given using the option `'q0'`. A brute-force search for an initial value can be requested by the option `'search'`. The simple Jacobian-transpose approach of Eq. 8.12 can be invoked using the option `'transpose'` along with the value of α.

8.7   Advanced Topics

8.7.1   Computing the Manipulator Jacobian Using Twists

In Sect. 7.1.2.2 we computed the forward kinematics as a product of exponentials based on the screws representing the joint axes in a zero-joint angle configuration. It is easy to differentiate the product of exponentials with respect to motion about each screw axis which leads to the Jacobian matrix

$$^{0}J^{\mathcal{V}} = \Big(S_1 \;\; \mathrm{Ad}\Big(e^{[S_1]q_1}\Big)S_2 \;\; \cdots \;\; \mathrm{Ad}\Big(e^{[S_1]q_1}\cdots e^{[S_{N-1}]q_{N-1}}\Big)S_N\Big)$$

for velocity in the world coordinate frame. The Jacobian is very elegantly expressed and can be easily built up column by column. Velocity in the end-effector coordinate frame is related to joint velocity by the Jacobian matrix

$$^{E}J^{\mathcal{V}} = \mathrm{Ad}\Big(^{E}\xi_0\Big)\,^{0}J^{\mathcal{V}}$$

where $\mathrm{Ad}\,(\cdot)$ is the adjoint matrix introduced in Sect. 3.1.2.

> However, compared to the Jacobian of Sect. 8.1, these Jacobians give the velocity of the end-effector as a *velocity twist*, not a spatial velocity as defined on page 63.

To obtain the Jacobian that gives spatial velocity as described in Sect. 8.1 we must apply a velocity transformation

$$^{0}J = \begin{pmatrix} I_{3\times3} & -\big[^{0}t_E\big]_{\times} \\ 0_{3\times3} & I_{3\times3} \end{pmatrix} \,^{0}J^{\mathcal{V}}$$

8.8   Wrapping Up

Jacobians are an important concept in robotics, relating changes in one space to changes in another. We previously encountered Jacobians for estimation in Chap. 6 and will use them later for computer vision and control.

In this chapter we have learned about the manipulator Jacobian which describes the relationship between the rate of change of joint coordinates and the spatial velocity of the end-effector expressed in either the world frame or the end-effector frame. We showed how the inverse Jacobian can be used to resolve desired Cartesian velocity into joint velocity as an alternative means of generating Cartesian paths for under- and over-actuated robots. For over-actuated robots we showed how null-space motions can be used to move the robot's joints without affecting the end-effector pose. The numerical properties of the Jacobian tell us about manipulability, that is how well the manipulator is able to move, or exert force, in different directions. At a singularity, indicated by linear dependence between columns of the Jacobian, the robot is unable to move in certain directions. We visualized this by means of the velocity and force ellipsoids.

We also created Jacobians to map angular velocity to roll-pitch-yaw or Euler angle rates, and these were used to form the analytic Jacobian matrix. The Jacobian transpose is used to map wrenches applied at the end-effector to joint torques, and also to map wrenches between coordinate frames. It is also the basis of numerical inverse kinematics for arbitrary robots and singular poses.

Further Reading

The manipulator Jacobian is covered by almost all standard robotics texts such as the robotics handbook (Siciliano and Khatib 2016), Lynch and Park (2017), Siciliano et al. (2009), Spong et al. (2006), Craig (2005), and Paul (1981). An excellent discussion of manipulability and velocity ellipsoids is provided by Siciliano et al. (2009), and the most common manipulability measure is that proposed by Yoshikawa (1984). Computing the manipulator Jacobian based on Denavit-Hartenberg parameters, as used in this Toolbox, was first described by Paul and Shimano (1978).

The resolved-rate motion control scheme was proposed by Whitney (1969). Extensions such as pseudo-inverse Jacobian-based control are reviewed by Klein and Huang (1983) and damped least-squares methods are reviewed by Deo and Walker (1995).

MATLAB and Toolbox Notes

The MATLAB Robotics System Toolbox™ describes a serial-link manipulator using an instance of the `RigidBodyTree` class. Jacobians can be computed using the class method `GeometricJacobian`.

Exercises

1. For the simple 2-link example (page 228) compute the determinant symbolically and determine when it is equal to zero. What does this mean physically?
2. For the Puma 560 robot can you devise a configuration in which three joint axes are parallel?
3. Derive the analytical Jacobian for Euler angles.
4. Velocity and force ellipsoids for the two link manipulator (page 234, 243). Perhaps using the interactive `teach` method with the `'callback'` option:
   a) What configuration gives the best manipulability?
   b) What configuration is best for throwing a ball in the positive x-direction?
   c) What configuration is best for carrying a heavy weight if gravity applies a force in the negative y-direction?
   d) Plot the velocity ellipse (x- and y-velocity) for the two-link manipulator at a grid of end-effector positions in its workspace. Each ellipsoid should be centered on the end-effector position.
5. Velocity and force ellipsoids for the Puma manipulator (page 235)
   a) For the Puma 560 manipulator find a configuration where manipulability is greater than at `qn`.
   b) Use the `teach` method with the `'callback'` option to interactively animate the ellipsoids. You may need to use the `'workspace'` option to `teach` to prevent the ellipsoid being truncated.
6. Resolved-rate motion control (page 235)
   a) Experiment with different Cartesian translational and rotational velocity demands, and combinations.
   b) Extend the Simulink system of Fig. 8.6 to also record the determinant of the Jacobian matrix to the workspace.
   c) In Fig. 8.6 the robot's motion is simulated for 5 s. Extend the simulation time to 10 s and explain what happens.
   d) Set the initial pose and direction of motion to mimic that of Sect. 7.3.4. What happens when the robot reaches the singularity?
   e) Replace the Jacobian inverse block in Fig. 8.5 with the MATLAB function `pinv`.
   f) Replace the Jacobian inverse block in Fig. 8.5 with a damped least squares function, and investigate the effect of different values of the damping factor.

g) Replace the Jacobian inverse block in Fig. 8.5 with a block based on the MATLAB function `lscov`.

7. The model `mdl_p8` describes an 8-joint robot (PPRRRRRR) comprising an *xy*-base (PP) carrying a Puma arm (RRRRRR).

a) Compute a Cartesian end-effector path and use numerical inverse kinematics to solve for the joint coordinates. Analyze how the motion is split between the base and the robot arm.

b) With the end-effector at a constant pose explore null-space control. Set a velocity for the mobile base and see how the arm configuration accomodates that.

c) Develop a null-space controller that keeps the last six joints in the middle of their working range by using the first two joints to position the base of the Puma. Modify this so as to maximize the manipulability of the P8 robot.

d) Consider now that the Puma robot is mounted on a nonholonomic robot, create a controller that generates appropriate steering and velocity inputs to the mobile robot (challenging).

e) For an arbitrary pose and end-point spatial velocity we will move six joints and lock two joints. Write an algorithm to determine which two joints should be locked.

8. The model `mdl_hyper3d(20)` is a 20-joint robot that moves in 3-dimensional space.

a) Explore the capabilities of this robot.

b) Compute a Cartesian end-effector trajectory that traces a circle on the ground, and use numerical inverse kinematics to solve for the joint coordinates.

c) Add a null-space control strategy that keeps all joint angles close to zero while it is moving.

d) Define an end-effector pose on the ground that the robot must reach after passing through two holes in vertical planes. Can you determine the joint configuration that allows this?

9. Write code to compute the Jacobian of a robot represented by a `SerialLink` object using twists as described in Sect. 8.7.1.

10. Consider the Puma 560 robot moving in the *xz*-plane. Divide the plane into 2-cm grid cells and for each cell determine if it is reachable, and if it is then determine the manipulability for the first three joints of the robot arm and place that value in the corresponding grid cell. Display a heat map of the robot's manipulability in the plane.

9

Dynamics and Control

In this chapter we consider the dynamics and control of a serial-link manipulator arm. The motion of the end-effector is the composition of the motion of each link, and the links are ultimately moved by *forces* and *torques* exerted by the joints. Section 9.1 describes the key elements of a robot joint control system that enables a single joint to follow a desired trajectory; and the challenges created by real-world effects such as friction, gravity load and varying inertia.

Each link in the serial-link manipulator is supported by a reaction force and torque from the preceding link, and is subject to its own weight as well as the reaction forces and torques from the links that it supports. Section 9.2 introduces the *rigid-body* equations of motion, a set of coupled dynamic equations, that describe the joint torques necessary to achieve a particular manipulator state. These equations can be factored into terms describing inertia, gravity load and gyroscopic coupling which provide insight into how the motion of one joint exerts a disturbance force on other joints, and how inertia and gravity load varies with configuration and payload. Section 9.3 introduces the forward dynamics which describe how the manipulator moves, that is, how its configuration evolves with time in response to forces and torques applied by the joints and by external forces such as gravity. Section 9.4 introduces control systems that compute the required joint forces based on the desired trajectory as well as the rigid-body dynamic forces. This enables improved control of the end-effector trajectory, despite changing robot configuration, as well as compliant motion. Section 9.5 covers an important application of what we have learned about joint control – series-elastic actuators for human-safe robots.

9.1 Independent Joint Control

A robot drive train comprises an actuator or motor, and a transmission to connect it to the link. A common approach to robot joint control is to consider each joint or axis as an independent control system that attempts to accurately follow its joint angle trajectory. However as we shall see, this is complicated by various *disturbance* torques due to gravity, velocity and acceleration coupling, and friction that act on the joint. A very common control structure is the nested control loop. The outer loop is responsible for maintaining position and determines the velocity of the joint that will minimize position error. The inner loop is responsible for maintaining the velocity of the joint as demanded by the outer loop.

9.1.1 Actuators

The vast majority of robots today are driven by rotary electric motors (Fig. 9.1). Large industrial robots typically use brushless servo motors while small laboratory or hobby robots use brushed DC motors or stepper motors. Manipulators for very large payloads as used in mining, forestry or construction are typically hydraulically driven using electrically operated hydraulic valves – electro-hydraulic actuation.

© Springer Nature Switzerland AG 2022
P. Corke, *Robotics and Control*, Springer Tracts in Advanced Robotics 141,
https://doi.org/10.1007/978-3-030-79179-7_9

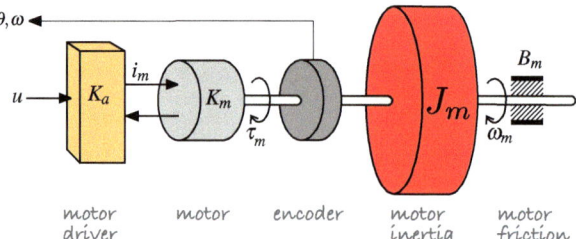

Fig. 9.1.
Key components of a robot-joint actuator. A demand voltage u controls the current i_m flowing into the motor which generates a torque τ_m that accelerates the rotational inertia J_m and is opposed by friction $B_m \, \omega_m$. The encoder measures rotational speed and angle

Electric motors can be either current or voltage controlled.▶ Here we assume current control where a motor driver or amplifier provides current

$$i_m = K_a u$$

that is linearly related to the applied control voltage u and where K_a is the transconductance of the amplifier with units of $\mathrm{A\,V^{-1}}$. The torque generated by the motor is proportional to current

$$\tau_m = K_m i_m$$

where K_m is the motor torque constant with units of $\mathrm{N\,m\,A^{-1}}$. The torque accelerates the rotational inertia J_m, due to the rotating part of the motor itself, which has a rotational velocity of ω. Frictional effects are modeled by B_m.

Current control is implemented by an electronic constant current source, or a variable voltage source with feedback of actual motor current. A variable voltage source is most commonly implemented by a pulse-width modulated (PWM) switching circuit. Voltage control requires that the electrical dynamics of the motor due to its resistance and inductance, as well as back EMF, must be taken into account when designing the control system.

9.1.2   Friction

Any rotating machinery, motor or gearbox, will be affected by friction – a force or torque that *opposes* motion. The net torque from the motor is

$$\tau' = \tau_m - \tau_f$$

where τ_f is the friction torque which is function of velocity

$$\tau_f = B\omega + \tau_C \tag{9.1}$$

where the slope $B > 0$ is the viscous friction coefficient and the offset is Coulomb friction. The latter is frequently modeled by the nonlinear function

$$\tau_C = \begin{cases} \tau_C^+ & \omega > 0 \\ 0 & \omega = 0 \\ \tau_C^- & \omega < 0 \end{cases} \tag{9.2}$$

In general the friction coefficients depend on the direction of rotation and this asymmetry is more pronounced for Coulomb than for viscous friction.

The total friction torque as a function of rotational velocity is shown in Fig. 9.2. At very low speeds, highlighted in grey, an effect known as stiction becomes evident. The applied torque must exceed the stiction torque before rotation can occur – a process known as *breaking stiction*. Once the machine is moving the stiction force rapidly decreases and viscous friction dominates.

There are several sources of friction *experienced* by the motor. The first component is due to the motor itself: its bearings and, for a brushed motor, the brushes rubbing on the commutator. The friction parameters are often provided in the motor manufacturer's data sheet. Other sources of friction are the gearbox and the bearings that support the link.

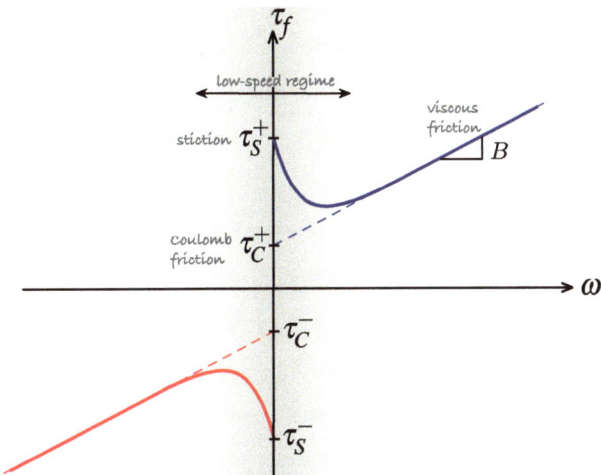

Fig. 9.2.
Typical friction versus speed characteristic. The *dashed lines* depict a simple piecewise-linear friction model characterized by slope (viscous friction) and intercept (Coulomb friction). The low-speed regime is *shaded* and shown in exaggerated fashion

Charles-Augustin de Coulomb (1736–1806) was a French physicist. He was born in Angoulême to a wealthy family and studied mathematics at the Collége des Quatre-Nations under Pierre Charles Monnier, and later at the military school in Méziéres. He spent eight years in Martinique involved in the construction of Fort Bourbon and there he contracted tropical fever.

Later he worked at the shipyards in Rochefort which he used as laboratories for his experiments in static and dynamic friction of sliding surfaces. His paper *Théorie des machines simples* won the Grand Prix from the Académie des Sciences in 1781. His later research was on electromagnetism and electrostatics and he is best known for the formula on electrostatic forces, named in his honor, as is the SI unit of charge. After the revolution he was involved in determining the new system of weights and measures.

9.1.3   Effect of the Link Mass

A motor in a robot arm does not exist in isolation, it is connected to a link as shown schematically in Fig. 9.3. The link has two obvious significant effects on the motor – it adds extra inertia and it adds a torque due to the weight of the arm and both vary with the configuration of the joint.

With reference to the simple 2-joint robot shown in Fig. 9.4 consider the first joint which is directly attached to the first link which is colored red. If we assume the mass of the red link is concentrated at its center of mass (CoM) the extra inertia of the link will be $m_1 r_1^2$. The motor will also experience the inertia of the blue link and this will depend on the value of q_2 – the inertia of the arm when it is straight is greater than the inertia when it is folded.

We also see that gravity acting on the center of mass of the red link will create a torque on the joint 1 motor which will be proportional to $\cos q_1$. Gravity acting on the center of mass of the blue link also creates a torque on the joint 1 motor, and this is more pronounced since it is acting at a greater distance from the motor – the *lever arm effect* is greater.

These effects are clear from even a cursory examination of Fig. 9.4 but the reality is even more complex. Jumping ahead to material we will cover in the next section, we can use the Toolbox◄ to determine the torque acting on each of the joints as a function of the position, velocity and acceleration of the joints

This requires the MATLAB Symbolic Math Toolbox™.

```
>> mdl_twolink_sym
>> syms q1 q2 q1d q2d q1dd q2dd real
>> tau = twolink.rne([q1 q2], [q1d q2d], [q1dd q2dd]);
```

and the result is a symbolic 2-vector, one per joint, with surprisingly many terms which we can summarize as:

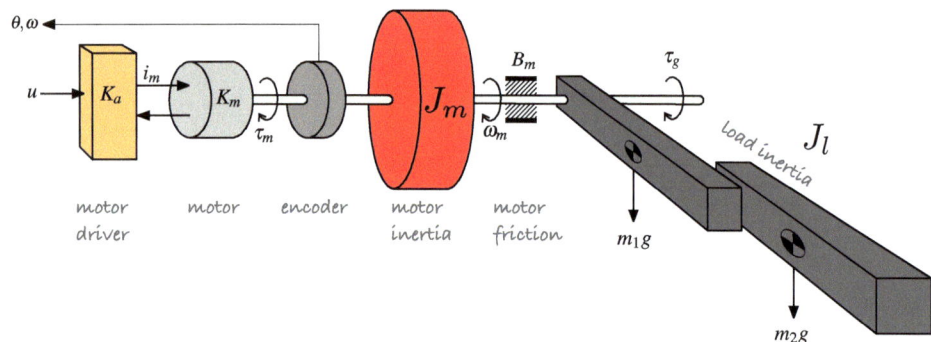

Fig. 9.3.
Robot joint actuator with at-tached links. The center of mass of each link is indicated by ◗

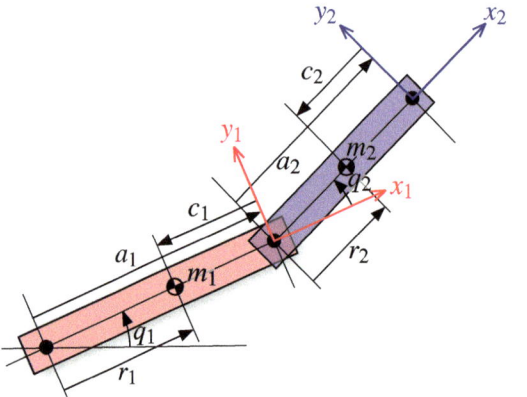

Fig. 9.4.
Notation for rigid-body dynam-ics of two-link arm showing link frames and relevant dimensions. The center of mass (CoM) of each link is indicated by ◗. The CoM is a distance of r_i from the axis of joint i, and c_i from the origin of frame $\{i\}$ as defined in Fig. 7.5 – therefore $r_i = a_i + c_i$

$$\tau_1 = M_{11}(q_2)\ddot{q}_1 + \underbrace{M_{12}(q_2)\ddot{q}_2 + C_1(q_2)\dot{q}_1\dot{q}_2 + C_2(q_2)\dot{q}_2^2 + g(q_1, q_2)}_{\text{disturbance}}$$

$$M_{11} = m_1\left(a_1^2 + 2a_1c_1 + c_1^2\right) + m_2\left(a_1^2 + (a_2 + c_2)^2 + (2a_1a_2 + 2a_1c_2)\cos q_2\right)$$

$$M_{12} = m_2(a_2 + c_2)(a_2 + c_2 + a_1\cos q_2)$$

$$C_1 = -2a_1m_2(a_2 + c_2)\sin q_2 \tag{9.3}$$

$$C_2 = -a_1m_2(a_2 + c_2)\sin q_2$$

$$g = (a_1m_1 + a_1m_2 + c_1m_1)\cos q_1 + (a_2m_2 + c_2m_2)\cos(q_1 + q_2)$$

We have already discussed the first and last terms in a qualitative way – the inertia is dependent on q_2 and the gravity torque g is dependent on q_1 and q_2. What is perhaps most surprising is that the torque applied to joint 1 depends on the velocity and the ac-celeration of q_2 and this will covered in more detail in Sect. 9.2.

In summary, the effect of joint motion in a series of mechanical links is nontrivial. The motion of any joint is affected by the motion of *all* the other joints and for a robot with many joints this becomes quite complex.

9.1.4   Gearbox

Electric motors are compact and efficient and can rotate at very high speed, but produce very low torque. Therefore it is common to use a reduction gearbox to tradeoff speed for increased torque. For a prismatic joint the gearbox might convert rotary motion to linear. The disadvantage of a gearbox is increased cost, weight, friction, backlash, mechanical noise and, for harmonic gears, torque ripple. Very high-performance robots, such as those used in high-speed electronic assembly, use expensive high-torque motors with a direct drive or a very low gear ratio achieved using cables or thin metal bands rather than gears.

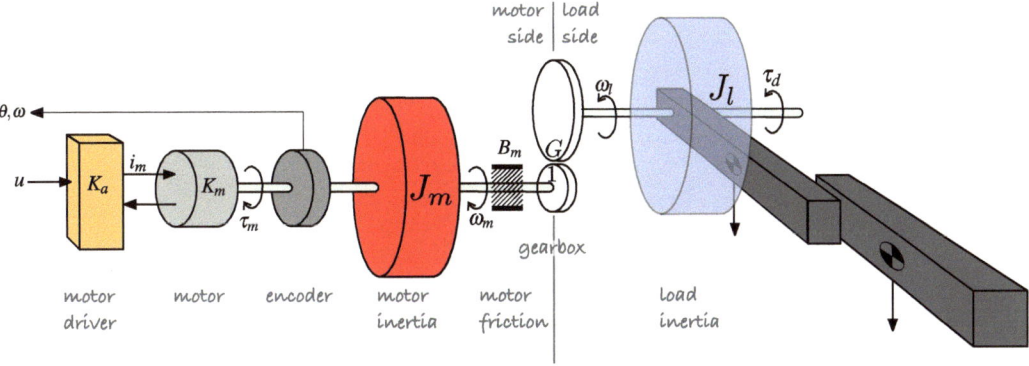

Fig. 9.5. Schematic of complete robot joint including gearbox. The effective inertia of the links is shown as J_l and the disturbance torque due to the link motion is τ_d

Figure 9.5 shows the complete drive train of a typical robot joint. For a $G\!:\!1$ reduction drive the torque at the link is G times the torque at the motor. For rotary joints the quantities measured at the link, reference frame l, are related to the motor referenced quantities, reference frame m, as shown in Table 9.1. The inertia of the load is reduced by a factor of G^2▸ and the disturbance torque by a factor of G.

For example if you turned the motor shaft by hand you would *feel* the inertia of the load through the gearbox but it would be reduced by G^2.

There are two components of inertia *seen* by the motor. The first is due to the rotating part of the motor itself, its rotor, which is denoted J_m. This is a constant intrinsic characteristic of the motor and the value is provided in the motor manufacturer's data sheet. The second component is the variable load inertia J_l which is the inertia of the driven link and all the other links that are attached to it. For joint j this is element M_{jj} of the configuration dependent inertia matrix of Eq. 9.3.

Table 9.1. Relationship between load and motor referenced quantities for reduction gear ratio G

$^l J = G^2\,{}^m J$
$^l B = G^2\,{}^m B$
$^l \tau_C = G\,{}^m \tau_C$
$^l \tau = G\,{}^m \tau$
$^l \omega = {}^m \omega / G$
$^l \dot{\omega} = {}^m \dot{\omega} / G$

9.1.5   Modeling the Robot Joint

The complete motor drive comprises the motor to generate torque, the gearbox to amplify the torque and reduce the effects of the load, and an encoder to provide feedback of position and velocity. A schematic of such a device is shown in Fig. 9.6.

Collecting the various equations above we can write the torque balance on the motor shaft as

$$K_m K_a u - B' \omega - \tau_C'(\omega) - \frac{\tau_d(\boldsymbol{q})}{G} = J' \dot{\omega} \tag{9.4}$$

where B', τ_C' and J' are the effective total viscous friction, Coulomb friction and inertia due to the motor, gearbox, bearings and the load

$$B' = B_m + \frac{B_l}{G^2}, \quad \tau_C' = \tau_{C,m} + \frac{\tau_{C,l}'}{G}, \quad J' = J_m + \frac{J_l}{G^2} \tag{9.5}$$

In order to analyze the dynamics of Eq. 9.4 we must first linearize it, and this can be done simply by setting all additive constants to zero

$$J' \dot{\omega} + B' \omega = K_m K_a u$$

and then applying the Laplace transformation

$$s J' \Omega(s) + B' \Omega(s) = K_m K_a U(s)$$

where $\Omega(s)$ and $U(s)$ are the Laplace transform of the time domain signals $\omega(t)$ and $u(t)$ respectively. This can be rearranged as a linear transfer function

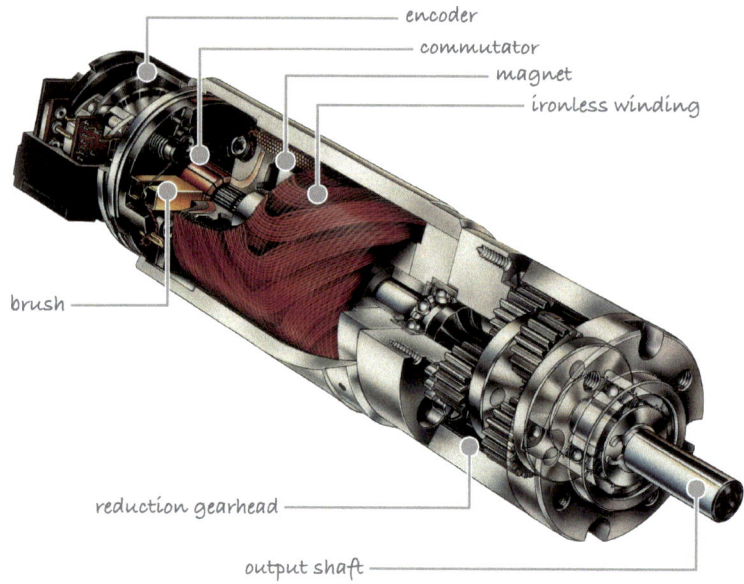

Fig. 9.6.
Schematic of an integrated motor-encoder-gearbox assembly (courtesy of maxon precision motors, inc.)

$$\frac{\Omega(s)}{U(s)} = \frac{K_m K_a}{J's + B'}$$

relating motor speed to control input, and has a single pole▶ at $s = -B'/J'$.   The mechanical pole.

We will use data for joint 2 – the shoulder – of the Puma 560 robot since its parameters are well known and are listed in Table 9.2. In the absence of other information we will take $B' = B_m$. The link inertia M_{22} experienced by the joint 2 motor as a function of configuration is shown in Fig. 9.16c and we see that it varies significantly – from 3.66 to 5.21 kg m². Using the mean value of the extreme inertia values, which is 4.43 kg m², the effective inertia is

$$J' = J_m + \frac{1}{G^2} M_{22}$$

$$= 200 \times 10^{-6} + \frac{4.43}{(107.815)^2}$$

$$= 200 \times 10^{-6} + 380 \times 10^{-6} = 580 \times 10^{-6} \text{ kg m}^2$$

and we see that the inertia of the link referred to the motor side of the gearbox is comparable to the inertia of the motor itself.

The Toolbox can automatically generate▶ a dynamic model suitable for use with   This requires the Control Systems Toolbox™.
the MATLAB control design tools

```
>> tf = p560.jointdynamics(qn);
```

is a vector of continuous-time linear-time-invariant (LTI) models, one per joint, computed for the particular pose qn. For the shoulder joint we are considering here that transfer function is

```
>> tf(2)
ans =
            1
   ----------------------
   0.0005797 s + 0.000817
Continuous-time transfer function.
```

which is similar to that above except that it does not account for K_m and K_a since these are not parameters of the Link object. Once we have a model of this form we can plot the step response and use a range of standard control system design tools.

Table 9.2.
Motor and drive parameters for Puma 560 shoulder joint with respect to the motor side of the gearbox (Corke 1996b)

Parameter	Symbol	Value	Unit
Motor torque constant	K_m	0.228	$N\,m\,A^{-1}$
Motor inertia	J_m	200×10^{-6}	$kg\,m^2$
Drive viscous friction	B_m	817×10^{-6}	$N\,m\,s\,rad^{-1}$
Drive Coulomb friction	τ_C^+	0.126	$N\,m$
	τ_C^-	−0.709	$N\,m$
Gear ratio	G	107.815	
Maximum torque	τ_{max}	0.900	$N\,m$
Maximum speed	$\dot{q}_{max}$	165	$rad\,s^{-1}$

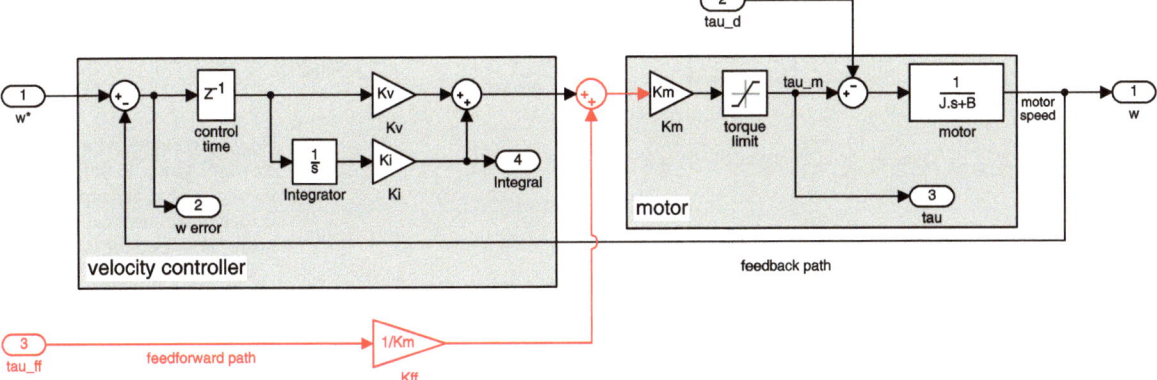

Fig. 9.7. Velocity control loop, Simulink model `vloop`

9.1.6 Velocity Control Loop

A very common approach to controlling the position output of a motor is the nested control loop. The outer loop is responsible for maintaining position and determines the velocity of the joint that will minimize position error. The inner loop – the velocity loop – is responsible for maintaining the velocity of the joint as demanded by the outer loop. Motor speed control is important for all types of robots, not just arms. For example it is used to control the speed of the wheels for car-like vehicles and the rotors of a quadrotor as discussed in Chap. 4.

The Simulink® model is shown in Fig. 9.7. The input to the motor driver is based on the error between the demanded and actual velocity.◀ A delay of 1 ms is included to model the computational time of the velocity loop control algorithm and a saturator models the finite maximum torque that the motor that can deliver.

We first consider the case of proportional control where $K_i = 0$ and

$$u^* = K_v\left(\dot{q}^* - \dot{q}\right) \tag{9.6}$$

To test this velocity controller we create a test harness

```
>> vloop_test
```

with a trapezoidal velocity demand which is shown in Fig. 9.8. Running the simulator

```
>> sim('vloop_test');
```

and with a little experimentation we find that a gain of $K_v = 0.6$ gives satisfactory performance as shown in Fig. 9.9. There is some minor overshoot at the discontinuity but less gain leads to increased velocity error and more gain leads to oscillation – as always control engineering is all about tradeoffs.

The motor velocity is typically computed by taking the difference in motor position at each sample time, and the position is measured by a shaft encoder. This can be problematic at very low speeds where the encoder tick rate is lower than the sample rate. In this case a better strategy is to measure the time between encoder ticks.

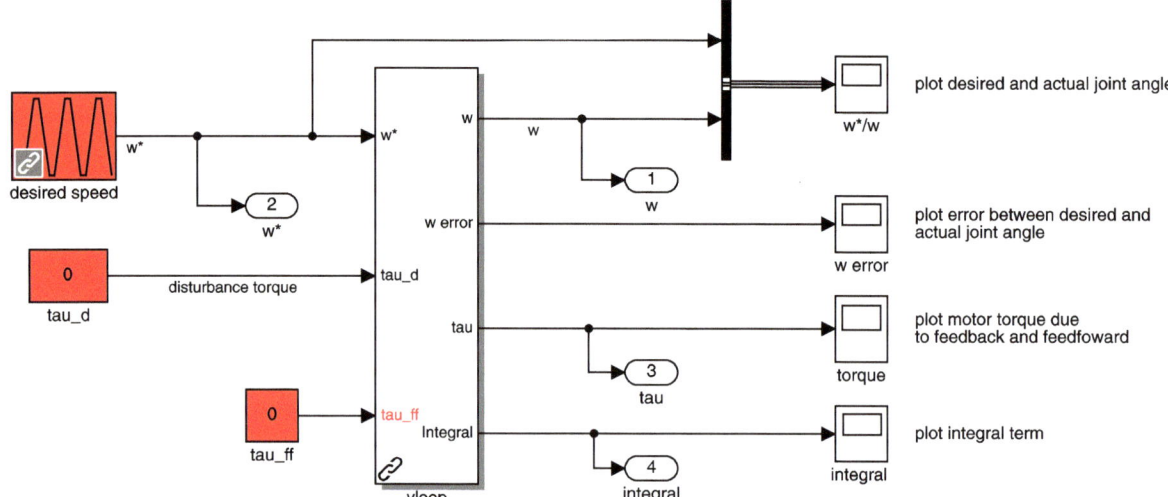

Fig. 9.8. Test harness for the velocity control loop, Simulink model `vloop_test`. The input `tau_d` is used to simulate a disturbance torque acting on the joint

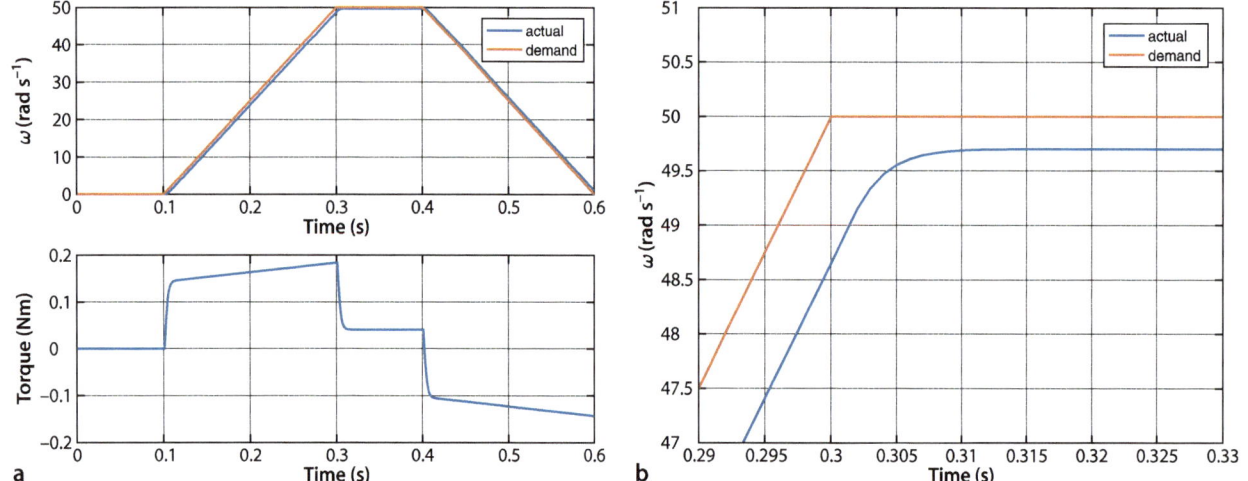

a

b

Fig. 9.9. Velocity loop with a trapezoidal demand. **a** Response; **b** closeup of response

We also observe a very slight steady-state error – the actual velocity is less than the demand at all times. From a classical control system perspective, the velocity loop contains no integrator block and is classified as a Type 0 system – a characteristic of Type 0 systems is they exhibit a finite error for a constant input. More intuitively we can argue that in order to move at constant speed the motor must generate a finite torque to overcome friction, and since motor torque is proportional to velocity error there must be a finite velocity error.

Now we will investigate the effect of inertia variation on the closed-loop response. Using Eq. 9.5 and the data from Fig. 9.16c we find that the minimum and maximum joint inertia at the motor are 515×10^{-6} and 648×10^{-6} kg m^2 respectively. Figure 9.10 shows the velocity tracking error using the control gains chosen above for various values of link inertia. We can see that the tracking error decays more slowly for larger inertia, and is showing signs of instability for the case of zero link inertia. For a case where the inertia variation is more extreme the gain should be chosen to achieve satisfactory closed-loop performance at both extremes.

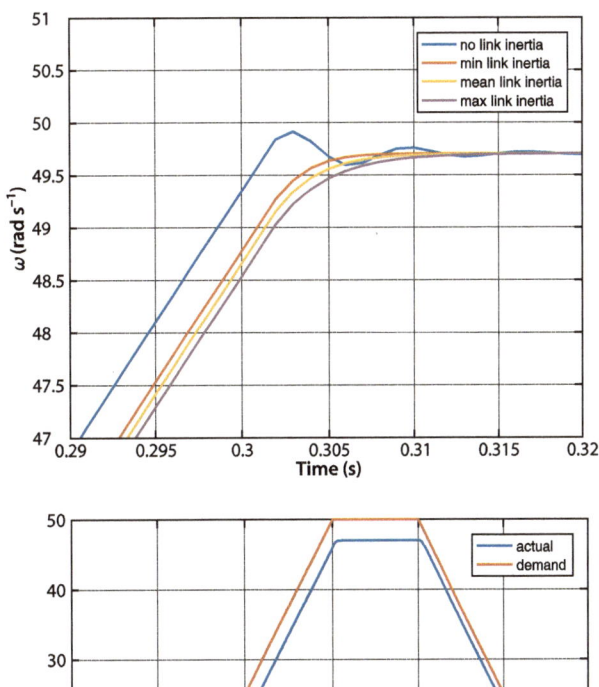

Fig. 9.10.
Velocity loop response with a trapezoidal demand for varying inertia M_{22}

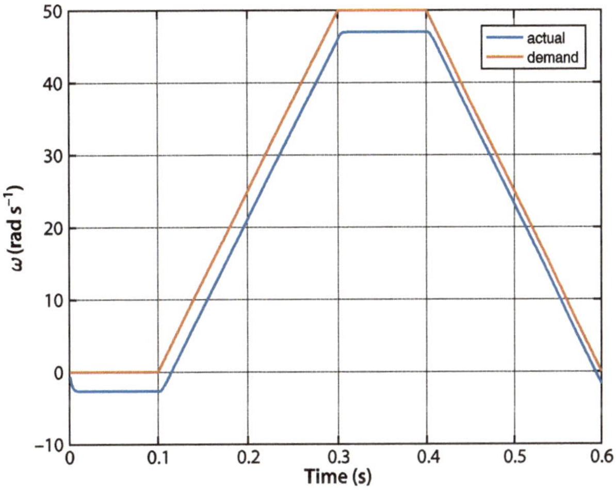

Fig. 9.11.
Velocity loop response to a trapezoidal demand with a gravity disturbance of 20 N m

Motor limits. Electric motors are limited in both torque and speed. The maximum torque is defined by the maximum current the drive electronics can provide. A motor also has a maximum rated current beyond which the motor can be damaged by overheating or demagnetization of its permanent magnets which irreversibly reduces its torque constant. As speed increases so does friction and the maximum speed is $\omega_{max} = \tau_{max} / B$.

The product of motor torque and speed is the mechanical output power and this also has an upper bound. Motors can tolerate some overloading, peak power and peak torque, for short periods of time but the sustained rating is significantly lower than the peak.

Figure 9.15a shows that the gravity torque on this joint varies from approximately -40 to 40 N m. We now add a disturbance torque equal to just half that maximum amount, 20 N m applied on the load side of the gearbox. We do this by setting a non-zero value in the `tau_d` block and rerunning the simulation. The results shown in Fig. 9.11 indicate that the control performance has been badly degraded – the tracking error has increased to more than 2 rad s^{-1}. This has the same root cause as the very small error we saw in Fig. 9.9 – a Type 0 system exhibits a finite error for a constant input or a constant disturbance.

There are three common approaches to counter this error. The first, and simplest, is to increase the gain. This will reduce the tracking error but push the system toward instability and increase the overshoot.

The second approach, commonly used in industrial motor drives, is to add integral action – adding an integrator changes the system to Type 1 which has zero error for a constant input or constant disturbance. We change Eq. 9.6 to a proportional-integral controller

$$u^* = \left(K_v + \frac{K_i}{s}\right)(\dot{q}^* - \dot{q}), \;\; K_i > 0$$

In the Simulink model of Fig. 9.7 this is achieved by setting `Ki` to a nonzero value. With some experimentation we find the gains $K_v = 1$ and $K_i = 10$ work well and the performance is shown in Fig. 9.12. The integrator state evolves over time to cancel out the disturbance term and we can see the error decaying to zero. In practice the disturbance varies over time and the integrator's ability to track it depends on the value of the integral gain K_i. In reality other disturbances affect the joint, for instance Coulomb friction and torques due to velocity and acceleration coupling. The controller needs to be well tuned so that these have minimal effect on the tracking performance.

As always in engineering there are some tradeoffs. The integral term can lead to increased overshoot so increasing K_i usually requires some compensating reduction of K_v. If the joint actuator is pushed to its performance limit, for instance the torque limit is reached, then the tracking error will grow with time since the motor acceleration will be lower than required. The integral of this increasing error will grow leading to a condition known as integral windup. When the joint finally reaches its destination the large accumulated integral keeps driving the motor forward until the integral decays – leading to large overshoot. Various strategies are employed to combat this, such as limiting the maximum value of the integrator, or only allowing integral action when the motor is close to its setpoint.

These two approaches are collectively referred to as disturbance rejection and are concerned with reducing the effect of an unknown disturbance. However if we think about the problem in its robotics context the gravity disturbance is not unknown. In Sect. 9.1.3 we showed how to compute the torque due to gravity that acts on each joint. If we know this torque, and the motor torque constant, we can *add* it to the output of the PI controller.▶

Even if the gravity load is known imprecisely this trick will reduce the magnitude of the disturbance.

The third approach is therefore to predict the disturbance and cancel it out – a strategy known as torque feedforward control. This is shown by the red wiring in Fig. 9.7 and can be demonstrated by setting the `tau_ff` block of Fig. 9.8 to the same, or approximately the same, value as the disturbance.

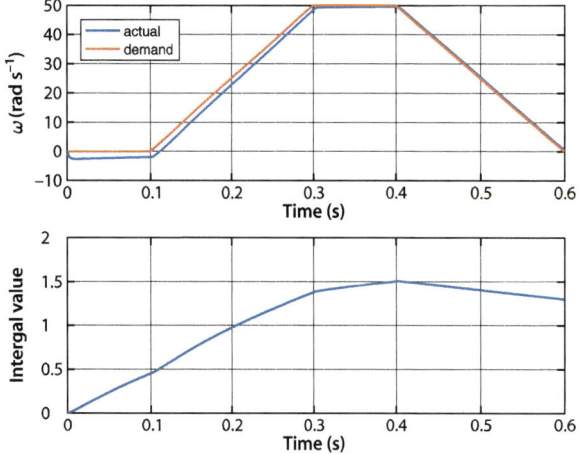

Fig. 9.12.
Velocity loop response to a trapezoidal demand with a gravity disturbance of 20 N m and proportional-integral control

Back EMF. A spinning motor acts like a generator and produces a voltage V_b called the back EMF which opposes the current flowing into the motor. Back EMF is proportional to motor speed $V_b = K_m \omega$ where K_m is the motor torque constant whose units can also be interpreted as V s rad^{-1}. When this voltage equals the maximum possible voltage from the drive electronics then no more current can flow into the motor and torque falls to zero. This provides a practical upper bound on motor speed, and torque at high speeds.

9.1.7   Position Control Loop

The outer loop is responsible for maintaining position and we use a proportional controller[◄] based on the error between actual and demanded position to compute the desired speed of the motor

Another common approach is to use a proportional-integral-derivative (PID) controller for position but it can be shown that the D gain of this controller is related to the P gain of the inner velocity loop.

$$\dot{q}^* = K_p \big(q^*(t) - q \big) \tag{9.7}$$

A Simulink model is shown in Fig. 9.13 and the position demand $q^*(t)$ comes from an LSPB trajectory generator that moves from 0 to 0.5 rad in 1 s with a sample rate of 1 000 Hz. Joint position is obtained by integrating joint velocity, obtained from the motor velocity loop via the gearbox. The error between the motor and desired position provides the velocity demand for the inner loop.

We load this control loop model

```
>> ploop_test
```

Fig. 9.13. Position control loop, Simulink model `ploop_test`. **a** Test harness for following an LSPB angle trajectory. **b** The position loop `ploop` which is a proportional controller around the inner velocity loop of Fig. 9.7

and its performance is tuned by adjusting the three gains: K_p, K_v, K_i in order to achieve good tracking performance along the trajectory. For $K_p = 40$ the tracking and error responses are shown in Fig. 9.14a. We see that the final error is zero but there is some tracking error along the path where the motor position lags behind the demand. The error between the demand and actual curves is due to the cumulative velocity error of the inner loop which has units of angle.

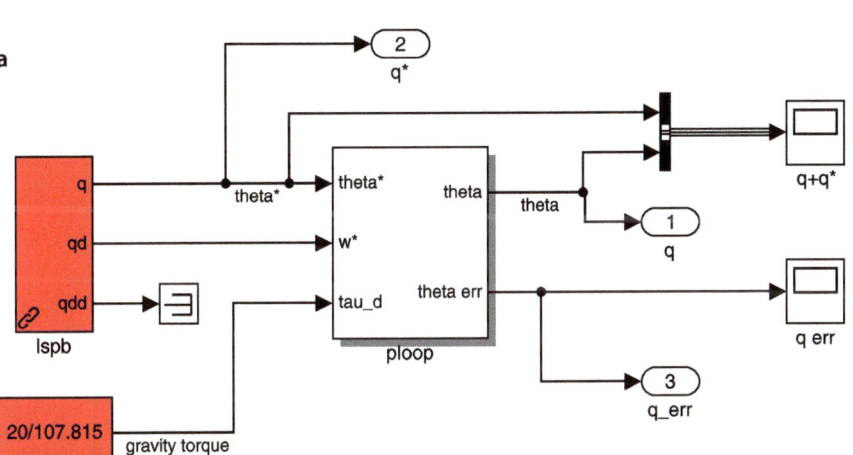

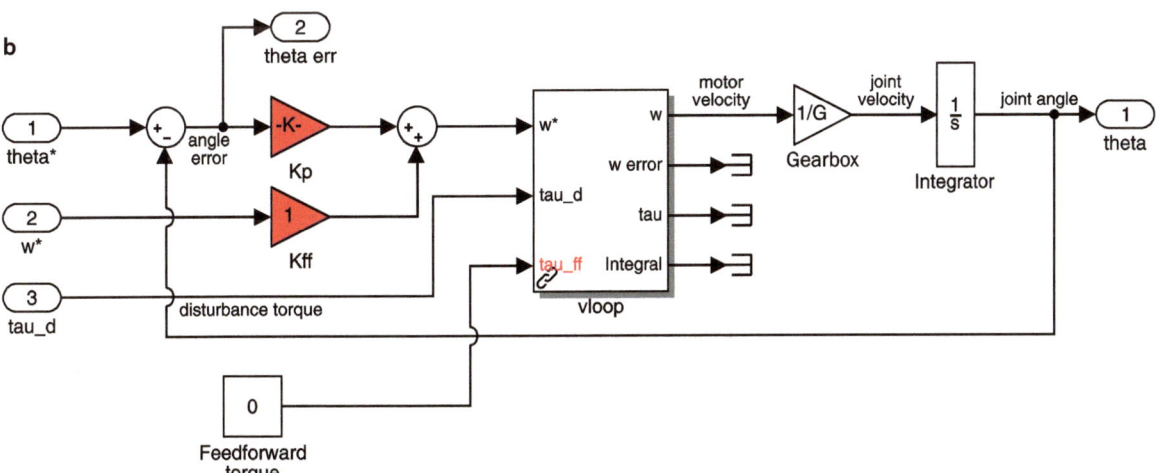

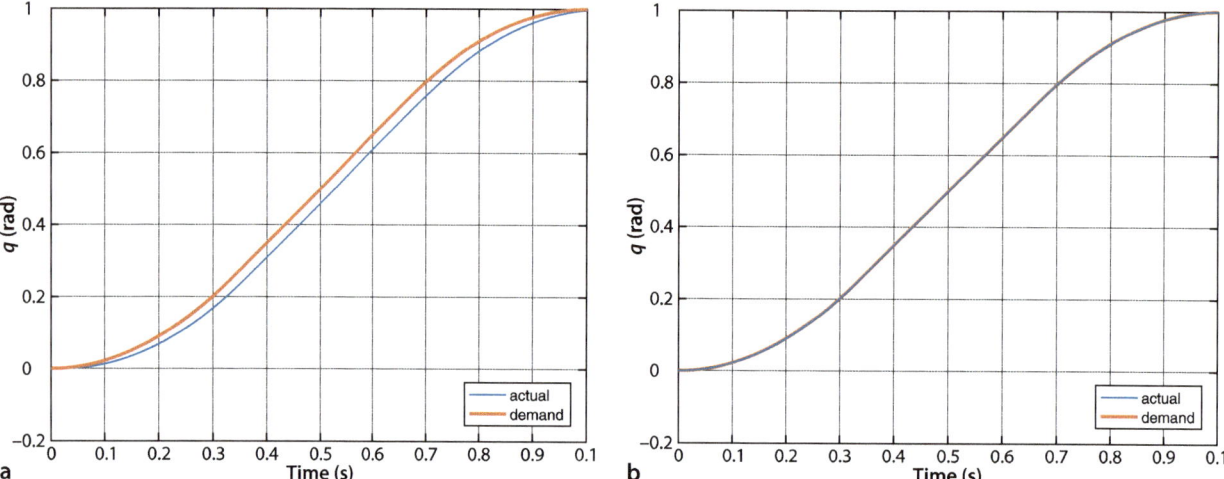

a

b

The position loop, like the velocity loop is based on classical negative feedback. Having zero position error while tracking a ramp would mean zero demanded velocity to the inner loop which is actually contradictory. More formally, we know that a Type 1 system◄ exhibits a constant error to a ramp input. If we care about reducing this tracking error there are two common remedies. We can add an integrator to the position loop – making it a proportional-integral controller but this gives us yet another parameter to tune. A simple and effective alternative is velocity feedforward control – we add the desired velocity to the output of the proportional control loop, which is the input to the velocity loop. The LSPB trajectory function computes velocity as a function of time as well as position. The time response with velocity feedforward is shown in Fig. 9.14b and we see that tracking error is greatly reduced.

Fig. 9.14. Position loop following an LSPB trajectory. **a** Proportional control only **b** proportional control plus velocity demand feedforward

Since the model contains an integrator after the velocity loop.

9.1.8   Independent Joint Control Summary

A common structure for robot joint control is the nested control loop. The inner loop uses a proportional or proportional-integral control law to generate a torque so that the actual velocity closely follows the velocity demand. The outer loop uses a proportional control law to generate the velocity demand so that the actual position closely follows the position demand. Disturbance torques due to gravity and other dynamic coupling effects impact the performance of the velocity loop as do variation in the parameters of the plant being controlled, and this in turn leads to errors in position tracking. Gearing reduces the magnitude of disturbance torques by $1 / G$ and the variation in inertia and friction by $1 / G^2$ but at the expense of cost, weight, increased friction and mechanical noise.

The velocity loop performance can be improved by adding an integral control term, or by feedforward of the disturbance torque which is largely predictable. The position loop performance can also be improved by feedforward of the desired joint velocity. In practice, control systems use both feedforward and feedback control. Feedforward is used to inject signals that we can compute, in this case the joint velocity, and in the earlier case the gravity torque. Feedback control compensates for all remaining sources of error including variation in inertia due to manipulator configuration and payload, changes in friction with time and temperature, and all the disturbance torques due to velocity and acceleration coupling. In general the use of feedforward allows the feedback gain to be reduced since a large part of the demand signal now comes from the feedforward.

9.2   Rigid-Body Equations of Motion

Consider the motor which actuates the jth revolute joint of a serial-link manipulator. From Fig. 7.5 we recall that joint j connects link $j-1$ to link j. The motor exerts a torque that causes the outward link, j, to rotationally accelerate but it also exerts a reaction torque on the inward link $j-1$. Gravity acting on the outward links j to N exert a weight force, and rotating links also exert gyroscopic forces on each other. The inertia that the motor *experiences* is a function of the configuration of the outward links.

The situation at the individual link is quite complex but for the *series* of links the result can be written elegantly and concisely as a set of coupled differential equations in matrix form

$$Q = M(q)\ddot{q} + C(q,\dot{q})\dot{q} + F(\dot{q}) + G(q) + J(q)^T W \tag{9.8}$$

where q, $\dot{q}$ and $\ddot{q}$ are respectively the vector of generalized joint coordinates, velocities and accelerations, M is the joint-space inertia matrix, C is the Coriolis and centripetal coupling matrix, F is the friction force, G is the gravity loading, and Q is the vector of generalized actuator forces associated with the generalized coordinates q. The last term gives the joint forces due to a wrench W applied at the end-effector and J is the manipulator Jacobian. This equation describes the manipulator rigid-body dynamics and is known as the inverse dynamics – given the pose, velocity and acceleration it computes the required joint forces or torques.

The recursive form of the inverse dynamics does not explicitly calculate the matrices M, C and G of Eq. 9.8. However we can use the recursive Newton-Euler algorithm to calculate these matrices and the Toolbox functions `inertia` and `coriolis` use Walker and Orin's (1982) 'Method 1'. While the recursive forms are computationally efficient for the inverse dynamics, to compute the coefficients of the individual dynamic terms (M, C and G) in Eq. 9.8 is quite costly – $O(N^3)$ for an N-axis manipulator.

These equations can be derived using any classical dynamics method such as Newton's second law and Euler's equation of motion, as discussed in Sect. 3.2.1, or a Lagrangian energy-based approach. A very efficient way for computing Eq. 9.8 is the recursive Newton-Euler algorithm which starts at the base and working outward adds the velocity and acceleration of each joint in order to determine the velocity and acceleration of each link. Then working from the tool back to the base, it computes the forces and moments acting on each link and thus the joint torques. The recursive Newton-Euler algorithm has $O(N)$ complexity and can be written in functional form as

$$Q = \mathcal{D}^{-1}(q,\dot{q},\ddot{q}) \tag{9.9}$$

Not all robot arm models in the Toolbox have dynamic parameters, see the "dynamics" tag in the output of the models() command, or use models('dyn') to list models with dynamic parameters. The Puma 560 robot is used for the examples in this chapter since its dynamic parameters are reliably known.

In the Toolbox it is implemented by the `rne` method of the `SerialLink` object. Consider the Puma 560 robot

```
>> mdl_puma560
```

at the nominal pose, and with zero joint velocity and acceleration. To achieve this state, the required generalized joint forces, or joint torques in this case, are

```
>> Q = p560.rne(qn, qz, qz)
Q =
   -0.0000   31.6399    6.0351    0.0000    0.0283         0
```

Since the robot is not moving (we specified $\dot{q}=\ddot{q}=0$) these torques must be those required to *hold the robot up* against gravity. We can confirm this by computing the torques required in the absence of gravity

```
>> Q = p560.rne(qn, qz, qz, 'gravity', [0 0 0])
ans =
     0     0     0     0     0     0
```

by overriding the object's default gravity vector.

Like most Toolbox methods `rne` can operate on a trajectory

```
>> q = jtraj(qz, qr, 10)
>> Q = p560.rne(q, 0*q, 0*q)
```

which has returned

```
>> about(Q)
Q [double] : 10x6 (480 bytes)
```

a 10×6 matrix with each row representing the generalized force required for the corresponding row of q. The joint torques corresponding to the fifth time step are

```
>> Q(5,:)
ans =
    0.0000   29.8883   0.2489        0        0        0
```

Consider now a case where the robot is moving. It is *instantaneously* at the nominal pose but joint 1 is moving at 1 rad s^{-1} and the acceleration of all joints is zero. Then in the absence of gravity, the required joint torques

```
>> p560.rne(qn, [1 0 0 0 0 0], qz, 'gravity', [0 0 0])
    30.5332    0.6280   -0.3607   -0.0003   -0.0000        0
```

are nonzero. The torque on joint 1 is that needed to overcome friction which always opposes the motion. More interesting is that torques need to be exerted on joints 2, 3 and 4. This is to oppose the gyroscopic effects (centripetal and Coriolis forces) – referred to as velocity coupling torques since the rotational velocity of one joint has induced a torque on several other joints.

The elements of the matrices *M*, *C*, *F* and *G* are complex functions of the link's kinematic parameters ($\theta_j, d_j, a_j, \alpha_j$) and inertial parameters. Each link has ten independent inertial parameters: the link mass m_j; the center of mass (COM) $r_j \in \mathbb{R}^3$ with respect to the link coordinate frame; and six second moments which represent the inertia of the link about the COM but with respect to axes aligned with the link frame {*j*}, see Fig. 7.5. We can view the dynamic parameters of a robot's link by

```
>> p560.links(1).dyn
Revolute(std): theta=q, d=0, a=0, alpha=1.5708, offset=0
  m    = 0
  r    = 0              0              0
  I    = | 0            0              0         |
         | 0            0.35           0         |
         | 0            0              0         |
  Jm   = 0.0002
  Bm   = 0.00148
  Tc   = 0.395        (+) -0.435       (-)
  G    = -62.61
  qlim = -2.792527 to 2.792527
```

which in order are: the kinematic parameters, link mass, COM position, link inertia matrix, motor inertia, motor friction, Coulomb friction, reduction gear ratio and joint angle limits.

The remainder of this section examines the various matrix components of Eq. 9.8.

9.2.1   Gravity Term

$$Q = M(q)\ddot{q} + C(q,\dot{q})\dot{q} + F(\dot{q}) + \boxed{G(q)} + J(q)^T W$$

We start our detailed discussion with the gravity term because it is generally the dominant term in Eq. 9.8 and is present even when the robot is stationary or moving slowly. Some robots use counterbalance weights▶ or even springs to reduce the *gravity* torque that needs to be provided by the motors – this allows the motors to be smaller and thus lower in cost.

In the previous section we used the rne method to compute the gravity load by setting the joint velocity and acceleration to zero. A more convenient approach is to use the gravload method

Counterbalancing will however increase the inertia associated with a joint since it adds additional mass at the end of a lever arm, and increase the overall mass of the robot.

```
>> gravload = p560.gravload(qn)
gravload =
  -0.0000   31.6399    6.0351    0.0000    0.0283         0
```

The SerialLink object contains a default gravitational acceleration vector which is initialized to the nominal value for Earth◄

The 'gravity' option for the SerialLink constructor can change this.

```
>> p560.gravity'
ans =
         0           0      9.8100
```

We could change gravity to the lunar value

```
>> p560.gravity = p560.gravity/6;
```

resulting in reduced joint torques

```
>> p560.gravload(qn)
ans =
   0.0000    5.2733    1.0059    0.0000    0.0047         0
```

or we could turn our lunar robot upside down

```
>> p560.base = SE3.Rx(pi);
>> p560.gravload(qn)
ans =
   0.0000   -5.2733   -1.0059   -0.0000   -0.0047         0
```

and see that the torques have changed sign. Before proceeding we bring our robot back to Earth and right-side up

```
>> mdl_puma560
```

The torque exerted on a joint due to gravity acting on the robot depends very strongly on the robot's pose. Intuitively the torque on the shoulder joint is much greater when the arm is stretched out horizontally

```
>> Q = p560.gravload(qs)
Q =
  -0.0000   46.0069    8.7722    0.0000    0.0283         0
```

than when the arm is pointing straight up

```
>> Q = p560.gravload(qr)
Q =
         0   -0.7752    0.2489         0         0         0
```

The gravity torque on the elbow is also very high in the first pose since it has to support the lower arm and the wrist. We can investigate how the gravity load on joints 2 and 3 varies with joint configuration by

```
1   [Q2,Q3] = meshgrid(-pi:0.1:pi, -pi:0.1:pi);
2   for i=1:numcols(Q2),
3       for j=1:numcols(Q3);
4       g = p560.gravload([0 Q2(i,j) Q3(i,j) 0 0 0]);
5       g2(i,j) = g(2);
6       g3(i,j) = g(3);
7       end
8   end
9   surfl(Q2, Q3, g2); surfl(Q2, Q3, g3);
```

Joseph-Louis Lagrange (1736–1813) was an Italian-born (Giuseppe Lodovico Lagrangia) French mathematician and astronomer. He made significant contributions to the fields of analysis, number theory, classical and celestial mechanics. In 1766 he succeeded Euler as the director of mathematics at the Prussian Academy of Sciences in Berlin, where he stayed for over twenty years, producing a large body of work and winning several prizes of the French Academy of Sciences. His treatise on analytical mechanics "Mécanique Analytique" first published in 1788, offered the most comprehensive treatment of classical mechanics since Newton and formed a basis for the development of mathematical physics in the nineteenth century. In 1787 he became a member of the French Academy, was the first professor of analysis at the École Polytechnique, helped drive the decimalization of France, was a member of the Legion of Honour and a Count of the Empire in 1808. He is buried in the Panthéon in Paris.

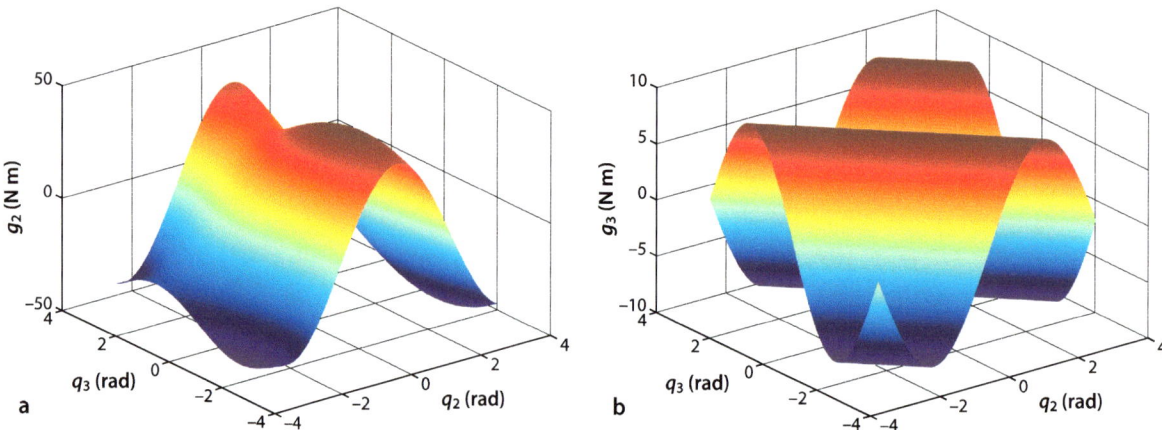

Fig. 9.15. Gravity load variation with manipulator pose. **a** Shoulder gravity load, $g_2(q_2, q_3)$; **b** elbow gravity load $g_3(q_2, q_3)$

and the results are shown in Fig. 9.15. The gravity torque on joint 2 varies between ± 40 N m and for joint 3 varies between ± 10 N m. This type of analysis is very important in robot design to determine the required torque capacity for the motors.

9.2.2   Inertia Matrix

$$Q = \boxed{M(q)\ddot{q}} + C(q, \dot{q})\dot{q} + F(\dot{q}) + G(q) + J(q)^T W$$

The joint-space inertia is a positive definite, and therefore symmetric, matrix►

The diagonal elements of this inertia matrix includes the motor armature inertias, multiplied by G^2.

```
>> M = p560.inertia(qn)
M =
    3.6594   -0.4044    0.1006   -0.0025    0.0000   -0.0000
   -0.4044    4.4137    0.3509    0.0000    0.0024    0.0000
    0.1006    0.3509    0.9378    0.0000    0.0015    0.0000
   -0.0025    0.0000    0.0000    0.1925    0.0000    0.0000
    0.0000    0.0024    0.0015    0.0000    0.1713    0.0000
   -0.0000    0.0000    0.0000    0.0000    0.0000    0.1941
```

which is a function of the manipulator configuration. The diagonal elements M_{jj} describe the inertia *experienced* by joint j, that is, $Q_j = M_{jj}\ddot{q}_j$. Note that the first two diagonal elements, corresponding to the robot's waist and shoulder joints, are large since motion of these joints involves rotation of the heavy upper- and lower-arm links. The off-diagonal terms $M_{ij} = M_{ji}$, $i \neq j$ are the products of inertia and represent coupling of acceleration from joint j to the generalized force on joint i.

We can investigate some of the elements of the inertia matrix and how they vary with robot configuration using the simple (but slow►) commands

```
1   [Q2,Q3] = meshgrid(-pi:0.1:pi, -pi:0.1:pi);
2   for i=1:numcols(Q2)
3       for j=1:numcols(Q3)
4           M = p560.inertia([0 Q2(i,j) Q3(i,j) 0 0 0]);
5           M11(i,j) = M(1,1);
6           M12(i,j) = M(1,2);
7       end
8   end
9   surfl(Q2, Q3, M11); surfl(Q2, Q3, M12);
```

Displaying the value of the robot object
```
>> p560
```
displays a tag `slowRNE` or `fastRNE`. The former indicates all calculations are done in MATLAB code. Build the MEX version, provided in the mex folder, to enable the `fastRNE` mode which is around 100 times faster.

The results are shown in Fig. 9.16 and we see significant variation in the value of M_{11} which changes by a factor of

```
>> max(M11(:)) / min(M11(:))
ans =
    2.1558
```

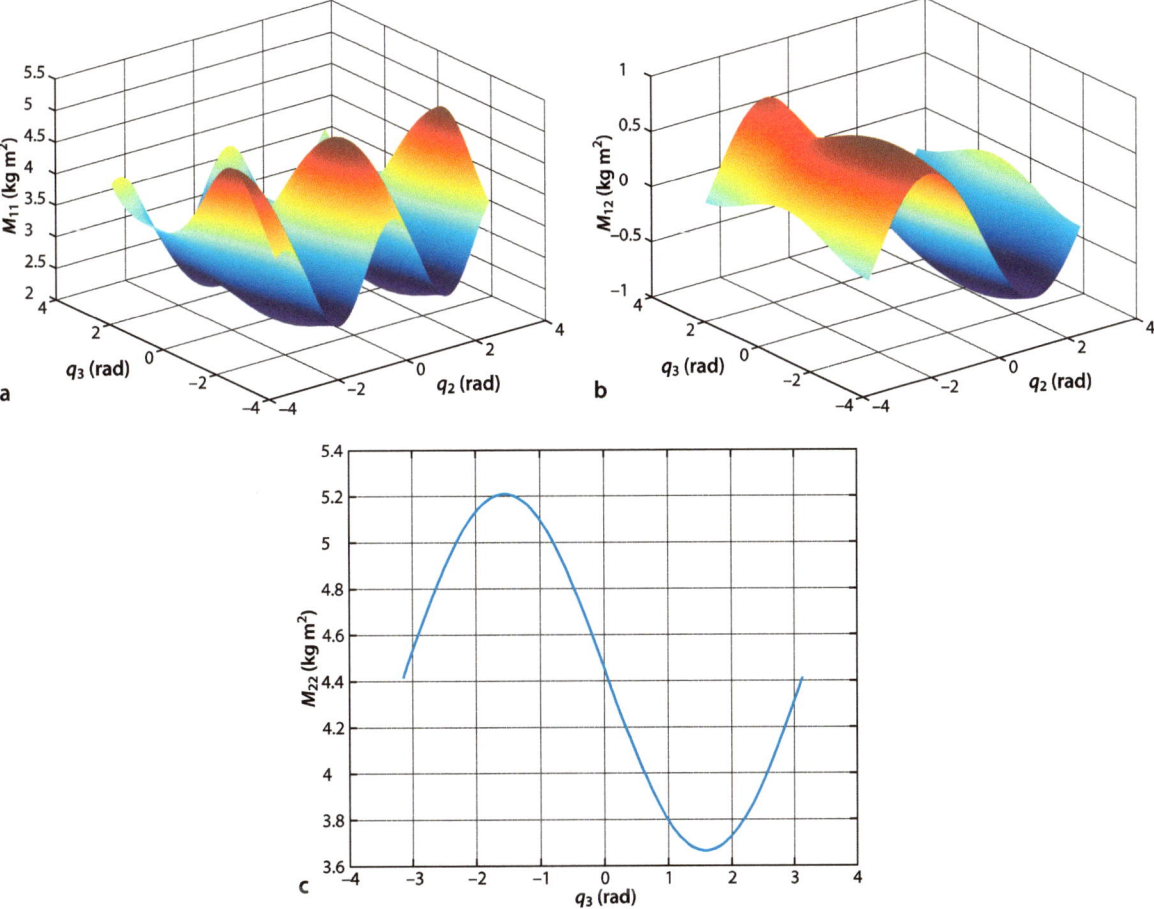

Fig. 9.16. Variation of inertia matrix elements as a function of manipulator pose. **a** Joint 1 inertia as a function of joint 2 and 3 angles $M_{11}(q_2, q_3)$; **b** product of inertia $M_{12}(q_2, q_3)$; **c** joint 2 inertia as a function of joint 3 angle $M_{22}(q_3)$

This is important for robot design since, for a fixed maximum motor torque, inertia sets the upper bound on acceleration which in turn effects path following accuracy.

The off-diagonal term M_{12} represents coupling between the angular acceleration of joint 2 and the torque on joint 1. That is, if joint 2 accelerates then a torque will be exerted on joint 1 and vice versa.

9.2.3 Coriolis Matrix

$$Q = M(q)\ddot{q} + \boxed{C(q, \dot{q})\dot{q}} + F(\dot{q}) + G(q) + J(q)^T W$$

The Coriolis matrix C is a function of joint coordinates and joint velocity. The centripetal torques are proportional to $\dot{q}_j^2$, while the Coriolis torques are proportional to $\dot{q}_i \dot{q}_j$. For example, at the nominal pose with the elbow joint moving at 1 rad s^{-1}

```
>> qd = [0 0 1 0 0 0];
```

the Coriolis matrix is

```
>> C = p560.coriolis(qn, qd)
C =
    0.8992   -0.2380   -0.2380    0.0005   -0.0375    0.0000
   -0.0000    0.9106    0.9106         0   -0.0036         0
    0.0000    0.0000   -0.0000         0   -0.0799         0
   -0.0559    0.0000    0.0000   -0.0000    0.0000   -0.0000
   -0.0000    0.0799    0.0799   -0.0000         0         0
    0.0000         0         0    0.0000         0         0
```

The off-diagonal terms $C_{i,j}$ represent coupling of joint j velocity to the generalized force acting on joint i. $C_{2,3} = 0.9106$ represents significant coupling from joint 3 velocity to torque on joint 2 – rotation of the elbow exerting a torque on the shoulder. Since the elements of this matrix represents a coupling from velocity to joint force they have the same dimensions as viscous friction or damping, however the sign can be positive or negative. The joint torques due to the motion of just this one joint are

```
>> C*qd'
ans =
   -0.2380
    0.9106
   -0.0000
    0.0000
    0.0799
         0
```

9.2.4   Friction

$$Q = M(q)\ddot{q} + C(q,\dot{q})\dot{q} + \boxed{F(\dot{q})} + G(q) + J(q)^T W$$

For most electric drive robots friction is the next most dominant joint force after gravity.▶

The Toolbox models friction within the `Link` object. The friction values are lumped and motor referenced, that is, they apply to the motor side of the gearbox. Viscous friction is a scalar that applies for positive and negative velocity.▶ Coulomb friction is a 2-vector comprising (Q_C^+, Q_C^-). The dynamic parameters of the Puma robot's first link are shown on page 262 as link parameters `Bm` and `Tc`. The online documentation for the `Link` class describes how to set these parameters.

For the Puma robot joint friction varied from 10 to 47% of the maximum motor torque for the first three joints (Corke 1996b).

In practice some mechanisms have a velocity dependent friction characteristic.

9.2.5   Effect of Payload

Any real robot has a specified maximum payload which is dictated by two dynamic effects. The first is that a mass at the end of the robot will increase the inertia *experienced* by the joint motors and which reduces acceleration and dynamic performance. The second is that mass generates a weight force which all the joints need to support. In the worst case the increased gravity torque component might exceed the rating of one or more motors. However even if the rating is not exceeded there is less torque available for acceleration which again reduces dynamic performance.

As an example we will add a 2.5 kg point mass to the Puma 560 which is its rated maximum payload. The center of mass of the payload cannot be at the center of the wrist coordinate frame, that is inside the wrist, so we will offset it 100 mm in the z-direction of the wrist frame. We achieve this by modifying the inertial parameters of the robot's last link▶

This assumes that the last link itself has no mass which is a reasonable approximation.

```
>> p560.payload(2.5, [0 0 0.1]);
```

The inertia at the nominal pose is now

```
>> M_loaded = p560.inertia(qn);
```

and the *ratio* with respect to the unloaded case, computed earlier, is

```
>> M_loaded ./ M
ans =
    1.3363    0.9872    2.1490   49.3960   80.1821    1.0000
    0.9872    1.2667    2.9191    5.9299   74.0092    1.0000
    2.1490    2.9191    1.6601   -2.1092   66.4071    1.0000
   49.3960    5.9299   -2.1092    1.0647   18.0253    1.0000
   83.4369   74.0092   66.4071   18.0253    1.1454    1.0000
    1.0000    1.0000    1.0000    1.0000    1.0000    1.0000
```

We see that the diagonal elements have increased, for instance the elbow joint inertia has increased by 66% which reduces the maximum acceleration by nearly 40%. Reduced acceleration impairs the robot's ability to accurately follow a high speed path. The inertia of joint 6 is unaffected since this added mass lies on the axis of this joint's rotation. Some off-diagonal terms have increased significantly, particularly in rows and columns four and five. This indicates that motion of joints 4 and 5, the wrist joints, which are swinging the offset mass give rise to large reaction forces that are *felt* by all the other robot joints.

The gravity load has also increased by some significant factors

```
>> p560.gravload(qn) ./ gravload
ans =
    0.3737    1.5222    2.5416    10.7826   86.8056        NaN
```

at the elbow and wrist. Note that the values for joints 1, 4 and 6 are invalid since they are each the quotient of numbers that are almost zero. We set the payload of the robot back to zero before proceeding

```
>> p560.payload(0)
```

9.2.6 Base Force

A moving robot exerts a wrench on its base – its weight as well as reaction forces and torques as the arm moves around. This wrench is returned as an optional output argument of the rne method, for example

```
>> [Q,Wb] = p560.rne(qn, qz, qz);
```

The wrench

```
>> Wb'
ans =
         0   -0.0000   230.0445   -48.4024   -31.6399   -0.0000
```

needs to be applied to the base to keep it in equilibrium. The vertical force of 230 N is the total weight of the robot which has a mass of

```
>> sum([p560.links.m])
ans =
   23.4500
```

There is also a moment about the x- and y-axes since the center of mass of the robot in this configuration is not over the origin of the base coordinate frame.

The base forces are important in situations where the robot does not have a rigid base such as on a satellite in space, on a boat, an underwater vehicle or even on a vehicle with soft suspension.

9.2.7 Dynamic Manipulability

In Sect. 8.2.2 we discussed a kinematic measure of manipulability, that is, how well configured the robot is to achieve velocity in any Cartesian direction. The force ellipsoid of Sect. 8.5.2 describes how well the manipulator is able to accelerate in different Cartesian directions but is based on the kinematic, not dynamic, parameters of the robot arm. Following a similar approach, we consider the set of generalized joint forces with unit norm

$$Q^T Q = 1$$

From Eq. 9.8 and ignoring gravity and assuming $\dot{q} = 0$ we write

$$Q = M\ddot{q}$$

Differentiating Eq. 8.2 and still assuming $\dot{q} = 0$ we write

$$\dot{\nu} = J(q)\ddot{q}$$

Combining these we write

$$\dot{\nu}^T \left(JM^{-1}M^{-T}J^T \right)^{-1} \dot{\nu} = 1$$

or more compactly

$$\dot{\nu}^T M_x^{-1} \dot{\nu} = 1$$

which is the equation of a hyperellipsoid in Cartesian acceleration space. For example, at the nominal pose

```
>> J = p560.jacob0(qn);
>> M = p560.inertia(qn);
>> Mx = (J * inv(M) * inv(M)' * J');
```

If we consider just the translational acceleration, that is the top left 3×3 submatrix of M_x

```
>> Mx = Mx(1:3, 1:3);
```

this is a 3-dimensional ellipsoid

```
>> plot_ellipse( Mx )
```

which is plotted in Fig. 9.17. The major axis of this ellipsoid is the direction in which the manipulator has maximum acceleration at this configuration. The radii of the ellipse are the square roots of the eigenvalues

```
>> sqrt(eig(Mx))
ans =
     0.4412
     0.1039
     0.1677
```

and the direction of maximum acceleration is given by the first eigenvector. The ratio of the minimum to maximum radius

```
>> min(ans)/max(ans)
ans =
     0.2355
```

is a measure of the nonuniformity of end-effector acceleration.▶ It would be unity for isotropic acceleration capability. In this case acceleration capability is good in the x- and z-directions, but poor in the y-direction.

The 6-dimensional ellipsoid has dimensions with different units: m s^{-2} and rad s^{-2}. This makes comparison of all 6 radii problematic.

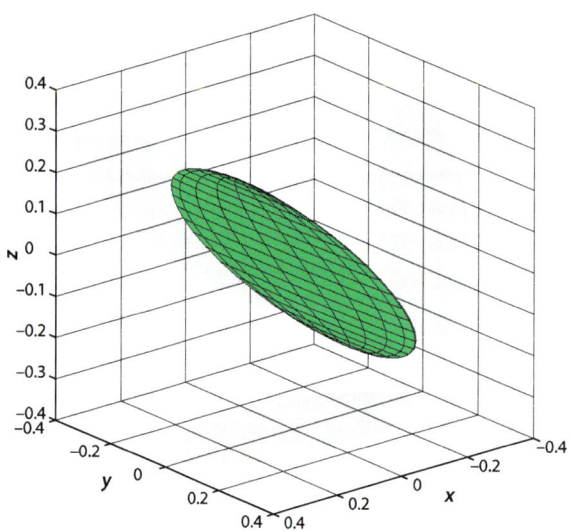

Fig. 9.17.
Spatial acceleration ellipsoid for Puma 560 robot in nominal pose

The scalar dynamic manipulability measure proposed by Asada is similar but considers the ratios of the eigenvalues of

$$J^{-T}MJ^{-1}$$

and returns a uniformity measure $m \in [0, 1]$ where 1 indicates uniformity of acceleration in all directions. For this example

```
>> p560.maniplty(qn, 'asada')
ans =
    0.2094
```

9.3   Forward Dynamics

To determine the motion of the manipulator in response to the forces and torques applied to its joints we require the forward dynamics or integral dynamics. Rearranging the equations of motion Eq. 9.8 we obtain the joint acceleration

$$\ddot{q} = M^{-1}(q)\Big(Q - C(q,\dot{q})\dot{q} - F(\dot{q}) - G(q) - J(q)^T W\Big) \tag{9.10}$$

and M is always invertible. This function is computed by the `accel` method of the `SerialLink` class

```
qdd = p560.accel(q, qd, Q)
```

given the joint coordinates, joint velocity and applied joint torques. This functionality is also encapsulated in the Simulink block `Robot` and an example of its use is

```
>> sl_ztorque
```

which is shown in Fig. 9.18. The torque applied to the robot is zero and the initial joint angles is set as a parameter of the `Robot` block, in this case to the *zero-angle pose*. The simulation is run

```
>> r = sim('sl_ztorque');
```

and the joint angles as a function of time are returned in the object `r`

```
>> t = r.find('tout');
>> q = r.find('yout');
```

We can show the robot's motion in animation

```
>> p560.plot(q)
```

and see it collapsing under gravity since there are no torques to counter gravity and hold in upright. The shoulder falls and swings back and forth as does the elbow, while the waist joint rotates because of Coriolis coupling. The motion will slowly decay as the energy is dissipated by viscous friction.

Fig. 9.18.
Simulink model `sl_ztorque` for the Puma 560 manipulator with zero joint torques. This model removes Coulomb friction in order to simplify the numerical integration

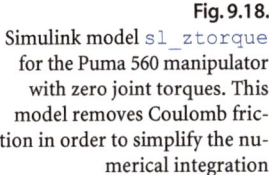
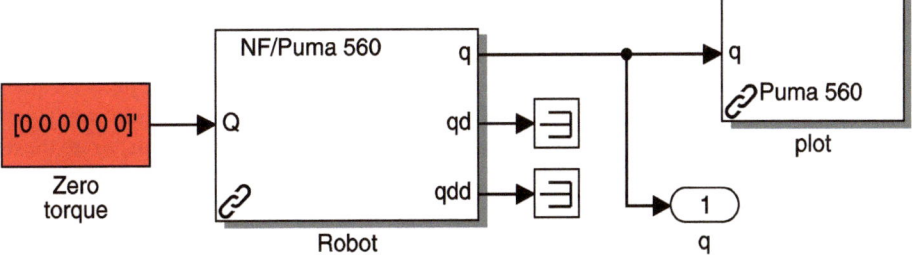

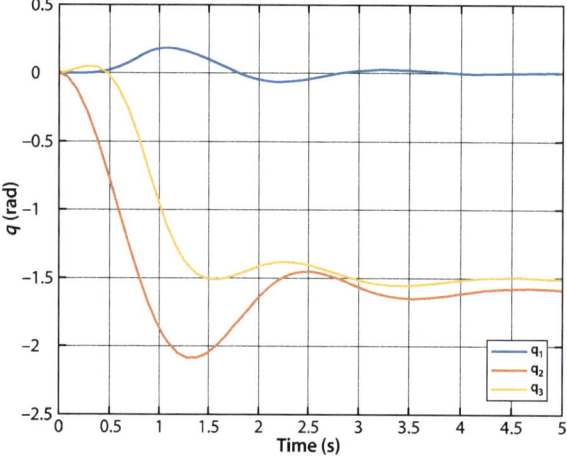

Fig. 9.19.
Joint angle trajectory for
Puma 560 robot with zero
Coulomb friction collapsing
under gravity from initial joint
configuration qz

Alternatively we can plot the joint angles as a function of time

```
>> plot(t, q(:,1:3))
```

and this is shown in Fig. 9.19. The method `fdyn` can be used as a nongraphical alternative to Simulink and is described in the online documentation.

This example is rather unrealistic and in reality the joint torques would be computed by some control law as a function of the actual and desired robot joint angles. This is the topic of the next section.

> Coulomb friction is a strong nonlinearity and can cause difficulty when using numerical integration routines to solve the forward dynamics. This is usually manifested by very long integration times. Fixed-step solvers tend to be more tolerant, and these can be selected through the Simulink `Simulation+Model Configuration Parameters+Solver` menu item.
>
> The default Puma 560 model, defined using `mdl_puma560`, has nonzero viscous and Coulomb friction parameters for each joint. Sometimes it is useful to zero the friction parameters for a robot and this can be achieved by
>
> ```
> >> p560_nf = p560.nofriction();
> ```
>
> which returns a copy of the robot object that is similar in all respects except that the Coulomb friction is zero. Alternatively we can set Coulomb *and* viscous friction coefficients to zero
>
> ```
> >> p560_nf = p560.nofriction('all');
> ```

9.4   Rigid-Body Dynamics Compensation

In Sect. 9.1 we discussed some of the challenges for independent joint control and introduced the concept of feedforward to compensate for the gravity disturbance torque. Inertia variation and other dynamic coupling forces were not explicitly dealt with and were left for the feedback controller to handle. However inertia and coupling torques can be computed according to Eq. 9.8 given knowledge of joint angles, joint velocities and accelerations, and the inertial parameters of the links. We can incorporate these torques into the control law using one of two *model-based* approaches: feedforward control, and computed torque control. The structural differences are contrasted in Fig. 9.20 and Fig. 9.21.

9.4.1   Feedforward Control

The torque feedforward controller shown in Fig. 9.20 is given by

$$Q^* = \underbrace{M(q^*)\ddot{q}^* + C(q^*,\dot{q}^*)\dot{q}^* + F(\dot{q}^*) + G(q^*)}_{\text{feedforward}} + \underbrace{\left\{ K_v(\dot{q}^* - \dot{q}^\#) + K_p(q^* - q^\#) \right\}}_{\text{feedback}}$$

$$= \mathcal{D}^{-1}(q^*,\dot{q}^*,\ddot{q}^*) + \left\{ K_v(\dot{q}^* - \dot{q}^\#) + K_p(q^* - q^\#) \right\} \tag{9.11}$$

where K_p and K_v are the position and velocity gain (or damping) matrices respectively, and $\mathcal{D}^{-1}(\cdot)$ is the inverse dynamics function. The gain matrices are typically diagonal. The feedforward term provides the joint forces required for the desired manipulator state $(q^*, \dot{q}^*, \ddot{q}^*)$ and the feedback term compensates for any errors due to uncertainty in the inertial parameters, unmodeled forces or external disturbances.

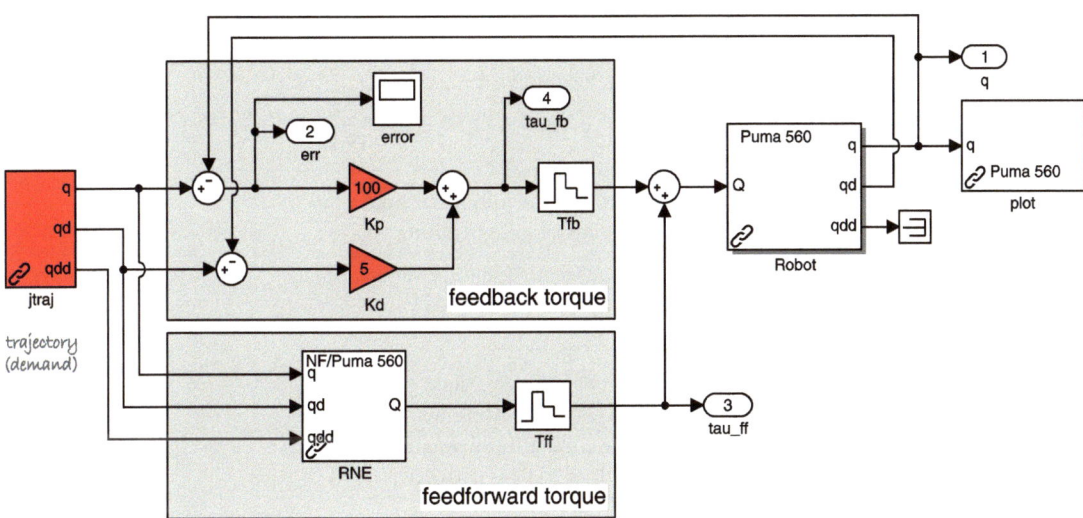

Fig. 9.20. The Simulink model `sl_fforward` for Puma 560 with torque feedforward control. The blocks with the staircase icons are zero-order holds

Fig. 9.21. Robotics Toolbox example `sl_ctorque`, computed torque control

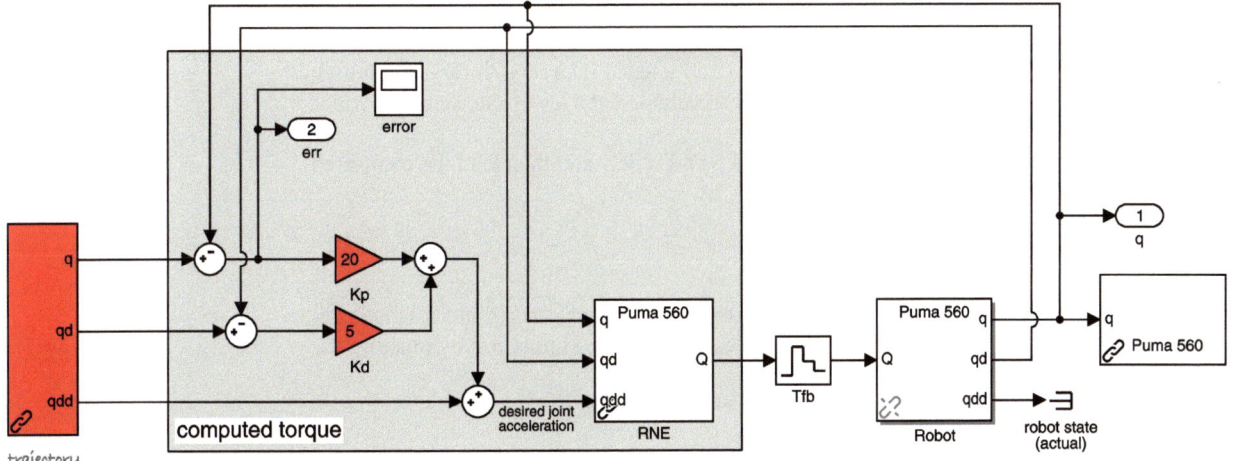

To test this controller using Simulink we first create a `SerialLink` object

```
>> mdl_puma560
```

and then load the torque feedforward controller model

```
>> sl_fforward
```

The feedforward torque is computed using the RNE block and added to the feedback torque computed from position and velocity error. The desired joint angles and velocity are generated using a `jtraj` block. Since the robot configuration changes relatively slowly the feedforward torque can be evaluated at a greater interval, T_{ff} than the error feedback loops, T_{fb}. In this example we use a zero-order hold block sampling at the relatively low sample rate of 20 Hz.

We run the simulation by pushing the Simulink play button or

```
>> r = sim('sl_fforward');
```

We can also consider that the feedforward term linearizes the nonlinear dynamics about the operating point $(q^*, \dot{q}^*, \ddot{q}^*)$. If the linearization is ideal then the dynamics of the error $e = q^* - q^{\#}$ can be obtained by combining Eq. 9.8 and 9.11

$$M\left(q^*\right)\ddot{e} + K_v\dot{e} + K_pe = 0 \tag{9.12}$$

For well chosen K_p and K_v the error will decay to zero but the joint errors are coupled▶ and their dynamics are dependent on the manipulator configuration.   Due to the nondiagonal matrix M.

9.4.2   Computed Torque Control

The computed torque controller is shown in Fig. 9.21. It belongs to a class of controllers known as inverse dynamic control. The principle is that the nonlinear system is cascaded with its inverse so that the overall system has a constant unity gain. In practice the inverse is not perfect so a feedback loop is required to deal with errors.

The computed torque control is given by

$$\begin{aligned}
Q &= M(q)\left\{\ddot{q}^* + K_v\left(\dot{q}^* - \dot{q}^{\#}\right) + K_p\left(q^* - q^{\#}\right)\right\} + C(q^*,\dot{q}^*)\dot{q}^* + F(\dot{q}^*) + G(q^*) \\
&= \mathcal{D}^{-1}\left(q^*, \dot{q}^*, \left(\ddot{q}^* + K_v\left(\dot{q}^* - \dot{q}^{\#}\right) + K_p\left(q^* - q^{\#}\right)\right)\right) \tag{9.13}
\end{aligned}$$

where K_p and K_v are the position and velocity gain (or damping) matrices respectively, and $\mathcal{D}^{-1}(\cdot)$ is the inverse dynamics function.

In this case the inverse dynamics must be evaluated at each servo interval, although the coefficient matrices M, C, and G could be evaluated at a lower rate since the robot configuration changes relatively slowly.

Using Simulink we first create a `SerialLink` object and then load the computed torque controller

```
>> mdl_puma560
>> sl_ctorque
```

The desired joint angles and velocity are generated using a `jtraj` block whose parameters are the initial and final joint angles. We run the simulation by pushing the Simulink play button or

```
>> r = sim('sl_ctorque');
```

Assuming ideal modeling and parameterization the error dynamics of the system are obtained by combining Eq. 9.8 and 9.13

$$\ddot{e} + K_v\dot{e} + K_p e = 0 \tag{9.14}$$

where $e = q^* - q^\#$. Unlike Eq. 9.12 the joint errors are uncoupled and their dynamics are therefore independent of manipulator configuration. In the case of model error there will be some coupling between axes, and the right-hand side of Eq. 9.14 will be a nonzero forcing function.

9.4.3 Operational Space Control

The control strategies so far have been posed in terms of the robot's joint coordinates – its configuration space. Equation 9.8 describes the relationship between joint position, velocity, acceleration and applied forces or torques. However we can also express the dynamics of the end-effector in the Cartesian operational space where we consider the end-effector as a rigid body with inertia that actuator and disturbance forces and torques act on. We can reformulate Eq. 9.8 in operational space as

$$\Lambda(x)\ddot{x} + \mu(x,\dot{x})\dot{x} + p(x) = W \tag{9.15}$$

where $x \in \mathbb{R}^6$ is the manipulator Cartesian pose◄ and Λ is the end-effector inertia which is subject to a gyroscopic and Coriolis force μ and gravity load p and an applied control wrench W. These operational space terms are related to those we have already discussed by

$$\Lambda(x) = J(q)^{-T} J(q)J(q)^{-1}$$
$$\mu(x,\dot{x}) = J(q)^{-T} C(q,\dot{q}) - \Lambda(q)\dot{J}(q)\dot{q}$$
$$p(x) = J(q)^{-T} g(q)$$

Comprising translation plus orientation represented using Euler or roll-pitch-yaw angles.

Imagine the task of wiping a table when the table's height is unknown and its surface is only approximately horizontal. The robot's z-axis is vertical, so to achieve the task we need to move the end-effector along a path in the xy-plane to achieve coverage and hold the wiper at a constant orientation about the z-axis. Simultaneously we maintain a constant force in the z-direction to hold the wiper against the table and a constant torque about the x- and y-axes in order to conform to the orientation of the table top. The first group of axes are position controlled, and the second group are force controlled. Each Cartesian degree of freedom can be either position or force controlled. The operational space control allows independent control of position and forces along and about the axes of the operational space coordinate frame.

A Simulink model of the controller and a simplified version of this scenario can be loaded by

```
>> sl_opspace
```

and is shown in Fig. 9.22. It comprises a position-control loop and a force-control loop whose results are summed together and used to drive the operational space robot model – details can be found by opening that block in the Simulink diagram. In this simulation the operational space coordinate frame is parallel to the end-effector coordinate frame. Motion is position controlled in the x- and y-directions and about the x-, y- and z-axes of this frame – the robot moves from its initial pose to a nearby pose using 5 out of the 6 Cartesian DOF.◄

The robot model and the compliance specification are set by the model's `InitFcn` *callback function. The setpoints are the red user adjustable boxes in the top-level diagram.*

Motion is force controlled in the z-direction with a setpoint of –5 N. To achieve this the controller moves the end-effector downward in order to decrease the force. It moves in free space until it touches the surface at $z = -0.2$ which is modeled as a stiffness of 100 N m^{-1}. Results in Fig. 9.23 show the x- and y-position moving toward the goal and the z-position decreasing and the simulated sensed force decreasing after contact. The controller is able to simultaneously satisfy both position and force constraints.

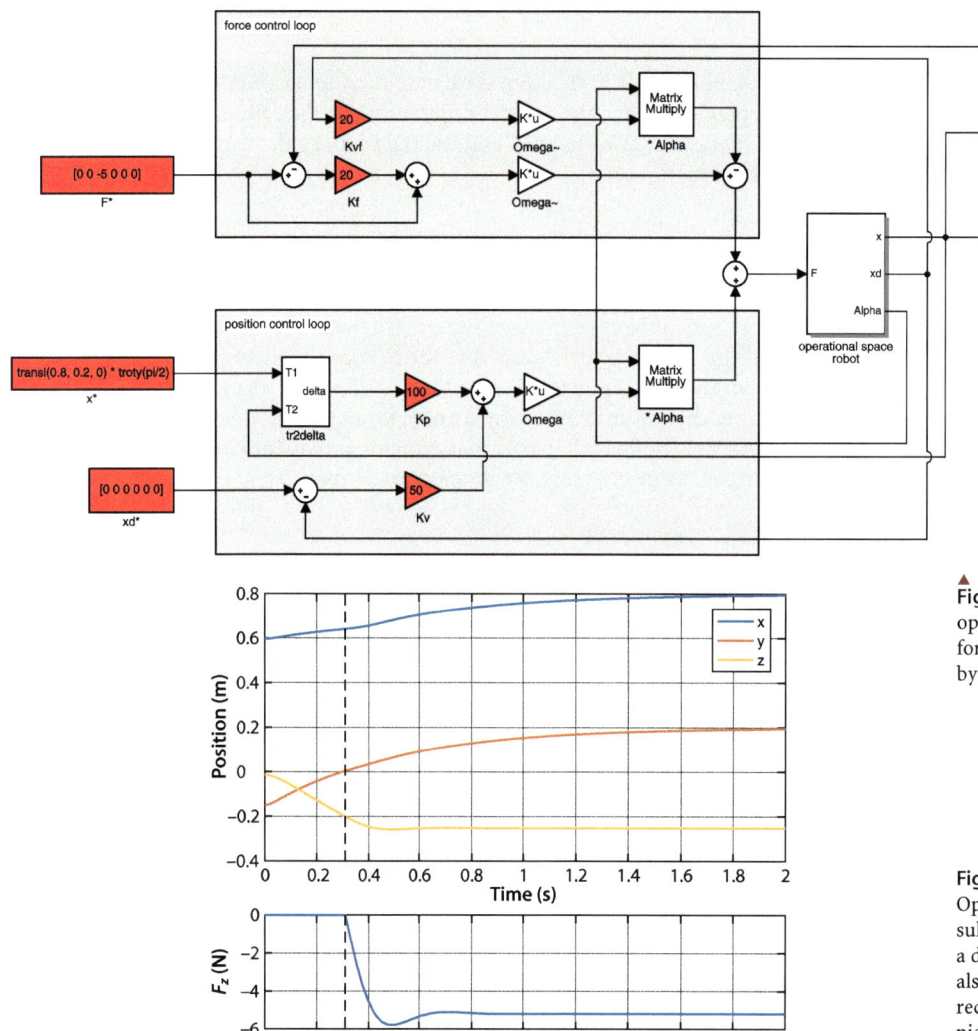

Fig. 9.22. Simulink model of an operational-space control system for a Puma 560 robot as described by (Khatib 1987)

Fig. 9.23.
Operational space controller results. The end-effector moves to a desired x- and y-position while also moving in the negative z-direction until it contacts the work piece and is able to exert the specified force of –5 N

9.5   Applications

9.5.1   Series-Elastic Actuator (SEA)

For high-speed robots the elasticity of the links and the joints becomes a significant dynamic effect which will affect path following accuracy. Joint elasticity is typically caused by elements of the transmission such as: longitudinal elasticity of a toothed belt or cable drive, a harmonic gearbox which is inherently elastic, or torsional elasticity of a motor shaft. In dynamic terms, as shown schematically in Fig. 9.24, the problem arises because the force is applied to one side of an elastic element and we wish to control the position of the other side – the actuator and sensor are not colocated. More complex still, and harder to analyze, is the case where the elasticity of the links must be taken into account.

However there are advantages in having some flexibility between the motor and the load. Imagine a robot performing a task that involves the gripper picking an object off a table whose height is uncertain.▶ A simple strategy to achieve this is to move down until the gripper touches the table, close the gripper and then lift up. However at the instant of contact a large and discontinuous force will be exerted on the robot which has the potential to damage the object or the robot. This is particularly problematic for robots with large

Or the robot is not very accurate.

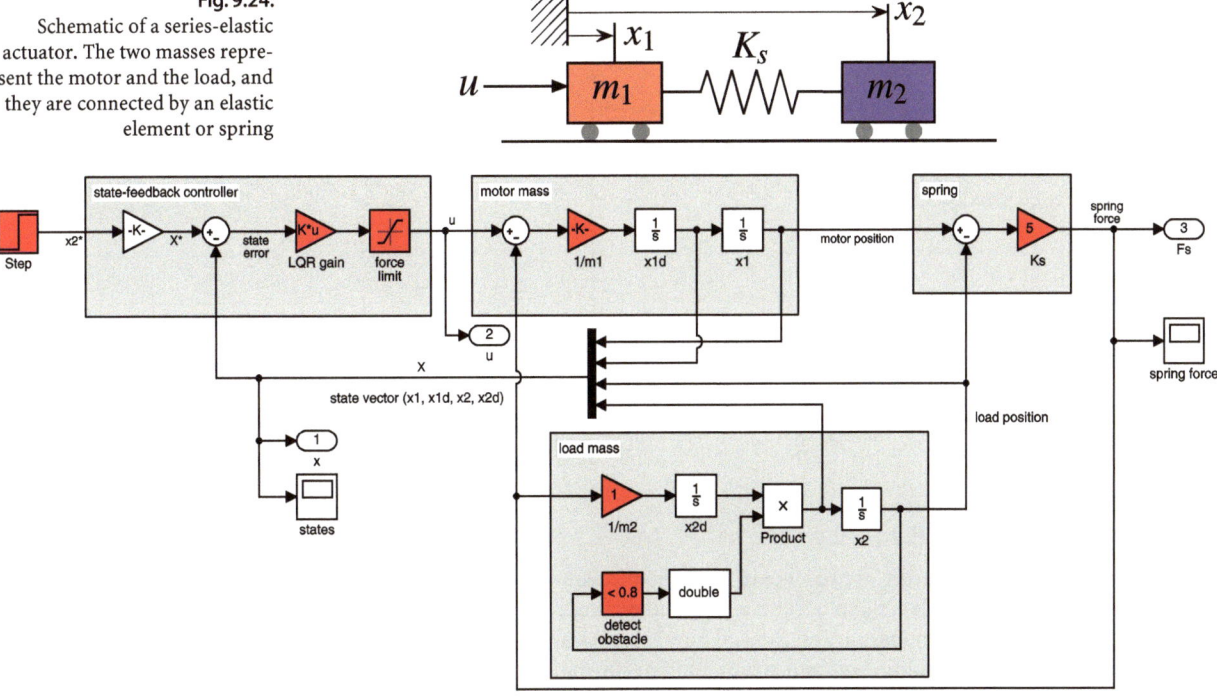

Fig. 9.24.
Schematic of a series-elastic actuator. The two masses represent the motor and the load, and they are connected by an elastic element or spring

Fig. 9.25. Simulink model `sl_sea` of a series-elastic actuator colliding with an obstacle

inertia that are moving quickly – the kinetic energy must be instantaneously dissipated. An elastic element – a spring – between the motor and the joint would help here. At the moment of contact the spring would start to compress and the kinetic energy is transferred to potential energy in the spring – the robot control system has time to react and stop or reverse the motors. We have changed the problem from a damaging hard impact to a soft impact. In addition to shock absorption, the deformation of the spring provides a means of determining the force that the robot is exerting. This capability is particularly useful for robots that interact closely with people since it makes the robot less dangerous in case of collision, and a spring is simple technology that cannot fail. For robots that must exert a force as part of their task, this is a simpler approach than the operational space controller introduced in Sect. 9.4.3. However position control is now more challenging because there is an elastic element between the motor and the load.

Consider the 1-dimensional case shown in Fig. 9.24 where the motor is represented by a mass m_1 to which a controllable force u is applied.◄ It is connected via a linear elastic element or spring to the load mass m_2. If we apply a positive force to m_1 it will move to the right and compress the spring, and this will exert a positive force on m_2 which will also move to the right. Controlling the position of m_2 is not trivial since this system has no friction and is marginally stable. It can be stabilized by feedback of position and velocity of the motor *and* of the load – all of which are potentially measurable.

In a real robot this is a rotary system with a torsional spring.

In robotics such a system, built into a robot joint, is known as a series-elastic actuator or SEA. The Baxter robot of Fig. 7.1b includes SEAs in some of its joints.

A Simulink model of an SEA system can be loaded by

```
>> sl_sea
```

and is shown in Fig. 9.25. A state-feedback LQR controller has been designed using MATLAB and requires input of motor and load position and velocity which form a vector x in the Simulink model. Fig. 9.26 shows a simulation of the model moving the load m_2 to a position $x_2^* = 1$. In the first case there is no obstacle and it achieves the goal with minimal overshoot, but note the complex force profile applied to m_1. In the second case the load mass is stopped at $x_2 = 0.8$ and the elastic force changes to accomodate this.

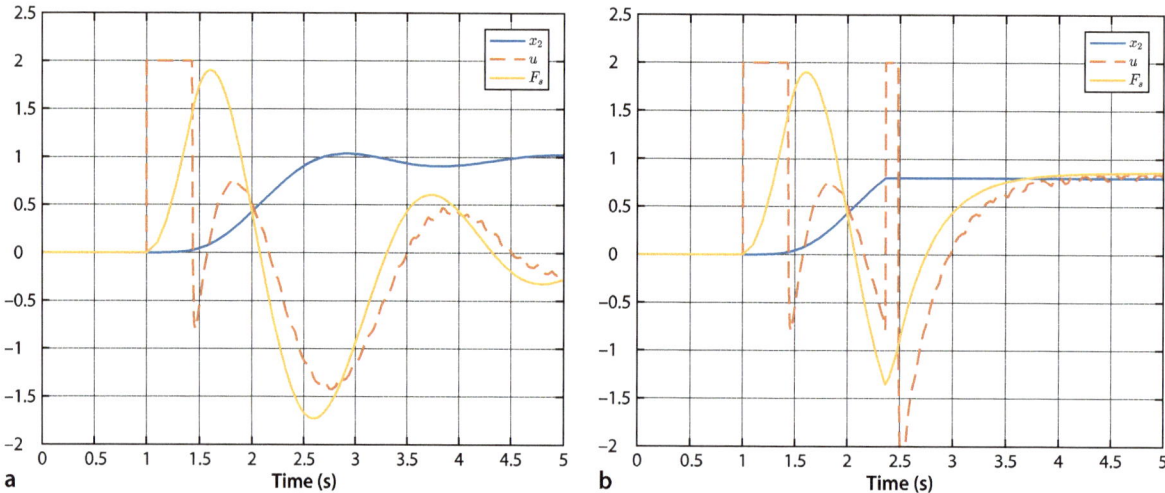

9.6   Wrapping Up

In this chapter we discussed approaches to robot manipulator control. We started with the simplest case of independent joint control, and explored the effect of disturbance torques and variation in inertia, and showed how feedforward of disturbances such as gravity could provide significant improvement in performance. We then learned how to model the forces and torques acting on the individual links of a serial-link manipulator. The equations of motion or inverse dynamics compute the joint forces required to achieve particular joint velocity and acceleration. The equations have terms corresponding to inertia, gravity, velocity coupling, friction and externally applied forces. We looked at the significance of these terms and how they vary with manipulator configuration and payload. The equations of motion provide insight into important issues such as how the velocity or acceleration of one joint exerts a disturbance force on other joints which is important for control design. We then discussed the forward dynamics which describe how the configuration evolves with time in response to forces and torques applied at the joints by the actuators and by external forces such as gravity. We extended the feedforward notion to full model-based control using torque feedforward, computed torque and operational-space controllers. Finally we discussed series-elastic actuators where a compliant element between the robot motor and the link enables force control and people-safe operation.

Further Reading

The engineering design of motor control systems is covered in mechatronics textbooks such as Bolton (2015). The dynamics of serial-link manipulators is well covered by all the standard robotics textbooks such as Paul (1981), Spong et al. (2006), Siciliano et al. (2009) and the Robotics Handbook (Siciliano and Khatib 2016). The efficient recursive Newton-Euler method we use today is the culmination of much research in the early 1980s and described in Hollerbach (1982). The equations of motion can be derived via a number of techniques, including Lagrangian (energy based), Newton-Euler, d'Alembert (Lee and Lee 1987; Lee et al. 1983) or Kane's method (Kane and Levinson 1983). However the computational cost of Lagrangian methods (Uicker 1965; Kahn 1969) is enormous, $O(N^4)$, which made it infeasible for real-time use on computers of that era and many simplifications and approximation had to be made. Orin et al. (1979) proposed an alternative approach based on the Newton-Euler (NE) equations of rigid-body motion applied to each link. Armstrong (1979) then showed how recursion could be applied resulting in

$O(N)$ complexity. Luh et al. (1980) provided a recursive formulation of the Newton-Euler equations with linear and angular velocities referred to link coordinate frames which resulted in a thousand-fold improvement in execution time making it practical to implement in real-time. Hollerbach (1980) showed how recursion could be applied to the Lagrangian form, and reduced the computation to within a factor of 3 of the recursive NE form, and Silver (1982) showed the equivalence of the recursive Lagrangian and Newton-Euler forms, and that the difference in efficiency was due to the representation of angular velocity.

The forward dynamics, Sect. 9.3, is computationally more expensive. An $O(N^3)$ method was proposed by Walker and Orin (1982) and is used in the Toolbox. Featherstone's (1987) articulated-body method has $O(N)$ complexity but for $N < 9$ is more expensive than Walker's method.

Critical to any consideration of robot dynamics is knowledge of the inertial parameters, ten per link, as well as the motor's parameters. Corke and Armstrong-Hélouvry (1994, 1995) published a meta-study of Puma parameters and provide a consensus estimate of inertial and motor parameters for the Puma 560 robot. Some of this data was obtained by painstaking disassembly of the robot and determining the mass and dimensions of the components. Inertia of components can be estimated from mass and dimensions by assuming mass distribution, or it can be measured using a bifilar pendulum as discussed in Armstrong et al. (1986).

Alternatively the parameters can be estimated by measuring the joint torques or the base reaction force and moment as the robot moves. A number of early works in this area include Mayeda et al. (1990), Izaguirre and Paul (1985), Khalil and Dombre (2002) and a more recent summary is Siciliano and Khatib (2016, § 6). Key to successful identification is that the robot moves in a way that is sufficiently exciting (Gautier and Khalil 1992; Armstrong 1989). Friction is an important dynamic characteristic and is well described in Armstrong's (1988) thesis. The survey by Armstrong-Hélouvry et al. (1994) is a very readable and thorough treatment of friction modeling and control. Motor parameters can be obtained directly from the manufacturer's data sheet or determined experimentally, without having to remove the motor from the robot, as described by Corke (1996a). The parameters used in the Toolbox Puma model are the best estimates from Corke and Armstrong-Hélouvry (1995) and Corke (1996a).

The discussion on control has been quite brief and has strongly emphasized the advantages of feedforward control. Robot joint control techniques are well covered by Spong et al. (2006), Craig (2005) and Siciliano et al. (2009) and summarized in Siciliano and Khatib (2016, § 8). Siciliano et al. have a good discussion of actuators and sensors as does the, now quite old, book by Klafter et al. (1989). The control of flexible joint robots is discussed in Spong et al. (2006). Adaptive control can be used to accomodate the time-varying inertial parameters and there is a large literature on this topic but some good early references include the book by Craig (1987) and key papers include Craig et al. (1987), Spong (1989), Middleton and Goodwin (1988) and Ortega and Spong (1989). The operational-space control structure was proposed in Khatib (1987). There has been considerable recent interest in series-elastic as well as variable stiffness actuators (VSA) whose position and stiffness can be independently controlled much like our own muscles – a good collection of articles on this technology can be found in the special issue by Vanderborght et al. (2008).

Dynamic manipulability is discussed in Spong et al. (2006) and Siciliano et al. (2009). The Asada measure used in the Toolbox is described in Asada (1983).

Historical and general. Newton's second law is described in his master work *Principia Nautralis* (mathematical principles of natural philosophy), written in Latin but an English translation is available online at http://www.archive.org/details/newtonspmathema00newtrich. His writing on other subjects, including transcripts of his notebooks, can be found online at http://www.newtonproject.sussex.ac.uk.

Exercises

1. Independent joint control (page 256ff)
   a) Investigate different values of `Kv` and `Ki` as well as demand signal shape and amplitude.
   b) Perform a root-locus analysis of `vloop` to determine the maximum permissible gain for the proportional case. Repeat this for the PI case.
   c) Consider that the motor is controlled by a voltage source instead of a current source, and that the motor's impedance is 1 mH and 1.6 Ω. Modify `vloop` accordingly. Extend the model to include the effect of back EMF.
   d) Increase the required speed of motion so that the motor torque becomes saturated. With integral action you will observe a phenomena known as integral windup – examine what happens to the state of the integrator during the motion. Various strategies are employed to combat this, such as limiting the maximum value of the integrator, or only allowing integral action when the motor is close to its setpoint. Experiment with some of these.
   e) Create a Simulink model of the Puma robot with each joint controlled by `vloop` and `ploop`. Parameters for the different motors in the Puma are described in Corke and Armstrong-Hélouvry (1995).
2. The motor torque constant has units of N m A^{-1} and is equal to the back EMF constant which has units of V s rad^{-1}. Show that these units are equivalent.
3. Simple two-link robot arm of Fig. 9.4
   a) Plot the gravity load as a function of both joint angles. Assume $m_1 = 0.45$ kg, $m_2 = 0.35$ kg, $r_1 = 8$ cm and $r_2 = 8$ cm.
   b) Plot the inertia for joint 1 as a function of q_2. To compute link inertia assume that we can model the link as a point mass located at the center of mass.
4. Run the code on page 263 to compute gravity loading on joints 2 and 3 as a function of configuration. Add a payload and repeat.
5. Run the code on page 264 to show how the inertia of joints 1 and 2 vary with payload?
6. Generate the curve of Fig. 9.16c. Add a payload and compare the results.
7. By what factor does this inertia vary over the joint angle range?
8. Why is the manipulator inertia matrix symmetric?
9. The robot exerts a wrench on the base as it moves (page 267). Consider that the robot is sitting on a frictionless horizontal table (say on a large air puck). Create a simulation model that includes the robot arm dynamics and the sliding dynamics on the table. Show that moving the arm causes the robot to translate and spin. Can you devise an arm motion that moves the robot base from one position to another and stops?
10. Overlay the dynamic manipulability ellipsoid on the display of the robot. Compare this with the force ellipsoid from Sect. 8.5.2.
11. Model-based control (page 271ff)
   a) Compute and display the joint tracking error for the torque feedforward and computed torque cases. Experiment with different motions, control parameters and sample rate T_{fb}.
   b) Reduce the rate at which the feedforward torque is computed and observe its effect on tracking error.
   c) In practice the dynamic model of the robot is not exactly known, we can only invert our best estimate of the rigid-body dynamics. In simulation we can model this by using the `perturb` method, see the online documentation, which returns a robot object with inertial parameters varied by plus and minus the specified percentage. Modify the Simulink models so that the RNE block is using a robot model with parameters perturbed by 10%. This means that the inverse dynamics are computed for a slightly different dynamic model to the robot under control and shows the effect of model error on control performance. Investigate the effects on error for both the torque feedforward and computed torque cases.

d) Expand the operational-space control example to include a sensor that measures all the forces and torques exerted by the robot.on an inclined table surface. Move the robot end-effector along a circular path in the xy-plane while exerting a constant downward force – the end-effector should move up and down as it traces out the circle. Show how the controller allows the robot tool to conform to a surface with unknown height and surface orientation.

12. Series-elastic actuator (page 274)

a) Experiment with different values of stiffness for the elastic element and control parameters. Try to reduce the settling time.

b) Modify the simulation so that the robot arm moves to touch an object at unknown distance and applies a force of 5 N to it.

c) Plot the frequency response function $X_2(s)/X_1(s)$ for different values of K_s, m_1 and m_2.

d) Simulate the effect of a collision between the load and an obstacle by adding a step to the spring force.

Appendices

Appendix A   **Installing the Toolbox**

Appendix B   **Linear Algebra Refresher**

Appendix C   **Geometry**

Appendix D   **Lie Groups and Algebras**

Appendix E   **Linearization, Jacobians and Hessians**

Appendix F   **Solving Systems of Equations**

Appendix G   **Gaussian Random Variables**

Appendix H   **Kalman Filter**

Appendix I   **Graphs**

A  Installing the Toolbox

The Toolbox is freely available from the book's home page

> http://www.petercorke.com/RVC

which also has a lot of additional information related to the book such as web links (all those printed in the book and more), code, figures, exercises and errata.

Downloading and Installing

One toolbox supports this book: the Robotics Toolbox (RTB). The relevant versions is RTB v10.

Toolboxes can be installed from .zip or .mltbx format files, with details below. Once the toolboxes are downloaded you can explore their capability using

```
>> rtbdemo
```

From .mltbx File

Since MATLAB® R2014b toolboxes can be packaged as, and installed from, files with the extension .mltbx.

1. Download the most recent version of `robot.mltbx` to your computer.
2. Using MATLAB navigate to the folder where you downloaded the file and double-click it (or right-click then select Install). The Toolbox will be installed within the local MATLAB file structure, and the paths will be appropriately configured for this, and future MATLAB sessions.
3. Other installed Toolboxes may interfere with the operation of this Toolbox. To check for potential problems run the command

```
>> rvccheck
```

and follow any instructions it suggests.

From .zip File

Follow these steps:

1. Download the most recent version of `robot.zip` to your computer.
2. Use your favorite unarchiving tool to unzip the files that you downloaded.
3. Add the Toolbox to your MATLAB path by running the commands

```
>> addpath RVCDIR
>> startup_rvc
```

© Springer Nature Switzerland AG 2022
P. Corke, *Robotics and Control*, Springer Tracts in Advanced Robotics 141,
https://doi.org/10.1007/978-3-030-79179-7

where RVCDIR is the full pathname of the directory where the folder rvctools was created when you unzipped the Toolbox files. The command startup_rvc adds various subfolders to your path and displays the version of the Toolbox.

4. Other installed Toolboxes may interfere with the operation of this Toolbox. To check for potential problems run the command

```
>> rvccheck
```

and follow any instructions it suggests.

You will need to run the startup_rvc command each time you start MATLAB. Alternatively, once you have run startup_rvc you can run

```
>> savepath
```

to save the path configuration for all future sessions.

For installation from zip files, the files for the Toolbox reside in a top-level directory called rvctools and beneath this are a number of subdirectories:

robot   The Robotics Toolbox.
common   Utility functions common to the Robotics and Machine Vision
               Toolboxes.
simulink   Simulink® blocks for robotics and vision, as well as examples.
contrib   Code written by third-parties.

MEX-Files

Some functions in the Toolbox are implemented as MEX-files, that is, they are written in C for computational efficiency but are callable from MATLAB just like any other function. Source code is provided in the mex folder along with instructions and scripts to build the MEX-files from inside MATLAB or from the command line. You will require a C-compiler in order to build these files, but prebuilt MEX-files for a limited number of architectures are included.

Contributed Code

A number of useful functions are provided by third-parties and wrappers have been written to make them consistent with other Toolbox functions. If you attempt to access a contributed function that is not installed you will receive an error message.

The contributed code contrib.zip can be downloaded, expanded and then added your MATLAB path. If you installed the Toolboxes from .zip files then expand contrib.zip inside the folder RVCDIR.

Many of these contributed functions are part of active software projects and the downloadable file is a snapshot that has been tested and works as described in this book.

Getting Help

A Google group at http://tiny.cc/rvcforum provides answers to frequently asked questions, and has a user forum for discussing questions, issues and bugs.

License

All the non-third-party code is released under the LGPL license. This means you are free to distribute it in original or modified form provided that you keep the license and authorship information intact.

The third-party code modules are provided under various open-source licenses. The Toolbox compatibility wrappers for these modules are provided under compatible licenses.

MATLAB Versions

The Toolbox software for this book has been developed and tested using MATLAB R2017b and R2018a under Mac OS X (10.13 High Sierra). MATLAB continuously evolves so older versions of MATLAB are increasingly unlikely to work. Please do not report bugs if you are using a MATLAB version older than R2016a.

Learning MATLAB

Resources for learning MATLAB are available at https://matlabacademy.mathworks. com. Some basic courses are free and open, while others require a MATLAB licence.

Octave

GNU Octave (www.octave.org) is free software that implements a language that is close to, but not the same as, MATLAB. There is support for some of the functionality of this Toolbox in Octave. For up-to-date details see http://petercorke.com/wordpress/toolboxes/other-languages.

Python

Python is a free programming language with a large and active community. The language has a very different syntax to MATLAB but well-supported libraries provide excellent support for linear algebra and graphics. Several packages provide some of the functionality of this Toolbox in Python. For up-to-date details see http://petercorke.com/wordpress/toolboxes/other-languages.

B

Linear Algebra Refresher

B.1   Vectors

A rank 1 tensor.

We will only consider real vectors◀ which are an ordered *n-tuple* of real numbers $v_1, v_2, \cdots v_n$ which is usually written as

$$\boldsymbol{v} = \begin{pmatrix} v_1 \\ v_2 \\ \vdots \\ v_n \end{pmatrix} \quad \text{or} \quad \boldsymbol{v} = (v_1, v_2, \cdots v_n)$$

which are a colum- and row-vector respectively. These are equivalent to an $n \times 1$ and a $1 \times n$ matrix respectively, and can be multiplied with a conforming matrix.

The numbers v_1, v_2 etc. are called the scalar components of $\boldsymbol{v}$, and v_i is called the ith component of $\boldsymbol{v}$. For a 3-vector we often write the elements as $\boldsymbol{v} = (v_x, v_y, v_z)$.

The symbol $\mathbb{R}^n$ represents the set of ordered n-tuples of real numbers, each vector is a point in this space, that is $\boldsymbol{v} \in \mathbb{R}^n$. The elements of $\mathbb{R}^2$ can be represented in a plane by a point or a directed line segment. The elements of $\mathbb{R}^3$ can be represented in a volume by a point or a directed line segment.

A vector space is an n-dimensional space whose elements are vectors *plus* the operations of addition and scalar multiplication. The addition of any two elements $\boldsymbol{a}, \boldsymbol{b} \in \mathbb{R}^n$ yields $(a_1 + b_1, a_2 + b_2 \cdots a_n + b_n)$ and $s\boldsymbol{a} = (sa_1, sa_2 \cdots sa_n)$. Both results are element of $\mathbb{R}^n$. The negative of a vector is obtained by negating each element of the vector $-\boldsymbol{a} = (-a_1, -a_2 \cdots -a_n)$.

We can use a vector to represent a point with coordinates $(x_1, x_2, \cdots x_n)$ which is called a coordinate vector. However we need to be careful because the operations of addition and scalar multiplication, while valid for vectors are meaningless for points. We can add a vector to the coordinate vector of a point to obtain the coordinate vector of another point, and we can subtract one coordinate vector from another, and the result is the is the displacement between the points.

The magnitude or length of a vector is a nonnegative scalar given by its *p*-norm

$$\|v\|_p = \left(\sum_{i=1}^{n} |v_i|^p \right)^{1/p}$$

The Euclidean length of a vector is given by $\|v\|_2$ which is also referred to as the L_2 norm and is generally assumed when p is omitted, for example $\|v\|$. A unit vector is one where $\|v\|_2 = 1$ and is denoted as $\hat{v}$. The L_1 norm is sum of the absolute value of the elements and is also known as the Manhattan distance, it is the distance traveled when confined to moving along the lines in a grid. The L_∞ norm is the maximum element of the vector.

The dot product of two column vectors is a scalar

$$\boldsymbol{a} \cdot \boldsymbol{b} = \boldsymbol{b} \cdot \boldsymbol{a} = \boldsymbol{a}^T \boldsymbol{b} = \boldsymbol{b}^T \boldsymbol{a} = \sum_{i=1}^{n} a_i b_i = \|a\|_2 \|b\|_2 \cos \theta$$

where θ is the angle between the vectors. $\boldsymbol{a} \cdot \boldsymbol{b} = 0$ when the vectors are orthogonal. For 3-vectors the cross product

$$\boldsymbol{a} \times \boldsymbol{b} = -\boldsymbol{b} \times \boldsymbol{a} = \det\begin{pmatrix} \hat{\boldsymbol{x}} & \hat{\boldsymbol{y}} & \hat{\boldsymbol{z}} \\ a_1 & a_2 & a_3 \\ b_1 & b_1 & b_3 \end{pmatrix} = [a]_\times \boldsymbol{b} = \|a\|_2 \|b\|_2 \sin\theta\hat{\boldsymbol{n}}$$

where $\hat{\boldsymbol{x}}$ is a unit-vector parallel to the x-axis etc., $[\cdot]_\times$ is a skew-symmetric matrix as described in the next section, and $\hat{\boldsymbol{n}}$ is a unit-vector normal to the plane containing $\boldsymbol{a}$ and $\boldsymbol{b}$. If the vectors are parallel $\boldsymbol{a} \times \boldsymbol{b} = 0$.

B.2   Matrices

A taxonomy of matrices is shown in Fig. B.1. In this book we are concerned only with real $m \times n$ matrices▶

> Real matrices are a subset of all matrices. For the general case of complex matrices the term Hermitian is the analog of symmetric, and unitary the analog of orthogonal. A^H denotes the Hermitian transpose, the complex conjugate transpose of the complex matrix A. Matrices are rank 2 tensors.

$$A = \begin{pmatrix} a_{1,1} & a_{1,2} & \cdots & a_{1,n} \\ a_{2,1} & a_{2,2} & \cdots & a_{2,n} \\ \vdots & \vdots & \ddots & \\ a_{m,1} & a_{n,2} & \cdots & a_{m,n} \end{pmatrix}, \quad A \in \mathbb{R}^{m \times n}$$

with m rows and n columns. If $n = m$ the matrix is square.

The transpose is

$$\boldsymbol{B} = \boldsymbol{A}^T, \quad b_{i,j} = a_{j,i} \quad \forall i, j$$

and it can be shown that

$$(\boldsymbol{AB})^T = \boldsymbol{B}^T\boldsymbol{A}^T, \quad (\boldsymbol{ABC})^T = \boldsymbol{C}^T\boldsymbol{B}^T\boldsymbol{A}^T, \quad \text{etc.}$$

Fig. B.1. Taxonomy of matrices. Classes of matrices that are always singular are shown in *red*, those that are never singular are shown in *blue*

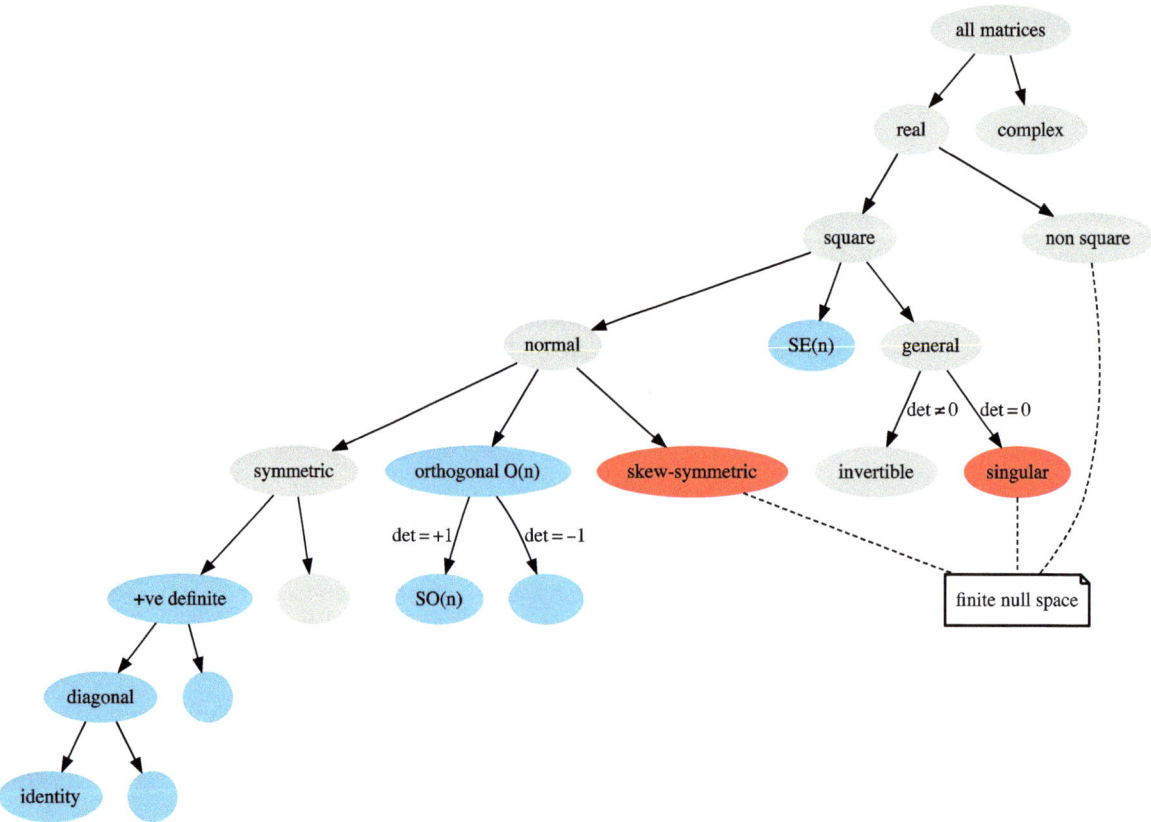

B.2.1 Square Matrices

A square matrix may have an inverse A^{-1} in which case

`Ai = inv(A)` $\qquad AA^{-1} = A^{-1}A = I_{n \times n}$

where

$$
I_{n \times n} = \begin{pmatrix} 1 & & & 0 \\ & 1 & & \\ & & \ddots & \\ 0 & & & 1 \end{pmatrix} \in \mathbb{R}^{n \times n}
$$

is the identity matrix, a unit diagonal matrix. The inverse exists provided that the matrix is nonsingular, that is, its determinant $\det(A) \neq 0$. The inverse can be computed from the matrix of cofactors. If A and B are square and nonsingular then

$$
(AB)^{-1} = B^{-1}A^{-1}, \quad (ABC)^{-1} = C^{-1}B^{-1}A^{-1}, \quad \text{etc.}
$$

and also

$$
\left(A^T \right)^{-1} = \left(A^{-1} \right)^T
$$

The inverse can be written as

$$
A^{-1} = \frac{1}{\det(A)} \text{adj}(A)
$$

where $\text{adj}(A)$ is the transpose of the matrix of cofactors and known as the adjugate or adjoint matrix and sometimes denoted by A^*. If $B = \text{adj}(A)$ then $A = \text{adj}(B)$. If A is nonsingular the adjugate can be computed by

$$
\text{adj}(A) = \det(A)A^{-1}
$$

For a square matrix if

$A = A^T$ ....the matrix is **symmetric**. The inverse of a symmetric matrix is also symmetric. Many matrices that we encounter in robotics are symmetric, for example covariance matrices and manipulator inertia matrices.

`S = skew(v)` $\quad A = -A^T$ ....the matrix is **skew-symmetric** or **anti-symmetric**. Such a matrix has a zero diagonal, is always singular and has the property that $[av]_\times = a[v]_\times$, $[Rv]_\times = R[v]_\times R^T$ and $v^T[v]_\times = [v]_\times v = 0, \forall v$. For the 3×3 case

$$
S = [v]_\times = \begin{pmatrix} 0 & -v_z & v_y \\ v_z & 0 & -v_x \\ -v_y & v_x & 0 \end{pmatrix} \tag{B.1}
$$

`v = vex(S)` $\qquad\qquad$ and the inverse operation is

$$
v = \text{vex}(S)
$$

$A^{-1} = A^T$ ....the matrix is **orthogonal**. The matrix is also known as orthonormal since its column vectors (and row vectors) must be of unit length and orthogonal to each other. The product of two orthogonal matrices of the same size is also an orthogonal matrix. The set of $n \times n$ orthogonal matrices forms a group $\mathbf{O}(n)$, known as the orthogonal group. The determinant of an orthogonal matrix is either $+1$ or -1. The subgroup $\mathbf{SO}(n)$ consisting of orthogonal matrices with determinant $+1$ is called the special orthogonal group. The columns (and rows) are orthogonal vectors, that is, their dot product is zero.

$A^T A = A A^T$ ......the matrix is **normal** and can be diagonalized by an orthogonal matrix U so that $U^T A U$ is a diagonal matrix. All symmetric, skew-symmetric and orthogonal matrices are normal matrices as are matrices of the form $A = B^T B = B B^T$ where B is an arbitrary matrix.

The square matrix $A \in \mathbb{R}^{n \times n}$ can be applied as a linear transformation to a vector $x \in \mathbb{R}^n$

$$x' = Ax$$

which results in another vector, generally with a change in its length and direction. However there are some important special cases. If $A \in \mathbf{SO}(n)$ the transformation is isometric and the vector's *length* is unchanged $\|x'\| = \|x\|$.

In 2-dimensions if x is the set of all points lying on a circle then x' defines points that lie on an ellipse. The MATLAB® builtin demonstration

```
>> eigshow
```

shows this very clearly as you interactively drag the tip of the vector x around the unit circle.

The eigenvectors of a square matrix are those vectors x such that

`[x,e] = eig(A)`

$$Ax = \lambda_i x \qquad (B.2)$$

that is, their direction is unchanged when transformed by the matrix. They are simply scaled by λ_i, the corresponding eigenvalue. The matrix has n eigenvalues (the *spectrum* of the matrix) which can be real or complex. For an orthogonal matrix the eigenvalues lie on a unit circle in the complex plane, $|\lambda_i| = 1$, and the eigenvectors are all orthogonal to one another.

The eigenvalues of a real symmetric matrix are all real and we classify the matrix according to the sign of its eigenvalues

- $\lambda_i > 0, \forall i$ positive definite
- $\lambda_i \geq 0, \forall i$ positive semi-definite
- $\lambda_i < 0, \forall i$ negative definite
- otherwise   indefinite

The inverse of a positive definite matrix is also positive definite.

The matrices $A^T A$ and $A A^T$ are always symmetric and positive semidefinite. This implies than any symmetric matrix A can be written as

$$A = L L^T$$

where L is the Cholesky decomposition of A.

`L = chol(A)`

The matrix R such that

$$A = RR$$

`R = sqrtm(A)`

is the square root of A or $A^{\frac{1}{2}}$.

If T is any nonsingular matrix then

$$A = TBT^{-1}$$

is known as a similarity transform and A and B are said to be similar, and it can be shown that the eigenvalues are unchanged by the transformation.

If A is nonsingular then the eigenvectors of A^{-1} are the same as A and the eigenvalues of A^{-1} are the reciprocal of those of A. The eigenvalues of A^T are the same as those of A but the eigenvectors are different.

The matrix form of Eq. B.2 is

$$AX = X\Lambda$$

where $X \in \mathbb{R}^{n \times n}$ is a matrix of eigenvectors of A, arranged column-wise, and Λ is a diagonal matrix of corresponding eigenvalues. If X is not singular we can rearrange this as

$$A = X\Lambda X^{-1}$$

which is the eigenvalue or spectral decomposition of the matrix. This implies that the matrix can be diagonalized by a similarity transform

$$\Lambda = X^{-1}AX$$

If A is symmetric then X is orthogonal and we can instead write

$$A = X\Lambda X^T \tag{B.3}$$

det(A)   The determinant of a square matrix $A \in \mathbb{R}^{n \times n}$ is the factor by which the transformation changes changes volumes in an n-dimensional space. For 2-dimensions imagine a shape defined by points x_i with an enclosed area a. The shape formed by the points Ax_i would have an enclosed area $a\det(A)$. If A is singular the points Ax_i would lie at a single point or along a line and have zero enclosed area. In a similar way for 3-dimensions, the determinant is a scale factor applied to the volume of a set of points mapped through the transformation A.

The determinant is equal to the product of the eigenvalues

$$\det(A) = \prod_{i=1}^{n} \lambda_i$$

thus a matrix with one or more zero eigenvalues will be singular. A positive definite
trace(A)   matrix, $\lambda_i > 0$, therefore has $\det(A) > 0$ and is not singular. The trace of a matrix is the sum of the diagonal elements

$$\text{tr}(A) = \sum_{i=1}^{n} A_{ii}$$

which is also the sum of the eigenvalues

$$\text{tr}(A) = \sum_{i=1}^{n} \lambda_i$$

The columns of $A = (c_1 c_2 \cdots c_n)$ can be considered as a set of vectors that define a space – the column space. Similarly, the rows of A can be considered as a set of vectors that define a space – the row space. The column rank of a matrix is the number of linearly independent columns of A. Similarly, the row rank is the number of linearly

independent rows of A. The column rank and the row rank are always equal and are simply   `rank(A)`
called the rank of A and the rank has an upper bound of $\min(m, n)$. The rank is the dimension of the largest nonsingular square submatrix that can be formed from A. A square matrix for which $\text{rank}(A) < n$ is said to be rank deficient or not of full rank. The rank shortfall $\min(m, n) - \text{rank}(A)$ is the nullity of A. In addition $\text{rank}(AB) \leq \min(\text{rank}(A), \text{rank}(B))$ and $\text{rank}(A + B) \leq \text{rank}(A) + \text{rank}(B)$. The matrix vv^T has rank 1 for all $v \neq 0$.

B.2.2   Nonsquare and Singular Matrices

For a nonsquare matrix $A \in \mathbb{R}^{m \times n}$ we can determine the left generalized inverse or pseudo inverse or Moore-Penrose pseudo inverse

$$A^+ A = I_{n \times n}$$

where $A^+ = (A^T A)^{-1} A^T$. The right generalized inverse is

$$AA^+ = I_{m \times m}$$

where $A^+ = A^T (AA^T)^{-1}$.

If the matrix A is not of full rank then it has a finite null space or kernel. A vector x lies in the null space of the matrix if

$$Ax = 0$$

More precisely this is the right-null space. A vector lies in the left-null space if

$$xA = 0$$

The left null space is equal to the right null space of A^T.

The null space is defined by a set of orthogonal basis vectors whose dimension is the   `null(A)`
nullity of A. Any linear combination of these null-space basis vectors lies in the null space.

For a nonsquare matrix $A \in \mathbb{R}^{m \times n}$ the analog to Eq. B.2 is

$$Av_i = \sigma_i u_i$$

where $u_i \in \mathbb{R}^m$ and $v_i \in \mathbb{R}^n$ are respectively the right- and left-singular vectors of A, and σ_i its singular values. The singular values are nonnegative real numbers that are the square root of the eigenvalues of AA^T and u_i are the corresponding eigenvectors. v_i are the eigenvectors of $A^T A$.

The singular value decomposition or SVD of the matrix A is   `[U,S,Vt] = svd(A)`

$$A = U\Sigma V^T$$

where $U \in \mathbb{R}^{m \times m}$ and $V \in \mathbb{R}^{n \times n}$ are both orthogonal matrices comprising, as columns, the corresponding singular vectors u_i and v_i. $\Sigma \in \mathbb{R}^{m \times n}$ is a diagonal matrix of the singular values

$$\Sigma = \begin{pmatrix} \sigma_1 & & & & \\ & \ddots & & & 0 \\ & & \sigma_r & & \\ & & & 0 & \\ 0 & & & & \ddots \\ & & & & & 0 \end{pmatrix}$$

where $r = \mathrm{rank}(A)$ is the rank of A and $\sigma_i \geq \sigma_{i+1}$. For the case where $r < n$ the diagonal will have zero elements as shown. Columns of V^T corresponding to the zero columns of Σ define the null space of A. The condition number of a matrix A is $\max \sigma_i / \min \sigma_i$ and a high value means the matrix is close to singular or "poorly conditioned".

cond(A)

The matrix quadratic form

$$s = \boldsymbol{x}^T A \boldsymbol{x} \tag{B.4}$$

is a scalar. If A is positive definite then $s = \boldsymbol{x}^T A \boldsymbol{x} > 0, \forall \boldsymbol{x} \neq 0$.

For the case that A is diagonal this can be written

$$s = \sum_{i=1}^{n} A_{ii} x_i^2$$

which is a weighted sum of squares. If A is symmetric then

$$s = \sum_{i=1}^{n} A_{ii} x_i^2 + 2 \sum_{i=1}^{n} \sum_{j=i+1}^{n} A_{ij} x_i x_j$$

the result also includes products or correlations between elements of $\boldsymbol{x}$.

The Mahalanobis distance is a weighted distance or norm

$$s = \sqrt{\boldsymbol{v}^T P^{-1} \boldsymbol{v}}$$

where $P \in \mathbb{R}^{n \times n}$ is a covariance matrix which down-weights components of $\boldsymbol{v}$ where uncertainty is high.

C   Geometry

Geometric concepts such as points, lines, ellipses and planes are critical to the fields of robotics and robotic vision. We briefly summarize key representations in both Euclidean and projective (homogeneous coordinate) space.

C.1   Euclidean Geometry

C.1.1   Points

A point in n-dimensional space is represented by an n-tuple, an ordered set of n numbers $(x_1, x_2 \cdots x_n)$ which define the coordinates of the point. The tuple can also be interpreted as a vector – a coordinate vector – from the origin to the point.

C.1.2   Lines

C.1.2.1   Lines in 2D

A line is defined by $\ell = (a, b, c)$ such that

$$ax + by + c = 0 \tag{C.1}$$

which is a generalization of the line equation we learned in school $y = mx + c$ but which can easily represent a vertical line by setting $a = 0$. $v = (a, b)$ is a vector parallel to the line, and $v = (-b, a)$ is a vector normal to the line. The line that joins two points is given by the solution to

$$\begin{pmatrix} x_1 & y_1 & 1 \\ x_2 & y_2 & 1 \end{pmatrix} \begin{pmatrix} a \\ b \\ c \end{pmatrix} = 0$$

which is found from the right-null space of the left-most term. The intersection point of two lines is

$$\begin{pmatrix} a_1 & b_1 \\ a_2 & b_2 \end{pmatrix} \begin{pmatrix} x \\ y \end{pmatrix} = -\begin{pmatrix} c_1 \\ c_2 \end{pmatrix}$$

which has no solution if the lines are parallel – the left-most term is singular.

We can also represent the line in polar form

$$\cos\theta x + \sin\theta y + \rho = 0$$

where θ is the angle from the x-axis to the line and ρ is the normal distance between the line and the origin.

C.1.2.2   Lines in 3D and Plücker Coordinates

We can define a line by two points, **p** and **q**, as shown in Fig. C.1, which would require a total of six parameters $\ell = (q_x, q_y, q_z, p_x, p_y, p_z)$. However since these points can be arbitrarily chosen there would be an infinite set of parameters that represent the same line making it hard to determine the equivalence of two lines.

There are advantages in representing a line as

$$\ell = (\boldsymbol{\omega} \times \boldsymbol{q}, \boldsymbol{p} - \boldsymbol{q}) = (\boldsymbol{v}, \boldsymbol{\omega}) \in \mathbb{R}^6$$

where $\boldsymbol{\omega}$ is the direction of the line and $\boldsymbol{v}$ is the moment of the line – a vector from the origin to a point on the line and normal to the line. This is a Plücker coordinate vector – a six dimensional quantity subject to two constraints: the coordinates are homogeneous and thus invariant to overall scale factor; and $\boldsymbol{v} \cdot \boldsymbol{\omega} = 0$. Lines therefore have 4 degrees-of-freedom▶ and the Plücker coordinates lie on a 4-dimensional manifold in 6-dimensional space. Lines with $\boldsymbol{\omega} = 0$ lie at infinity and are known as ideal lines.▶

This is not intuitive but consider two parallel planes and an arbitrary 3D line passing through them. The line can be described by the 2-dimensional coordinates of its intersection point on each plane – a total of four coordinates.

Ideal as in imaginary, not as in perfect.

In MATLAB® we will first define two points as column vectors

```
>> P = [2 3 4]';   Q = [3 5 7]';
```

and then create a Plücker line object

```
>> L = Plucker(P, Q)
L =
{ 1  -2  1; -1  -2  -3 }
```

which displays the v and w components. These can be accessed as properties

```
>> L.v'
ans =
     1    -2     1
>> L.w'
ans =
    -1    -2    -3
```

A Plücker line can also be represented as a skew-symmetric matrix

```
>> L.skew
ans =
     0     1     2    -1
    -1     0     1    -2
    -2    -1     0    -3
     1     2     3     0
```

which can also be formed by $\tilde{\boldsymbol{p}}\tilde{\boldsymbol{q}}^T - \tilde{\boldsymbol{q}}\tilde{\boldsymbol{p}}^T$.

To plot this line we first define a region of 3D space▶ then plot it in blue

Since lines lines are infinite we need to specify a finite volume in which to draw it.

```
>> axis([-5 5  -5 5  -5 5]);
>> L.plot('b');
```

The line is the set of all points

$$\boldsymbol{p}(\lambda) = \frac{\boldsymbol{v} \times \boldsymbol{\omega}}{\boldsymbol{\omega} \cdot \boldsymbol{\omega}} + \lambda \boldsymbol{\omega}, \ \ \lambda \in \mathbb{R}$$

which can be generated parametrically in terms of a scalar parameter

```
>> L.point([0 1 2])
ans =
    0.5714     0.3042     0.0369
    0.1429    -0.3917    -0.9262
   -0.2857    -1.0875    -1.8893
```

where the columns are points on the line corresponding to $\lambda = 0, 1, 2$.

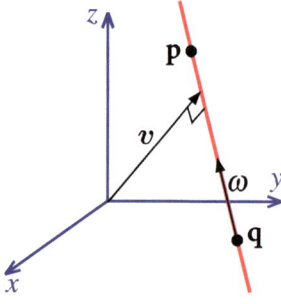

Fig. C.1. Describing a line in 3-dimensions

A point x is closest to the line when

$$\lambda = \frac{(x - q) \cdot \omega}{\omega \cdot \omega}$$

For the point $(1, 2, 3)$ the closest point on the line, and its distance, is given by

```
>> [x, d] = L.closest([1 2 3]')
x =
    1.5714
    2.1429
    2.7143
d =
    0.6547
```

The line intersects the plane $n^T x + d = 0$ at the point coordinate

$$x = \frac{v \times n - d\omega}{\omega \cdot n}$$

For the xy-plane the line intersects at

```
>> L.plane_intersect([0 0 1 0])'
ans =
    0.6667    0.3333         0
```

Two lines can be identical, coplanar or skewed. Identical lines have linearly dependent Plücker coordinates, that is, $\ell_1 = \lambda \ell_2$. If coplanar they can be parallel or intersecting and if skewed can be intersecting or not. If lines have $\omega_1 \times \omega_2 = 0$ they are parallel otherwise they are skewed.

The minimum distance between two lines is

$$d = \omega_1 \cdot v_2 + \omega_2 \cdot v_1$$

and is zero if they intersect.

The side operator is a permuted dot product

$$\text{side}\left(\ell^1, \ell^2\right) = \ell_1^1 \ell_5^2 + \ell_2^1 \ell_6^2 + \ell_3^1 \ell_4^2 + \ell_4^1 \ell_3^2 + \ell_5^1 \ell_1^2 + \ell_6^1 \ell_2^2$$

which is zero if the lines intersect or are parallel and is computed by the `side` method.

We can transform a Plücker line by the adjoint of a rigid-body motion.

$$\ell' = \text{Ad}(\xi)\ell$$

Julius Plücker (1801–1868) was a German mathematician and physicist who made contributions to the study of cathode rays and analytical geometry. He was born at Elberfeld and studied at Düsseldorf, Bonn, Heidelberg and Berlin and went to Paris in 1823 where he was influenced by the French geometry movement. In 1825 he returned to the University of Bonn, was made professor of mathematics in 1828, and professor of physics in 1836. In 1858 he proposed that the lines of the spectrum, discovered by his colleague Heinrich Geissler (of Geissler tube fame), were characteristic of the chemical substance which emitted them. In 1865, he returned to geometry and invented what was known as line geometry. He was the recipient of the Copley Medal from the Royal Society in 1866, and is buried in the Alter Friedhof (Old Cemetery) in Bonn.

C.1.3   Planes

A plane is defined by a 4-vector $\pi = (a, b, c, d)$ such that

$$ax + by + cz + d = 0$$

which can be written in point-normal form as

$$n^T(x - p) = 0$$

for a plane containing a point with coordinate p and a normal n, or more generally as

$$n^T x + d = 0$$

A plane can be defined by 3 points

$$\begin{pmatrix} x_1 & y_1 & z_1 & 1 \\ x_2 & y_2 & z_2 & 1 \\ x_3 & y_3 & z_3 & 1 \end{pmatrix} \pi = 0$$

and solved for using the right-null space of the left-most term, or by two nonparallel lines

$$\pi = \left(\boldsymbol{\omega}_1 \times \boldsymbol{\omega}_2, \, \boldsymbol{v}_1 \cdot \boldsymbol{\omega}_2 \right)$$

or by a line and a point with coordinate r

$$\pi = \left(\boldsymbol{\omega} \times \boldsymbol{r} - \boldsymbol{v}, \, \boldsymbol{v} \cdot \boldsymbol{r} \right)$$

A point is defined as the intersection point of three planes

$$\begin{pmatrix} a_1 & b_1 & c_1 \\ a_2 & b_2 & c_2 \\ a_3 & b_3 & c_3 \end{pmatrix} \begin{pmatrix} x \\ y \\ z \end{pmatrix} = - \begin{pmatrix} d_1 \\ d_2 \\ d_3 \end{pmatrix}$$

The Plücker line formed by the intersection of two planes is

$$\ell = \left(\boldsymbol{n}_1 \times \boldsymbol{n}_2, \, d_2 \boldsymbol{n}_1 - d_1 \boldsymbol{n}_2 \right)$$

Fig. C.2. Ellipses. **a** Canonical ellipse centered at the origin and aligned with the x- and y-axes; **b** general form of ellipse

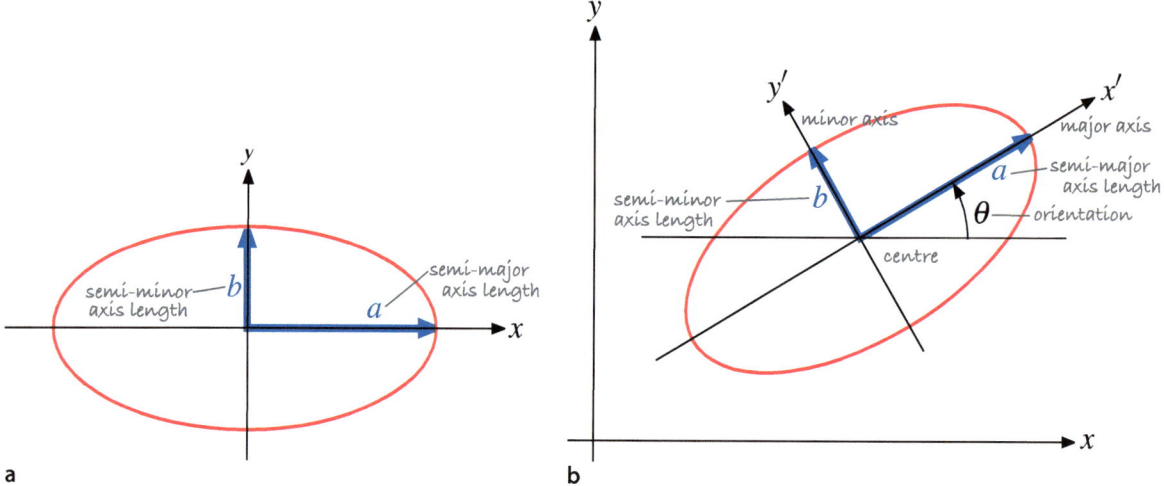

a

b

C.1.4 Ellipses and Ellipsoids

An ellipse belongs to the family of planar curves known as conics. The simplest form of an ellipse is defined implicitly

$$\frac{x^2}{a^2} + \frac{y^2}{b^2} = 1$$

and is shown in Fig. C.2a. This canonical ellipse is centered at the origin and has its major and minor axes aligned with the x- and y-axes. The radius in the x-direction is a and in the y-direction is b. The longer of the two radii is known as the semi-major axis length and the other is the semi-minor axis length.

We can write the ellipse in matrix quadratic form Eq. B.4 as

$$\begin{pmatrix} x & y \end{pmatrix} \begin{pmatrix} 1/a^2 & 0 \\ 0 & 1/b^2 \end{pmatrix} \begin{pmatrix} x \\ y \end{pmatrix} = 1$$

$$\boldsymbol{x}^T \begin{pmatrix} a^2 & 0 \\ 0 & b^2 \end{pmatrix}^{-1} \boldsymbol{x} = 1 \tag{C.2}$$

$$\boldsymbol{x}^T \boldsymbol{E}^{-1} \boldsymbol{x} = 1 \tag{C.3}$$

In the most general form E is a symmetric matrix

$$E = \begin{pmatrix} A & C \\ C & B \end{pmatrix} \tag{C.4}$$

and its determinant $\det(E) = AB - C^2$ defines the type of conic

$$\det(E) \begin{cases} > 0 & \text{ellipse} \\ = 0 & \text{parabola} \\ < 0 & \text{hyperbola} \end{cases}$$

An ellipse is therefore represented by a positive definite symmetric matrix E. Conversely any positive definite symmetric matrix, such as an inertia matrix or covariance matrix, can be represented by an ellipse.

Nonzero values of C change the orientation of the ellipse. The ellipse can be arbitrarily centered at $\boldsymbol{x}_c$ by writing it in the form

$$\left(\boldsymbol{x} - \boldsymbol{x}_c \right)^T \boldsymbol{E}^{-1} \left(\boldsymbol{x} - \boldsymbol{x}_c \right) = 1$$

which leads to the general ellipse shown in Fig. C.2b.

Since E is symmetric it can be diagonalized by Eq. B.3

$$E = X \Lambda X^T$$

where X is an orthogonal matrix comprising the eigenvectors of E. The inverse is

$$E^{-1} = X \Lambda^{-1} X^T$$

so the quadratic form becomes

$$\boldsymbol{x}^T X \Lambda^{-1} X^T \boldsymbol{x} = 1$$
$$\left(X^T \boldsymbol{x} \right)^T \Lambda^{-1} \left(X^T \boldsymbol{x} \right) = 1$$
$$\boldsymbol{x}'^T \Lambda^{-1} \boldsymbol{x}' = 1$$

This is similar to Eq. C.3 but with the ellipse defined by the diagonal matrix $\mathbf{\Lambda}$ with respect to the rotated coordinated frame $\mathbf{x}' = \mathbf{X}^T\mathbf{x}$. The major and minor ellipse axes are aligned with the eigenvectors of $\mathbf{E}$. The squared radii of the ellipse are the eigenvalues of $\mathbf{E}$ or the diagonal elements of $\mathbf{\Lambda}$.

For the general case of $\mathbf{E} \in \mathbb{R}^{n \times n}$ the result is an ellipsoid in n-dimensional space. The Toolbox function `plot_ellipse` will draw an ellipse for the $n = 2$ case and an ellipsoid for the $n = 3$ case.

Alternatively the ellipse can be represented in polynomial form by writing as

$$\big(\mathbf{x} - (x_0, \, y_0)\big)^T \begin{pmatrix} a & c \\ c & b \end{pmatrix} \big(\mathbf{x} - (x_0, \, y_0)\big) = 1$$

and expanding to obtain

$$e_1 x^2 + e_2 y^2 + e_3 xy + e_4 x + e_5 y + e_6 = 0$$

where $e_1 = a$, $e_2 = b$, $e_3 = 2c$, $e_4 = -2(ax_0 + cy_0)$, $e_5 = -2(by_0 + cx_0)$ and $e_6 = ax_0^2 + by_0^2 + 2cx_0 y_0 - 1$. The ellipse has only five degrees of freedom, its center coordinate and the three unique elements in $\mathbf{E}$. For a nondegenerate ellipse where $e_1 \neq 0$ we can rewrite the polynomial in normalized form

$$x^2 + E_1 y^2 + E_2 xy + E_3 x + E_4 y + E_5 = 0 \qquad (C.5)$$

with five unique parameters.

C.1.4.1   Properties

The area of an ellipse is πab and its eccentricity is

$$\varepsilon = \frac{\sqrt{a^2 - b^2}}{a}$$

The eigenvectors of $\mathbf{E}$ define the principal directions of the ellipse and the square root of the eigenvalues are the corresponding radii.

Consider the ellipse

$$\mathbf{x} \begin{pmatrix} 2 & -1 \\ -1 & 1 \end{pmatrix}^{-1} \mathbf{x} = 1$$

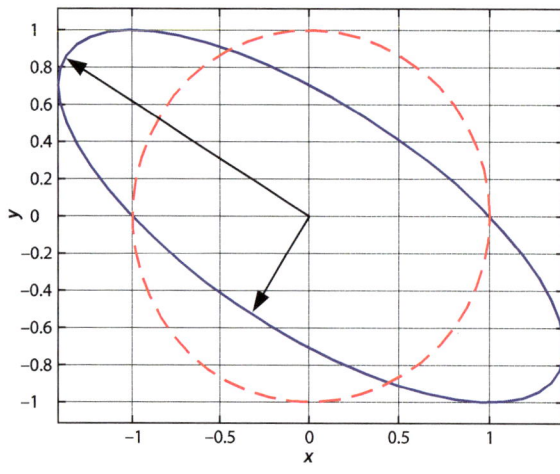

Fig. C.3.
Ellipse corresponding to a symmetric 2×2 matrix, and the unit circle shown in *red*. The arrows indicate the major and minor axes of the ellipse

which is represented in MATLAB by

```
>> E = [2 -1; -1 1];
```

We can plot this by

```
>> plot_ellipse(E)
```

which is shown in Fig. C.3.

The eigenvectors and eigenvalues of E are

```
>> [x,e] = eig(E)
x =
    -0.5257   -0.8507
    -0.8507    0.5257
e =
     0.3820         0
          0    2.6180
```

and the ellipse radii are

```
>> r = sqrt(diag(e))
r =
    0.6180
    1.6180
```

which correspond to b and a respectively. If either radius is equal to zero the ellipse is degenerate and becomes a line. If both radii are zero the ellipse is a point.

The eigenvectors are unit vectors in the minor- and major-axis directions and we will scale them by the radii to yield radius vectors which we can plot

```
>> arrow([0 0]', x(:,1)*r(1));
>> arrow([0 0]', x(:,2)*r(2));
```

The orientation of the ellipse is the angle of the major-axis with respect to the horizontal axis and is

$$\theta = \tan^{-1}\frac{x_y}{x_x}$$

For our example this is

```
>> atan2(x(2,2), x(1,2)) * 180/pi
ans =
   148.2825
```

in units of degrees.

The ellipse area is $\pi r_1 r_2$ and the ellipsoid volume is $\tfrac{4}{3}\pi\, r_1 r_2 r_3$ where the radii $r_i = \sqrt{\lambda_i}$ where λ_i are the eigenvalues of E. Since $\det(E) = \Pi \lambda_i$ the area or volume is proportional to $\sqrt{\det(E)}$.

C.1.4.2 Drawing an Ellipse

In order to draw an ellipse we first define a point coordinate $y = [x, y]^T$ on the unit circle

$$y^T y = 1$$

and rewrite Eq. C.3 as

$$x^T E^{-\frac{1}{2}} E^{-\frac{1}{2}} x = 1$$

where $E^{\frac{1}{2}}$ is the matrix square root (MATLAB function `sqrtm`). Equating these two equations we can write

$$\boldsymbol{x}^T E^{-\frac{1}{2}} E^{-\frac{1}{2}} \boldsymbol{x} = \boldsymbol{y}^T \boldsymbol{y}$$

It is clear that

$$\boldsymbol{y} = E^{-\frac{1}{2}} \boldsymbol{x}$$

which we can rearrange as

$$\boldsymbol{x} = E^{\frac{1}{2}} \boldsymbol{y}$$

which transforms a point on the unit circle to a point on an ellipse. If the ellipse is centered at $\boldsymbol{x}_c$ rather than the origin we can perform a change of coordinates

$$\left(\boldsymbol{x} - \boldsymbol{x}_c\right)^T E^{-\frac{1}{2}} E^{-\frac{1}{2}} \left(\boldsymbol{x} - \boldsymbol{x}_c\right) = 1$$

from which we write the transformation as

$$\boldsymbol{x} = E^{\frac{1}{2}} \boldsymbol{y} + \boldsymbol{x}_c$$

Continuing the MATLAB example above

```
>> E = [2 -1; -1 1];
```

We define a set of points on the unit circle

```
>> th = linspace(0, 2*pi, 50);
>> y = [cos(th); sin(th)];
```

which we transform to points on the perimeter of the ellipse

```
>> x = (sqrtm(E) * y)';
>> plot(x(:,1), x(:,2));
```

which is encapsulated in the Toolbox function

```
>> plot_ellipse(E, [0 0])
```

An ellipsoid is described by a positive-definite symmetric 3×3 matrix. Drawing an ellipsoid is tackled in an analogous fashion and `plot_ellipse` is also able to display a 3-dimensional ellipsoid.

C.2   Homogeneous Coordinates

A point in homogeneous coordinates, or the projective space $\mathbb{P}^n$, is represented by a coordinate vector $\tilde{\boldsymbol{x}} = (\tilde{x}_1, \tilde{x}_2 \cdots \tilde{x}_{n+1})$. The Euclidean coordinates are related to the projective coordinates by

$$x_i = \frac{\tilde{x}_i}{\tilde{x}_{n+1}}, \quad i = 1 \cdots n$$

Conversely a homogeneous coordinate vector can be constructed from a Euclidean coordinate vector by

$$\tilde{\boldsymbol{x}} = \left(x_1, x_2 \cdots x_n, 1\right)$$

and the tilde is used to indicate that the quantity is homogeneous.

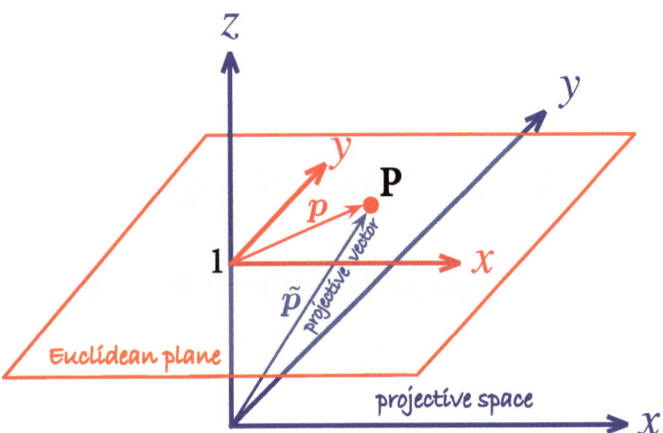

The extra *degree of freedom* offered by projective coordinates has several advantages. It allows points and lines at infinity, known as ideal points and lines, to be represented using only real numbers. It also means that scale is unimportant, that is $\tilde{x}$ and $\tilde{x}' = \alpha\tilde{x}$ both represent the same Euclidean point for all $\alpha \neq 0$. We express this as $\tilde{x} \simeq \tilde{x}'$. Points in homogeneous form can also be rotated with respect to a coordinate frame and translated simply by multiplying the homogeneous coordinate by an $(n+1) \times (n+1)$ homogeneous transformation matrix.

Homogeneous vectors are important in computer vision when we consider points and lines that exist in a plane – a camera's image plane. We can also consider that the homogeneous form represents a ray in Euclidean space as shown in Fig. C.4. The relationship between points and rays is at the core of the projective transformation.

C.2.1   Two Dimensions

In two dimensions there is a duality between points and lines. In $\mathbb{P}^2$ a line is defined by a 3-tuple, $\tilde{\ell} = (\ell_1, \ell_2, \ell_3)^T$, not all zero, and the equation of the line is the set of all points

$$\tilde{\ell}^T \tilde{x} = 0$$

which expands to $\ell_1 x + \ell_2 y + \ell_3 = 0$ and can be manipulated into the more familiar representation of a line. Note that this form can represent a vertical line, parallel to the y-axis, which the familiar form $y = mx + c$ cannot. This is the point equation of a line. The nonhomogeneous vector (ℓ_1, ℓ_2) is a parallel to the line, and $(-\ell_2, \ell_1)$ is normal to the line.

A point is defined by the intersection of two lines. If we write the point equations for two lines $\tilde{\ell}_1^T \tilde{p} = 0$ and $\tilde{\ell}_2^T \tilde{p} = 0$ their intersection is the point with coordinates

$$\tilde{p} = \tilde{\ell}_1 \times \tilde{\ell}_2$$

and is known as the line equation of a point. Similarly, a line joining two points $\tilde{p}_1$ and $\tilde{p}_2$ is given by the cross-product

$$\tilde{\ell}_{12} = \tilde{p}_1 \times \tilde{p}_2$$

Consider the case of two parallel lines at 45° to the horizontal axis

```
>> l1 = [1 -1 0]';
>> l2 = [1 -1 -1]';
```

which we can plot

```
>> plot_homline(l1, 'b')
>> plot_homline(l2, 'r')
```

The intersection point of these parallel lines is

```
>> cross(l1, l2)
ans =
        1        1        0
```

This is an *ideal point* since the third coordinate is zero – the equivalent Euclidean point would be at infinity. Projective coordinates allow points and lines at infinity to be simply represented and manipulated without special logic.

The distance from a point with coordinates $\tilde{p}$ to a line $\tilde{\ell}$ is

$$d = \frac{\tilde{\ell}^T \tilde{p}}{p_3 \sqrt{\ell_1^2 + \ell_2^2}} \qquad \text{(C.6)}$$

C.2.2   Three Dimensions

In three dimensions there is a duality between points and planes.

C.2.2.1   Lines

Using the homogeneous representation of the two points $\tilde{p}$ and $\tilde{q}$ we can form a 4×4 skew-symmetric matrix

$$\begin{aligned} L &= \tilde{q}\tilde{p}^T - \tilde{p}\tilde{q}^T \\ &= \begin{pmatrix} 0 & v_3 & -v_2 & -\omega_1 \\ -v_3 & 0 & v_1 & -\omega_2 \\ v_2 & -v_1 & 0 & -\omega_3 \\ \omega_1 & \omega_2 & \omega_3 & 0 \end{pmatrix} \end{aligned}$$

whose 6 unique elements comprise the Plücker coordinate vector. This matrix is rank 2 and the determinant is a quadratic in the Plücker coordinates – a 4-dimensional quadric hypersurface known as the Klein quadric. All points that lie on this manifold are valid lines. Many of the relationships in Sect. C.1.2.2 (between lines and points and planes) can be expressed in terms of this matrix. This matrix is returned by the `L` method of the `Plucker` class.

For a perspective camera with a camera matrix C the 3-dimensional Plücker line represented as a 4×4 skew-symmetric matrix L is projected onto the image plane as

$$\ell = CLC^T$$

which is a homogeneous line in $\mathbb{P}^2$. This is computed automatically if a `Plucker` object is passed to the `project` method of a `CentralCamera` object.

C.2.2.2   Planes

The plane described by $\pi\tilde{x} = 0$ can be defined by a line and a point

$$\pi = L\tilde{p}$$

The join and incidence relationships are more complex than the cross products used for the 2-dimensional case. Three points define a plane and the join relationship is

$$
\begin{pmatrix} \tilde{\boldsymbol{p}}_1^T \\ \tilde{\boldsymbol{p}}_2^T \\ \tilde{\boldsymbol{p}}_3^T \end{pmatrix} \tilde{\pi} = 0
$$

and the solution is found from the right-null space of the matrix. The incidence of three planes is the dual

$$
\begin{pmatrix} \tilde{\pi}_1^T \\ \tilde{\pi}_2^T \\ \tilde{\pi}_3^T \end{pmatrix} \tilde{\boldsymbol{p}} = 0
$$

and is an ideal point, zero last component, if the planes do not intersect at a point.

D Lie Groups and Algebras

We cannot go very far in the study of rotations or rigid-body motion without coming across the terms Lie groups, Lie algebras or Lie brackets – all named in honor of the Norwegian mathematician Sophus Lie. Rotations and rigid-body motion in 2- and 3-dimensions can be represented by matrices which form Lie groups and which have Lie algebras.

We will start simply by considering the set of all real 2×2 matrices $A \in \mathbb{R}^{2 \times 2}$

$$A = \begin{pmatrix} a_{11} & a_{12} \\ a_{21} & a_{22} \end{pmatrix}$$

which we could write as a linear combination of basis matrices

$$A = a_{11} \begin{pmatrix} 1 & 0 \\ 0 & 0 \end{pmatrix} + a_{12} \begin{pmatrix} 0 & 1 \\ 0 & 0 \end{pmatrix} + a_{21} \begin{pmatrix} 0 & 0 \\ 1 & 0 \end{pmatrix} + a_{22} \begin{pmatrix} 0 & 0 \\ 0 & 1 \end{pmatrix}$$

where each basis matrix represents a *direction* in a 4-dimensional space of 2×2 matrices. That is, the four axes of this space are *parallel* with each of these basis matrices. Any 2×2 matrix can be represented by a point in this space – this particular matrix is a point with the coordinates $(a_{11}, a_{12}, a_{21}, a_{22})$.

All proper rotation matrices, those belonging to $\mathbf{SO}(2)$, are a subset of points within the space of all 2×2 matrices. For this example the points lie in a 1-dimensional subset, a closed curve, in the 4-dimensional space. This is an instance of a manifold, a lower-dimensional smooth *surface* embedded within a space.

The notion of a curve in the 4-dimensional space makes sense when we consider that the $\mathbf{SO}(2)$ rotation matrix

$$A = \begin{pmatrix} \cos\theta & \sin\theta \\ -\sin\theta & \cos\theta \end{pmatrix}$$

has only one free parameter, and varying that parameter moves the point along the manifold.

Sophus Lie (1842–1899) (surname pronounced lee) was a Norwegian mathematician who obtained his Ph.D. from the University of Christiania in Oslo in 1871. He spent time in Berlin working with Felix Klein, and later contributed to Klein's Erlangan program to characterize geometries based on group theory and projective geometry. On a visit to Milan during the Franco-Prussian war he was arrested as a German spy and spent one month in prison. He is best known for his discovery that continuous transformation groups (now called Lie groups) can be understood by linearizing them and studying their generating vector spaces. He is buried in the Vår Frelsers gravlund in Oslo. (Photograph by Ludwik Szacinski)

Invoking mathematical formalism we say that rotations $\mathbf{SO}(2)$ and $\mathbf{SO}(3)$, and rigid-body motions $\mathbf{SE}(2)$ and $\mathbf{SE}(3)$ are matrix Lie groups and this has two implications. Firstly, they are an *algebraic group*, a mathematical structure comprising elements and a single operator. In simple terms, a group $\mathbf{G}$ has the following properties:

1. if g_1 and g_2 are elements of the group, that is $g_1, g_2 \in \mathbf{G}$, then the result of the group's operator $\circ$ is also an element of the group: $g_1 \circ g_2 \in \mathbf{G}$. In general, groups are not commutative so $g_1 \circ g_2 \neq g_2 \circ g_1$. For rotations and rigid-body motions the group operator $\circ$ represents composition. ◥

2. the group operator is associative, that is, $(g_1 \circ g_2) \circ g_3 = g_1 \circ (g_2 \circ g_3)$.

3. for $g \in \mathbf{G}$ there is an identity element $I \in \mathbf{G}$ such that $g \circ I = I \circ g = g$. ◥

4. for every $g \in \mathbf{G}$ there is a unique inverse $h \in \mathbf{G}$ such that $g \circ h = h \circ g = I$. ▶

In this book's notation the $\oplus$ operator is the group operator.

In this book's notation the identity is denoted by 0 (implying null motion) so we can say that $\xi \oplus 0 = 0 \oplus \xi = \xi$.

In this book's notation we use the operator $\ominus \xi$ to form the inverse.

The second implication of being a Lie group is that there is a smooth (differentiable) manifold structure. At any point on the manifold we can construct tangent vectors. The set of all tangent vectors at that point form a vector space – the tangent space. This is the multidimensional equivalent to a tangent line on a curve, or a tangent plane on a solid. We can think of this as the set of all possible derivatives of the manifold at that point.

The tangent space *at the identity* is described by the Lie algebra of the group, and the basis directions of the tangent space are called the generators of the group. Points in this tangent space map to elements of the group via the exponential function. If $\mathbf{g}$ is the Lie algebra for group $\mathbf{G}$ then

$$\forall X \in \mathbf{g} \Rightarrow e^X \in \mathbf{G}$$

where the elements of $\mathbf{g}$ and $\mathbf{G}$ are matrices of the same size and which each have a specific structure.

The surface of a sphere is a manifold in 3-dimensional space and at any point on that surface we can create a tangent vector. In fact we can create an infinite number of them and they lie within a plane which is a 2-dimensional vector space – the tangent space. We can choose a set of basis directions and establish a 2-dimensional coordinate system and we can map points on the plane to points on the sphere's surface.

Now consider an arbitrary real 3×3 matrix $A \in \mathbb{R}^{3 \times 3}$

$$A = \begin{pmatrix} a_{11} & a_{12} & a_{13} \\ a_{21} & a_{22} & a_{23} \\ a_{31} & a_{32} & a_{33} \end{pmatrix}$$

which we could write as a linear combination of basis matrices

$$A = a_{11} \begin{pmatrix} 1 & 0 & 0 \\ 0 & 0 & 0 \\ 0 & 0 & 0 \end{pmatrix} + a_{12} \begin{pmatrix} 0 & 1 & 0 \\ 0 & 0 & 0 \\ 0 & 0 & 0 \end{pmatrix} \cdots + a_{33} \begin{pmatrix} 0 & 0 & 0 \\ 0 & 0 & 0 \\ 0 & 0 & 1 \end{pmatrix}$$

where each basis matrix represents a *direction* in a 9-dimensional space of 3×3 matrices. Every possible 3×3 matrix is represented by a point in this space.

Not all matrices in this space are proper rotation matrices belonging to $\mathbf{SO}(3)$, but those that do lie on a manifold since $\mathbf{SO}(3)$ is a Lie group. The null rotation, represented by the identity matrix, is one point in this space. At that point we can construct a tangent space which has only 3 dimensions. Every point in the tangent space – the derivatives of the manifold – can be expressed as a linear combination of basis matrices

$$\Omega = \omega_1 \underbrace{\begin{pmatrix} 0 & 0 & 0 \\ 0 & 0 & -1 \\ 0 & 1 & 0 \end{pmatrix}}_{G_1} + \omega_2 \underbrace{\begin{pmatrix} 0 & 0 & 1 \\ 0 & 0 & 0 \\ -1 & 0 & 0 \end{pmatrix}}_{G_2} + \omega_3 \underbrace{\begin{pmatrix} 0 & -1 & 0 \\ 1 & 0 & 0 \\ 0 & 0 & 0 \end{pmatrix}}_{G_3} \tag{D.1}$$

which is the Lie algebra of the SO(3) group. The bases of this space: G_1, G_2 and G_3 are called the generators of SO(3) and belong to so(3).◄

The equivalent algebra is denoted using lower case letters and is a set of matrices.

Equation D.1 can be written as a skew-symmetric matrix parameterized by the vector $\omega = (\omega_1, \omega_2, \omega_3) \in \mathbb{R}^3$

$$\Omega = [\omega]_\times = \begin{pmatrix} 0 & -\omega_3 & \omega_2 \\ \omega_3 & 0 & -\omega_1 \\ -\omega_2 & \omega_1 & 0 \end{pmatrix} \in so(3)$$

and this reflects the 3 degrees of freedom of the SO(3) group embedded in the space of all 3×3 matrices. The 3DOF is consistent with our intuition about rotations in 3D space and also Euler's rotation theorem.

Mapping between vectors and skew-symmetric matrices is frequently required and the following shorthand notation will be used

$$[\cdot]_\times : \mathbb{R} \mapsto so(2),\ \mathbb{R}^3 \mapsto so(3)$$

$$\vee_\times(\cdot): so(2) \mapsto \mathbb{R},\ so(3) \mapsto \mathbb{R}^3$$

The first mapping is performed by the Toolbox function `skew` and the second by `vex` (which is named after the $\vee_\times$).

The exponential of *any* matrix in so(3) is a valid member of SO(3)

$$R(\theta, \hat{\omega}) = e^{[\hat{\omega}]_\times \theta} \in SO(3)$$

and an efficient closed-form solution is given by Rodrigues' rotation formula

$$R(\theta, \hat{\omega}) = I + \sin\theta [\hat{\omega}]_\times + (1 - \cos\theta)[\hat{\omega}]_\times^2$$

Finally, consider an arbitrary real 4×4 matrix $A \in \mathbb{R}^{4\times4}$

$$A = \begin{pmatrix} a_{11} & a_{12} & a_{13} & a_{14} \\ a_{21} & a_{22} & a_{23} & a_{24} \\ a_{31} & a_{32} & a_{33} & a_{34} \\ a_{41} & a_{42} & a_{43} & a_{44} \end{pmatrix}$$

which we could write as a linear combination of basis matrices

$$A = a_{11}\begin{pmatrix} 1 & 0 & 0 & 0 \\ 0 & 0 & 0 & 0 \\ 0 & 0 & 0 & 0 \\ 0 & 0 & 0 & 0 \end{pmatrix} + a_{12}\begin{pmatrix} 0 & 1 & 0 & 0 \\ 0 & 0 & 0 & 0 \\ 0 & 0 & 0 & 0 \\ 0 & 0 & 0 & 0 \end{pmatrix} \cdots + a_{44}\begin{pmatrix} 0 & 0 & 0 & 0 \\ 0 & 0 & 0 & 0 \\ 0 & 0 & 0 & 0 \\ 0 & 0 & 0 & 1 \end{pmatrix}$$

where each basis matrix represents a *direction* in a 16-dimensional space of all possible 4×4 matrices. Every 4×4 matrix is represented by a point in this space.

Not all matrices in this space are proper homogeneous transformation matrices belonging to SE(3), but those that do lie on a smooth manifold. The null motion (zero rotation and translation), which is represented by the identity matrix, is one point in this space. At that point we can construct a tangent space, which has 6 dimensions in this case, and points in the tangent space can be expressed as a linear combination of basis matrices

$$\Sigma = \omega_1 \begin{pmatrix} 0 & 0 & 0 & 0 \\ 0 & 0 & -1 & 0 \\ 0 & 1 & 0 & 0 \\ 0 & 0 & 0 & 0 \end{pmatrix} + \omega_2 \begin{pmatrix} 0 & 0 & 1 & 0 \\ 0 & 0 & 0 & 0 \\ -1 & 0 & 0 & 0 \\ 0 & 0 & 0 & 0 \end{pmatrix} + \omega_3 \begin{pmatrix} 0 & -1 & 0 & 0 \\ 1 & 0 & 0 & 0 \\ 0 & 0 & 0 & 0 \\ 0 & 0 & 0 & 0 \end{pmatrix}$$

$$+ v_1 \begin{pmatrix} 0 & 0 & 0 & 1 \\ 0 & 0 & 0 & 0 \\ 0 & 0 & 0 & 0 \\ 0 & 0 & 0 & 0 \end{pmatrix} + v_2 \begin{pmatrix} 0 & 0 & 0 & 0 \\ 0 & 0 & 0 & 1 \\ 0 & 0 & 0 & 0 \\ 0 & 0 & 0 & 0 \end{pmatrix} + v_3 \begin{pmatrix} 0 & 0 & 0 & 0 \\ 0 & 0 & 0 & 0 \\ 0 & 0 & 0 & 1 \\ 0 & 0 & 0 & 0 \end{pmatrix}$$

and these generator matrices belong to the Lie algebra of the group **SE**(3) and are denoted **se**(3). This can be written in general form as

$$\Sigma = [S] = \left(\begin{array}{ccc|c} 0 & -\omega_3 & \omega_2 & v_1 \\ \omega_3 & 0 & -\omega_1 & v_2 \\ -\omega_2 & \omega_1 & 0 & v_3 \\ \hline 0 & 0 & 0 & 0 \end{array} \right) \in \mathbf{se}(3)$$

which is an augmented skew symmetric matrix parameterized by $S = (v, \omega) \in \mathbb{R}^6$ which is referred to as a twist and has physical interpretation in terms of a screw axis direction and position. The sparse matrix structure and this concise parameterization reflects the 6 degrees of freedom of the **SE**(3) group embedded in the space of all 4×4 matrices. We extend our earlier shorthand notation

$$[\cdot] \colon \mathbb{R}^3 \mapsto \mathbf{se}(2),\ \mathbb{R}^6 \mapsto \mathbf{se}(3)$$

$$\vee(\cdot) \colon \mathbf{se}(2) \mapsto \mathbb{R}^3,\ \mathbf{se}(3) \mapsto \mathbb{R}^6$$

We can use these operators to convert between a twist representation which is a 6-vector and a Lie algebra representation which is a 4×4 augmented skew-symmetric matrix. We convert the Lie algebra to the Lie group representation using

$$\boldsymbol{T}(\theta, S) = \mathrm{e}^{[S]\theta} \in \mathbf{SE}(3)$$

or the inverse using the matrix logarithm. The exponential and the logarithm each have an efficient closed form solution.

Transforming a Twist – the Adjoint Representation

We have seen that rigid-body motions can be described by a twist which represents motion in terms of a screw axis direction and position, for example in Fig. D.1 the twist S_A can be used to transform points on the body. If the screw is rigidly attached to the body which undergoes some motion in **SE**(3) the new twist is

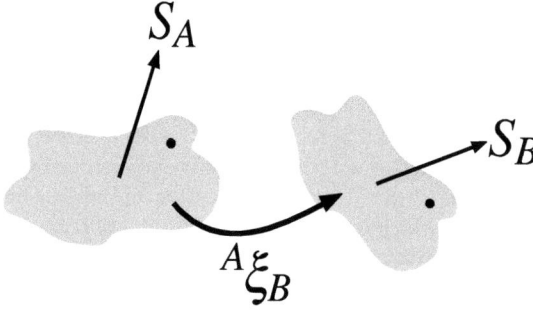

Fig. D.1.
Points in the body (*grey cloud*) can be transformed by the twist S_A. If the body and the screw axis undergo a rigid-body transformation ${}^A\xi_B$ the new twist is S_B

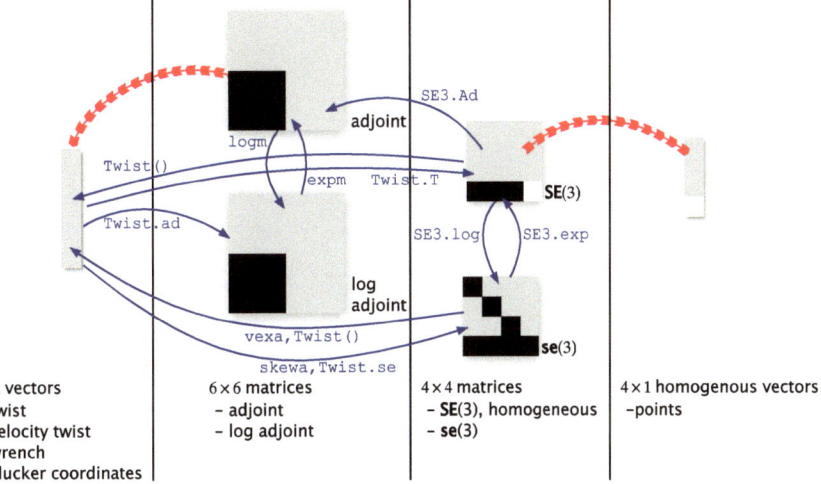

Fig. D.2.
The menagerie of **SE**(3) related quantities. Matrix values are coded as: 0 (*black*), 1 (*white*), other values (*grey*). Transformations between types are indicated by *blue arrows* with the relevant class plus method name. Operations are indicated by *red arrows*: the tail-end object operates on the head-end object and results in another object of the head-end type

6 × 1 vectors
– twist
– velocity twist
– wrench
– Plücker coordinates

6 × 6 matrices
– adjoint
– log adjoint

4 × 4 matrices
– **SE**(3), homogeneous
– **se**(3)

4 × 1 homogenous vectors
– points

$$^{B}S = \mathrm{Ad}\!\left(^{B}\xi_{A}\right){}^{A}S$$

where

$$\mathrm{Ad}(\xi) = \begin{pmatrix} R & [t]_{\times} R \\ 0 & R \end{pmatrix} \in \mathbb{R}^{6\times 6} \tag{D.2}$$

is the adjoint representation of the rigid-body motion. Alternatively we can write

$$\mathrm{Ad}\!\left(e^{[S]}\right) = e^{\mathrm{ad}(S)}$$

where ad(S) is the logarithm of the adjoint and defined in terms of the twist parameters as

$$\mathrm{ad}(S) = \begin{pmatrix} [\boldsymbol{\omega}]_{\times} & [\boldsymbol{v}]_{\times} \\ 0 & [\boldsymbol{\omega}]_{\times} \end{pmatrix} \in \mathbb{R}^{6\times 6}$$

The relationship between the various mathematical objects discussed are shown in Fig. D.2.

E

Linearization, Jacobians and Hessians

In robotics and computer vision the equations we encounter are often nonlinear. To apply familiar and powerful analytic techniques we must work with linear or quadratic approximations to these equations. The principle is illustrated in Fig. E.1 for the 1-dimensional case, and the analytical approximations shown in red are made at $x = x_0$. The approximation equals the nonlinear function at x_0 but is increasing inaccurate as we move away from that point. This is called a local approximation since it is valid in a region local to x_0 – the size of the valid region depends on the severity of the nonlinearity. This approach can be extended to an arbitrary number of dimensions.

Scalar Function of a Scalar

The function $f \colon \mathbb{R} \mapsto \mathbb{R}$ can be expressed as a Taylor series

$$f(x_0 + \Delta) = f(x_0) + \frac{\mathrm{d}f}{\mathrm{d}x}\Delta + \tfrac{1}{2}\frac{\mathrm{d}^2 f}{\mathrm{d}x^2}\Delta^2 + \cdots$$

which we truncate to form a first-order or linear approximation

$$f'(\Delta) \approx f(x_0) + J(x_0)\Delta$$

or a second-order approximation

$$f'(\Delta) \approx f(x_0) + J(x_0)\Delta + \tfrac{1}{2}H(x_0)\Delta^2$$

where $\Delta \in \mathbb{R}$ is an infinitesimal change in x relative to the linearization point x_0, and the first and second derivatives are given by $J(x_0) = \mathrm{d}f/\mathrm{d}x|_{x_0}$ and $H(x_0) = \mathrm{d}^2 f/\mathrm{d}x^2|_{x_0}$ respectively.

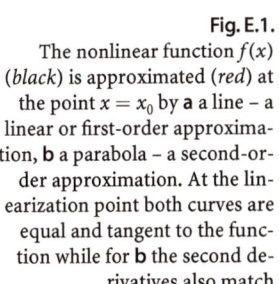

Fig. E.1.
The nonlinear function $f(x)$ (*black*) is approximated (*red*) at the point $x = x_0$ by **a** a line – a linear or first-order approximation, **b** a parabola – a second-order approximation. At the linearization point both curves are equal and tangent to the function while for **b** the second derivatives also match

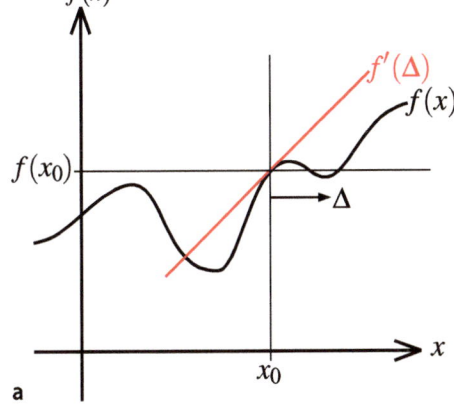

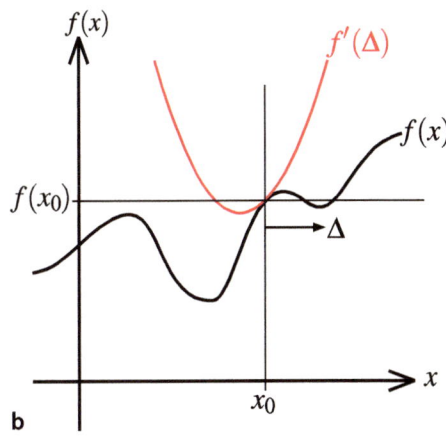

Scalar Function of a Vector

The scalar field $f(\boldsymbol{x})\colon \mathbb{R}^n \mapsto \mathbb{R}$ can be expressed as a Taylor series

$$f(\boldsymbol{x}_0 + \boldsymbol{\Delta}) = f(\boldsymbol{x}_0) + J(\boldsymbol{x}_0)\boldsymbol{\Delta} + \tfrac{1}{2}\boldsymbol{\Delta}^T H(\boldsymbol{x}_0)\boldsymbol{\Delta} + \cdots$$

which we can truncate to form a first-order or linear approximation

$$f'(\boldsymbol{\Delta}) \approx f(\boldsymbol{x}_0) + J(\boldsymbol{x}_0)\boldsymbol{\Delta}$$

or a second-order approximation

$$f'(\boldsymbol{\Delta}) \approx f(\boldsymbol{x}_0) + J(\boldsymbol{x}_0)\boldsymbol{\Delta} + \tfrac{1}{2}\boldsymbol{\Delta}^T H(\boldsymbol{x}_0)\boldsymbol{\Delta}$$

Ludwig Otto Hesse (1811–1874) was a German mathematician, born in Königsberg, Prussia, who studied under Jacobi (p. 230) and Bessel at the University of Königsberg. He taught at Königsberg, Halle, Heidelberg and finally at the newly established Polytechnic School in Munich. In 1869 he joined the Bavarian Academy of Sciences.

where $\boldsymbol{\Delta} \in \mathbb{R}^n$ is an infinitesimal change in $\boldsymbol{x} \in \mathbb{R}^n$ relative to the linearization point $\boldsymbol{x}_0$, $J \in \mathbb{R}^{1 \times n}$ is the vector version of the first derivative, and $H \in \mathbb{R}^{n \times n}$ is the Hessian – the matrix version of the second derivative.

The derivative of the function $f(\cdot)$ with respect to the vector $\boldsymbol{x}$ is

$$J(\boldsymbol{x}) = \nabla f(\boldsymbol{x})^T = \begin{pmatrix} \frac{\partial f}{\partial x_1} & \frac{\partial f}{\partial x_2} & \cdots & \frac{\partial f}{\partial x_n} \end{pmatrix}$$

and is itself a vector that points in the direction at which the function $f(\boldsymbol{x})$ has maximal increase. It is often written as $\nabla_x f$ to make explicit that the differentiation is with respect to $\boldsymbol{x}$.

The Hessian is an $n \times n$ symmetric matrix of second derivatives

$$H = \begin{pmatrix} \frac{\partial^2 f}{\partial x_1^2} & \frac{\partial^2 f}{\partial x_1 \partial x_2} & \cdots & \frac{\partial^2 f}{\partial x_1 \partial x_n} \\ \frac{\partial^2 f}{\partial x_1 \partial x_2} & \frac{\partial^2 f}{\partial x_2^2} & \cdots & \frac{\partial^2 f}{\partial x_2 \partial x_n} \\ \vdots & \vdots & \ddots & \vdots \\ \frac{\partial^2 f}{\partial x_1 \partial x_n} & \frac{\partial^2 f}{\partial x_2 \partial x_n} & \cdots & \frac{\partial^2 f}{\partial x_n^2} \end{pmatrix}$$

The function is at a critical point when $\|J\| = 0$. If the Hessian is positive definite then the function is at a local minimum, if negative definite then a local maximum, and if indefinite then the function is at a saddle point.

For functions which are quadratic in $\boldsymbol{x}$, as is the case for least-squares problems, it can be shown that the Hessian is

$$H(\boldsymbol{x}) = J(\boldsymbol{x})^T J(\boldsymbol{x}) + \sum_{i=1}^{m} f_i(\boldsymbol{x}) \frac{\partial^2 f_i}{\partial \boldsymbol{x}^2} \approx J(\boldsymbol{x})^T J(\boldsymbol{x})$$

which is frequently approximated by just the first term and this is key to Gauss-Newton least-squares optimization discussed in Sect. F.2.2.

Vector Function of a Vector

The vector field $\boldsymbol{f}(\boldsymbol{x})\colon \mathbb{R}^n \mapsto \mathbb{R}^m$ can be expressed as a Taylor series which can also be written as a set of scalar functions

$$\boldsymbol{f}(\boldsymbol{x}) = \begin{pmatrix} f_1(\boldsymbol{x}) \\ f_2(\boldsymbol{x}) \\ \vdots \\ f_n(\boldsymbol{x}) \end{pmatrix}$$

where $f_i:\mathbb{R}^m \to \mathbb{R}$ for $i \in \{1, 2, \cdots n\}$. The derivative of f with respect to the vector x can be expressed in matrix form as a Jacobian matrix

$$J = \begin{pmatrix} \frac{\partial f_1}{\partial x_1} & \cdots & \frac{\partial f_1}{\partial x_n} \\ \vdots & \ddots & \vdots \\ \frac{\partial f_m}{\partial x_1} & \cdots & \frac{\partial f_m}{\partial x_n} \end{pmatrix}$$

which can also be written as

$$J(x) = \begin{pmatrix} \nabla f_1^T \\ \nabla f_2^T \\ \vdots \\ \nabla f_n^T \end{pmatrix}$$

This derivative is also known as the tangent map of f, denoted Tf, or the differential of f denoted Df. To make explicit that the differentiation is with respect to x this can be denoted as J_x, $T_x f$, $D_x f$ or even $\partial f / \partial x$. The function is at a critical point when the Jacobian is not full rank.

The Hessian in this case is $H \in \mathbb{R}^{n \times m \times n}$ which is a 3-dimensional array called a cubix.

Deriving Jacobians

Jacobians of functions are required for many optimization algorithms as well as for the extended Kalman filter, and can be evaluated numerically or symbolically.

Consider Eq. 6.8 for the range and bearing angle of a landmark given the pose of the vehicle and the position of the landmark. We can express this as the very simple MATLAB® anonymous function

```
>> zrange = @(xi, xv, w) ...
    [ sqrt((xi(1)-xv(1))^2 + (xi(2)-xv(2))^2) + w(1);
      atan((xi(2)-xv(2))/(xi(1)-xv(1)))-xv(3) + w(2) ];
```

To estimate the Jacobian $H_{xv} = \partial h / \partial x_v$ for $x_v = (1, 2, \frac{\pi}{3})$ and $x_i = (10, 8)$ we can compute a first-order numerical difference

```
>> xv = [1, 2, pi/3]; xi = [10, 8]; w= [0,0];
>> h0 = zrange(xi, xv, w)
h0 =
   10.8167
   -0.4592
>> d = 0.001;
>> Hxv = [ zrange(xi, xv+[1,0,0]*d, w)-h0 ...
           zrange(xi, xv+[0,1,0]*d, w)-h0, ...
           zrange(xi, xv+[0,0,1]*d,w)-h0]  / d
Hxv =
   -0.8320    -0.5547          0
    0.0513    -0.0769    -1.0000
```

which shares the characteristic last column with the Jacobian shown in Eq. 6.14. Note that in computing this Jacobian we have set the measurement noise w to zero. The principal difficulty with this approach is choosing d, the difference used to compute the finite-difference approximation to the derivative. Too large and the results will be quite inaccurate if the function is nonlinear, too small and numerical problems will lead to reduced accuracy.

Alternatively we can perform the differentiation symbolically. This particular function is relatively simple and the derivatives can be determined easily using differential calculus. The numerical derivative can be used as a quick check for correctness. To avoid the possibility of error, or for more complex functions we can perform the differentiation symbolically using any of a large number of computer algebra packages. Using the MATLAB Symbolic Math Toolbox™ we can declare some symbolic variables

```
>> syms xi yi xv yv thetav wr wb
```

and then evaluate the same function as above

```
>> z = zrange([xi yi], [xv yv thetav], [wr wb])
z =
         wr + ((xi - xv)/(yi - yv)^2)^(1/2)
  wb - thetav + atan((yi - yv)/(xi - xv))
```

which is simply Eq. 6.8 in MATLAB symbolic form. The Jacobian is computed by a Symbolic Math Toolbox™ function

```
>> Hxv = jacobian(z, [xv yv thetav])
Hxv =
[ -(2*xi - 2*xv)/(2*((xi - xv)^2 + (yi - yv)^2)^(1/2)),↵
    -(2*yi - 2*yv)/(2*((xi - xv)^2 + (yi - yv)^2)^(1/2)),   0]
[ (yi - yv)/((xi - xv)^2*((yi - yv)^2/(xi - xv)^2 + 1)),↵
    -1/((xi - xv)*((yi - yv)^2/(xi - xv)^2 + 1)), -1]
```

which has the required dimensions

```
>> about(Hxv)
Hxv [sym] : 2x3 (112 bytes)
```

and the characteristic last column. We could cut and paste this code into our program or automatically create a MATLAB callable function

```
>> HxvFunc = matlabFunction(Hxv);
```

where `HxvFunc` is a MATLAB function handle. We can evaluate the Jacobian at the operating point given above

```
>> xv = [1,· 2, pi/3]; xi = [10, 8]; w = [0,0];
>> HxvFunc( xi(1), xv(1), xi(2), xv(2) )
ans =
   -0.8321   -0.5547         0
    0.0513   -0.0769   -1.0000
```

which is similar to the approximation above obtained numerically. The function `matlabFunction` can also write the function to an M-file. The functions `ccode` and `fcode` generate C and Fortran representations of the Jacobian.

Another interesting approach is the package ADOL-C which is an open-source tool for the automatic differentiation of C and C++ programs, that is, given a function written in C it will return a Jacobian function written in C. It is available at http://www.coin-or.org/projects/ADOL-C.xml.

F Solving Systems of Equations

Solving systems of linear and nonlinear equations, particularly over-constrained systems, is a common problem in robotics and computer vision.

F.1 Linear Problems

F.1.1 Nonhomogeneous Systems

These are equations of the form

$$Ax = b$$

where we wish to solve for the unknown vector $x \in \mathbb{R}^n$ and $A \in \mathbb{R}^{m \times n}$ and $b \in \mathbb{R}^m$ are constants.

If $n = m$ then A is square, and if A is nonsingular then the solution is obtained using the matrix inverse

$$x = A^{-1}b$$

In practice we often encounter systems where $m > n$, that is there are more equations than unknowns. In general there will not be an exact solution but we can attempt to find the *best* solution, in a least-squares sense, which is

$$x^* = \arg\min_x \|Ax - b\|$$

That solution is given by

$$x^* = \left(A^T A\right)^{-1} A^T b = A^+ b$$

which is known as the pseudo inverse or more formally the left-generalized inverse.◄
Using SVD where $A = U\Sigma V^T$ this is

$$x = V\Sigma^{-1}U^T b$$

Since the inverse left multiplies b.

where Σ^{-1} is simply the element-wise inverse of the diagonal elements of Σ^T.

If the matrix is singular, or the system is under constrained $n < m$, then there are infinitely many solutions. We can again use the SVD approach

$$x = V\Sigma^{-1}U^T b$$

where this time Σ^{-1} is the element-wise inverse of the *nonzero* diagonal elements of Σ, all other zeros are left in place.

In MATLAB all these problems can be solved using the backslash operator

```
>> x = A\b
```

For the problem

$$RP = Q$$

where R is an unknown rotation matrix in $\mathbf{SO}(n)$, and $P = \{p_1 \cdots p_m\} \in \mathbb{R}^{n \times m}$ and $Q = \{q_1 \cdots q_m\} \in \mathbb{R}^{n \times m}$ comprise column vectors for which $q_i = Rp_i$. We first compute the moment matrix

$$M = \sum_{i=1}^{N} q_i p_i^T$$

and then compute the SVD $M = U\Sigma V^T$. The least squares estimate of the rotation matrix is

$$R = UV^T$$

and is guaranteed to be an orthogonal matrix.

F.1.2   Homogeneous Systems

These are equations of the form

$$Ax = 0$$

and always have the trivial solution $x = 0$. If A is square and nonsingular this is the only solution. Otherwise, if A is not of full rank, that is the matrix is nonsquare, or square and singular then there are an infinite number of solutions which are linear combinations of vectors in the right null space of A which is computed by the MATLAB function `null`.

F.2   Nonlinear Problems

Many problems in robotics and computer vision involves sets of nonlinear equations. Solution of these problems requires linearizing the equations about an estimated solution, solving for an improved solution and iterating. Linearization is discussed in Appendix E.

F.2.1   Finding Roots

Consider a set of equations expressed in the form

$$f(x) = 0$$

where $f : \mathbb{R}^n \mapsto \mathbb{R}^m$. This is a nonlinear version of the homogeneous system described above. We first linearize the equation about our best estimate of the solution x_0

$$f(x_0 + \Delta) = f(x_0) + J(x_0)\Delta + \tfrac{1}{2}\Delta^T H(x_0)\Delta + \cdots \tag{F.1}$$

where $\Delta \in \mathbb{R}^n$ is an infinitesimal change in x relative to x_0. We truncate this to form a linear approximation

$$f'(\Delta) \approx f_0 + J\Delta \tag{F.2}$$

where $\boldsymbol{f}_0 = \boldsymbol{f}(\boldsymbol{x}_0)$ is the function value and $J = J(\boldsymbol{x}_0) \in \mathbb{R}^{m \times n}$ the Jacobian, both evaluated at the linearization point. Now we solve an approximation of our original problem $\boldsymbol{f}'(\Delta) = 0$

$$\boldsymbol{f}_0 + J\Delta = 0 \Rightarrow \Delta = -J^{-1}\boldsymbol{f}_0$$

If $n \neq m$ then J is nonsquare and we can use the pseudo-inverse or the MATLAB backslash operator $-\text{J}\backslash\text{f0}$. The computed step Δ is based on an approximation to the original nonlinear function so $\boldsymbol{x}_0 + \Delta$ will generally not be the solution but it will be closer. This leads to an iterative solution – the Newton-Raphson method:

1   **while** $\left\| \boldsymbol{f}(\boldsymbol{x}_0) \right\| > \varepsilon$ **do**

2   compute $\boldsymbol{f}_0 = \boldsymbol{f}(\boldsymbol{x}_0), J(\boldsymbol{x}_0)$

3       $\Delta = -J^{-1}\boldsymbol{f}_0$

4       $\boldsymbol{x}_0 \leftarrow \boldsymbol{x}_0 + \Delta$

5   **end**

F.2.2   Nonlinear Minimization

A very common class of problems involves finding the *minimum* of a scalar function $f(\boldsymbol{x}) : \mathbb{R}^n \mapsto \mathbb{R}$ which can be expressed as

$$\boldsymbol{x}^* = \arg\min_{\boldsymbol{x}} f(\boldsymbol{x})$$

The derivative of the linearized system Eq. F.2 is

$$\frac{\mathrm{d}f'}{\mathrm{d}\Delta} = J$$

and if we consider the function to be a multi-dimensional surface then $J(\boldsymbol{x}_0)$ is vector indicating the direction and magnitude of the *slope* at $\boldsymbol{x} = \boldsymbol{x}_0$ so an update of

$$\Delta = -\beta J$$

will move the estimate *down hill* toward the minimum. This leads to an iterative solution called gradient descent:

1   **repeat**

2   compute $J = J(\boldsymbol{x}_0)$

3       $\Delta = -\beta J$

4       $\boldsymbol{x}_0 \leftarrow \boldsymbol{x}_0 + \Delta$

5   **until** $\left\| \Delta \right\| < \varepsilon$

and the challenge is to choose the appropriate step size β.

If we include the second-order term from Eq. F.1 the approximation becomes

$$f'(\Delta) \approx \boldsymbol{f}_0 + J\Delta + \tfrac{1}{2}\Delta^T H(\boldsymbol{x}_0)\Delta$$

and to find its minima we take the derivative and set it to zero

$$\frac{\mathrm{d}f'}{\mathrm{d}\Delta} = 0 \Rightarrow J + H\Delta = 0$$

and the update is

$$\Delta = -H^{-1}J$$

This leads to another iterative solution – Newton's method. The challenge is determining the Hessian of the nonlinear system, either by numerical approximation or symbolic manipulation.

F.2.3   Nonlinear Least Squares Minimization

Very commonly the scalar function we wish to optimize is a quadratic cost function

$$F(\boldsymbol{x}) = \|\boldsymbol{f}(\boldsymbol{x})\|^2 = \boldsymbol{f}(\boldsymbol{x})^T \boldsymbol{f}(\boldsymbol{x})$$

where $\boldsymbol{f}(\boldsymbol{x})\colon \mathbb{R}^n \mapsto \mathbb{R}^m$ is some vector-valued nonlinear function which we can linearize as

$$\boldsymbol{f}'(\boldsymbol{\Delta}) \approx \boldsymbol{f}_0 + J\boldsymbol{\Delta}$$

and the scalar cost is

$$
\begin{aligned}
F(\boldsymbol{\Delta}) &\approx \left(\boldsymbol{f}_0 + J\boldsymbol{\Delta}\right)^T \left(\boldsymbol{f}_0 + J\boldsymbol{\Delta}\right) \\
&\approx \boldsymbol{f}_0^T \boldsymbol{f}_0 + \boxed{\boldsymbol{f}_0^T J\boldsymbol{\Delta} + \boldsymbol{\Delta}^T J^T \boldsymbol{f}_0} + \boldsymbol{\Delta}^T J^T J\boldsymbol{\Delta} \\
&\approx \boldsymbol{f}_0^T \boldsymbol{f}_0 + 2\boldsymbol{f}_0^T J\boldsymbol{\Delta} + \boldsymbol{\Delta}^T J^T J\boldsymbol{\Delta}
\end{aligned}
$$

One term is the transpose of the other, but since both result in a scalar transposition doesn't matter.

where $J^T J \in \mathbb{R}^{n \times n}$ is the *approximate* Hessian from page 318.

To minimize the error of this linearized least squares system we take the derivative with respect to $\boldsymbol{\Delta}$ and set it to zero

$$\frac{\mathrm{d}F}{\mathrm{d}\boldsymbol{\Delta}} = 0 \Rightarrow 2\boldsymbol{f}_0^T J + 2J^T J\boldsymbol{\Delta} = 0$$

which we can solve for the locally optimal update

$$\boldsymbol{\Delta} = -\left(J^T J\right)^{-1} J^T \boldsymbol{f}_0 \tag{F.3}$$

where we can recognize the pseudo or left generalized-inverse of J. Once again we iterate to find the solution – a Gauss-Newton iteration.

Numerical Issues

When solving Eq. F.3 we may find that the Hessian $J^T J$ is poorly conditioned or singular and this can be remedied by adding a damping term

$$\boldsymbol{\Delta} = -\left(J^T J + \lambda I\right)^{-1} J^T \boldsymbol{f}_0$$

which makes the system more positive definite. Since $J^T J + \lambda I$ is effectively in the denominator, increasing λ will decrease $\|\boldsymbol{\Delta}\|$ and slow convergence.

How do we choose λ? We can experiment with different values but a better way is the Levenberg-Marquardt algorithm (Algorithm F.1) which adjusts λ to ensure convergence. If the error increases compared to the last step then the step is repeated with increased λ to reduce the step size. If the error decreases then λ is reduced to increase the convergence rate. The updates vary continuously between Gauss-Newton (low λ) and gradient descent (high λ).

For problems where n is large inverting the $n \times n$ approximate Hessian is expensive. Typically $m < n$ which means the Jacobian is not square and Eq. F.3 can be rewritten as

$$\begin{aligned}
&\text{1}\quad \text{initialize } \lambda \\
&\text{2}\quad \textbf{repeat} \\
&\text{3}\quad \quad \text{compute } \boldsymbol{f}_0 = \boldsymbol{f}(\boldsymbol{x}_0), J = J(\boldsymbol{x}_0), H = J^T J \\
&\text{4}\quad \quad \Delta = -(H + \lambda I)^{-1} J^T \boldsymbol{f}_0 \\
&\text{5}\quad \quad \textbf{if } F(\boldsymbol{x}_0 + \Delta) < F(\boldsymbol{x}_0) \textbf{ then} \\
&\text{6}\quad \quad \quad - \textit{error decreased: reduce damping} \\
&\text{7}\quad \quad \quad \boldsymbol{x}_0 \leftarrow \boldsymbol{x}_0 + \Delta \\
&\text{8}\quad \quad \quad \lambda \leftarrow \lambda/c \\
&\text{9}\quad \quad \textbf{else} \\
&\text{10}\quad \quad \quad - \textit{error increased: discard and raise damping} \\
&\text{11}\quad \quad \quad \lambda \leftarrow c\lambda \\
&\text{12}\quad \quad \textbf{end} \\
&\text{13}\quad \textbf{until } \|\Delta\| < \varepsilon
\end{aligned}$$

Algorithm F.1.
Levenberg-Marquardt algorithm, c is typically chosen in the range 2 to 10

$$\Delta = -J^T \left(JJ^T \right)^{-1} \boldsymbol{f}_0$$

which is the right pseudo-inverse and involves inverting a smaller matrix. We can reintroduce a damping term

$$\Delta = -J^T \left(JJ^T + \lambda I \right)^{-1} \boldsymbol{f}_0$$

and if λ is large this becomes simply

$$\Delta \approx -\beta J^T \boldsymbol{f}_0$$

but exhibits very slow convergence.

If $\boldsymbol{f}_k(\cdot)$ has additive noise that is zero mean, normally distributed and time invariant we have a maximum likelihood estimator of $\boldsymbol{x}$. Outlier data has a significant impact on the result since errors are squared. Robust estimators minimize the effect of outlier data, and in an M-estimator

$$F(\boldsymbol{x}) = \rho\big(\boldsymbol{f}_k(\boldsymbol{x})\big)$$

the squared norm is replaced by a loss function $\rho(\cdot)$ which models the likelihood of its argument. Unlike the squared norm these functions flatten off for large values, and some common examples include the Huber loss function and the Tukey biweight function.

F.2.4   Sparse Nonlinear Least Squares

For a large class of problems the overall cost is the sum of quadratic costs

$$F(\boldsymbol{x}) = \sum_k \big\| \boldsymbol{f}_k(\boldsymbol{x}) \big\|^2 = \sum_k \boldsymbol{f}_k(\boldsymbol{x})^T \boldsymbol{f}_k(\boldsymbol{x}) \tag{F.4}$$

Consider the problem of fitting a model $z = \phi(w; \boldsymbol{x})$ where $\phi \colon \mathbb{R}^p \mapsto \mathbb{R}^m$ with parameters $\boldsymbol{x} \in \mathbb{R}^n$ to a set of data points (w_k, z_k). The error vector associated with the kth data point is

$$\boldsymbol{f}_k(\boldsymbol{x}) = z_k - \phi\big(w_k; \boldsymbol{x}\big) \in \mathbb{R}^m$$

and minimizing Eq. F.4 gives the optimal model parameters $\boldsymbol{x}$.

Another example is pose-graph optimization as used for pose-graph SLAM and bundle adjustment. Edge k in the graph connects vertices i and j and has an associated cost $\boldsymbol{f}_k(\cdot)\colon \mathbb{R}^n \mapsto \mathbb{R}^m$

$$\boldsymbol{f}_k(\boldsymbol{x}) = \hat{e}_k(\boldsymbol{x}) - \boldsymbol{e}_k^{\#} \tag{F.5}$$

where $\boldsymbol{e}_k^{\#}$ is the observed value of the edge parameter and $\hat{e}_k(\boldsymbol{x})$ is the estimate based on the state $\boldsymbol{x}$ of the pose graph. This is linearized

$$\boldsymbol{f}_k'(\boldsymbol{\Delta}) \approx \boldsymbol{f}_{0,k} + J_k \boldsymbol{\Delta}$$

and the squared error for the edge is

$$F_k(\boldsymbol{x}) = \boldsymbol{f}_k(\boldsymbol{x})^T \Omega_k \boldsymbol{f}_k(\boldsymbol{x})$$

where $\Omega_k \in \mathbb{R}^{m \times m}$ is a positive-definite constant matrix► which we combine as

> This can be used to specify the significance of the edge det Ω_k with respect to other edges, as well as the relative significance of the elements of $\boldsymbol{f}_k(\cdot)$.

$$\begin{aligned}
F_k(\boldsymbol{\Delta}) &\approx \left(\boldsymbol{f}_{0,k} + J_k \boldsymbol{\Delta}\right)^T \Omega_k \left(\boldsymbol{f}_{0,k} + J_k \boldsymbol{\Delta}\right) \\
&\approx \boldsymbol{f}_{0,k}^T \Omega_k \boldsymbol{f}_{0,k} + \boldsymbol{f}_{0,k}^T \Omega_k J_k \boldsymbol{\Delta} + \boldsymbol{\Delta}^T J_k^T \Omega_k \boldsymbol{f}_{0,k} + \boldsymbol{\Delta}^T J_k^T \Omega_k J_k \boldsymbol{\Delta} \\
&\approx c_k + 2 b_k^T \boldsymbol{\Delta} + \boldsymbol{\Delta}^T H_k \boldsymbol{\Delta}
\end{aligned}$$

where $b_k^T = \boldsymbol{f}_{0,k}^T \Omega_k J_k$ and $H_k = \Sigma_k J_k^T \Omega_k J_k$. The total cost is the sum of all edge costs

$$\begin{aligned}
F(\boldsymbol{\Delta}) &= \sum_k F_k(\boldsymbol{\Delta}) \\
&\approx \sum_k \left(c_k + 2 b_k^T \boldsymbol{\Delta} + \boldsymbol{\Delta}^T H_k \boldsymbol{\Delta}\right) \\
&\approx \sum_k c_k + 2\left(\sum_k b_k^T\right)\boldsymbol{\Delta} + \boldsymbol{\Delta}^T \left(\sum_k H_k\right)\boldsymbol{\Delta} \\
&\approx c + 2 b^T \boldsymbol{\Delta} + \boldsymbol{\Delta}^T H \boldsymbol{\Delta}
\end{aligned}$$

where $b^T = \Sigma_k \boldsymbol{f}_{0,k}^T \Omega_k J_k$ and $H = \Sigma_k J_k^T \Omega_k J_k$ are summations over the edges of the graph. Once they are computed we proceed as previously, taking the derivative with respect to $\boldsymbol{\Delta}$ and setting it to zero, solving for the update $\boldsymbol{\Delta}$ and iterating using Algorithm F.1.

State Vector

The state vector is a concatenation of all poses and coordinates in the optimization problem. For pose-graph SLAM it takes the form

$$\boldsymbol{x} = \left\{\xi_1, \xi_2 \cdots \xi_N\right\} \in \mathbb{R}^n$$

Poses must be represented in a vector form and preferably one that is compact and singularity free. For **SE**(2) this is quite straightforward and we use $\xi \sim (x, y, \theta) \in \mathbb{R}^3$. For **SE**(3) we will use $\xi \sim (t, r) \in \mathbb{R}^6$ which comprises translation $t \in \mathbb{R}^3$ and rotation $r \in \mathbb{R}^3$. The latter can be triple angles (Euler or roll-pitch-yaw), axis-angle, exponential coordinates or the vector part of a unit-quaternion. The state vector has structure, comprising a sequence of subvectors one per pose. We denote the ith subvector of $\boldsymbol{x}$ as $\boldsymbol{x}_i \in \mathbb{R}^{N_\xi}$, where $N_\xi = 3$ for **SE**(2) and $N_\xi = 6$ for **SE**(3).

For pose-graph SLAM with landmarks, or bundle adjustment the state vector comprises poses and coordinate vectors

$$x = \{\xi_1, \xi_2 \cdots \xi_N | P_1, P_2 \cdots P_M\} \in \mathbb{R}^n$$

and the ith and j^{th} subvectors of x are denoted $x_i \in \mathbb{R}^{N_\xi}$ and $x_j \in \mathbb{R}^{N_P}$ and correspond to ξ_i and P_j respectively.

Inherent Structure

A key observation is that the error vector $f_k(x)$ for edge k depends only on the associated vertices i and j, and this means that the Jacobian

$$J_k = \frac{\partial f_k(x)}{\partial x} \in \mathbb{R}^{m \times m}$$

is mostly zeros

$$J_k = \left(0 \cdots A_i \cdots B_j \cdots 0\right), \ A_i = \frac{\partial f_k(x)}{\partial x_i}, \ B_j = \frac{\partial f_k(x)}{\partial x_j}$$

where $A_i \in \mathbb{R}^{m \times N_\xi}$ and $B_j \in \mathbb{R}^{m \times N_\xi}$ or $B_j \in \mathbb{R}^{m \times N_P}$ according to the state vector structure.

This sparse block structure means that the vector b_k and the Hessian $J_k^T \Omega_k J_k$ also have a sparse block structure as shown in Fig. F.1. The Hessian has just four small nonzero blocks so rather than compute the product $J_k^T \Omega_k J_k$, which involves many multiplications by zero, we can just compute the four nonzero blocks and add them into the Hessian for the least squares system. All blocks in a row have the same height, and in a column have the same width. For pose-graph SLAM with landmarks, or bundle adjustment the blocks are of different sizes as shown in Fig. F.1b.

If the value of an edge represents pose then Eq. F.5 must be replaced with $f_k(x) = \hat{e}_k(x) \ominus e_k^\#$. We generalize this with the $\boxminus$ operator to indicate that the use of $-$ or $\ominus$ as appropriate. Similarly when updating the state vector at the end of an iteration the poses must be compounded $x_0 \leftarrow x_0 \oplus \Delta$ and we generalize this to the $\boxplus$ operator. The pose-graph optimization is solved by the iteration in Algorithm F.2.

1	**repeat**
2	$H \leftarrow 0, \ b \leftarrow 0$
3	**for each** k **do**
4	$f_{0,k}(x_0) = \hat{e}_k(x) \boxminus e_k^\#$
5	$(i, j) = \text{vertices}(k)$
6	compute $A_i(x_i), \ B_j(x_j)$
7	$b_i \leftarrow b_i + f_{0,k}^T \Omega_k A_i$
8	$b_j \leftarrow b_j + f_{0,k}^T \Omega_k B_j$
9	$H_{i,i} \leftarrow H_{i,i} + A_i^T \Omega_k A_i$
10	$H_{i,j} \leftarrow H_{i,j} + A_i^T \Omega_k B_j$
11	$H_{j,i} \leftarrow H_{j,i} + B_j^T \Omega_k A_i$
12	$H_{j,j} \leftarrow H_{j,j} + B_j^T \Omega_k B_j$
13	**end**
14	$\Delta = -H^{-1} b$
15	$x_0 \leftarrow x_0 \boxplus \Delta$
16	**until** $\|\Delta\| < \varepsilon$

Algorithm F.2. Pose graph optimization. For Levenberg-Marquardt optimization replace line 14 with lines 4–12 from Algorithm F.1

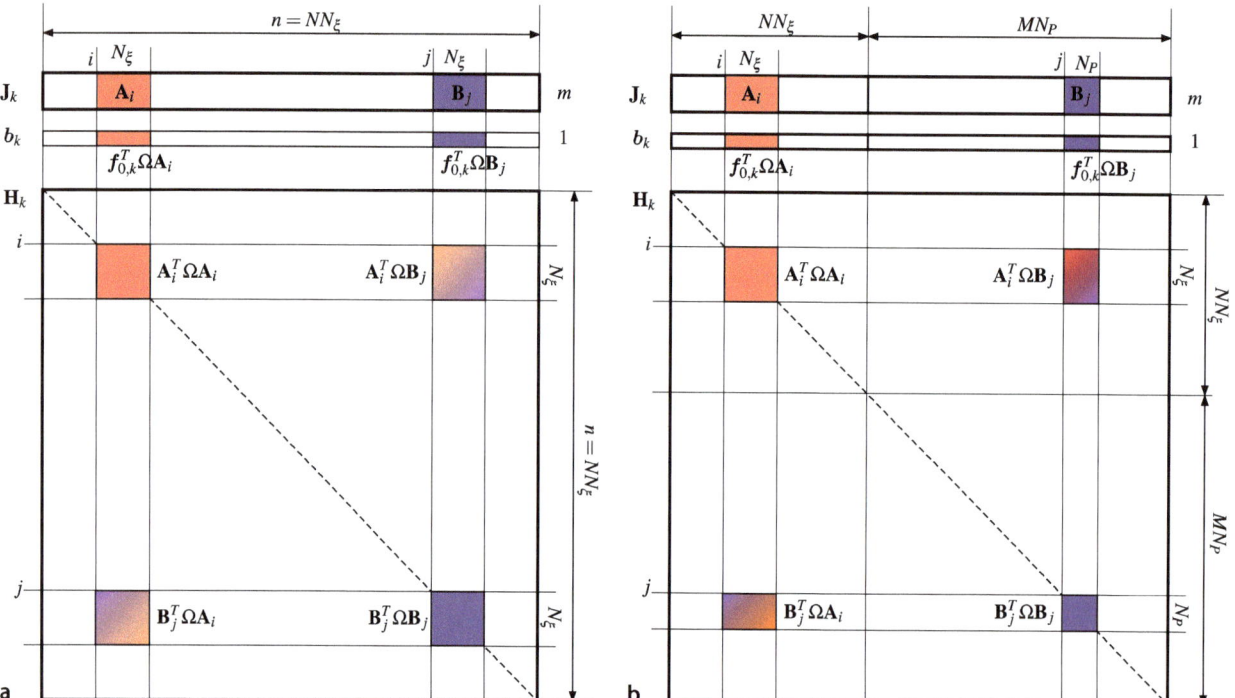

Large Scale Problems

For pose-graph SLAM with thousands of poses or bundle adjustment with thousands of cameras and millions of landmarks the Hessian matrix will be massive leading to computation and storage challenges. The overall Hessian is the summation of many edge Hessians structured as shown in Fig. F.1 and the total Hessian for two problems we have discussed are shown in Fig. F.2. They have clear structure which we can exploit.

Firstly, in both cases the Hessian is sparse – that is, it contains mostly zeros. MATLAB has built-in support for such matrices and instead of storing all those zeros (at 8 bytes each) it simply keeps a list of the nonzero elements. All the standard matrix operations employ efficient algorithms for manipulating sparse matrices.

Secondly, for the bundle adjustment case we see that the Hessian has two block diagonal submatrices so we partition the system as

$$\begin{pmatrix} B & E \\ E^T & C \end{pmatrix} \begin{pmatrix} \Delta_\xi \\ \Delta_P \end{pmatrix} = \begin{pmatrix} b_\xi \\ b_P \end{pmatrix}$$

where B and C are block diagonal.▶ The subscripts ξ and P denote the blocks of Δ and b associated with camera poses and landmark positions respectively. We solve first for the camera pose updates Δ_ξ

$$S\Delta_\xi = b_\xi - EC^{-1}b_P$$

where $S = B - EC^{-1}E^T$ is the Schur complement which is a symmetric positive-definite matrix that is also block diagonal. Then we solve for the update to landmark positions

$$\Delta_P = C^{-1}\left(b_P + E^T\Delta_\xi\right)$$

More sophisticated techniques exploit the fine-scale block structure to further reduce computational time, for example GTSAM (https://bitbucket.org/gtborg/gtsam) and SLAM++ (https://sourceforge.net/projects/slam-plus-plus).

Fig. F.1. Inherent structure of the error vector, Jacobian and Hessian matrices for graph-based least-squares problems. **a** Pose-graph SLAM with N nodes representing robot pose as $\mathbb{R}^{N_\xi}$; **b** bundle adjustment with N nodes representing camera pose as $\mathbb{R}^{N_\xi}$ and M nodes representing landmark position as $\mathbb{R}^{N_P}$. The indices i and j denote the ith and j^{th} block not the ith and j^{th} row or column. *White* indicates zero values

A block diagonal matrix is inverted by simply inverting each of the nonzero blocks along the diagonal.

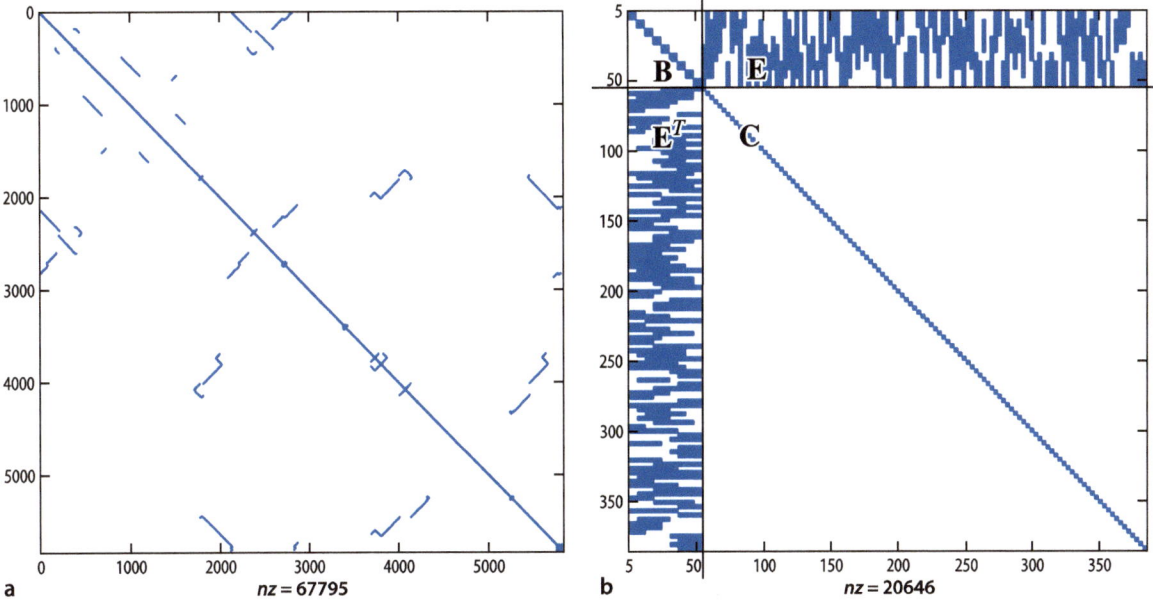

Fig. F.2. Hessian sparsity maps produced using the MATLAB `spy` function, the number of nonzero elements is shown beneath the plot. **a** Hessian for the pose-graph SLAM problem of Fig. 6.17, the diagonal elements represent pose constraints between successive nodes due to odometry, the off-diagonal terms represent constraints due to revisiting locations (loop closures); **b** Hessian for a bundle adjustment problem with 10 cameras and 110 landmarks (`vision/examples/bademo.m`)

Anchoring

Optimization provides a solution where the *relative* poses and positions give the lowest overall cost, and the solution will have an arbitrary transformation with respect to a global reference frame. To obtain absolute poses and positions we must anchor or fix some nodes – assign them values with respect to the global frame and prevent the optimization from adjusting them. The appropriate way to achieve this is to remove from H and b the rows and columns corresponding to the anchored poses and positions. We then solve a lower dimensional problem for Δ' which will be shorter than x and careful book keeping is required to correctly match the subvectors of Δ' with those of x for the update.

G Gaussian Random Variables

The 1-dimensional Gaussian function

$$g(x) = \frac{1}{\sqrt{\sigma^2 2\pi}} e^{-\frac{1}{2\sigma^2}(x-\mu)^2} \tag{G.1}$$

is described by the position of its peak μ and its width σ. The total area under the curve is unity and $g(x) > 0, \forall x$. The function can be plotted using the Toolbox function `gaussfunc`

```
>> x = linspace(-6, 6, 500);
>> plot(x, gaussfunc(0, 1, x), 'r' )
>> hold on
>> plot(x, gaussfunc(0, 2^2, x), '--b' )
```

and Fig. G.1 shows two Gaussians with zero mean and $\sigma = 1$ and $\sigma = 2$. Note that the second argument to `gaussfunc` is the variance not standard deviation.

If the Gaussian is considered to be a probability density function (PDF) then this is the well known normal distribution and the peak position μ is the mean value and the width σ is the standard deviation. A random variable drawn from a normal distribution is often written as $X \sim N(\mu, \sigma^2)$, and $N(0, 1)$ is referred to as the standard normal distribution – the MATLAB function `randn` draws random numbers from this distribution. To draw one hundred Gaussian random numbers with mean `mu` and standard deviation `sigma` is

```
>> g = sigma * randn(100) + mu;
```

The probability that a random value falls within an interval $x \in [x_1, x_2]$ is obtained by integration

$$P = \int_{x_1}^{x_2} g(x)\mathrm{d}x = \Phi(x_2) - \Phi(x_1) \quad \text{where} \quad \Phi(x) = \int_{-\infty}^{x} g(x)\mathrm{d}x$$

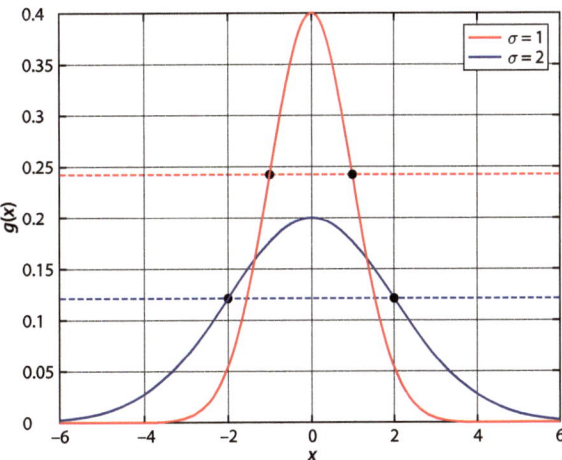

Fig. G.1.
Two Gaussian functions, both with with mean $\mu = 0$, and with standard deviation $\sigma = 1$, and $\sigma = 2$. The markers indicate the points $x = \mu \pm 1\sigma$. The *blue curve* is wider but less tall, since the total area under the curve is unity

or evaluation of the cumulative normal distribution function $\Phi(x)$. The marked points in Fig. G.1 at $\mu \pm 1\sigma$ delimit the 1σ confidence interval. The area under the curve over this interval is 0.68, so the probability of a random value being drawn from this interval is 68%.

The Gaussian can be extended to an arbitrary number of dimensions. The n-dimensional Gaussian, or multivariate normal distribution, is

$$g(\boldsymbol{x}) = \frac{1}{\sqrt{\det(\boldsymbol{P})(2\pi)^n}} e^{-\frac{1}{2}(\boldsymbol{x}-\boldsymbol{\mu})^T \boldsymbol{P}^{-1}(\boldsymbol{x}-\boldsymbol{\mu})} \tag{G.2}$$

and compared to the scalar case of Eq. G.1 $\boldsymbol{x} \in \mathbb{R}^n$ and $\boldsymbol{\mu} \in \mathbb{R}^n$ have become vectors, the squared term in the exponent has been replaced by a matrix quadratic form, and σ^2, the variance, has become a positive-definite (and hence symmetric) covariance matrix $\boldsymbol{P} \in \mathbb{R}^{n \times n}$. The diagonal elements represent the variance of x_i and the off-diagonal elements P_{ij} are the correlationss between x_i and x_j. If the variables are independent or uncorrelated the matrix P would be diagonal. The covariance matrix is symmetric and positive definite.

We can plot a 2-dimensional Gaussian

```
>> [x,y] = meshgrid(-5:0.1:5, -5:0.1:5);
>> P = diag([1 2^2]);
>> surfc(x, y, gaussfunc([0 0], P, x, y))
```

as a surface which is shown in Fig. G.2. In this case $\boldsymbol{\mu} = (0, 0)$ and $\boldsymbol{P} = \mathrm{diag}(1^2, 2^2)$ which corresponds to uncorrelated variables with standard deviation of 1 and 2 respectively. Figure G.2 also shows a number of elliptical contours – contours of constant probability density. If this 2-dimensional probability density function represents the position of a robot in the xy-plane the most likely position for the robot is at $(0, 0)$ and the size of the ellipse says something about our spatial certainty. A particular contour indicates the boundary of a region within which the robot is located with a particular probability. A large ellipse indicates we know, with that probability, that the robot is somewhere inside a large area – we have low certainty about the robot's position. Conversely, a small ellipse means that we know the robot, with the same probability, is somewhere within a much smaller area.

The contour lines are ellipses and in this example the radii in the y- and x-directions are in the ratio $2:1$ as defined by the ratio of the standard deviations. For higher order Gaussians, $n > 2$, the corresponding confidence interval is the surface of an ellipsoid in n-dimensional space.

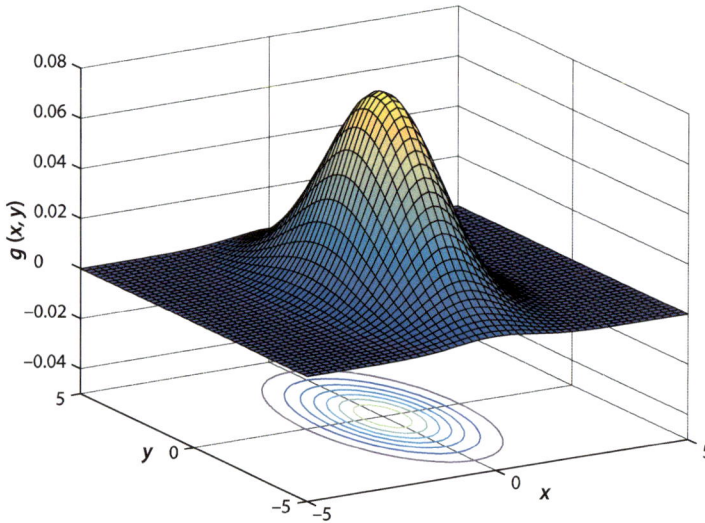

Fig. G.2.
The 2-dimensional Gaussian with covariance $\boldsymbol{P} = \mathrm{diag}(1^2, 2^2)$. Contours lines of constant probability density are shown beneath

The connection between Gaussian probability density functions and ellipses can be found in the quadratic exponent of Eq. G.2 which is the equation of an ellipse or ellipsoid◄. All the points that satisfy

It is also the definition of Mahalanobis distance, the covariance weighted distance between x and μ.

$$(\boldsymbol{x} - \boldsymbol{\mu})^T \boldsymbol{P}^{-1}(\boldsymbol{x} - \boldsymbol{\mu}) = s$$

result in a constant probability density value, that is, a contour of the 2-dimensional Gaussian. s is related to the probability by

$$s = \chi_n^2(p)$$

If we draw a vector of length n from the multivariate Gaussian each element is normally distributed. The sum of squares of independent normally distributed values is known to be distributed according to a χ^2 (chi-squared) distribution with n degrees of freedom.

which is the χ^2 distribution◄ with n degrees of freedom, 2 in this case, and p is the probability that the point x lies on the ellipse. For example the 50% confidence interval is

```
>> s = chi2inv(0.5, 2)
s =
    1.3863
```

This function requires the MATLAB Statistics and Machine Learning Toolbox™. The Robotics Toolbox provides chi2inv_rtb which is an approximation for the case $n = 2$.

where the first argument is the probability and the second is the number of degrees of freedom►.

If the covariance matrix is diagonal then the ellipse is aligned with the x- and y-axes as we saw in Sect. C.1.4. This indicates that the two variables are independent and have zero correlation. Conversely a rotated ellipse indicates that the covariance is not diagonal and the two variables are correlated.

To draw a covariance ellipse we use the general approach for ellipses outlined in Sect. C.1.4 but the right-hand side of the ellipse equation is s not 1, and $\boldsymbol{E} \equiv \boldsymbol{P}$.

H

Kalman Filter

All models are wrong.
Some models are useful.
George Box

Consider the system shown in Fig. H.1. The physical robot is a "black box" which has a true state or pose x that evolves over time according to the applied inputs. We cannot directly measure the state, but sensors on the robot have outputs which are a function of that true state. Our challenge is: given the system inputs and sensor outputs *estimate* the unknown true state x and how certain we are of that estimate.

At face value this might seem hard, or even impossible, but there are quite a lot of things we know about system that will help us. Firstly, we know how the state evolves over time as a function of the inputs – this is the state transition◄ model $f(\cdot)$, and we know the inputs to the system u. Our model is unlikely to be perfect▲ and it is common to represent this uncertainty by an imaginary random number generator which is corrupting the system state – process noise. Secondly, we know how the sensor output depends on the state – this is the sensor model $h(\cdot)$ and its uncertainty is also modeled by an imaginary random number generator – sensor noise.

The imaginary random number sources v and w are *inside* the black box so the random numbers are also unknowable. However we can describe the characteristics of these random numbers – their *distribution* which tells us how likely it is that we will *draw* a random number with a particular value. A lot of noise in physical systems can be modeled well by the Gaussian (aka normal) distribution $\mathcal{N}(\mu, \sigma^2)$ which is characterized by a mean μ and a standard deviation σ. There are infinitely many possible distributions◄ but the Gaussian distribution has some nice mathematical properties that we will rely on. However we should never assume that noise is Gaussian – we should attempt to determine the distribution by understanding the physics of the process and the sensor, or from careful measurement and analysis.

Often called the process or motion model.

For example wheel slippage on a mobile ground robot or wind gusts for a UAV.

Which can be nonsymmetrical or have multiple peaks.

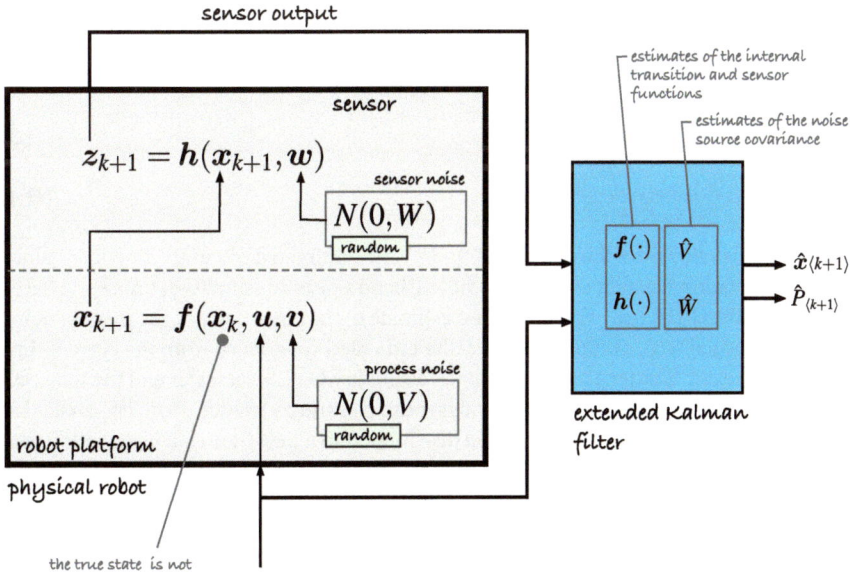

Fig. H.1.
The physical robot on the left has a true state that cannot be directly measured, however we gain a clue from the sensor output which is a function of this unknown true state

In general terms, the problem we wish to solve is:

> given a model of the system $f(\cdot)$, $h(\cdot)$, $\hat{V}$ and $\hat{W}$; the known inputs applied to the system u; and some noisy sensor measurements z, find an estimate $\hat{x}$ of the system state and our uncertainty $\hat{P}$ in that estimate.

In a robotic localization context x is the unknown position or pose of the robot, u is the commands sent to the motors and z is the output of various sensors on the robot. For a ground robot x would be the pose in $SE(2)$ and u would be the motor commands and z might be the measured odometry or range and bearing to landmarks. For a flying robot x would be the pose in $SE(3)$ and u are the known forces applied to the airframe and z might be the measured accelerations and angular velocities.▶

The state is a vector and there are many approaches to mapping pose to a vector, especially the rotational component – Euler angles, quaternions, and exponential coordinates are commonly used.

H.1   Linear Systems – Kalman Filter

Consider the transition model described as a discrete-time linear time-invariant system

$$x\langle k{+}1\rangle = Fx\langle k\rangle + Gu\langle k\rangle + v\langle k\rangle \tag{H.1}$$

$$z\langle k\rangle = Hx\langle k\rangle + w\langle k\rangle \tag{H.2}$$

where k is the time step, $x \in \mathbb{R}^n$ is the state vector, and $u \in \mathbb{R}^m$ is a vector of inputs to the system at time k, for example a velocity command, or applied forces and torques. The matrix $F \in \mathbb{R}^{n \times n}$ describes the dynamics of the system, that is, how the states evolve with time. The matrix $G \in \mathbb{R}^{n \times m}$ describes how the inputs are coupled to the system states. The vector $z \in \mathbb{R}^p$ represents the outputs of the system as measured by sensors. The matrix $H \in \mathbb{R}^{p \times n}$ describes how the system states are mapped to the system outputs which we can observe.

To account for errors in the motion model (F and G) or unmodeled disturbances we introduce a Gaussian random variable $v \in \mathbb{R}^n$ termed the process noise. $v\langle k\rangle \sim N(0, V)$, that is, it has zero mean and covariance $V \in \mathbb{R}^{n \times n}$. Covariance is a matrix quantity which is the variance for a multi-dimensional distribution – it is a positive definite matrix and therefore symmetric. The sensor measurement model H is not perfect either and this is modeled by sensor measurement noise, a Gaussian random variable $w \in \mathbb{R}^p$, $w\langle k\rangle \sim N(0, W)$ and covariance $W \in \mathbb{R}^{p \times p}$.

The Kalman filter is an optimal estimator for the case where the process and measurement noise are zero-mean Gaussian noise. The filter has two steps: prediction and update. The prediction is based on the previous state and the inputs that were applied

$$\hat{x}^{+}\langle k{+}1\rangle = F\hat{x}\langle k\rangle + Gu\langle k\rangle \tag{H.3}$$

$$\hat{P}^{+}\langle k{+}1\rangle = \underbrace{F\hat{P}\langle k\rangle F^{T}} + \hat{V} \tag{H.4}$$

where $\hat{x}$ is the estimate of the state and $\hat{P} \in \mathbb{R}^{n \times n}$ is the estimated covariance, or uncertainty, in $\hat{x}$. The notation $^{+}$ makes explicit that the left-hand side is an estimate at time $k + 1$ based on information from time k. $\hat{V}$ is our best estimate of the covariance of the process noise.

The indicated term in Eq. H.4 *projects* the estimated covariance from the current time step to the next. Consider a one dimensional example where F is a scalar and the state estimate $\hat{x}\langle k\rangle$ has a PDF which is Gaussian with a mean $\bar{x}\langle k\rangle$ and a variance $\sigma^2\langle k\rangle$. The prediction equation maps the state and its Gaussian distribution to a new Gaussian distribution with a mean $F\bar{x}\langle k\rangle$ and a variance $F^2\sigma^2\langle k\rangle$. The term $FP\langle k\rangle F\langle k\rangle^{T}$ is the matrix form of this since

$$\mathrm{cov}(Fx) = F\,\mathrm{cov}(x)F^{T} \tag{H.5}$$

which scales the covariance appropriately.

The prediction of $\hat{P}$ involves the addition of two positive-definite matrices so the uncertainty will increase – this is to be expected since we have used an uncertain model to predict the future value of an already uncertain estimate. $\hat{V}$ must be a reasonable estimate of the covariance of the actual process noise. If we overestimate it, that is our estimate of process noise is larger than it really is, then we will have a large increase in uncertainty at this step, a pessimistic estimate of our certainty.

To counter this growth in uncertainty we need to introduce new information such as measurements made by the sensors since they depend on the state. The difference between what the sensors measure and what the sensors are predicted to measure is

$$\nu = \boldsymbol{z}^{\#}\langle k{+}1\rangle - \boldsymbol{H}\hat{\boldsymbol{x}}^{+}\langle k{+}1\rangle \in \mathbb{R}^{p}$$

Some of this difference is due to noise in the sensor, the measurement noise, but the remainder provides valuable information related to the error between the actual and the predicted value of the state. Rather than considering this as error we refer to it more positively as *innovation* – new information.

The second step of the Kalman filter, the *update* step, maps the innovation into a correction for the predicted state, optimally tweaking the estimate based on what the sensors observed

$$\hat{\boldsymbol{x}}\langle k{+}1\rangle = \hat{\boldsymbol{x}}^{+}\langle k{+}1\rangle + \boldsymbol{K}\nu \tag{H.6}$$

$$\hat{\boldsymbol{P}}\langle k{+}1\rangle = \hat{\boldsymbol{P}}^{+}\langle k{+}1\rangle - \boldsymbol{K}\boldsymbol{H}\hat{\boldsymbol{P}}^{+}\langle k{+}1\rangle \tag{H.7}$$

Uncertainty is now *decreased* or *deflated*, since new information, from the sensors, is being incorporated. The matrix

$$\boldsymbol{K} = \boldsymbol{P}^{+}\langle k{+}1\rangle \boldsymbol{H}^{T}\left[\underbrace{\boldsymbol{H}\boldsymbol{P}^{+}\langle k{+}1\rangle \boldsymbol{H}^{T} + \hat{\boldsymbol{W}}}\right]^{-1} \in \mathbb{R}^{n\times p} \tag{H.8}$$

is known as the Kalman gain. The term indicated is the estimated covariance of the innovation and comprises the uncertainty in the state and the estimated measurement noise covariance. If the innovation has high uncertainty in some dimensions then the Kalman gain will be correspondingly small, that is, if the new information is uncertain then only small changes are made to the state vector. The term $\boldsymbol{H}\boldsymbol{P}^{+}\langle k{+}1\rangle \boldsymbol{H}^{T}$ in Eq. H.13 *projects* the covariance of the state estimate into the space of sensor values.

The covariance matrix must be positive-definite but after many updates the accumulated numerical errors may cause this matrix to be no longer symmetric. The positive-definite structure can be enforced by using the Joseph form of Eq. H.7

$$\hat{\boldsymbol{P}}\langle k{+}1\rangle = \left(\boldsymbol{I}_{n\times n} - \boldsymbol{K}\boldsymbol{H}\right)\hat{\boldsymbol{P}}^{+}\langle k{+}1\rangle \left(\boldsymbol{I}_{n\times n} - \boldsymbol{K}\boldsymbol{H}\right)^{T} + \boldsymbol{K}\hat{\boldsymbol{V}}\boldsymbol{K}^{T}$$

but this is computationally more costly.

The equations above constitute the classical Kalman filter which is widely used in robotics, aerospace and econometric applications. The filter has a number of important characteristics. Firstly it is optimal, but only if the noise is truly Gaussian with zero mean and time invariant parameters. This is often a good assumption but not always. Secondly it is recursive, the output of one iteration is the input to the next. Thirdly, it is asynchronous. At a particular iteration if no sensor information is available we just perform the prediction step and not the update. In the case that there are different sensors, each with their own $\boldsymbol{H}$, and different sample rates, we just apply the update with the appropriate $\boldsymbol{z}$ and $\boldsymbol{H}$. The filter must be initialized with some reasonable value of $\hat{\boldsymbol{x}}$ and $\hat{\boldsymbol{P}}$, as well as good choices of the covariance estimates $\hat{\boldsymbol{V}}$ and $\hat{\boldsymbol{W}}$. As the filter runs the estimated covariance $\|\hat{\boldsymbol{P}}\|$ decreases but never reaches zero – the minimum value can be shown to be a function of $\hat{\boldsymbol{V}}$ and $\hat{\boldsymbol{W}}$. The Kalman-Bucy filter is a continuous-time version of this filter.

The covariance matrix $\hat{P}$ is rich in information. The diagonal elements $\hat{P}_{ii}$ are the variance, or uncertainty, in the state x_i. The off-diagonal elements $\hat{P}_{ij}$ are the correlations between states x_i and x_j and indicate that the errors are not independent. The correlations are critical in allowing any piece of new information to *flow through* to adjust all the states that affect a particular process output.

H.2   Nonlinear Systems – Extended Kalman Filter

For the case where the system is not linear it can be described generally by two functions: the state transition (the motion model in robotics) and the sensor model

$$x\langle k{+}1\rangle = f\big(x\langle k\rangle, u\langle k\rangle, v\langle k\rangle\big) \tag{H.9}$$

$$z\langle k\rangle = h\big(x\langle k\rangle, w\langle k\rangle\big) \tag{H.10}$$

and as before we represent model uncertainty, external disturbances and sensor noise by Gaussian random variables v and w.

We linearize the state transition function about the current state estimate $\hat{x}_k$ as shown in Fig. H.2 resulting in

$$x'\langle k{+}1\rangle \approx F_x x'\langle k\rangle + F_u u\langle k\rangle + F_v v\langle k\rangle \tag{H.11}$$

$$z'\langle k\rangle \approx H_x x'\langle k\rangle + H_w w\langle k\rangle \tag{H.12}$$

where $F_x = \partial f / \partial x \in \mathbb{R}^{n\times n}, F_u = \partial f / \partial u \in \mathbb{R}^{n\times m}, F_v = \partial f / \partial v \in \mathbb{R}^{n\times n}, H_x = \partial h / \partial x \in \mathbb{R}^{p\times n}$ and $H_w = \partial h / \partial w \in \mathbb{R}^{p\times p}$ are Jacobians of the functions $f(\cdot)$ and $h(\cdot)$. Equating coefficients between Eq. H.1 and Eq. H.11 gives $F \sim F_x, G \sim F_u$ and $v\langle k\rangle \sim F_v v\langle k\rangle$; and between Eq. H.2 and Eq. H.12 gives $H \sim H_x$ and $w\langle k\rangle \sim H_w w\langle k\rangle$.

Taking the prediction equation Eq. H.9 with $v\langle k\rangle = 0$, and the covariance equation Eq. H.4 with the linearized terms substituted we can write the prediction step as

$$\hat{x}^+\langle k{+}1\rangle = f\big(\hat{x}\langle k\rangle, u\langle k\rangle\big)$$
$$\hat{P}^+\langle k{+}1\rangle = F_x \hat{P}\langle k\rangle F_x^T + F_v \hat{V} F_v^T$$

and the update step as

$$\hat{x}\langle k{+}1\rangle = \hat{x}^+\langle k{+}1\rangle + K\nu$$
$$\hat{P}\langle k{+}1\rangle = \hat{P}^+\langle k{+}1\rangle - K H_x \hat{P}^+\langle k{+}1\rangle$$

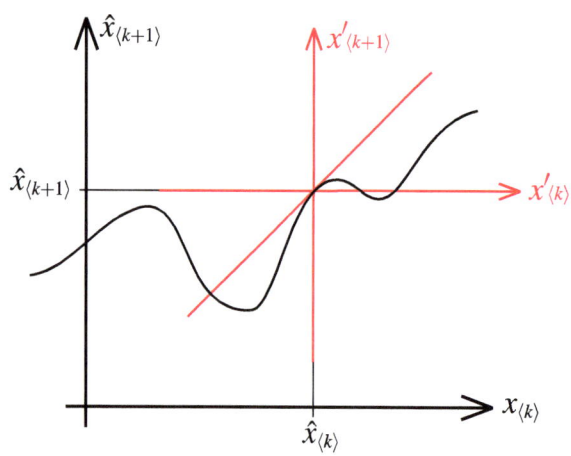

Fig. H.2.
One dimensional example illustrating how the nonlinear state transition function $f: x_k \mapsto x_{k+1}$ shown in *black* is linearized about the point $(\hat{x}\langle k\rangle, \hat{x}\langle k{+}1\rangle)$ shown in *red*

Procedure EKF

Input : $\hat{\boldsymbol{x}}\langle k\rangle \in \mathbb{R}^n$, $\hat{\boldsymbol{P}}\langle k\rangle \in \mathbb{R}^{n\times n}$, $\boldsymbol{u}\langle k\rangle \in \mathbb{R}^m$, $\boldsymbol{z}\langle k+1\rangle \in \mathbb{R}^p$, $\hat{\boldsymbol{V}} \in \mathbb{R}^{n\times n}$, $\hat{\boldsymbol{W}} \in \mathbb{R}^{p\times p}$

Output: $\hat{\boldsymbol{x}}\langle k+1\rangle \in \mathbb{R}^n$, $\hat{\boldsymbol{P}}\langle k+1\rangle \in \mathbb{R}^{n\times n}$

$-$ *linearize about* $\boldsymbol{x} = \hat{\boldsymbol{x}}\langle k\rangle$

compute Jacobians: $\boldsymbol{F}_x \in \mathbb{R}^{n\times n}$, $\boldsymbol{F}_v \in \mathbb{R}^{n\times n}$, $\boldsymbol{H}_x \in \mathbb{R}^{p\times n}$, $\boldsymbol{H}_w \in \mathbb{R}^{p\times p}$

$-$ *the prediction step*

$\qquad \hat{\boldsymbol{x}}^+\langle k+1\rangle = \boldsymbol{f}\big(\hat{\boldsymbol{x}}\langle k\rangle, \boldsymbol{u}\langle k\rangle\big)$ // *predict state at next time step*

$\qquad \hat{\boldsymbol{P}}^+\langle k+1\rangle = \boldsymbol{F}_x\hat{\boldsymbol{P}}\langle k\rangle\boldsymbol{F}_x^T + \boldsymbol{F}_v\hat{\boldsymbol{V}}\boldsymbol{F}_v^T$ // *predict cov ariance at next time step*

$-$ *the update step*

$\qquad \nu = \boldsymbol{z}\langle k+1\rangle - \boldsymbol{h}\big(\hat{\boldsymbol{x}}^+\langle k+1\rangle\big)$ // *innovation : measured $-$ predicted sensor value*

$\qquad \boldsymbol{K} = \boldsymbol{P}^+\langle k+1\rangle\boldsymbol{H}_x^T\Big[\boldsymbol{H}_x\boldsymbol{P}^+\langle k+1\rangle\boldsymbol{H}_x^T + \boldsymbol{H}_w\hat{\boldsymbol{W}}\boldsymbol{H}_w^T\Big]^{-1}$ // *Kalman gain*

$\qquad \hat{\boldsymbol{x}}\langle k+1\rangle = \hat{\boldsymbol{x}}^+\langle k+1\rangle + \boldsymbol{K}\nu$ // *update state estimate*

Algorithm H.1.
Procedure EKF

$\qquad \hat{\boldsymbol{P}}\langle k+1\rangle = \hat{\boldsymbol{P}}^+\langle k+1\rangle - \boldsymbol{K}\boldsymbol{H}_x\hat{\boldsymbol{P}}^+\langle k+1\rangle$ // *update covariance estimate*

where the Kalman gain is now

$$\boldsymbol{K} = \boldsymbol{P}^+\langle k+1\rangle\boldsymbol{H}_x^T\Big(\boldsymbol{H}_x\boldsymbol{P}^+\langle k+1\rangle\boldsymbol{H}_x^T + \boldsymbol{H}_w\hat{\boldsymbol{W}}\boldsymbol{H}_w^T\Big)^{-1} \tag{H.13}$$

Properly these matrices should be denoted as depending on the time step, i.e. $F_x\langle k\rangle$ but this has been dropped in the interest of readability.

These equations are only valid at the linearization point $\hat{x}\langle k\rangle$ – the Jacobians F_x, F_v, H_x, H_w must be computed at every iteration.◄ The full procedure is summarized in Algorithm H.1.

A fundamental problem with the extended Kalman filter is that PDFs of the random variables are no longer Gaussian after being operated on by the nonlinear functions $f(\cdot)$ and $h(\cdot)$. We can easily illustrate this by considering a nonlinear scalar function $y = (x + 2)^2 / 4$. We will draw a million Gaussian random numbers from the normal distribution $\mathcal{N}(5, 4)$ which has a mean of 5 and a standard deviation of 2

```
>> x = 2*randn(1000000,1) + 5;
```

and map them through our function

```
>> y = (x+2).^2 / 4;
```

and plot the probability density function of y

```
>> histogram(y, 'Normalization', 'pdf');
```

Fig. H.3.
PDF of the state x (*red*) which is Gaussian $\mathcal{N}(5, 4)$ and the PDF of the nonlinear function $y = (x + 2)^2 / 4$ (*black*). The peak and the mean of the nonlinear distribution are shown by *blue solid* and *dashed vertical lines* respectively

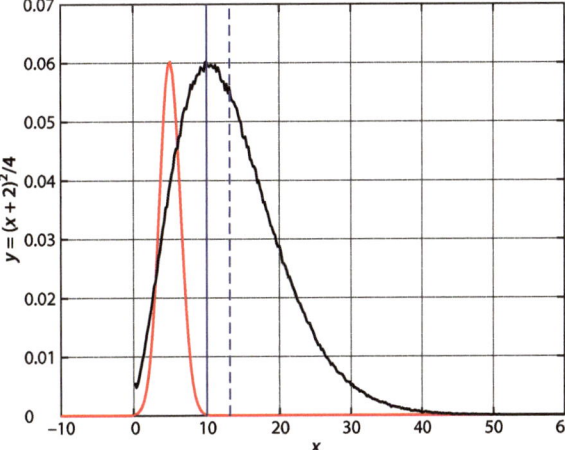

which is shown in Fig. H.3. We see that the PDF of y is substantially changed and no longer Gaussian. It has lost its symmetry so the mean value is greater than the mode. The Jacobians that appear in the EKF equations appropriately scale the covariance but the resulting non-Gaussian distribution breaks the assumptions which guarantee that the Kalman filter is an optimal estimator. Alternatives include the iterated EKF described by Jazwinski (2007) or the Unscented Kalman Filter (UKF) (Julier and Uhlmann 2004) or the sigma-point filter which uses discrete sample points (sigma points) to approximate the PDF.

Graphs

A graph is an abstract representation of a set of objects connected by links and depicted graphically as shown in Fig. I.1. Mathematically a graph is denoted $G(V, E)$ where V are the vertices or nodes, and E are the links that connect pairs of vertices and are called edges or arcs. Edges can be directed (*arrows*) or undirected as in this case. Edges can have an associated weight or cost associated with moving from one vertex to another. A sequence of edges from one vertex to another is a path, and a sequence that starts and ends at the same vertex is a cycle. An edge from a vertex to itself is a loop. Graphs can be used to represent transport, communications or social networks, and this branch of mathematics is graph theory.

The Toolbox provides a MATLAB® graph class called `PGraph` that supports embedded graphs where the vertices are associated with a point in an n-dimensional space.◀ To create a new graph

This class is used other Toolbox classes such as `PRM`, `Lattice`, `RRT`, `PoseGraph` and `BundleAdjust`. MATLAB 2015b introduced a built in `graph` class to represent graphs.

```
>> g = PGraph()
g =
   2 dimensions
   0 vertices
   0 edges
   0 components
```

and by default the nodes of the graph exist in a 2-dimensional space. We can add nodes to the graph

```
>> g.add_node( rand(2,1) );
>> g.add_node( rand(2,1) );
>> g.add_node( rand(2,1) );
>> g.add_node( rand(2,1) );
>> g.add_node( rand(2,1) );
```

and each has a random coordinate. The `add_node` method returns an integer identifier for the node just added. A summary of the graph is given with its display method

```
>> g
g =
   2 dimensions
   5 vertices
   0 edges
   0 components
```

and shows that the graph has 5 nodes but no edges. The nodes are numbered 1 to 5 and we add edges between pairs of nodes

```
>> g.add_edge(1, 2);
>> g.add_edge(1, 3);
>> g.add_edge(1, 4);
>> g.add_edge(2, 3);
>> g.add_edge(2, 4);
>> g.add_edge(4, 5);
>> g
g =
   2 dimensions
   5 vertices
   6 edges
   1 components
```

By default the distance between the nodes is the Euclidean distance between the vertices but this can be overridden by a third argument to `add_edge`. The methods `add_node` and `add_edge` return an integer that uniquely identifies the node or edge just created. The graph has one component, that is all the nodes are connected into one network. The graph can be plotted by

```
>> g.plot('labels')
```

as shown in Fig. I.1. The vertices are shown as blue circles, and the option `'labels'` displays the vertex index next to the circle. Edges are shown as black lines joining vertices. Many options exist to change default plotting behavior. Note that only graphs embedded in 2- and 3-dimensional space can be plotted.

The neighbors of vertex 2 are

```
>> g.neighbours(2)
ans =
     3     4     1
```

which are vertices connected to vertex 2 by edges. Each edge has a unique index and the edges connecting to vertex 2 are

```
>> e = g.edges(2)
e =
     4     5     1
```

The cost or length of these edges is

```
>> g.cost(e)
ans =
    0.9597    0.3966    0.6878
```

and clearly edge 5 has a lower cost than edges 4 and 1. Edge 5

```
>> g.vertices(5)'
ans =
     2     4
```

joins vertices 2 and 4, and vertex 4 is clearly the closest neighbor of vertex 2. Frequently we wish to obtain a node's neighboring vertices and their distances at the same time, and this can be achieved conveniently by

```
>> [n,c] = g.neighbours(2)
n =
     3     4     1
c =
    0.9597    0.3966    0.6878
```

Concise information about a node can be obtained by

```
>> g.about(1)
Node 1 #1@ (0.814724 0.905792 )
  neighbours:    >-o-> 2 3 4
  edges:    >-o-> 1 2 3
```

Arbitrary data can be attached to any node or edge by the methods `setvdata` and `setedata` respectively and retrieved by the methods `vdata` and `edata` respectively.

The vertex closest to the coordinate (0.5, 0.5) is

```
>> g.closest([0.5, 0.5])
ans =
     4
```

and the vertex closest to an interactively selected point is given by `g.pick`.

The minimum cost path between any two nodes in the graph can be computed using well known algorithms such as A* (Nilsson 1971)

```
>> g.Astar(3, 5)
ans =
     3     2     4     5
```

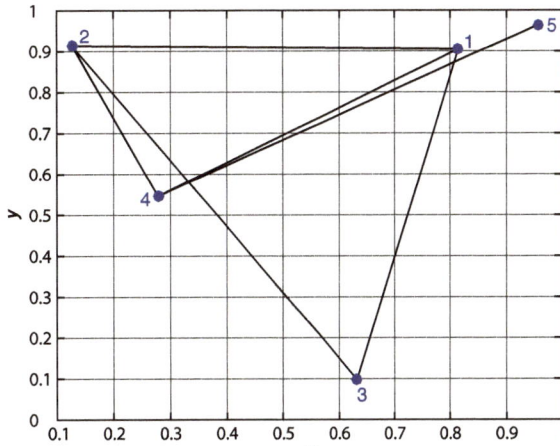

Fig. I.1.
An example graph generated by
the PGraph class

or the earlier method by Dijstrka (1959). By default the graph is treated as undirected, that is, the edges have no preferred direction. The 'directed' option causes edges to be treated as directed, and the path will only traverse edges in their specified direction which is from the first to the second argument of the method add_edge.

Methods exist to compute various other representations of the graph such as adjacency, incidence, degree and Laplacian matrices.

Bibliography

Achtelik MW (2014) Advanced closed loop visual navigation for micro aerial vehicles. Ph.D. thesis, ETH Zurich

Agarwal P, Burgard W, Stachniss C (2014) Survey of geodetic mapping methods: Geodetic approaches to mapping and the relationship to graph-based SLAM. IEEE Robot Autom Mag 21(3):63–80

Albertos P, Mareels I (2010) Feedback and control for everyone. Springer-Verlag, Berlin Heidelberg

Altmann SL (1989) Hamilton, Rodrigues, and the quaternion scandal. Math Mag 62(5):291–308

Alton K, Mitchell IM (2006) Optimal path planning under defferent norms in continuous state spaces. In: Proceedings of the IEEE International Conference on Robotics and Automation (ICRA). pp 866–872

Antonelli G (2014) Underwater robots: Motion and force control of vehicle-manipulator systems, 3rd ed. Springer Tracts in Advanced Robotics, vol 2. Springer-Verlag, Berlin Heidelberg

Arkin RC (1999) Behavior-based robotics. MIT Press, Cambridge, Massachusetts

Armstrong WW (1979) Recursive solution to the equations of motion of an N-link manipulator. In: Proceedings of the 5th World Congress on Theory of Machines and Mechanisms, Montreal, Jul, pp 1343–1346

Armstrong BS (1988) Dynamics for robot control: Friction modelling and ensuring excitation during parameter identification. Stanford University

Armstrong B (1989) On finding exciting trajectories for identification experiments involving systems with nonlinear dynamics. Int J Robot Res 8(6):28

Armstrong B, Khatib O, Burdick J (1986) The explicit dynamic model and inertial parameters of the Puma 560 Arm. In: Proceedings of the IEEE International Conference on Robotics and Automation (ICRA), vol 3. pp 510–518

Armstrong-Hélouvry B, Dupont P, De Wit CC (1994) A survey of models, analysis tools and compensation methods for the control of machines with friction. Automatica 30(7):1083–1138

Asada H (1983) A geometrical representation of manipulator dynamics and its application to arm design. J Dyn Syst-T ASME 105:131

Bailey T (n.d.) Software resources. University of Sydney. http://www-personal.acfr.usyd.edu.au/tbailey

Bailey T, Durrant-Whyte H (2006) Simultaneous localization and mapping: Part II. IEEE Robot Autom Mag 13(3):108–117

Ball RS (1876) The theory of screws: A study in the dynamics of a rigid body. Hodges, Foster & Co., Dublin

Ball RS (1908) A treatise on spherical astronomy. Cambridge University Press, New York

Bar-Shalom Y, Fortmann T (1988) Tracking and data association. Mathematics in science and engineering, vol 182. Academic Press, London Oxford

Bar-Shalom Y, Rong Li X, Thiagalingam Kirubarajan (2001) Estimation with applications to tracking and navigation. John Wiley & Sons, Inc., Chichester

Bertozzi M, Broggi A, Cardarelli E, Fedriga R, Mazzei L, Porta P (2011) VIAC expedition: Toward autonomous mobility. IEEE Robot Autom Mag 18(3):120–124

Biber P, Straßer W (2003) The normal distributions transform: A new approach to laser scan matching. In: Proceedings of the IEEE/RSJ International Conference on intelligent robots and systems (IROS), vol 3. pp 2743–2748

Blewitt M (2011) Celestial navigation for yachtsmen. Adlard Coles Nautical, London

Bolton W (2015) Mechatronics: Electronic control systems in mechanical and electrical engineering, 6th ed. Pearson, Harlow

Borenstein J, Everett HR, Feng L (1996) Navigating mobile robots: Systems and techniques. AK Peters, Ltd. Natick, MA, USA, Out of print and available at http://www-personal.umich.edu/~johannb/Papers/pos96rep.pdf

Borgefors G (1986) Distance transformations in digital images. Comput Vision Graph 34(3):344–371

Bostrom N (2016) Superintelligence: Paths, dangers, strategies. Oxford University Press, Oxford, 432 p

Brady M, Hollerbach JM, Johnson TL, Lozano-Pérez T, Mason MT (eds) (1982) Robot motion: Planning and control. MIT Press, Cambridge, Massachusetts

© Springer Nature Switzerland AG 2022
P. Corke, *Robotics and Control*, Springer Tracts in Advanced Robotics 141,
https://doi.org/10.1007/978-3-030-79179-7

Braitenberg V (1986) Vehicles: Experiments in synthetic psychology. MIT Press, Cambridge, Massachusetts

Bray H (2014) You are here: From the compass to GPS, the history and future of how we find ourselves. Basic Books, New York

Brockett RW (1983) Asymptotic stability and feedback stabilization. In: Brockett RW, Millmann RS, Sussmann HJ (eds) Progress in mathematics. Differential geometric control theory, vol 27. pp 181–191

Brooks RA (1989) A robot that walks: Emergent behaviors from a carefully evolved network. MIT AI Lab, Memo 1091

Brynjolfsson E, McAfee A (2014) The second machine age: Work, progress, and prosperity in a time of brilliant technologies. W.W. Norton & Co., New York

Buehler M, Iagnemma K, Singh S (eds) (2007) The 2005 DARPA grand challenge: The great robot race. Springer Tracts in Advanced Robotics, vol 36. Springer-Verlag, Berlin Heidelberg

Buehler M, Iagnemma K, Singh S (eds) (2010) The DARPA urban challenge. Tracts in Advanced Robotics, vol 56. Springer-Verlag, Berlin Heidelberg

Censi A (2008) An ICP variant using a point-to-line metric. In: Proceedings of the IEEE International Conference on Robotics and Automation (ICRA). pp 19–25

Choset HM, Lynch KM, Hutchinson S, Kantor G, Burgard W, Kavraki LE, Thrun S (2005) Principles of robot motion. MIT Press, Cambridge, Massachusetts

Corke PI (1996a) In situ measurement of robot motor electrical constants. Robotica 14(4):433–436

Corke PI (1996b) Visual control of robots: High-performance visual servoing. Mechatronics, vol 2. Research Studies Press (John Wiley). Out of print and available at http://www.petercorke.com/bluebook

Corke PI (2007) A simple and systematic approach to assigning Denavit-Hartenberg parameters. IEEE T Robotic Autom 23(3):590–594

Corke PI, Armstrong-Hélouvry BS (1994) A search for consensus among model parameters reported for the PUMA 560 robot. In: Proceedings of the IEEE International Conference on Robotics and Automation (ICRA). San Diego, pp 1608–1613

Corke PI, Armstrong-Hélouvry BS (1995) A meta-study of PUMA 560 dynamics: A critical appraisal of literature data. Robotica 13(3):253–258

Corke P, Lobo J, Dias J (2007) An introduction to inertial and visual sensing. Int J Robot Res 26(6):519–535

Craig JJ (1987) Adaptive control of mechanical manipulators. Addison-Wesley Longman Publishing Co., Inc. Boston

Craig JJ (2005) Introduction to robotics: Mechanics and control, 3rd ed. Pearson/Prentice Hall, Upper Saddle River, New Jersey

Craig JJ, Hsu P, Sastry SS (1987) Adaptive control of mechanical manipulators. Int J Robot Res 6(2):16–28

Dellaert F, Kaess M (2006) Square root SAM: Simultaneous localization and mapping via square root information smoothing. Int J Robot Res 25(12):1181–1203

Denavit J, Hartenberg RS (1955) A kinematic notation for lower-pair mechanisms based on matrices. J Appl Mech-T ASME 22(1):215–221

Deo AS, Walker ID (1995) Overview of damped least-squares methods for inverse kinematics of robot manipulators. J Intell Robot Syst 14(1):43–68

Dickmanns ED (2007) Dynamic vision for perception and control of motion. Springer-Verlag, London

Dickmanns ED, Graefe V (1988b) Dynamic monocular machine vision. Mach Vision Appl 1(4):223–240

Dickmanns ED, Zapp A (1987) Autonomous high speed road vehicle guidance by computer vision. In: Tenth Triennial World Congress of the International Federation of Automatic Control, vol 4. Munich, pp 221–226

Durrant-Whyte H, Bailey T (2006) Simultaneous localization and mapping: Part I. IEEE Robot Autom Mag 13(2):99–110

Engelberger JF (1980) Robotics in practice. Management and applications of industrial robots. Springer-Verlag, Berlin Heidelberg

Engelberger JF (1989) Robotics in service. MIT Press, Cambridge, Massachusetts

Everett HR (1995) Sensors for mobile robots: Theory and application. AK Peters Ltd., Wellesley

Featherstone R (1987) Robot dynamics algorithms. Kluwer Academic, Dordrecht

Ferguson D, Stentz A (2006) Using interpolation to improve path planning: The Field D* algorithm. J Field Robotics 23(2):79–101

Ford M (2015) Rise of the robots: Technology and the threat of a jobless future. Basic Books, New York

Friedman DP, Felleisen M, Bibby D (1987) The little LISPer. MIT Press, Cambridge, Massachusetts

Funda J, Taylor RH, Paul RP (1990) On homogeneous transforms, quaternions, and computational efficiency. IEEE T Robotic Autom 6(3):382–388

Gautier M, Khalil W (1992) Exciting trajectories for the identification of base inertial parameters of robots. Int J Robot Res 11(4):362

Geraerts R, Overmars MH (2004) A comparative study of probabilistic roadmap planners. In: Boissonnat J-D, Burdick J, Goldberg K, Hutchinson S (eds) Springer tracts in advanced robotics. Algorithmic Foundations of Robotics V, vol 7. Springer-Verlag, Berlin Heidelberg, pp 43–58

Glover A, Maddern W, Warren M, Reid S, Milford M, Wyeth G (2012) OpenFABMAP: An open source toolbox for appearance-based loop closure detection. In: Proceedings of the IEEE International Conference on Robotics and Automation (ICRA). pp 4730–4735

Grey CGP (2014) Humans need not apply. YouTube video, www.youtube.com/watch?v=7Pq-S557XQU

Grisetti G (n.d.) Teaching resources. Sapienza University of Rome. http://www.dis.uniroma1.it/~grisetti/teaching.html

Groves PD (2013) Principles of GNSS, inertial, and multisensor integrated navigation systems, 2nd ed. Artech House, Norwood, USA

Hamel T, Mahony R, Lozano R, Ostrowski J (2002) Dynamic modelling and configuration stabilization for an X4-flyer. IFAC World Congress 1(2), p 3

Hartenberg RS, Denavit J (1964) Kinematic synthesis of linkages. McGraw-Hill, New York, available online at http://kmoddl.library.cornell.edu/bib.php?m=23

Hirata T (1996) A unified linear-time algorithm for computing distance maps. Inform Process Lett 58(3):129–133

Hirt C, Claessens S, Fecher T, Kuhn M, Pail R, Rexer M (2013) New ultrahigh-resolution picture of Earth's gravity field. Geophys Res Lett 40:4279–4283

Hoag D (1963) Consideration of Apollo IMU gimbal lock. MIT Instrumentation Laboratory, E–1344, http://www.hq.nasa.gov/alsj/e-1344.htm

Hollerbach JM (1980) A recursive Lagrangian formulation of manipulator dynamics and a comparative study of dynamics formulation complexity. IEEE T Syst Man Cyb 10(11):730–736, Nov

Hollerbach JM (1982) Dynamics. In: Brady M, Hollerbach JM, Johnson TL, Lozano-Pérez T, Mason MT (eds) Robot motion – Planning and control. MIT Press, Cambridge, Massachusetts, pp 51–71

Howard TM, Green CJ, Kelly A, Ferguson D (2008) State space sampling of feasible motions for high-performance mobile robot navigation in complex environments. J Field Robotics 25(6–7):325–345

Hua M-D, Ducard G, Hamel T, Mahony R, Rudin K (2014) Implementation of a nonlinear attitude estimator for aerial robotic vehicles. IEEE T Contr Syst T 22(1):201–213

Izaguirre A, Paul RP (1985) Computation of the inertial and gravitational coefficients of the dynamics equations for a robot manipulator with a load. In: Proceedings of the IEEE International Conference on Robotics and Automation (ICRA). Mar, pp 1024–1032

Jarvis RA, Byrne JC (1988) An automated guided vehicle with map building and path finding capabilities. In: Robotics Research: The Fourth international symposium. MIT Press, Cambridge, Massachusetts, pp 497–504

Jazwinski AH (2007) Stochastic processes and filtering theory. Dover Publications, Mineola

Julier SJ, Uhlmann JK (2004) Unscented filtering and nonlinear estimation. P IEEE 92(3):401–422

Kaess M, Ranganathan A, Dellaert F (2007) iSAM: Fast incremental smoothing and mapping with efficient data association. In: Proceedings of the IEEE International Conference on Robotics and Automation (ICRA). pp 1670–1677

Kahn ME (1969) The near-minimum time control of open-loop articulated kinematic linkages. Stanford University, AIM-106

Kálmán RE (1960) A new approach to linear filtering and prediction problems. J Basic Eng-T Asme 82(1):35–45

Kane TR, Levinson DA (1983) The use of Kane's dynamical equations in robotics. Int J Robot Res 2(3):3–21

Karaman S, Walter MR, Perez A, Frazzoli E, Teller S (2011) Anytime motion planning using the RRT*. In: Proceedings of the IEEE International Conference on Robotics and Automation (ICRA). pp 1478–1483

Kavraki LE, Svestka P, Latombe JC, Overmars MH (1996) Probabilistic roadmaps for path planning in high-dimensional configuration spaces. IEEE T Robotic Autom 12(4):566–580

Kelly A (2013) Mobile robotics: Mathematics, models, and methods. Cambridge University Press, New York

Khalil W, Creusot D (1997) SYMORO+: A system for the symbolic modelling of robots. Robotica 15(2):153–161

Khalil W, Dombre E (2002) Modeling, identification and control of robots. Kogan Page Science, London

Khatib O (1987) A unified approach for motion and force control of robot manipulators: The operational space formulation. IEEE T Robotic Autom 3(1):43–53

King-Hele D (2002) Erasmus Darwin's improved design for steering carriages and cars. Notes and Records of the Royal Society of London 56(1):41–62

Klafter RD, Chmielewski TA, Negin M (1989) Robotic engineering – An integrated approach. Prentice Hall, Upper Saddle River, New Jersey

Klein CA, Huang CH (1983) Review of pseudoinverse control for use with kinematically redundant manipulators. IEEE T Syst Man Cyb 13:245–250

Klein G, Murray D (2007) Parallel tracking and mapping for small AR workspaces. In: Sixth IEEE and ACM International Symposium on Mixed and Augmented Reality (ISMAR 2007). pp 225–234

Koenig S, Likhachev M (2005) Fast replanning for navigation in unknown terrain. IEEE T Robotic Autom 21(3):354–363

Kuipers JB (1999) Quaternions and rotation sequences: A primer with applications to orbits, aeroespace and virtual reality. Princeton University Press, Princeton, New Jersey

Kümmerle R, Grisetti G, Strasdat H, Konolige K, Burgard W (2011) g^2o: A general framework for graph optimization. In: Proceedings of the IEEE International Conference on Robotics and Automation (ICRA). pp 3607–3613

Lamport L (1994) LATEX: A document preparation system. User's guide and reference manual. Addison-Wesley Publishing Company, Reading

LaValle SM (1998) Rapidly-exploring random trees: A new tool for path planning. Computer Science Dept., Iowa State University, TR 98–11

LaValle SM (2006) Planning algorithms. Cambridge University Press, New York

LaValle SM (2011a) Motion planning: The essentials. IEEE Robot Autom Mag 18(1):79–89

LaValle SM (2011b) Motion planning: Wild frontiers. IEEE Robot Autom Mag 18(2):108–118

Lee CSG, Lee BH (1987) Development of generalized d'Alembert equations of motion for robot manipulators. IEEE T Syst Man Cy 17:311–325

Lee CSG, Lee BH, Nigham R (1983) Development of the generalized d'Alembert equations of motion for mechanical manipulators. In: Proceedings of the 22nd CDC, San Antonio, Texas. pp 1205–1210

Li T, Bolic M, Djuric P (2015) Resampling methods for particle filtering: Classification, implementation, and strategies. IEEE Signal Proc Mag 32(3):70–86

Lloyd J, Hayward V (1991) Real-time trajectory generation using blend functions. In: Proceedings of the IEEE International Conference on Robotics and Automation (ICRA). Seoul, pp 784–789

Lovell J, Kluger J (1994) Apollo 13. Coronet Books

Lu F, Milios E (1997) Globally consistent range scan alignment for environment mapping. Auton Robot 4:333–349

Luh JYS, Walker MW, Paul RPC (1980) On-line computational scheme for mechanical manipulators. J Dyn Syst-T ASME 102(2):69–76

Lumelsky V, Stepanov A (1986) Dynamic path planning for a mobile automaton with limited information on the environment. IEEE T Automat Contr 31(11):1058–1063

Lynch KM, Park FC (2017) Modern robotics: Mechanics, planning, and control. Cambridge University Press, New York

Magnusson M, Lilienthal A, Duckett T (2007) Scan registration for autonomous mining vehicles using 3D-NDT. J Field Robotics 24(10):803–827

Magnusson M, Nuchter A, Lorken C, Lilienthal AJ, Hertzberg J (2009) Evaluation of 3D registration reliability and speed – A comparison of ICP and NDT. In: Proceedings of the IEEE International Conference on Robotics and Automation (ICRA). pp 3907–3912

Mahony R, Kumar V, Corke P (2012) Multirotor aerial vehicles: Modeling, estimation, and control of quadrotor. IEEE Robot Autom Mag (19):20–32

Martins FN, Celeste WC, Carelli R, Sarcinelli-Filho M, Bastos-Filho TF (2008) An adaptive dynamic controller for autonomous mobile robot trajectory tracking. Control Eng Pract 16(11):1354–1363

Matarić MJ (2007) The robotics primer. MIT Press, Cambridge, Massachusetts

Matthews ND, An PE, Harris CJ (1995) Vehicle detection and recognition for autonomous intelligent cruise control. Technical Report, University of Southampton

Mayeda H, Yoshida K, Osuka K (1990) Base parameters of manipulator dynamic models. IEEE T Robotic Autom 6(3):312–321

Merlet JP (2006) Parallel robots. Kluwer Academic, Dordrecht

Mettler B (2003) Identification modeling and characteristics of miniature rotorcraft. Kluwer Academic, Dordrecht

Middleton RH, Goodwin GC (1988) Adaptive computed torque control for rigid link manipulations. Syst Control Lett 10(1):9–16

Mindell DA (2008) Digital Apollo. MIT Press, Cambridge, Massachusetts

Montemerlo M, Thrun S (2007) FastSLAM: A scalable method for the simultaneous localization and mapping problem in robotics, vol 27. Springer-Verlag, Berlin Heidelberg

Montemerlo M, Thrun S, Koller D, Wegbreit B (2003) FastSLAM 2.0: An improved particle filtering algorithm for simultaneous localization and mapping that provably converges. In: Proceedings of the 18th International Joint Conference on Artificial Intelligence. Morgan Kaufmann, San Francisco, pp 1151–1156

Muja M, Lowe DG (2009) Fast approximate nearest neighbors with automatic algorithm configuration. International Conference on Computer Vision Theory and Applications (VISAPP), Lisbon, Portugal (Feb 2009), pp 331–340

Murray RM, Sastry SS, Zexiang L (1994) A mathematical introduction to robotic manipulation. CRC Press, Inc., Boca Raton

NASA (1970) Apollo 13: Technical air-to-ground voice transcription. Test Division, Apollo Spacecraft Program Office, http://www.hq.nasa.gov/alsj/a13/AS13_TEC.PDF

Neilson S (2011) Robot nation: Surviving the greatest socio-economic upheaval of all time. Eridanus Press, New York, 124 p

Neira J, Tardós JD (2001) Data association in stochastic mapping using the joint compatibility test. IEEE T Robotic Autom 17(6):890–897

Neira J, Davison A, Leonard J (2008) Guest editorial special issue on Visual SLAM. IEEE T Robotic Autom 24(5):929–931

Nethery JF, Spong MW (1994) Robotica: A mathematica package for robot analysis. IEEE T Robotic Autom 1(1):13–20

Newman P (n.d.) C4B mobile robots and estimation resources. Oxford University. http://www.robots.ox.ac.uk/~pnewman/Teaching/C4CourseResources/C4BResources.html

Ng J, Bräunl T (2007) Performance comparison of bug navigation algorithms. J Intell Robot Syst 50(1):73–84

Nilsson NJ (1971) Problem-solving methods in artificial intelligence. McGraw-Hill, New York

Olson E (2011) AprilTag: A robust and flexible visual fiducial system. In: Proceedings of the IEEE International Conference on Robotics and Automation (ICRA). pp 3400–3407

Orin DE, McGhee RB, Vukobratovic M, Hartoch G (1979) Kinematics and kinetic analysis of open-chain linkages utilizing Newton-Euler methods. Math Biosci 43(1/2):107–130

Ortega R, Spong MW (1989) Adaptive motion control of rigid robots: A tutorial. Automatica 25(6):877–888

Paul R (1972) Modelling, trajectory calculation and servoing of a computer controlled arm. Ph.D. thesis, technical report AIM-177, Stanford University

Paul R (1979) Manipulator Cartesian path control. IEEE T Syst Man Cyb 9:702–711

Paul RP (1981) Robot manipulators: Mathematics, programming, and control. MIT Press, Cambridge, Massachusetts

Paul RP, Shimano B (1978) Kinematic control equations for simple manipulators. In: IEEE Conference on Decision and Control, vol 17. pp 1398–1406

Paul RP, Zhang H (1986) Computationally efficient kinematics for manipulators with spherical wrists based on the homogeneous transformation representation. Int J Robot Res 5(2):32–44

Pivtoraiko M, Knepper RA, Kelly A (2009) Differentially constrained mobile robot motion planning in state lattices. J Field Robotics 26(3):308–333

Pomerleau D, Jochem T (1995) No hands across America Journal. http://www.cs.cmu.edu/~tjochem/nhaa/Journal.html

Pomerleau D, Jochem T (1996) Rapidly adapting machine vision for automated vehicle steering. IEEE Expert 11(1):19–27

Pounds P (2007) Design, construction and control of a large quadrotor micro air vehicle. Ph.D. thesis, Australian National University

Pounds P, Mahony R, Gresham J, Corke PI, Roberts J (2004) Towards dynamically-favourable quadrotor aerial robots. In: Proceedings of the Australasian Conference on Robotics and Automation. Canberra

Pounds P, Mahony R, Corke PI (2006) A practical quad-rotor robot. In: Proceedings of the Australasian Conference on Robotics and Automation. Auckland

Press WH, Teukolsky SA, Vetterling WT, Flannery BP (2007) Numerical recipes, 3rd ed. Cambridge University Press, New York

Prouty RW (2002) Helicopter performance, stability, and control. Krieger, Malabar FL

Pynchon T (2006) Against the day. Jonathan Cape, London

Rekleitis IM (2004) A particle filter tutorial for mobile robot localization. Technical report (TR-CIM-04-02), Centre for Intelligent Machines, McGill University

Russell S, Norvig P (2009) Artificial intelligence: A modern approach, 3rd ed. Prentice Hall Press, Upper Saddle River, NJ

Scaramuzza D, Fraundorfer F (2011) Visual odometry [tutorial]. IEEE Robot Autom Mag 18(4):80–92

Sharp A (1896) Bicycles & tricycles: An elementary treatise on their design an construction; With examples and tables. Longmans, Green and Co., London New York Bombay

Sheridan TB (2003) Telerobotics, automation, and human supervisory control. MIT Press, Cambridge, Massachusetts, 415 p

Shoemake K (1985) Animating rotation with quaternion curves. In: Proceedings of ACM SIGGRAPH, San Francisco, pp 245–254

Siciliano B, Khatib O (eds) (2016) Springer handbook of robotics, 2nd ed. Springer-Verlag, New York

Siciliano B, Sciavicco L, Villani L, Oriolo G (2009) Robotics: Modelling, planning and control. Springer-Verlag, Berlin Heidelberg

Siegwart R, Nourbakhsh IR, Scaramuzza D (2011) Introduction to autonomous mobile robots, 2nd ed. MIT Press, Cambridge, Massachusetts

Silver WM (1982) On the equivalance of Lagrangian and Newton-Euler dynamics for manipulators. Int J Robot Res 1(2):60–70

Sobel D (1996) Longitude: The true story of a lone genius who solved the greatest scientific problem of his time. Fourth Estate, London

Spong MW (1989) Adaptive control of flexible joint manipulators. Syst Control Lett 13(1):15–21

Spong MW, Hutchinson S, Vidyasagar M (2006) Robot modeling and control, 2nd ed. John Wiley & Sons, Inc., Chichester

Stachniss C, Burgard W (2014) Particle filters for robot navigation. Foundations and Trends in Robotics 3(4):211–282

Stentz A (1994) The D* algorithm for real-time planning of optimal traverses. The Robotics Institute, Carnegie-Mellon University, CMU-RI-TR-94-37

Stewart A (2014) Localisation using the appearance of prior structure. Ph.D. thesis, University of Oxford

Sussman GJ, Wisdom J, Mayer ME (2001) Structure and interpretation of classical mechanics. MIT Press, Cambridge, Massachusetts

Taylor RA (1979) Planning and execution of straight line manipulator trajectories. IBM J Res Dev 23(4):424–436

Thrun S, Burgard W, Fox D (2005) Probabilistic robotics. MIT Press, Cambridge, Massachusetts

Titterton DH, Weston JL (2005) Strapdown inertial navigation technology. IEE Radar, Sonar, Navigation and Avionics Series, vol 17, The Institution of Engineering and Technology (IET), 576 p

Uicker JJ (1965) On the dynamic analysis of spatial linkages using 4 by 4 matrices. Dept. Mechanical Engineering and Astronautical Sciences, NorthWestern University

Vanderborght B, Sugar T, Lefeber D (2008) Adaptable compliance or variable stiffness for robotic applications. IEEE Robot Autom Mag 15(3):8–9

Walker MW, Orin DE (1982) Efficient dynamic computer simulation of robotic mechanisms. J Dyn Syst-T ASME 104(3):205–211

Walter WG (1950) An imitation of life. Sci Am 182(5):42–45

Walter WG (1951) A machine that learns. Sci Am 185(2):60–63

Walter WG (1953) The living brain. Duckworth, London

Whitney DE (1969) Resolved motion rate control of manipulators and human prostheses. IEEE T Man Machine 10(2):47–53

Wiener N (1965) Cybernetics or control and communication in the animal and the machine. MIT Press, Cambridge, Massachusetts

Yoshikawa T (1984) Analysis and control of robot manipulators with redundancy. In: Brady M, Paul R (eds) Robotics research: The first international symposium. MIT Press, Cambridge, Massachusetts, pp 735–747

Zarchan P, Musoff H (2005) Fundamentals of Kalman filtering: A practical approach. Progress in Astronautics and Aeronautics, vol 208. American Institute of Aeronautics and Astronautics

Ziegler J, Bender P, Schreiber M, Lategahn H, Strauss T, Stiller C, Thao Dang, Franke U, Appenrodt N, Keller CG, Kaus E, Herrtwich RG, Rabe C, Pfeiffer D, Lindner F, Stein F, Erbs F, Enzweiler M, Knöppel C, Hipp J, Haueis M, Trepte M, Brenk C, Tamke A, Ghanaat M, Braun M, Joos A, Fritz H, Mock H, Hein M, Zeeb E (2014) Making Bertha drive – An autonomous journey on a historic route. IEEE Intelligent Transportation Systems Magazine 6(2):8–20

Index

Index of People

A

Ackermann, Rudolph **99**
Asimov, Issac 1

B

Ball, Sir Robert 50, **53**
Bayes, Reverend Thomas **155**
Black, Harold 4
Bode, Henrik 4
Braitenberg, Valentino **124**
Bryan, George 35

C

Čapek, Karel 1, 3
Cardano, Gerolamo **35**
Chasles, Michel **51**
Clifford, William 53
Cook, Captain James 150, 165
Coriolis, Gaspard-Gustave de **67**
Coulomb, Charles-Augustin de **251**

D

Delaunay, Boris 135
Denavit, Jacques **196**
Descartes, René **17**
Devol, George C. Jr. 1, **2**
Draper, Charles Stark (Doc) 79, **80**, 156

E

Einstein, Albert 68
Engelberger, Joseph F. **2**
Euclid of Alexandria **16**
Euler, Leonhard **34**, 66, 263

G

Gauss, Carl Friedrich 59
Goetz, Raymond 7

H

Hall, Edwin **83**
Hamilton, Sir William Rowan **42**, 53, 58, 59

H

Harrison, John 150
Hartenberg, Richard **196**
Hershey, Allen V. 218
Hesse, Ludwig Otto **314**

I

Ilon, Bengt 110

J

Jacobi, Carl Gustav Jacob **230**

K

Kálmán, Rudolf 155

L

Lagrange, Joseph-Louis **263**
Lazzarini, Mario 172
Leclerc, Georges-Louis 172
Lie, Sophus **307**

M

Markov, Andrey 135
McCarthy, John 4
McCulloch, Warren 4
Metropolis, Nicholas 172
Minsky, Marvin 4
Moler, Cleve 9

N

Newell, Allen 4
Newton, Sir Isaac **65–67**, 277
Nyquist, Harold 4

P

Pitts, Walter 4
Plücker, Julius **297**
Price, Richard 155

R

Rodrigues, Olinde **40**, 59

© Springer Nature Switzerland AG 2022
P. Corke, *Robotics and Control*, Springer Tracts in Advanced Robotics 141,
https://doi.org/10.1007/978-3-030-79179-7

S

Scheinman, Victor 193
Schmidt, Stanley F. 156
Shannon, Claude 4
Simon, Herbert 4

T

Tait, Peter 35, 59
Tesla, Nikola 5
Turing, Alan 4

U

Ulam, Stanislaw 172

V

von Kármán, Theodore 196
von Neumann, John 172
Voronoy, Georgy Feodosevich 135

W

Walter, William Grey 4, **124**
Wiener, Norbert 4

Index of Functions, Classes and Methods

Classes are shown in **bold**, Simulink® models in *italics*, and methods are prefixed by a dot. All others are Toolbox functions.

A

about 75, 116, 143, 202, 211, 262, 316
angdiff 101
angles 46
angvec2r 40
angvec2tr 59
animate 74, 75
AprilTag 162
apriltags **162**, 182
atan2 301

B

Bicycle 98, 99, 107, 109
Bicycle 161, 164, 166, 174, 183
BinaryOccupancyGrid 146
bug 127
bug.plot 127
bug2 127
    –, .path 127, 129
BundleAdjust 337

C

ccode 316
ctraj 8, 76, 212, 213

D

delta2tr 65
DHFactor 216, 220
    –, .dh.command 220
dim 161
Dstar 132
    –, .costmap 132
    –, .modify_cost 133
    –, .niter 133
    –, .plan 132, 133
    –, .query 133

DXform 129
    –, .plan 129, 130
    –, .query 129
    –, .visualize 129
    –, .visualize3d 130

E

e2h 27
eig 39
EKF 156, 158, 161, 164–166
eps 48
ETS2 192, 225
    –, .fkine 192
    –, .plot 193
    –, .Rz 192, 194
    –, .structure 193, 194
    –, .teach 192
    –, .Tx 192, 194
ETS3 225
    –, .fkine 194
    –, .Ry 194
    –, .Rz 194
    –, .Tx 194
    –, .Ty 194
    –, .Tz 194
eul2jac 231
eul2r 34, 35
eul2tr 46
eval 216, 220
ExampleHelperRobot Simulator 121
exp 24, 41
expm 24, 41, 45, 50, 52, 59
eye 27, 216

F

fcode 316
fkine 212
fminsearch 205

G

gait 223
gaussfunc 327, 328
GeometricJacobian 246
Graph 212

H

h2e 27

I

icp 178
InitFcn 273
interp 75, 76
InverseKinematics 225
ithin 134

J

jacobian 170, 228, 316
Jacobian 235
jsingu 232
jtraj 202, 210, 212, 214, 261,
    272, 273

L

LandmarkMap 158, 161
Lattice 139, 337
Link 196, 198, 215, 254, 266
    –, .A 197
    –, .a 197
    –, .offset 197
    –, .RP 197
log 23, 40
logm 23, 40, 52
lscov 246
lspb 70, 71, 76, 210

M

makemap 129, 146
matlabFunction 316
mdl_puma560 199, 205, 261, 263,
    272
mdl_quadrotor 116
mdl_twolink 251
meshgrid 263, 264
model 101
models 198
MonteCarloLocalization 183
mplot 236
mstraj 218, 222
mtraj 71, 74, 210

N

Navigation 128
null 318
numcols 79, 87, 88, 137, 218, 263,
    264
numrows 129, 218

O

oa2r 38

P

ParticleFilter 175, 183
pcregrigid 181
PGraph 337
    –, .add_edge 337, 338
    –, .add_node 337
    –, .closest 338
    –, .cost 338
    –, .edges 338
    –, .neighbours 338
    –, .plot 338
pinv 238, 239, 241
ploop 259, 278
ploop_test 259
plot 177
plot_homline 304
plot_point 26
Plucker 52, 304
    –, .L 304
    –, .side 297
pol2cart 177
Polygon 147
PoseGraph 170, 171, 177, 337
    –, .optimize 171
    –, .plot 170, 171
    –, .plotoccgrid 179
    –, .scan 177
    –, .scanmap 179
    –, .scanxy 177
    –, .time 178
PRM 136, 146, 337
    –, .plan 136
    –, .query 137
    –, .visualize 137

Q

q.animate 74
qplot 211
Quaternion 42

R

rand 137
randinit 136, 137
randn 137, 327
RandomPath 155, 164, 174
RangeBearingSensor 159, 160, 162, 164–166,
    175, 184
    –, .h 160
    –, .H_w 160
    –, .H_x 160
    –, .reading 159
Revolute 196
RevoluteMDH 217
RigidBodyTree 225, 246
RNE 272, 278
roblocks 101, 109
Robot 269

rotx 32, 33, 40, 41, 64
roty 32–34
rotz 32, 34
rpy2r 39, 60
rpy2tr 43
RRT 142, 337
    –, .path 143
    –, .plan 142
    –, .visualize 142
running 101

S

SE2 55, 170, 199
    –, .Rx 209
SE3 55, 201, 202, 206, 208, 209, 212–214, 219, 222, 231
    –, .Rx 214
    –, .Ry 213
    –, .Rz 222
    –, .torotvec 231
Sensor 159, 164
    –, .H_xf 164
sensorfield 125, 146
SerialLink 198, 204, 210, 222, 225, 247, 263
    –, .accel 269
    –, .base 201, 202, 263
    –, .coriolis 261
    –, .edit 198
    –, .fdyn 270
    –, .fellipse 243
    –, .fkine 198, 201, 202, 205, 207, 211, 228, 236
    –, .gravity 263
    –, .gravload 262, 263, 267
    –, .ikcon 225
    –, .ikine 206, 208, 213, 214, 222, 225, 244
    –, .ikine6s 205–209, 212–214, 219, 225
    –, .ikinesym 204
    –, .inertia 261, 264, 266, 268
    –, .jacob0 229–234, 242, 268
    –, .jacobn 230
    –, .jtraj 210
    –, .links 267
    –, .maniplty 213, 234, 269
    –, .motordynamics 254
    –, .nofriction 270
    –, .plot 201, 207, 211, 214, 219, 225, 269
    –, .plot3d 225
    –, .rne 261, 262, 267
    –, .teach 225, 233, 243, 246
    –, .tool 201
    –, .vellipse 233
shortest 74
sigma 327
simplify 23, 60, 170
skew 309
sl_bicycle 99
sl_braitenberg 124, **125**
sl_ctorque **271**, 272
sl_driveline 102, **103**
sl_drivepoint **101**, 102
sl_drivepose **106**, 107
sl_fforward **271**, 272
sl_jspace 212, **213**
sl_lanechange 99, **100**

sl_opspace 273
sl_pursuit **104**
sl_quadcopter **116**
sl_quadrotor 115, 116
sl_rrmc 235, **236**
sl_rrmc2 236, **237**
sl_sea **275**
sl_ztorque **269**
SO2 55, 72
SO3 55, 74
spy 325
sqrt 157, 268
sqrtm 302

T

T1.torpy 71
t2r 45
T2xyz 212
tags 162
tau-d 257
tau_ff 258
torpy 75
tpoly 69, 70, 76, 89, 90, 210
tr2angvec 39
tr2delta 65
tr2eul 34, 35
tr2rotvec 231
tr2rpy 36
Tracking Controller 109
traj 218
tranimate 32, 33, 59, 60
transl 8, 45, 201, 202, 206, 208, 211, 213, 221, 222, 236
transl2 25, 26
trexp 41, 50, 52, 59
trinterp 76
tripleangle 36, 60
triplepoint 134
trlog 41, 52, 231
trot2 25
trotx 45, 46, 202, 220, 263
troty 213
trotz 220
trplot 25, 33, 45, 59
trplot2 26, 59
Ts
    –, .t 212
    –, .torpy 212
Twist 28, 52, 199
    –, .expm 52
    –, .line 52
    –, .S 52
    –, .T 28, 52, 199

U

Unicycle 109, 121, 183
UnitQuaternion 43, 44, 48, 66, 74, 79
    –, .animate 79
    –, .dot 62
    –, .dotb 62
    –, .omega 79
    –, .plot 66
    –, .torpy 79

V

Vehicle 154, 156, 158, 183
  –, .Fv 156
  –, .Fx 156
  –, .step 155
vex 23, 24, 40, 41, 309

vloop **255**, 278
vloop_test **256**
VREP_class 185

X, Y, Z

xv 316

General Index

Symbols

\-operator 69, 317, 319

A

A* search 132, 137, 140, **339**
absorption shock 275
acceleration 79, 80, 85, 118, 249, 273
  –, angular 66
  –, centripetal **68**
  –, Coriolis **68**, 89
  –, discontinuity 76
  –, Euler 68
  –, gravitational 68, 81
  –, inertial 81
  –, proper 81
  –, sensor 81, **85**
  –, specific 81
accelerometer 37, 39, **79–81**, 85
  –, triaxial 81, 85
Ackermann steering **99**
actuation 118
  –, electric 254
  –, electro-hydraulic 249
actuator 118, 249
  –, joint 250
  –, saturation 116
  –, series-elastic (SEA) **274, 275**
addition, vector 287
adjoint
  –, logarithm of **311**
  –, matrix 63, 67, 199, 245, 297, **311**
adjugate 289
adjustment, bundle 182
AHRS (see *attitude and heading reference system*)
aircraft 117, 119
algebra 307
algebraic group 308
algorithm
  –, Bresenham 179
  –, *bug* **126–128**
  –, D* 132
  –, FastSLAM (see also *Rao-Blackwellized SLAM*) 181
  –, ICP (iterated closest point) 181
  –, Levenberg-Marquardt **244, 320**, 321
  –, Newton-Euler 261
  –, rapidly exploring 143
  –, resolved-rate motion control 235
  –, RRT (rapidly-exploring random tree) 143
  –, skeletonization 134, 135
  –, thinning 134, 135
  –, velocity loop control 255

ampullae 81
anaglyph, stereo glasses 33
analysis, root-locus 278
analytical Jacobian **231**
angle
  –, Cardan 30, 36
  –, declination 83
  –, elevation 150
  –, Euler **34–36**, 38, 57, 73, 194, 230, 231, 245
    –, singularity 37
  –, heading 85
  –, inclination 83
  –, joint 5, 12, **196**
  –, nautical 36
  –, representation 34
  –, roll-pitch-yaw **35, 36**, 76, **210–212**, 230
    –, rate 74
    –, singularity 36
  –, rotation 23, 24, 29, 33, 35, **37**, 41
  –, steering 99, 100, 139, 143
  –, Tait-Bryan 36
  –, trajectory
    –, joint 270
    –, LSPB (linear segment with parabolic blend) **70**, 259, 260
  –, XYZ sequence 36
angle-axis representation 39, 43
angular
  –, acceleration 66
  –, momentum 66, 77, 78
  –, rate 86
  –, uncertainty 157
  –, velocity 48, 50, **62**, 66, 68, 77, **78**, 153, 231, 332
anthropomorphic 145, 200, **201**
anti-symmetric matrix **289**
Apollo 13 36, 38
  –, Lunar Module 37, 79
approach vector 38, **39**, 208, 209
April tag 162
architecture, subsumption 125
ArduCopter (software project) 120
artificial intelligence **4**
Asimo humanoid robot **6**
ASV (see *autonomous surface vehicle*)
attitude and heading reference system (AHRS) 85
automata 126
automated guided vehicle **94**
autonomous surface vehicle (ASV) **94**
autonomous underwater vehicle (AUV) 94, 118, 119
axis
  –, instantaneous 62
  –, of motion **71**
  –, optical 38

–, rotation 30, 37, 39, 41, 46, 48, 61, 66
  –, Earth 83
–, screw 45, **50**

B

back
  –, EMF (electromotive force) 250, **258**
  –, end 168
ballbot 110
base
  –, force 267
  –, transform **197**
Baxter robot 209, 275
behavior-based robot **125**
Beidou (satellite navigation system) **151**
bi-quaternion (see *dual quaternion*)
bias 86
bicycle model **98, 105, 142, 143**
bifilar pendulum 277
blade flapping **113**
blend **70**
  –, parabolic 70
body
  –, acceleration estimation 81
  –, moving 66
body-fixed frame 37, 53, **68**, 77, 113
Braitenberg vehicle **124**
breaking, stiction 250
Bresenham algorithm 179
Buffon's needle problem **172**
bug algorithm **126**
bundle adjustment 182

C

C-space 54
calibration
  –, camera 9
  –, sensor 86
camera 168
  –, pose 173
car 117–119
Cardan angle sequence **34**
Cartesian
  –, coordinate system **20**
  –, geometry 17
  –, motion 75, 209, **212**, 236
  –, plane 17
  –, point 177
  –, trajectory 89, **212**, 222
celestial navigation 150
center of mass 62, 66, 113, 251, 262
centripetal
  –, acceleration **68**
  –, force 262
Chasles theorem 50
chi-squared distribution 158, 329
Cholesky decomposition **290**
circle 74
city block distance 128
closed-form solution 203
clothoid **99**
CML (see *concurrent mapping and localization*)

coefficient
  –, Coulomb 270
  –, viscous friction 250, 270
column space **291**
compass 39, **83**, 106, 149, 151, 153, 162
compensation, gravity 116
computed torque control **272**
concurrent mapping and localization (CML) **165**
condition number (see *matrix condition number*)
confidence test 162
configuration
  –, change **214**, 215
  –, kinematic 196, 206, 207, 213, 214, 236
  –, of a system **53**
  –, space 53, **54**, 112, 117, 119, 143, **196**, 199, 208, 209
  –, zero-angle 195
connected component, graph 137, 338
constraint
  –, nonholonomic 99, 109
  –, rolling 119
control
  –, feedback 260
  –, feedforward 116, 258, 260, 270, **271**
  –, flexible transmission 12
  –, force **273**
  –, independent joint 249
  –, integral
    –, action 257
    –, windup 278
  –, joint **249, 260**
  –, loop, nested 249
  –, mobile robot 100–107
  –, model-based 190
  –, proportional 101, 102, 104, 255
    –, derivative 114–116
    –, integral 116, 258, 259
  –, resolved-rate motion **235**, 246
  –, shared 7
  –, space, operational space 273, **274**
  –, torque 270
    –, computed **272**
    –, feedforward 271
  –, traded 6
  –, velocity 100, 255, 259
coordinate
  –, frame **15**, 16, **20**
    –, 2-dimensional 17
    –, 3-dimensional 17
    –, end-effector 192
    –, global 179
    –, moving 66
    –, right-handed 29
    –, velocity 66
  –, generalized **53**, 98, 107, 111, 117, 118, 192, 261
  –, homogeneous 302
  –, joint **196**, 216, 227, 261
  –, Plücker 50, 52, **296**
  –, point 20, 24, 45, 49
  –, random 337
  –, system **17**
  –, vector 15–17, 287, 295, 302
Coriolis
  –, acceleration **68**, 89
  –, force 261, 262, 265, 273

correlation, covariance 152, 328, 334
correspondence, point 178
cost map **132**
Coulomb friction **250**, 251, 253
covariance
 –, correlation 152, 328, 334
 –, ellipse **158**, 164, 329
 –, matrix 152, 154, 156, 158, 159, 161, 163, 165, 167, 168, 174, **328**
 –, extending 163
curvature 139
cybernetics 1, **4**, 124, 145

D

D* **132**
d'Alembert force 67
damped inverse 238
data
 –, association 162
  –, error **151**, **162**
 –, laser scan 177
 –, type 55, 56
dead reckoning 95, **149**, 153
declination
 –, angle 83
 –, magnetic **83**
decomposition
 –, Cholesky 290
 –, spectral **291**
Deep Phreatic Thermal Explorer (DEPTHX, AUV) 118, 119
definition
 –, eigenvalue, eigenvector 39
 –, frame 68
 –, Mahalanobis distance 329
 –, robot 5, 124, 128
degree of freedom (DOF) 37, **54**, 71, 112, 118, 119, 189, 191, 193, 206, 208, 229, 232, 234, 238–240
Denavit-Hartenberg
 –, notation 194, 195, 215, 216, 219, 227
  –, modified 216
 –, parameter 195, 198, 225
DEPTHX (see *Deep Phreatic Thermal Explorer*)
derivative
 –, orientation **62**, 66, 116
 –, pose 61, 62
 –, quaternion **62**
 –, **time 61**
detector, SURF (speeded up robust feature) 250
determinant 47, 233, 238, **291**
differential, kinematics **227**
Dijkstra method 130
dimension 15
 –, curved 15
direction 307–309
displacement
 –, rigid body 50, 51
 –, spatial 65, 243
distance 162
 –, Euclidean **16**, 128, 338
 –, Mahalanobis 162, **293**, 329
 –, Manhattan 128, **287**
 –, threshold 137
 –, transform 128, 132, 133, 135

distortion, iron 85
distribution
 –, chi-squared 329
 –, von Mises 154
DOF (see *degree of freedom*)
down hill 319
drag, aerodynamic 113
dual
 –, number **53**
 –, quaternion **53**
Dubbins path 99
dynamics **249**
 –, error 272
 –, forward 114, 116, 249, **269**, 270
 –, inverse 261, 271, 272
 –, quadrotor 113, **114**
 –, rigid-body **261**, **270**

E

Earth
 –, diameter 79
 –, gravity 80
 –, shape 79
 –, surface 68, 77
east-north-up (ENU) 77
eccentricity **300**
effect, Eötvös 89
effective inertia 254
EGNOS (satellite network) 151
eigenvalue 39, 158, 234, 268, 269, **290**
eigenvector 39, **290**
EISPACK project **9**
EKF (see *extendet Kalman filter* and *Kalman filter*)
EKF SLAM (see *Kalman filter, extended, SLAM*)
elasticity, joint **274**
ellipse 157, **299**, 329
 –, canonical 298, 299
 –, confidence 165, 166
 –, covariance **158**, 164, 329
 –, drawing 301
 –, equation 329
 –, error **158**, 161
 –, rotated 329
 –, size 328
 –, velocity 233, 242
ellipsoid 268, **299**, 300
 –, equation 329
 –, force 242, **243**
 –, shape 234
 –, surface 233, 243, 328
 –, velocity 242
  –, rotational 234
 –, volume 234, 301
 –, wrench 243
Elsie (robot) **93**, 123
encoder 253, 254
end-effector **191**
 –, coordinate frame 230
 –, force 242
 –, inertia 273
 –, torque 242
 –, velocity 227, 228
ENU (see *east-north-up*)

Eötvös, effect 89
ephemeris **150**
equation
  –, differential 49
  –, Eulers rotation **66**
  –, line 295, 303
  –, motion 99, 109, 269
    –, Euler 114, 261
    –, rigid-body 249, **261**
  –, nonhomogeneous **315**
  –, solving system 317
equivalence principle 68
error 47, 48, 167, 168
  –, cumulative 168
  –, edge 170
  –, ellipse 157, 164, 166
  –, ICP (iterated closest point) 180
  –, position 249
  –, vector 324
estimation **152**
  –, Monte-Carlo 155, 173, 181
  –, pose 81
ethics 7
Euclidean
  –, coordinate 27, 302
  –, distance **16**, 128, 338
  –, geometry 16, 17, 20, **295**
  –, group **19**, **25**, 44
  –, length **287**
  –, line 295
  –, plane 17, 303
  –, point 27, 295, 303, 304
  –, space 17, 53, 295, 303
Euler
  –, acceleration **68**
  –, angle **34–36**, 38, 57, 73, 194, 230, 231, 245
    –, singularity 37
  –, force **68**
  –, motion equation **66**, 114, 261
  –, rotation theorem **30**, **31**, 33–35, 309
explicit complementary filter 86, 87
exponential
  –, coordinate **41**, 231, 322
    –, rate 231
  –, mapping 48, 50
  –, matrix 23, 24, 41, 49
  –, product of 194, 198, 199
extended Kalman filter (EKF, see also *Kalman filter*) 86, 88, **155**, 167, 315, 334
exteroceptive sensor **5**, 168

F

FastSLAM (see also *SLAM* and *Rao-Blackwellized SLAM*) **167**
feature map 161, 166
feedback control 116, 258–260
feedforward control **116**, 258, 260, 270, **271**
fibre-optic gyroscope (FOG) 78
fictitious force **67**, 81
field
  –, magnetic, intensity 84, 85
  –, robot 3, 94
file 170
  –, MEX **284**

filter
  –, complementary explicit 86, 87
  –, Kalman 88, 89, **155**, 160–162, 167, 173, 180, 182, **332**
    –, extended (EKF) 86, 88, 155, 167, 315, **334**
    –, unscented (UKF) 182
  –, Kalman-Bucy 333
  –, particle 167, 173–176
flow, current 83
flux line, magnetic 83
FOG (see *fibre-optic gyroscope*)
font, Hershey 218
force 50, 66, **242**, 249
  –, apparent 67
  –, control **273**
  –, Coriolis 261, 262, 265, 273
  –, d'Alembert 67
  –, ellipsoid 242, **243**
  –, fictitious **67**, 81
  –, gyroscopic 273
  –, inertial 67
  –, pseudo 67
  –, translational 67
form, homogeneous 27
formula, Rodrigues rotation 35, **40**, 41, 50, 51, 59, 64, 309
forward
  –, dynamics 114, **269**
  –, kinematics **191**, 192, 199, 202, 228
    –, instantaneous 229
frame
  –, body-fixed 53, **68**, 77
  –, coordinate **15**, 16, **20**
  –, reference **67**
    –, inertial 66, 67, 77, 81
    –, noninertial 68
  –, right-handed coordinate 29
  –, world coordinate 16, 77
friction 249–251, 260, 261, 266
  –, aerodynamic 113
  –, Coulomb **250**, 251, 253, 266, 270
  –, stiction **250**
  –, viscous 244, **250**, 251, 253, 266, 269, 270
front end **168**
function
  –, Gaussian **327**
  –, Huber loss 321
  –, observation 162
  –, probability density (PDF) **151**, 158, 159, 173, 327, 328
    –, Gaussian 173
  –, scalar 313
  –, Tukey biweight 321
fusion, sensor **85**, 86, 161

G

gait pattern **223**
Galileo (satellite navigation system) **151**
gantry robot 189
Gaussian
  –, distribution 331, 332
  –, function **327**, 329
  –, multivariate **328**
  –, noise 155, **158**, 162, 332, 333
  –, probability 158, 162, 329
  –, random variable **327**, 332, 334
gearbox 252–254

generalized
 –, coordinate 53, 98, 107, 111, 117, 118, 192, 261
  –, joint **196**, 216, 261
 –, forces **261**
  –, joint 242, 244, 261, 262, 264, 266, 267
 –, matrix inverse **292**
 –, Voronoi diagram 134
generator matrix 308, 310
Genghis (robot) 145
geomagnet 83
geometric Jacobian **229**
geometry
 –, algebraic 48
 –, analytic 17
 –, Cartesian 17
 –, Euclidean 16, 17, 20, **295**
gimbal 203
 –, lock **36**, 206, 213, 232
 –, low-friction 78
Global Hawk unmanned aerial vehicle (UAV) 4, 112
Global Positioning System (GPS) 5, 115, 149, **151**, 163
 –, differential 151
 –, multi-pathing 151
 –, RTK **151**
 –, selective availability **151**
GLONASS (satellite navigation system) **151**
goal seeking 126
gradient, descent **319**, 320
graph 134, 137, **337**
 –, A* search 132, 137, 140, 339
 –, embedded **337**
gravity 68, 82, 113, 249, 251
 –, compensation 116
 –, disturbance 258
 –, load 249, 258, 261–263, 269
 –, term **262**
 –, torque 252, 262
 –, vector 82, 261
great circle **74**
ground effect 113
group
 –, algebraic 308
 –, Euclidean 19
 –, Lie **307**
 –, orthogonal **22**, 32, **290**
gyroscope 36, 77, 85, 99, 153
 –, fibre-optic (FOG) 78
 –, ring-laser (RLG) 78
 –, strapdown 78
 –, triaxial 78

H

Hall effect 83
 –, sensor 83
hard-iron distortion 85
heading 83
 –, angle 85
 –, rate (see *yaw rate*)
helicopter 119
Hershey font 218
Hessian 313, **314**, 320
 –, approximate 314, 320
 –, matrix 314
histogram, cumulative 174

holonomic constraint 54
homogeneous
 –, equation 318
 –, form **25**
 –, transformation **25**, **44**, 51, 52, 75, 197, 201, 303
  –, normalization 48
homography 162
hovercraft 117–119
Huber loss function 321
humanoid robot 3, 6
hybrid, trajectory **70**
hypersurface, quadric 304

I

ICP (see *iterated closest point*)
ICR (see *instantaneous center of rotation*)
ideal
 –, line **303**
 –, point **303**, 304
identity quaternion 43
image
 –, plane 303
 –, processing 128, 134
IMU (see *inertial measurement unit*)
inclination
 –, angle 83
 –, magnetic **83**, 84
incremental replanning 132
inertia 251–253
 –, effective 254
 –, end-effector 273
 –, load 253
 –, matrix 114, 264
 –, motor 253
inertial
 –, force 67
 –, measurement unit (IMU) 37, 85
 –, navigation system (INS) 77, 85, 115
 –, reference frame 66, **67**, 77, 81
 –, sensor 85
inflation, obstacle 130
innovation 87, **160**, 168, 333
INS (see *inertial navigation system*)
instantaneous center of rotation (ICR) **98**, 107
integral
 –, dynamics **269**
 –, windup 258
intelligence, artificial 12
intensity
 –, light 123
 –, magnetic field 83, 85
interpolation
 –, linear 73
 –, orientation **73**
 –, quaternion 58, **74**
 –, rotational 74
 –, scalar 210
 –, unit-quaternion 74, 75
inverse
 –, dynamic control 272
 –, dynamics **261**, 271, 272
 –, left-generalized 317
 –, pseudo 238, 240, **292**, **317**
iterated closest point (ICP) 177, 180, 181

J

Jacobian, Jacobian matrix 213, 216, **227**, **228**, 245, 313, **315**
-, analytical **230**, 231
-, condition **232**
-, damped inverse 238
-, end-effector coordinate frame **230**
-, geometric 229
-, insertion 163, 165
-, manipulability **232**, **233**
-, manipulator **227**, 229, **245**, 261
-, matrix 156, 170, 190, 213, **227**, **228**
-, numerical approximation **315**
-, over-actuated robot **240**
-, singularity **232**, **238**
-, transpose 227, 243, 244
-, under-actuated robot **239**
jerk 68
Johns Hopkins Beast (robot) 145
joint
-, actuator 250
-, angle 5, 12, **196**
-, control, independent 249
-, elasticity **274**
-, position 273
-, prismatic 191, 193
-, revolute 191
-, sliding 191
-, space **196**, 210, **242**
-, velocity 227, 228
Joseph form 333

K

Kalman filter 88, 89, **155**, 160–162, 167, 173, 180, 182, **332**
-, extended (EKF) 86, 88, 155, 167, 315, **334**
-, SLAM (EKF SLAM) **167**
-, gain **333**
-, unscented (UKF) 182
kernel, density approach 181
kinematic
-, configuration 196, 206, 207, 213, 214, 236
-, model **99**, 105, 109, 112, 141, 143, 200
kinematics **191**
-, differential **227**
-, forward **191**, 192, 199, 202, 228
-, instantaneous 229
-, symbolic 204, 228
-, inverse
-, closed form **203**
-, numerical 204, 207, 243
-, velocity **227**
Klein quadric 304

L

landmark **150**, 162, 167, 180
-, identity 162
-, navigation 149
-, observation 159
laser
-, odometry 177
-, rangefinder **176**, 177, 179
-, noise 178
-, scanner 168

lateral motion **98**
lattice planner 138
law
-, Newton
-, first **67**
-, second **66**, **68**, 80, 113, 261, 277
-, of robotics 1
least squares problem 238, 239, 244, 317
-, nonlinear 169, 314, **320**, **321**
-, rotation matrix **318**
Levenberg-Marquardt
-, algorithm **244**, **320**, 321
-, optimization 244, 323
lever arm effect 251
Lie
-, algebra 51, 52, **307–310**
-, group 23, 48, 307, **307–310**
light intensity 123
line 304
-, 2D 295
-, 3D 296
-, equation 295, 303
-, Euclidean 295
-, ideal **303**
-, of no motion **98**
-, Plücker **296–298**
-, projection **304**
linear segment with parabolic blend (LSPB) trajectory **70**, 259, 260
linearization 313
-, general 313
link 250
-, effect 251
-, elasticity 274
-, mass 251, 262
LINPACK project **9**
load 275
-, gravity 249, 258, 261–263, 269
-, inertia 253
localization 9, **149**, 165, 179
-, CML (concurrent mapping and localization) 165
-, error 151
-, laser-based **180**
-, Monte-Carlo **173**
-, problem 151, 152
-, SLAM (simultaneous localization and mapping) 165, 167–169, 173
longitude problem **150**
longitudinal motion **98**
LORAN (radio-based localization system) **151**
LORD MicroStrain 77
LSPB (see *linear segment with parabolic blend*)

M

magnetic
-, declination **83**
-, field 84, 85
-, flux 83
-, inclination **83**, 84
-, north **83**, 85
-, pole 83, 84
magnetometer **83**, 85
Mahalanobis distance 162, **293**, 329
Manhattan distance 128, **287**
manifold **307–309**

manipulability 213, **232–235**
 –, dynamic **267**, 269
manipulator (see *also robot*) 189
 –, Jacobian **229**, 242, 261
 –, kinematics 227
 –, over-actuated 54, **238, 240**
 –, serial-link, dynamics 249
 –, under-actuated 54, **208, 238, 239**
manoeuvre 118, 119
manufacturing robot 3
map 158, 162, 167
 –, building, laser-based 179
 –, feature 161, 166
 –, obstacle **129**
mapping **165**
 –, CML (concurrent mapping and localization) 165
 –, exponential 48, 50
 –, point 54
 –, PTAM (parallel tracking and mapping) 173
 –, SLAM (simultaneous localization and mapping) **165**, 167–169, 173
Mars rover 4, 6
mass 66, 275
 –, center of 62, 66, 113, 251, 262
 –, distribution 66
 –, link 251, 262
 –, payload 266
 –, proof 80
mathematical morphology 134
MATLAB®
 –, code 9
 –, command prompt 9
 –, matrix xxv
 –, MEX-file **284**
 –, object 8
 –, software 8
 –, Toolbox conventions **xxv**
matrix **288**
 –, adjoint 63, 67, 199, 245, 297, **311**
 –, adjugate **289**
 –, angular velocity 64
 –, anti-symmetric 289
 –, condition number 233, **293**
 –, covariance 152, 154, 156, 158, 159, 161, 163, 165, 167, 168, 174, **328**
  –, diagonal 159
  –, extending 163
  –, odometry 158
  –, sensor 159
 –, definite
  –, negative 314
  –, positive 314, 322
 –, diagonalization **291**
 –, estimation 9
 –, exponential 23, 24, 41, 49
 –, exponentiation 48
 –, generator 308, 310
 –, Hessian 314
 –, homography 9
 –, identity 64
 –, indefinite 314
 –, inertia 114, 264
 –, inverse
  –, damped 238
  –, pseudo 238–240

 –, Jacobian 170, **227, 228**
 –, logarithm 23
 –, MATLAB® xxv
 –, normalization 47
 –, orthogonal 22
 –, orthonormal 32, 47
 –, rank 232, **292**
 –, rotation 22, 33, 40, 48, 64
  –, determinant 47
  –, normalization 65
  –, product 23
 –, singular value decomposition **292**, 318
 –, skew-symmetric **23**, 40, **41**, 35, 41, 48, 49, 61, 64, 289, 309
  –, augmented **310**
 –, sparse 324
 –, transformation, homogeneous 50, 62
MAV (see *micro air vehicle*)
maximum
 –, torque **257**
 –, velocity 70
measurement
 –, odometry 154
 –, random 154
 –, strapdown inertial 85
 –, unit, inertial (IMU) 38, 85
mecanum wheel **110**
MEMS (see *micro-electro-mechanical system*)
method
 –, Newton's **320**
 –, Newton-Raphson 319
 –, roadmap **134**
MEX-file **284**
micro-electro-mechanical system (MEMS) 78
micro air vehicle (MAV) 112
Mikrokopter (software project) 120
minimization, nonlinear 319
minimum-norm solution 208, 213, 240
mobile robot 3, **93, 97**
mobility 119
model
 –, bicycle **98, 105, 142, 143**
 –, kinematic **99**, 105, 109, 112, 141, 143, 200
 –, motion 97, 107, 110, 112, 113, 138, 142, 153, 269, 331, 332
 –, nonlinear 86
 –, process 331
 –, quadrotor 113
 –, screw 46
 –, unicycle 105, 109
 –, vehicle 105
model-based control 270
moment 50
 –, image
  –, line 296
  –, matrix **318**
 –, of inertia **66**, 262
 –, torque 66, 113, 114, 242, 267
 –, vector 28, 45, **50**, 296
momentum, angular 66, 77
Monte-Carlo
 –, estimation 155, 173, 181
 –, localization 173
Moore-Penrose pseudo inverse **292**
morphology (see *mathematical morphology*)

motion **61**, 82
–, axis of 71
–, Cartesian **75**, 209, **212**, 236
–, complex 12
–, control, resolved-rate 232, 236, 237
–, discontinuity 76
–, end-effector 236
–, equation **66**, 99, 109, 114, 249, **261**, 269
–, inertial frame 82
–, joint-space **209**, 214
–, lateral 110
–, longitudinal 98
–, model 97, 107, 110, 112, 113, 138, 142, 153, 269, 331, 332
–, multi-dimensional 71
–, null-space 12
–, omnidirectional 97, 110, 126, 138
–, planner 103
–, resolved-rate 12
–, rigid-body 25, 44, 45, 52, 307, 308
  –, incremental **65**
–, rotational 49, 50, 66
–, screw 45, 46
–, segment 72
–, sickness 81
–, singularity **213**
–, straight-line 212
–, translational 28, 29, 49, 51, 66
motor 253, 254, 275
–, DC 249
–, high-torque 252
–, inertia 253
–, limit 257
–, servo 249
–, stepper 249
–, torque 250
multi-pathing **151**
multi-segment trajectory **72**

N

nautical
–, angle 36
–, chronometer 150
–, mile **149**
navigation 95, 120, **123**, 149
–, aerospace 42
–, algorithm 129
–, Beidou (satellite navigation system) 151
–, chart 151
–, dead reckoning 149
–, Galileo (satellite navigation system) 151
–, GLONASS (satellite navigation system) 151
–, GPS (Global Positioning System) 5, 115, 149, **151**, 163
–, inertial 61, 64, **77**, 85, 115
–, landmark 149
–, map-based 123
–, marine 165
–, radio 77
–, reactive 123, **124**
–, satellite 5, 115, 149, **151**, 163
–, spacecraft 36, 78
–, system 77, 85, 115
Navlab project 120
NED (see *north-east-down*)
nested control loop **249**

Newton's
–, first law **67**
–, method **320**
–, second law **66**, 68, 80, 113, 261, 277
Newton-Euler method 261, 276, 277
Newton-Raphson method 319
node, graph 18, 137, 139, 142, 168, **337**
noise 86, 154, 178
–, Gaussian 155, **158**, 162, 332, 333
–, odometry 154, 156, **331**
–, random 86, 154, 175
–, scanning laser rangefinder 178
–, sensor 160, 173
nonholonomy, nonholonomic 97
–, constraint 99, 109
–, system **119**
normalization
–, homogeneous transformation 48
–, rotation matrix **47**
normal matrix **290**
north
–, magnetic **83**, 85
–, true **83**
north-east-down (NED) **77**
null space of matrix 240, **292**, 318
number
–, denominate **15**
–, dual **53**
–, random 137, 172, 331

O

observation 159
obstacle
–, inflation 128
–, map **129**
occupancy grid 126, 128, **129**, 179
odometer **153**
odometry **153**, 154, 168
–, differential 153
–, laser 177
–, noise 154, 156, **331**
–, wheel 153
omnidirectional
–, motion 97, 110, 126, 138
–, vehicle 110
–, wheel 110
OmniSTAR satellite network 151
operational space **53**
–, control **273**, **274**
operator 69
–, associative binary 19
–, asterisc 79
–, backslash 69, 317, 319
–, group 308
–, inverse 65
–, multiplication 52
–, spatial displacement 65
optical axis 38
optimization 171, 173, 180
–, algorithm 244
–, graph 173
–, Levenberg-Marquardt 244, 323
–, pose graph 170–172, 181
–, problem 169, 204

orientation 15
  –, 2-dimensional 21
  –, 3-dimensional 30
  –, derivative 62, 66, 116
  –, end-effector 194
  –, error 86
  –, estimation 78, 82, 87
  –, interpolation 73
  –, vector 38
  –, vehicle 99, 106
origin 15
orthogonal or orthonormal matrix 32, 289, 292
over-actuated robot 54, 238, 240
over-actuation 119, 238

P

parabolic blend 70
parallel tracking and mapping (PTAM) system 173
parallel-link robot 189
parameter
  –, Denavit-Hartenberg 195, 198, 225
  –, ellipse 288
particle filter 167, 173
path 68, 129, 132
payload 12, 249, 260
  –, effect 266
  –, lift capability 113
  –, mass 266
PDF (see probability density function)
peak 151
  –, velocity 70
pendulum, bifilar 277
phototaxis 124
pitch
  –, angle 35
  –, screw 45, 50
planar
  –, robot 203
  –, surface 95, 117
  –, transformation 29
plane 298, 304
  –, Cartesian 17
  –, Euclidean 17, 303
  –, image 303
planning
  –, algorithm 133
  –, map-based 128
  –, robot path 128, 132
  –, trajectory 145
Plücker
  –, coordinate 50, 52, 296
  –, line 296–298
point 15
  –, 3D 29
  –, Cartesian 177
  –, cloud 179, 182
  –, coordinate 24
    –, homogeneous 49
    –, vector 20, 45
  –, corresponding, correspondence 178
  –, equation
    –, ellipsoid surface 233, 243
    –, line 303
  –, Euclidean 27, 295, 304
  –, homogeneous form 27
  –, ideal 303, 304
  –, instantaneous center of rotation (ICR) 98, 107
  –, line equation 303
  –, mapping 54
  –, moving to 100
  –, task space 54
  –, tool center (TCP) 201
  –, transformation 22
  –, vector xxv, 15, 20
  –, velocity, angular 62
polar-coordinate robot arm 194
pole
  –, magnetic 83, 84
  –, rotational 28
polynomial
  –, ellipse 300
  –, function of time 69
  –, matrix approximation 50
  –, trajectory 69
pose 15, 53, 58, 168
  –, 2D 55
  –, 3D 56
  –, camera 173
  –, change 61
  –, derivative 61, 62
  –, end-effector 191, 227
  –, error 168, 243
  –, estimation 81
  –, graph 168, 169
    –, optimization 170–172, 181
    –, SLAM (simultaneous localization and mapping) 165, 167–169, 173
  –, robot (see also manipulator) 177, 179
  –, singular 232
  –, trajectory 75
position 15
positive definite 290
posterior probability 155
power series 50
probabilistic roadmap (PRM) 135
probability 10, 35, 152, 155, 172
  –, conditional 155
  –, density function (PDF) 151, 158, 159, 173, 327, 328
  –, Gaussian 158, 162, 329
  –, posterior 155
  –, prior 155
process noise 154, 332
product
  –, of exponential 198, 199
  –, of inertia 66
projection, line 304
Prometheus Project 120
proof mass 80
proprioceptive sensor 5
pseudo
  –, force 67
  –, inverse 238, 240, 292, 317
    –, Moore-Penrose 292
  –, random numbers 172
PTAM (see parallel tracking and mapping)
Puma 560 robot 194, 200, 254, 274
pure
  –, pursuit 103
  –, quaternion 43, 53, 62

Q

quadric 304
- -, hypersurface 304
- -, Klein 304

quadrotor 54, 95, 97, **112**, 118
- -, control system 115
- -, dynamics 113, **114**
- -, model 113

quaternion 42
- -, computational efficiency **43**
- -, conjugate **43**
- -, convert to rotation matrix 43
- -, derivative **62**
- -, double cover **42**
- -, dual **53**
- -, identity **43**
- -, interpolation 58, 74
- -, pure **43**, 53, 62
- -, unit **42**, 43, 45, 48, 53, 56, **62**, 74

quintic polynomial 69

R

radio navigation 77, 151
radius, turning 139
random
- -, coordinate 337
- -, measurement 154
- -, noise 86, 154, 175
- -, number 137, 172, 331
- -, sampling 137, 143
- -, variable 327
  - -, Gaussian **327**, 332, 334

rangefinder
- -, remission 177
- -, scanning laser **176–179**

rank, matrix 232, **292**
Rao-Blackwellized SLAM (see also *FastSLAM*) 167
rapidly-exploring random tree (RRT) **142**, 143
rate
- -, angular 86
- -, exponential coordinate 231
- -, roll-pitch-yaw angle 74, 116, 231
- -, rotation matrix 62

ratio 266
- -, gear 252, 262

recursive Newton-Euler 261
redundant robot 54, **208**, 224, 238
Reeds-Shepp path **99**
reference
- -, frame **67**
  - -, inertial 66, **67**, 77, 81
  - -, noninertial 68
- -, system, attitude and heading (AHRS) 85

reflectance, reflectivity 177, 178
- -, specular 178

remission 177
renormalization 53
replanning, incremental 132
representational singularity 231
resampling 174
resectioning **150**
resolved-rate motion control 232, **235**

response
- -, position loop 260
- -, velocity loop 256–258

right-hand rule **29**
rigid-body
- -, displacement 44, 50, 51
- -, dynamics **261**, 270
- -, motion 25, 44, 45, 52, **65**, 307, 308

ring-laser gyroscope (RLG) 78
roadmap 134
robot (see also *manipulator*) 189
- -, arm 119
  - -, model 198
  - -, planar 192, 243
  - -, polar-coordinate **194**
  - -, PUMA 193
  - -, serial-link 194
  - -, SCARA (Selective Compliance Assembly Robot Arm) **189**, 193, 208
  - -, Stanford 193
- -, Asimo humanoid **6**
- -, base transform **201**, 216
- -, Baxter 209, 275
- -, behavior-based **125**
- -, definition of 5, 124, 128
- -, DEPTHX (Deep Phreatic Thermal Explorer, AUV) 118, 119
- -, Elsie 94
- -, end-effector 190
- -, field 3, 94
- -, gantry 189
- -, high-speed 274
- -, humanoid 3, 6
- -, joint
  - -, modelling 253
  - -, structure 193
- -, kidnapped 176
- -, law 1
- -, manipulability 213, 234
- -, manufacturing 3
- -, maximum payload 266
- -, mobile 3, **93**, **97**
- -, over-actuated 54, **240**
- -, path planning 129, 132
- -, parallel-link 189
- -, planar 203
- -, pose 177, 179
- -, Puma 560 194, **200**, 254, 274
- -, redundant 54, **208**, 224, 238
- -, Shakey **93**
- -, service 3
- -, singularity 206, 213
- -, tele- **5**
- -, tool transform **201**, 202, 216, 220
- -, tortoise **93**
- -, trajectory 167
- -, under-actuated 54, **208**, **238**, **239**
- -, walking **219**
- -, wrist 194, 213

Rodrigues
- -, rotation formula 35, **40**, 50, 51, 59, 64, 309
- -, vector **40**

roll angle 35
roll-pitch-yaw angle 35, 36, 38, 230, 231
- -, rate 74, 116, 231

–, singularity 36
–, XYZ **35**, **36**, 212, 230
–, ZYX **35**
rolling, constraint 119
root, finding 318
Rossum's Universal Robots (RUR) 3
rotation, rotational 45, 48, 52
    –, angle 23, 24, 29, 33, 35, **37**, 41
    –, axis 30, 37, 39, 41, 46, 48, 61, 66
    –, direction 74
    –, formula 35, **40**, 50, 51, 59, 64, 309
    –, incremental 64
    –, inertia 66
    –, interpolation 74
    –, matrix **22**, 33, 34, 38, 40, 43, 48, 64, 230
        –, determinant 47
        –, estimating 318
        –, least squares problem **318**
        –, normalization 65
        –, product 23
        –, reading 33
    –, motion 49, 50, 66
    –, pole 28
    –, rate 62
    –, theorem, Euler's **30**, 31, 33–35, 309
    –, torque 67
    –, twist 28
    –, vector 28
    –, velocity 61, 63, 67
row space **291**
RRT (see *rapidly-exploring random tree*)
RTK GPS (see *Global Positioning System (GPS), RTK*)
rule, right-hand 29
RUR (see *Rossum's Universal Robots*)

S

saccule 81
sampling
    –, importance 174
    –, probabilistic 145
    –, random 137, 143
satellite navigation
    –, system 5, 115, 149, **151**, 163
    –, network 151
saturation, actuator 116
scalar 15, 52
    –, field 314
    –, function 313, 314
    –, interpolation 210
    –, multiplication 287
scale factor 86
scanning laser rangefinder **176**, 177, 179
    –, noise 178
SCARA (see *Selective Compliance Assembly Robot Arm*)
Schur complement 324
screw 45, 50
    –, axis 45, **50**
    –, model 46
    –, motion 45, 46
    –, pitch **45**, 50
    –, theory **50**
SE(2) **25**, **32**
se(3) 51, 52, **310**

SE(3) **44**, 46, 51, 52, 71, 75, 310, **311**, 322
SEA (see *series-elastic actuator*)
selective availability **151**
Selective Compliance Assembly Robot Arm (SCARA) **189**, 193, 208
sensor 168
    –, acceleration 81, **85**
    –, bias 86
    –, calibration 86
    –, drift 86
    –, error 168
    –, fusion 86, **161**
    –, Hall effect 83
    –, inertial 85
    –, noise 160, 173
    –, range and bearing 159
serial-link manipulator 191
series-elastic actuator (SEA) **274**, **275**
Shakey (robot) **93**
shape
    –, change 233, 243
    –, Earth 79
    –, ellipsoid 234
shared control **7**
similarity transform, transformation **291**
similar matrix **291**
Simulink 10, 270
    –, block 99
        –, library 109
    –, kinematics 212
simultaneous localization and mapping (SLAM) **165**
    –, back end 168, 172, 173
    –, EKF (extended Kalman filter) 167
    –, Fast 167
    –, front end 168, 172
    –, pose graph **165**, 167–169, 173
    –, Rao-Blackwellized 167
    –, system, vision-based 173
singular
    –, pose 232
    –, value **292**
        –, decomposition **292**
    –, vector **292**
singularity 35, **36**, 206, 213
    –, angle
        –, Euler 37
        –, roll-pitch-yaw 36
    –, Jacobian **232**, **238**
    –, motion **213**
    –, representational 231
    –, three angle representation 36
    –, wrist 206, 213
singular value decomposition (SVD) 292, 317, 318
skeleton 135, 201
    –, topological 134
skeletonization 134, 135
skew-symmetric matrix **23–25**, 40, **41**, 48, 49, 61, 64, 88, **289**, 304, 309
    –, augmented **310**
skid steering 109
SLAM (see *simultaneous localization and mapping*)
SO(2) **22**, 307, **308**
so(3) 52, 231, **309**
SO(3) **32**, 66, 71, 73, 79, 308, 309

soft-iron distortion 85
solution
  –, closed-form 203
  –, minimum-norm 240
  –, numerical 204
solving system 317
space
  –, configuration 53, **54**, 112, 117, 119, 143, **196**, 199, 208, 209
  –, control 273, 274
  –, Euclidean 17, 53, 295, 303
  –, inertial reference equipment (SPIRE) 77
  –, joint **196**, 210, **242**
  –, operational 53
    –, control **273, 274**
  –, task **53**, 54, 208, 209
  –, vector 287
sparse matrix 324
spatial
  –, displacement 65, 243
  –, operator 65
  –, velocity 62, 63, 67, **229**, 230, 237
    –, vector **62**
special
  –, Euclidean group **19, 25**, 44
  –, orthogonal group **22**, 32, **290**
spectral decomposition **291**
specular reflection 178
speeded up robust feature (SURF) detector 250
spherical
  –, linear interpolation **74**
  –, wrist 197, **203, 205**
SPIRE (see *space inertial reference equipment*)
spring 80, 275
  –, torsional 275
Stanford, robot arm 193
steering
  –, Ackermann 99, 121
  –, angle 99, 100, 139, 143
  –, mechanism 97
  –, skid 109
stereo glasses 33
stiction 250
straight-line motion 212
strapdown
  –, configuration **78**
  –, gyroscope 78
  –, inertial measurement 85
subsumption architecture **125**
surface
  –, 3D 130
  –, Earth 68, 77
  –, ellipsoid 233, 243, 328
  –, hypersphere 233
  –, planar 95, 117
  –, polished 178
  –, writing on 218
SVD (see *singular value decomposition*)
Swedish wheel 110
symmetric matrix 264, **289**
system
  –, attitude and heading reference (AHRS) 85
  –, configuration **53**
  –, coordinate **17**
  –, homogeneous 318

  –, inertial navigation (INS) 77, 85, 115
  –, nonholonomic **119**
  –, nonhomogeneous 317
  –, nonintegrable 119
  –, nonlinear 334
  –, under-actuated 118
  –, vestibular 78, 81

T

tag, April 162
Tait-Bryan angle 36
tangent space 308
task space 53, 54, 208, 209
taxis **124**
Taylor series 313
TCP (see *tool center point*)
telerobot **5**
temperature drift 86
tensor 287
theorem
  –, Chasles 50
  –, Euler's rotation 30
theory
  –, Lie group 23
  –, screw **50**
thinning (also *skeletonization*) 134, 135
threshold distance 137
thrust 113
time **61**
  –, derivative 61
  –, series xxv
  –, varying pose 61, **68**
tool
  –, center point (TCP) **201**
  –, transform **197, 201**, 202, 216, 220
toolbox
  –, functions 55–57
  –, obtaining 283
topological skeleton 134
topology, algebraic 48
torque 249, 251, 252, 273
  –, control 270
    –, computed 270, **272**
    –, feedforward 258, 270, 271
  –, disturbance 249
  –, end-effector 242
  –, gravity 252, 262
  –, maximum **257**
  –, moment 66, 113, 114, 242, 267
  –, motor 250
  –, rotational 67
trace of matrix 291
traded control **6**
trajectory **68**, 72, 74–76, 88, 137, 167, 207, 209, 221, 223, 249, 261
  –, Cartesian 89, **212**, 222
  –, continuous 72, 218
  –, end-effector 249
  –, following **103**, 138
  –, hybrid **70**
  –, joint-space 210–212, 214
  –, lane-changing 100
  –, leg 219
  –, multi-axis 71

–, multi-segment 72
–, planning 145
–, polynomial **69**
–, pose **75**
–, robot 167
transconductance 250
transform
   –, base **197**
   –, distance 128, 132, 133, 135
   –, planar 29
   –, SE(2) 29
   –, tool **197**, **201**, 202, 216, 220
   –, homogeneous **25**, **44**, 51, 52, 75, 197, 201
   –, matrix 50, 62
   –, point 22
   –, SE(2) **25**
   –, SE(3) **44**
   –, wrench 242
translation 44, 51, 52
transmission 249, 274
   –, flexible 12
   –, mechanical 107
transpose, Jacobian 244
trapezoidal trajectory **70**
traversability **128**, 132
triangulation 150
triaxial
   –, accelerometer 81, 85
   –, gyroscope 78
   –, magnetometer 83
triple point 134
true north **83**
Tukey biweight function 321
turning radius **98**, 139
twist 28, 46, **50**, 51, 198, 245, **310**
   –, axis 45
   –, Jacobian computing 245
   –, nonunit 29, 46
   –, rotational 28
   –, transforming 310
   –, unit 28, 46, 50, 52
   –, vector 28, 29, 45
   –, velocity 63, 245

U

UAV (see *unmanned aerial vehicle*)
UKF (see *unscented Kalman Filter*)
uncertainty 158, 159, 161
under-actuated 54, 97, 118, 119, 193, 227
   –, robot, manipulator 54, **208**, **238**, **239**
   –, system **118**
unicycle, model 109
Unimation Inc. 2
unit
   –, inertial measurement (IMU) 38, 85
   –, quaternion **42**, 43, 45, 53, 56
      –, derivative **62**
      –, interpolation 74
      –, normalization 48
   –, twist 28, 46, 50, 52
unmanned aerial vehicle (UAV) **112**
unscented Kalman Filter (UKF) 182
utricle 81

V

VaMoRs system (autonomous van) 120
variable, Gaussian random 332
Vaucanson's duck 1
vector 15, **287**
   –, addition 287
   –, approach 38
   –, bound 15
   –, coordinate 15–17, 287, 295, 302
   –, error 324
   –, field 315
   –, gravity 82, 261
   –, moment 28, 45, **50**, 296
   –, normal 38
   –, orientation **38**
   –, point xxv, 15, 20
   –, Rodrigues **40**
   –, rotation 28
   –, scalar function of 314
   –, singular **292**
   –, space 287
   –, twist 28, 29, 45
   –, vector function of 314
   –, velocity **62**, 228
vehicle
   –, aerial 119
   –, autonomous 6, 94
      –, surface (ASV) 94
      –, underwater (AUV) 94
   –, Braitenberg 124
   –, car-like 97, 98
   –, configuration 98
   –, coordinate system 98
   –, differentially-steered 97, **107**
   –, frame 98
   –, micro air (MAV) 112
   –, mobile robot 3, **93**, **97**
   –, model 105
   –, omnidirectional 110
   –, orientation 99, 106
   –, path 101, 103, 107
   –, underwater 119
   –, unmanned aerial (UAV) 94, **112**
   –, velocity 99
   –, wheeled 95, 97
velocity 249, 273
   –, angular 48, 50, **62**, 66, 68, 77, **78**, 153, 231, 332
      –, time-varying 66
      –, vector 64
   –, control 100, 255, 259
      –, feedforward 260
      –, loop 255, 259
   –, coupling torque **262**
   –, discontinuity 76
   –, ellipse, ellipsoid 233, 234, 242
   –, end-effector 227, 228
   –, joint 227, 228
   –, kinematics **227**
   –, linear 50, 66
   –, maximum 70
   –, peak 70
   –, rotational 61, 63, 67

–, spatial **62**, 63, 67, **229**, 230, 237
–, translational 61, 63, 67
–, twist 63, 245
–, vector **62**, 228
–, vehicle 99
vestibular system 78, 81
via point 72
viscous friction coefficient **250**
visual simultaneous localization and mapping (VSLAM)
182
von Mises distribution 154
Voronoi
–, cell 135
–, diagram 134, 135
–, roadmap 135
–, tessellation **135**
VSLAM (see *visual simultaneous localization and mapping*)

W

WAAS (see *wide area augmentation system*)
walking robot **219**
waypoint 155
Wide Area Augmentation System (WAAS) **151**
world coordinate frame **16**, 77

wrench 63, **67**, **242**, 243, 261, 267
–, ellipsoid 243
–, end-effector 242, 243
–, transformation 242
wrist 206
–, coordinate frame 201
–, robot 194, 213
–, singularity 206, 213
–, spherical 197, **203**, **205**

X

XYZ, roll-pitch-yaw angle **36**, 212, 230

Y

yaw angle 35
yaw rate **99**, 161
Yoshikawa's manipulability measure 234

Z

zero-angle configuration 195
ZYX roll-pitch-yaw angle 35
ZYZ Euler angles **34**